Waves and Particles in Light and Matter

Waves and Particles in Light and Matter

Edited by

Alwyn van der Merwe

University of Denver
Denver, Colorado

and

Augusto Garuccio

University of Bari
Bari, Italy

Springer Science+Business Media, LLC

Library of Congress Cataloging-in-Publication Data

Waves and particles in light and matter / edited by Alwyn van der Merwe and Augusto Garuccio.
p. cm.
Includes bibliographical references and index.
ISBN 978-1-4613-6088-9 ISBN 978-1-4615-2550-9 (eBook)
DOI 10.1007/978-1-4615-2550-9
1. Wave-particle duality--Congresses. 2. Wave-motion, Theory of--Congresses. 3. Quantum theory--Congresses. I. Van der Merwe, Alwyn. II. Garuccio, Augusto. III. Workshop on "Waves and Particles in Light and Matter" (1992 Trani, Italy)
QC476.W38W43 1994
530.1'2--dc20 94-26744
CIP

ISBN 978-1-4613-6088-9

Originally published by Plenum Press, New York in 1994
Softcover reprint of the hardcover 1st edition 1994

PREFACE

From September 24 through 30, 1992 the Workshop on "Waves and Particles in Light and Matter" was held in the Italian city of Trani in celebration of the centenary of Louis de Broglie's birth.

As is well known, the relationship between quantum theory and objective reality was one of the main threads running through the researches of this French physicist. It was therefore in a fitting tribute to him on his 90th birthday that ten years ago an international conference on the same subject was convened in Perugia. On that occasion, physicists from all over the world interested in the problematics of wave-particle duality engaged in thoughtful debates (the proceedings of which were subsequently published) on recent theoretical and experimental developments in our understanding of the foundations of quantum mechanics.

This time around, about 120 scientists, coming from 5 continents, in the warm and pleasant atmosphere of Trani's Colonna Conference Center focussed their discussions on recent results concerned with the EPR paradox, matter-interferometry, reality of de Broglie's waves, photon detection, macroscopic quantum coherence, alternative theories to usual quantum mechanics, special relativity, state reduction, and other related topics.

The workshop was organized in plenary sessions, round tables, and poster sessions, and the present volume collects most—but not all—of the presented papers.

A number of acknowledgements are due. We thank, first of all, the contributors, without whose constant dedication this volume could not have been published.

Thanks are also due to the City of Trani for its hospitality in putting the Colonna Conference Center at our disposal.

We are deeply indebted to Prof. Franco Selleri, Chairman of the Organizing Committee, for his indispensable contributions to the success of the workshop, and we pay tribute to the secretarial staff, Ms. Rossella Colmoyer, Mr. Francesco Minerva, and Ms. Gabriella Pugliese, for their caring assistance.

We finally owe a debt of gratitude for financial support to the Commission of the European Communities and especially to Prof. P. Fasella, Dr. A. Pozzo, and Dr. W. Hebel. We are also grateful for financial support from the Istituto Nazionale di Fisica Nucleare and the Università di Bari.

The Editors

CONTENTS

THE GREAT VEIL, REALITY, AND LOUIS DE BROGLIE: PERSONAL MEMORIES

O. Costa de Beauregard

76 rue Murger
77780 Bourron-Marlotte, France

A history-making insight: Louis de Broglie as a realist, his reticences towards formalism.

Key words: deep admiration but differences in philosophical approach.

1. "He has lifted a corner of the great veil," so wrote Einstein in his letter dated 16 December 1924 to Langevin, who had sent him a copy of Louis de Broglie's Ph.D. Thesis [1]. And, indeed, this discovery of the matter wave, resulting from a reading and combining of equations in a fresh approach, ranks among the most impressive predictions of mathematical physics. Also, de Broglie's elegantly thought out and written short paper is a masterpiece and a landmark in the world's scientific literature.

The paper opens with a short historical survey of the competing concepts of waves and particle in optics, up to the new days of the quantum and of Einstein's photon—so openly manifested in Maurice de Broglie's "X-Ray Laboratory." The relevance of the relativity theory is mentioned.

Chapter 1 starts with the matter-wave hypothesis, the riddle of a faster than light propagating phase, and its resolution, via the group velocity concept, with crucial help from relativity. Chapter 2 identifies the mechanical and optical extremum principles, and relates an electron's 4-frequency to its kinetic-plus-potential energy-momentum via the Planck-Einstein quantum rule. All this occurs at a quick pace.

Chapters 3 and 4 then liken the Bohr-Sommerfeld quantization rule to the self-resonance of the matter wave.

Chapter 5 proposes that the photon may be a true particle, with an exceedingly small rest mass. Chapter 6 discusses the scattering of x and γ rays.

Waves and Particles in Light and Matter, Edited by
A. van der Merwe and A. Garuccio, Plenum Press, New York, 1994

Chapter 7, entitled "Statistical Mechanics and the Quantum," discusses the kinetic theory of gases and blackbody radiation; it should capture the attention of historians of science and be re-read in the light of Einstein's later work in these same fields.

2. *Statistics and the quantum* are words with an ominous ring, and to this I will come back. But first I must recall how Louis de Broglie came to theoretical physics, because it is a rather unusual story.

3. *At the 1911 Solvay Council* the great luminaries of fundamental physics were all present. Maurice de Broglie, the experimentalist, was one of the secretaries; on his return to Paris, he brought with him the papers for the Proceedings. At that time, Louis, his younger brother by 18 years, was studying history and law. Maurice sought his help in handling the papers, and Louis became fascinated by the nature of the discussions. From history he knew that Newton had viewed light as a stream of particles and had used a periodicity argument for explaining the interferences he had observed. Later, of course, expanding the opposite Huygens and Young paradigm, Fresnel established the wave theory of light.

Abundant evidence of the Planck-Einstein energy-frequency equivalence was found in the Solvay Council papers, and an intuition flashed in Louis' mind: Could it be that energy quantization of oscillators is a sort of resonance phenomenon? Could it be that Newton's and Fresnel's pictures of light are both true? And could it then not be that a wave-particle dualism exists for matter also? Of course, in Maurice de Broglie's "X-Ray Laboratory," waves and particles, particles and waves, were words ceaselessly spelled out and concepts repeatedly playing with each other.

Thus Louis de Broglie, aged 19, leaving aside everything else, converted to fundamental physics. He enrolled at the Faculty of Sciences of the Paris University in order to become a full member of the fraternity. And there he found more than he had expected: the formal analogy between geometrical optics and analytical mechanics (noticed by Hamilton), with the riddle however of the phase velocity in the denominator of Fermat's extremum law and the particle velocity in the numerator of Hamilton's law. To de Broglie this meant that his guess did make sense.

From 1914 to 1918 he served at the Eiffel tower under General Ferrié in military radio. Just an anecdote: It so happened that he was the first person to receive news of the German capitulation.

The year 1918 brought him back to physics, reading or listening to Poincaré and Langevin, following what was going on at his brother's laboratory. Whence started what I mention above and ends with the words "statistics and the quantum."

4. *In 1926* (it is incredible how fast things went in these days) Born, soon followed by Jordan, proposed "the statistical interpretation of wave-particle dualism," an efficient calculation recipe, but the subject of disputes fiercely going on ever since these days. As soon as 1927, at the Fifth Solvay Council, where Louis de Broglie [2] presented his realistic interpretation of "wave mechanics," Einstein [3] perceived the gravity of the challenge. His and de Broglie's feelings were extremely close on this point, and to this I can testify.

De Broglie's [4] little fable of one ball and two boxes, very similar to Einstein's [3] Solvay Council argument, is this: If in Berlin one ball is placed in one of two boxes sent to Atlanta and Cape Town, respectively, a recipient opening his box at any one of these locations will infer straightaway what the other recipient finds. Although his reasoning zigzags in spacetime along ABC or CBA, physical causality proceeds forward to A and C from B where, so to speak, "the dice are cast."

But the Born-Jordan recipe says something else, that is paradoxical: At B a superposition of amplitudes, not a sum of probabilities, is at stake, so that what shows up at A and C cannot have pre-existed at B! So, Born's seemingly innocent assumption that the wave's *intensity* measures the particle's position *probability density* unexpectedly challenges the very concept of a physical reality; still worse, perhaps, it hints at retrocausation! The 1935 EPR [5] argument anew raises these points.

"Nonseparability" is the wording fashionable today for this phenomenology, which can be concisely expressed in a Lorentz and CPT-invariant form [6], and has been tested in "delayed choice experiments."

Often I wonder how Einstein or de Broglie would have reacted to the delayed choice experiments.

5. *Like Einstein*, de Broglie was a dedicated realist, as quotations from him will make clear. These are quite consonant with the ideas he had expressed at the 1927 Solvay Council. Here are these 1963 and 1964 quotations:

> "Since more than thirty years most theoretical physicists have adhered to an interpretation of quantum physics issuing from Bohr's ideas... This interpretation seems perfectly adapted to the elegant and precise formalisms presently used, the predictions of which are usually quite well vindicated experimentally.
>
> "Although after my initial works on wave mechanics I had expressed, concerning the wave-particle dualism, ideas quite different from those the Copenhagen School was beginning to spread, I was soon stopped by great difficulties, so I finally accepted the interpretation that has become orthodox, and taught

> it. But since some ten years, returning to my initial conceptions, I became convinced that today's formalism, although seemingly rigorous and usually yielding exact conclusions, does not give a deep and truly explanatory view of the physical reality at the micro-level." [7]
>
> "Since some twelve years I have come back to a tentative interpretation of quantum mechanics which, under the names of "guiding wave theory" or "double solution theory," I had unsuccessfully proposed in 1927–1928 soon after my Ph.D. Thesis. Long thinking on the matter leads me now to state that the presently accepted interpretation of quantum mechanics does not yield a truly reasonable explanation of some basic and unquestionable experimental facts, so that it must be up-dated by re-establishing a permanent localization of the particle inside space during the course of time, and by restoring a physical reality to the accompanying wave, also assuming an appropriate binding between wave and particle." [8].

Duly taking into consideration what has happened since these days, mainly of course the advent of Bell's [9] theorem, I feel sure that to many of those here present these words have a welcome and familiar ring; and not only to them, but also to some of the people I have recently met at the Paris MAXENT 1992 Conference. Perhaps, even "Up There," Einstein, Schrödinger, and Bell also rejoice at this sound.

Most of you know that my "religion" is a different one, and so I seize this occasion to heartily thank my good friend Franco Selleri for, nonetheless, having invited me at this Centenary Celebration.

My personal comments on these two quotations are these: Like Einstein, de Broglie considers it unquestionable that the particle has a permanent reality, while it seems to me that formalism and experimentation would rather liken its "second quantization" to that of drops from a drip tube. As for the wave, I would rather see it as an information-carrying entity propagating covariantly throughout what we formalize as spacetime. I can testify that to de Broglie purely formal arguments were *a priori* suspect, and that to him explicit spacetime covariance was much less important than an unshakable faith in the *micro-reality* assumption. As you may have noticed, his words: "localization in space during the course of time" are not of the style used in, say, the Schwinger or the Feynman covariant formalizations. I can testify that the advent of these concepts, and their similarities with those I had been defending with not much success, was to him quite unexpected.

6. Anyhow, when I began working with Louis de Broglie in November 1940, he professed "The Copenhagen interpretation of quantum mechanics," which he understood in depth. It was then expressed in the non-relativistic Schrödinger-Heisenberg formalism, favoring time and energy over space and momentum. This de Broglie took as evidencing an essential "conflict between relativity and the quantum," those two ideas which his 1924 "wave mechanics" had so harmoniously united. Curiously, he resisted strongly the idea of trying once more what had then succeeded—and was going to succeed again: rely on, and stick to, explicit spacetime invariance.

Perhaps one reason for this reticence was (as it turned out in discussions we had on the EPR correlations) that a covariant rendering of "interference of probabilities" leads straight to the concept of retrocausation—which to both Einstein and Louis de Broglie was simply unthinkable, as quotations would prove. To both of them, realism and retarded causation were equally beyond question.

7. *A Brief Survey of de Broglie's Other Works.* Louis de Broglie's [10] massive-photon theory antedates the Proca and the Petiau-Duffin-Kemmer spin-1 formalisms. His [11] textbook on the matter makes very interesting reading. In quantum electrodynamics a transitory use of a photon rest mass helps in discussing self-energy problems [12]. Also, it allows use of the Lorentz condition in operator form [11,12].

De Broglie builds his spin-1 photon via "fusion" of two spin" 1/2 particles; he later extended this method to higher-spin states.

In 1951 Vigier's [13] and Bohm's [14] works persuaded de Broglie that his early thoughts, as expressed at the 5th Solvay Council, might well be physically sound. Changing his mind, he devoted since then much thinking along this line, either in the form of the "guiding wave theory" or in the more sophisticated form of the singularity-loaded wave. John Bell [15] has repeatedly stated his affinity for this approach, for details of which I refer to de Broglie's [16–19] synthetic presentations.

7. *Personal Memories.* Louis de Broglie was Professor of Theoretical Physics at the Institut Henri Poincaré, teaching a regular course in quantum mechanics, the subject matter of which changed each year. Often this material was subsequently published as a textbook, which is where much of what I know about his views can be found.

At our Institute Louis de Broglie had an office where he received us, either alone or in small groups, for discussions. We submitted papers to him, which he read very carefully, before handing them back with written annotations. As he was thoroughly versed in physics, both classical and recent, and had devoted much thought to every aspect of the subject, we thus learned a lot of physics, while receiving also much help in bringing our own research into publishable form.

He directed of course a seminar, on Tuesday afternoons, with speakers either from outside or inside his group.

I feel much gratitude for the great care he devoted to each of us and for the high quality of, and the insights gained from, the teaching we thus received.

Of course, from time to time a disagreement would turn up. If it so happened that you were right, it could take a long and step by step discussion to persuade Louis de Broglie of that— something about which I do have some typical memories. And if it turned out that the respective positions were irreconcilable, then you were in a difficult position: All scientific dialog, even on other matters, would become impossible. This is what happened to me in 1951, when Louis de Broglie openly came back to the hidden-determinism quantum mechanical paradigm, while I persisted in deriving the logical consequences of an explicitly quantum-relativistic formalism. Nevertheless, he offered me copies of his new books with kind handwritten signatures; it is from these that I have borrowed the preceding quotations.

This is the sort of peripeteia happening inevitably in scientific research, in the face of which one must remain faithful to his own working hypotheses as long as they are not disproven.

On the whole, having been Louis de Broglie's student has been for me a wonderful experience, and I remember these days with deep emotion.

REFERENCES

1. L. de Broglie, "Recherches sur la Théorie des Quanta," *Ann. Phys. (Paris)* 22–128 (1925); reprinted in *Ann. Fond. Louis de Broglie*, 1992.
2. L. de Broglie, "Nouvelle Dynamique des Quanta," in *Electrons et Photons: Rapports et Discussions du Cinquième Conseil Solvay* (Gauthier-Villars, Paris, 1928), p. 105.
3. A. Einstein, oral intervention, Ref. 2.
4. L. de Broglie, Ref. 17, pp. 28–30.
5. A. Einstein, B. Podolsky, and N. Rosen, *Phys. Rev.* **47**, 777 (1935).
6. O. Costa de Beauregard, *Phys. Rev. Lett.* **50**, 867 (1983).
7. L. de Broglie, Ref. 17, p. vii.
8. L. de Broglie, Ref. 18, p. v.
9. J. S. Bell, *Physics* **1**, 195 (1965).
10. L. de Broglie, *Une Nouvelle Conception de la Lumière* and *Nouvelles Recherches sur la Lumière* (Herman, Paris, 1934, 1936).

11. L. de Broglie, *Mécanique Ondulatoire du Photon et Théorie Quantique des Champs* (Gauthiers-Villars, Paris, 1949).
12. J. M. Jauch and F. Rohrlich, *The Theory of Photons and Electrons* (Addison-Wesley, Cambridge, Mass., 1955).
13. D. Bohm and J. P. Vigier, *Phys. Rev.* **96**, 208 (1956).
14. D. Bohm *Phys. Rev.* **85**, 166, 180 (1952).
15. J. S. Bell, *Speakable and Unspeakable in Quantum Mechanics* (Cambridge University Press, Cambridge, 1987).
16. L. de Broglie, *Une Tentative d'Interprétation Causale et Non Linéaire de la Mécanique Ondulatoire* (Gautheir-Villars, Paris, 1953).
17. L. de Broglie, *Etude Critique des Bases de l'Interprétation Actuelle de la Mécanique Ondulatoire* (Gauthier-Villars, Paris, 1963).
18. L. de Broglie, *La Thermodynamique de la Particule Isolée* (Gauthier-Villars, Paris, 1964).
19. L. de Broglie, *La Réinterprétation de la Mécanique Ondulatoire* (Gauthier- Villars, Paris, 1971).

THE FALLACY OF THE ARGUMENTS AGAINST LOCAL REALISM IN QUANTUM PHENOMENA

A. O. Barut

Department of Physics
University of Colorado
Boulder, Colorado 80309

The arguments refuting local realistic theories, including those invoking Bell inequalities, are all based on a tacit, preconceived, already quantized (and as it turns out inadmissible) notion of what constitutes a "single event." More appropriate models of single events with internal, classical (hidden) parameters reproduce all the quantum correlations after averaging.

Key words: single events, interference, spin correlations.

1. INTRODUCTION

The problem of the existence of a realistic picture underlying the probabilistic statements of quantum theory reduces, as we shall see, to the questions: what is *an* "electron," and what is *a* "photon"? I mean here, a single, individual electron or photon, because the concepts of an "individual event" in a single experiment, and a "typical event" in repeated experiments are quite different.

Physics has made a lot of progress in sharpening and understanding the structure of the electron. Coining a word for an entity does not mean that we completely understand it. We had in this century, in succession, the electron according to Lorentz, and to de Broglie, according to Heisenberg and Schrödinger, the Dirac electron, and the electron concept in quantum electrodynamics with its renormalized self-energy effects. And we still need an improved notion, because precisely of the perturbative nature of infinite renormalization of the radiative processes. But now many physicists working in the foundations of quantum theory claim that there are limits to further progress and knowledge, and proclaim that "there is no objective reality in quantum phenomena."

Waves and Particles in Light and Matter, Edited by
A. van der Merwe and A. Garuccio, Plenum Press, New York, 1994

The main basis of this assertion seems to be the Bell's theorem, or its corollaries, which states that "the objective description of an *individual* quantum system by certain intrinsic parameters λ is impossible."

The program of realism, on the other hand, is to look for a deterministic description of single events such that the experimental results will be obtained by a proper averaging over the parameters λ, i.e., over repeated events, as the experimenter in fact is always doing.

The apparent inevitability of the above impossibility is based on the claim that the theorem is proved without any assumptions on the nature of λ, hence on the nature of the individual events.

There are, however, major *tacit* and, as it will turn out, inadmissible assumptions as to what constitutes a *single* event. These are taken usually for granted, or obvious, so that remarkably enough, they are never stated even as assumptions.

We can demonstrate that they are only assumptions, and not obvious facts, by simply modifying them. This allows us indeed to obtain the correct experimental results, or the correct quantum results after averaging, whereas the so-called "obvious" assumptions fail to do so. Therefore, the discussion will shift now to the nature and description of the single events.

What are the standard tacit assumptions about a single electron and a single light event? They are:

(1) "If we lower the intensity of light sufficiently, we end up with single indivisible photons, which are either detected or not, either go one way or another, for each λ, stochastically. We have yes or no possibilities only, and this indeterministically. In other words single events are fundametnally indeterministic."

(2) "In the case of the electron, an individual spin has, for each λ, projections in *every* direction $\hat{a}$ equal to $\pm 1/2$ *simultaneously*, again the outcome being stochastic and indeterministic."

In these, and other similar situations, one attributes to the individual single system intrinsically the same discrete dichotomic property that we know only to be true on the average for repeated events in an ensemble as determined by idealized experimental arrangements. In other words, the single events are already "quantum" objects. But this assumption already contradicts the search for realism. We must look for parameters λ, and for a deterministic behavior of the individual result as a function of λ, such that only after averaging we obtain the observed probabilities or regularities. In this sense, the use of probabilities in quantum theory is the same as in any other classical situation with repeated events.

If we assign an intrinsic dichotomic property to every single event, we cannot go any further. Hence the ultimate reality would be indeterminate. However, the example of a coin toss might show that even here there is a deterministic description of single events. Let the position of the coin with

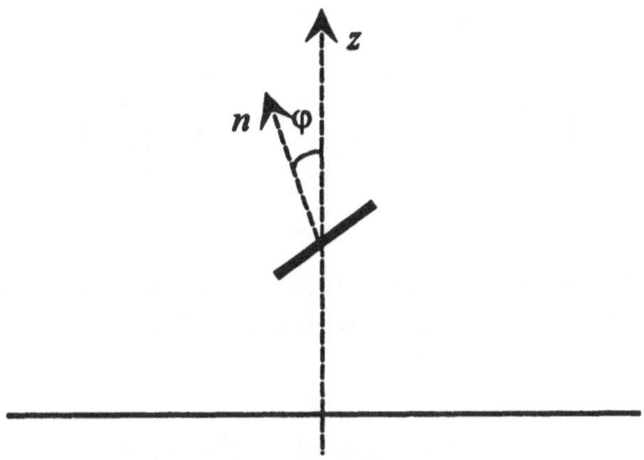

Fig. 1. Dynamics of coin toss.

normal $\hat{n}$ be specified by z and φ (Fig. 1). A simple model of motion is to use the gravitational force in z direction and a constant angular frequency $\dot{\varphi} = \omega$. Let the angle when one point of the coin touches the table be φ_f. This is a "hidden" variable for a single event, determined by the initial conditions of the toss; φ_f is continous and the single event is deterministic. Now for $2n\pi < \varphi_f < (2n+1)\pi$ we have H(ead), for $(2n+1)\pi < \varphi_f < (2n+2)\pi$, we have T(ail); $n = 1, 2 \ldots$. For repeated events, under certain initial conditions, we may have probabilities for H and T equal to 1/2, but not always, and certainly, the individual events do not have any ***intrinsic*** dichotomic property. The measurement of H or T on the table discretizes the result, but we could in principle also measure the continuous variable φ_f and work with it to answer more general questions than just head or tail.

As a second example, in a counting situation, consider a source S emitting particles between two hemispheres (Fig. 2). The currents in the two hemispheres are collected by the detectors D_+ and D_-. We could say that each event intrinscially makes a choice with probabilities P_+, P_-, i.e., we have a dichotomic stochastic variable X with values ± 1. Or, we can say that each event is deterministic with (hidden) parameters (θ, φ), such that when we integrate over the two hemispheres we obtain the probabilties P_+, P_-. The deterministic variable here is $X = \frac{1}{2}\cos\varphi$.

We thus see that we can make deterministic models of hidden variables. This goes beyond the standard quantum theory, which cannot make any statements about individual events. The standard quantum theory with its ensemble interpretation is incomplete. It tells us, as we said above, the behavior of a *typical* electron in repeated experiments, but not the behavior of an *individual* electron in a single experimental event. We have therefore to

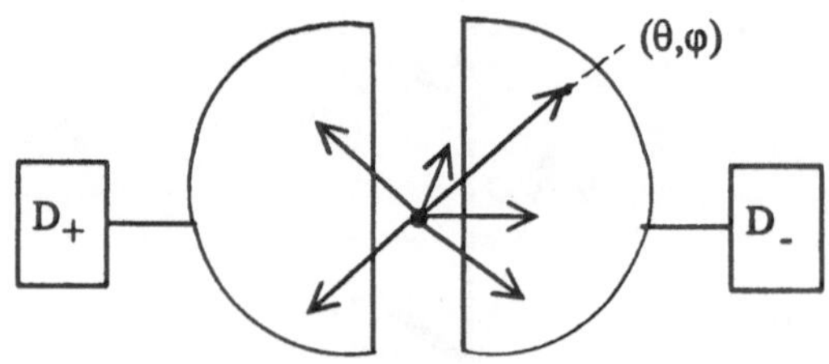

Fig. 2. Individual events (θ, φ) and their discretized integrated counts at detectors D_+ and D_-.

go beyond and find the correct models of single events which will reproduce *exactly* the experiments after averaging. This is the goal of the "Quantum theory of single events." [1]

What is then the origin of the usual tacit assumptions about dichotomic hidden variables?

There are two:

(A) For the electron spin it is the Stern-Gerlach experiment, rather the idealization of it. It is often stated that the atomic beam after the Stern-Gerlach magnet splits into two sharp lines, hence each individual atom must have exactly the spin projection $\pm 1/2$. However, both experiment and theory gives two broad bumps, with the tail of one bump reaching the other side. In a hidden-variable model of spin as a direction (θ, φ)—that we shall also use later—the motion of each spin must be individually calculated and the two maxima can be accounted for by the adiabatic effect of the fringe fields.

(B) The origin of the discrete "photon" concept is really the quantization of free fields into plane wave-normal modes with $\omega/c = k$, or $E = \hbar\omega = cp$. It is perhaps not generally recognized, that the plane wave representing a single event is already a statistical concept. It is used in QED calculations in the probabilistic sense. The probability of a process is calculated by the emission, absorption, or exchange of a plane wave. So here single events have no hidden parameters but are intrinsically probabilistic. A single light pulse emitted by a single atom is not a plane wave. It may be a localized lump which depends on the nature and state of the atom and have internal parameters λ. In fact there is no such unique physical object called "photon" in the laboratory. We know that the properties of light emitted by atoms depend on its preparation, there are squeezed lights, coherent lights of various degrees, incoherent light, etc.

Furthermore, the concept of photon is dispensable. There is a new formulation of quantum-electrodynamics which does not quantize the electromagnetic field, nor the electron field for that matter. The electromagnetic field is expressed in terms of the current of the source, the self-field, which acts back on the electron. Thus it has been possible to calculate all the radiative processes, beginning with Planck's law, photoeffect and Compton effect, then spontaneous emission, $g-2$, Lamb shift, scattering processes, in fact the whole of QED without ever mentioning photons.

We focussed our attention on the notion of single events and presented a critique of the usual tacit assumptions about individual electrons and photons. All the so-called "quantum paradoxes" stem from the assumption of discrete dichotomic nature of single events. We next present the alternative, namely classical, continuous and deterministic models of single events.

We first discuss the diffraction and interference experiments, then the spin correlation experiments (EPR) using deterministic single events with hidden parameters. These are indeed the fundamental quintessential quantum phenomena.

2. DIFFRACTION AND INTERFERENCE

For the discussion of interference phenomena a model of a single event with internal parameters is an exact localized oscillating solution of the wave equation, which I called a "de Broglie wavelet." The inital position $\boldsymbol{x}_0$ and the velocity $\boldsymbol{v}$ (which is smaller than c) are the parameters individualizing the quantum particle. Such solutions have been discussed recently extensively in the relativistic and electromagnetic case. Here we shall use only wavelet solutions of the Schödinger equation. The wavelet defines a rest frame and moves like a massive particle, even though it is a solution of the massless wave equation. For slow motion relative to the rest frame, the Schrödinger equation can in fact be derived from the massless wave equation $\Box\varphi = 0$.

The nonrelativistc limit of wavelets has the form

$$\psi(\boldsymbol{x},t;\boldsymbol{x}_0,\boldsymbol{v}) = F_\omega(\boldsymbol{x}-\boldsymbol{x}_0-\boldsymbol{v}-t)e^{i(\boldsymbol{k}\cdot\boldsymbol{x}-\omega t)}\,, \tag{1}$$

where $\boldsymbol{k} = m\boldsymbol{v}/\hbar$ and $\omega = \Omega + \frac{\hbar k^2}{2m}$, with Ω being the internal frequency, and are the solutions of the Schrödinger equation. The simplest spherically symmetric form of the localization function F is

$$F = C_0\sqrt{\frac{2}{\pi}}\frac{\sin\left(\sqrt{2m\frac{\Omega}{\hbar}}\rho\right)}{\sqrt{\frac{2m\Omega}{\hbar}}}, \quad \rho = |\boldsymbol{x}-\boldsymbol{x}_0-\boldsymbol{v}t|. \tag{2}$$

In the limit $\Omega \to \infty$, F approaches a δ function; then we have the classical particle limit, and in the limit $\Omega \to 0$, F approaches unity, and we get

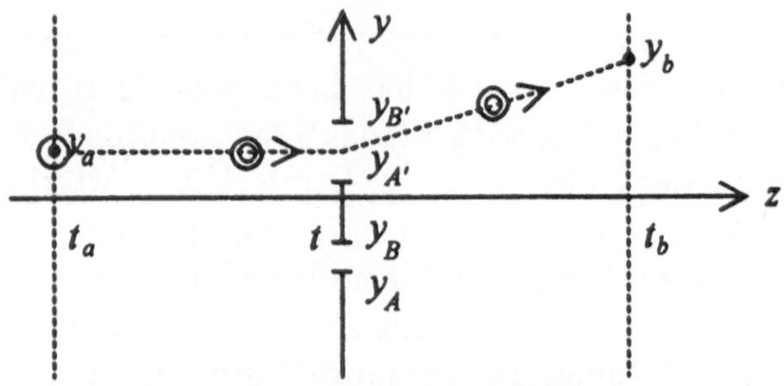

Fig. 3. Geometry of aperture for calculating the propagator of wavelets.

the plane-wave solution of the wave equation which has no longer internal parameters. The plane wave solution is also obtained from averaging of (1) over $\boldsymbol{x}_0$ or $(\boldsymbol{x}_0 - \boldsymbol{v}t)$. Thus wavelets form a bridge between classical particle dynamics and standard quantum mechanics. More general solutions are rotating waves with discrete quantum numbers (ℓ, m).

The wavelets, being solutios of the wave equation, propagate with the propagator of the wave equation, which in our case is

$$K(\boldsymbol{x}_b t_b, \boldsymbol{x}_a t_a) = \left(\frac{m}{2\pi i\hbar(t_b - t_a)}\right)^{3/2} \exp\left[\frac{im}{2\hbar}\frac{(\boldsymbol{x}_b - \boldsymbol{x}_a)^2}{(t_b - t_a)}\right] \qquad (3)$$

When applied to a wavelet (1) according to

$$\psi(\boldsymbol{x}_b, t_b) = \int K(\boldsymbol{x}_b t_b, \boldsymbol{x}_a t_a)\psi(\boldsymbol{x}_a t_a) d\boldsymbol{x}_a, \qquad (4)$$

it reproduces in fact the translated wavelet at $(\boldsymbol{x}_b, t_b)$ without spreading. For a single or double slit arrangement (Fig. 3) the propagator is changed into

$$K(\boldsymbol{x}_b t_b, \boldsymbol{x}_a t_a) = \left(\frac{m}{2\pi i\hbar(t_b - t_a)}\right)^{3/2} \exp\left[\frac{im}{2\hbar}\frac{(\boldsymbol{x}_b - \boldsymbol{x}_a)^2}{(t_b - t_a)}\right] \frac{1}{i\sqrt{\pi}} \int_{\zeta_A}^{\zeta_B} d\zeta e^{i\zeta^2}, \qquad (5)$$

where

$$\zeta = \sqrt{\frac{m}{2\hbar}} \sqrt{\frac{(t_b - t_a)}{(t_b - t)(t - t_a)}} y - \frac{1}{\sqrt{(t_b - t_a)}} \left(y_b \sqrt{\frac{t - t_a}{t_b - t}} + y_a \sqrt{\frac{t_b - t}{t - t_a}} \right)$$

and ζ_A, ζ_B are the position (or positions) of the slit (or slits) at time t.

In the Frauenhofer limit, $(t - t_a) \to \infty, (z - z_a) \to \infty$, with $(t_b - t) = \tau, \zeta$ becomes independent of y_a, and when we apply the propagator (5) to the wavelet, (1), i.e.,

$$\psi(\boldsymbol{x}_a, t_a) = F(\boldsymbol{x}_a - x_0 - vt_a)e^{i(\boldsymbol{k}\cdot\boldsymbol{x}_a - \omega t_a)},$$

according to Eq. (4), the wavelet reproduces itself, in the Frauenhofer limit, multiplied however by the Fresnel factor

$$\psi(\boldsymbol{x}_b t_b) \cong \boldsymbol{F}(\boldsymbol{x}_b - \boldsymbol{x}_0 - vt_b)e^{i(\boldsymbol{k}\cdot\boldsymbol{x}_b - \omega t_b)}\frac{1}{\sqrt{i\pi}}\int_{\zeta_A} d\zeta e^{i\zeta^2}, \tag{6}$$

where

$$\zeta_A = \sqrt{\frac{m}{2\hbar t}}(y_A - y_B).$$

Thus, as in the case of free propagation, the wave at the screen is

$$\psi(\boldsymbol{x}_b t_b) \cong F(\boldsymbol{x}_b - \boldsymbol{x}_0 - \boldsymbol{v}t_B)\Psi(\boldsymbol{x}_b, t_b), \tag{7}$$

where F is the localization functions depending on the individual paramters $\boldsymbol{x}_0$ and $\boldsymbol{v}$ and Ψ is the standard plane wave solution of the diffraction problem used as a probability amplitude. We have further for a single slit of width d, for example,

$$\int d\zeta e^{i\zeta^2} = \sqrt{\frac{m}{2\hbar\tau}}e^{i\frac{m}{2\hbar\tau}y_b^2}\frac{\sin\left(\frac{my_b}{2\hbar\tau}d\right)}{\left(\frac{my_b}{2\hbar\tau}\right)d}. \tag{8}$$

We see from Eq. (7) that, because F is localized, a single event samples effectively a localized portion of Ψ; ψ or $|\psi|^2$ is maximum where F is and where Ψ (or $|\Psi|^2$) is. In particular, no single events will be seen at positions where Ψ is zero or very small. But, when we average over the parameters $\boldsymbol{x}_0$ or $\boldsymbol{v}$ of the single event, we recover Ψ. Thus this theory completely accounts for interference experiments made by single events one after another, where the obesrvations on the screen are collected in a memory.

3. SPIN CORRELATIONS

Finally, I discuss briefly the case of the so-called discrete spin or polarization experiments, which has been treated in detail elsewhere. As mentioned at the beginning, the quantal assumption that the individual spin is a dichotomoic variable should be replaced by a classical continuous variable. The discreteness comes after averaging and by an idealization of the experiments. We take as the model of an individual spin the vector

$$\boldsymbol{S} = S(\cos\varphi\sin\theta, \sin\varphi\sin\theta, \cos\theta), \tag{8'}$$

which can also be thought as the expectation value of quantum spin in a spin-coherent state. Here θ and φ are the "hidden" parameters of the single event. Like the spatially localized wavelets, we obtain the "spin-localized" waves in the direction (θ, φ) in a magnetic field $B = (B_0 - \beta z)\hat{k} + \beta y \hat{\jmath}$, for example, from the wave equation

$$i\hbar \frac{\partial \psi}{\partial t} = -\frac{\hbar^2}{2m}\Delta\psi + g\frac{\mu_0}{\hbar}S\left(\beta y \sin\varphi \sin\theta + (\beta_0 - \beta z)\cos\theta\right)\psi, \qquad (9)$$

where θ and φ are fixed for a single event. For a large constant field B_0, the rapid oscillations $\exp\left(ig\mu_0 \frac{B_0 S}{\hbar} t\right)$ can be factorized. Now we average over θ for positive and negative values (upper and lower hemisphere for the spin direction ($\int_0^{\pi/2} \cos\theta d\theta = 1$, $\int_{\pi/2}^{\pi} \cos\theta = -1$, similarly for $\sin\varphi$ and $\sin\theta$) and obtain the two equations for the averages $\bar{\psi}_\pm$:

$$i\hbar \frac{\partial \bar{\psi}_\pm}{\partial t} = -\frac{\hbar^2}{2m}\Delta\psi_\pm + \left(g\mu_0 \frac{S\beta}{\hbar}\right)\begin{pmatrix} -z \\ +z \end{pmatrix}\bar{\psi}_\pm, \qquad (10)$$

which are the well-known discrete quantum equations for each spin component. Even then, the solutions for $\psi_\pm$, which are Airy functions on the screen, do not give two sharp lines, but two maxima in agreement with experiments. This shows that the usual idealization of spin counting as $\pm$ or, in popular expositions of the discussion of spin correlations, as two red and green lights, is not correct.

In the EPR-type spin correlation experiments with two spins in a singlet state ψ_s, one measures the correlation function

$$E(A,B) = \frac{\langle AB\rangle - \bar{A}\bar{B}}{\sqrt{\langle A^2\rangle\langle B^2\rangle}}, \qquad (11)$$

where $A = S_1 \cdot \hat{a}, B = S_2 \cdot \hat{b}$ are the projections of two spins in the $\hat{a}$ and $\hat{b}$ directions, respectively, and $< A >$ means in quantum mechanics $\langle \psi_S \left| \frac{1}{2}\sigma \cdot \hat{a} \right| \psi_S \rangle$, but in our case it means the integration over θ and φ: $\langle A\rangle = \int d\theta \sin\theta d\varphi S_1(\theta,\varphi)\cdot\hat{a}$, using (8). We have therefore

$$E(A,B) = \frac{\int \sin\theta d\theta d\varphi \left(S_1(\theta,\varphi)\cdot\hat{a}\right)\left(S_2(\theta,\varphi)\cdot\hat{b}\right)}{\left[\int \sin\theta d\theta d\varphi \left(S_1\cdot\hat{a}\right)^2 \int \sin\theta d\theta d\varphi \left(S_2\cdot\hat{b}\right)^2\right]^{1/2}}.$$

For $S_1 = -S_2$, the correlation function so calculated gives exactly the same result as quantum calculation. Similarly for triplet and other states.

One might say that experiment actually measures discrete correlations $(A,B) = (++,+-,-+,--)$. But in reality one has to choose a time interval $\Delta\tau$ within which one decides to call two events to be in coincidence. And there are also single events without a partner on the other side. These single events are not usually counted in experiment, and they are due to the tails of the two bumps in the Stern-Gerlach measurement that we talked about. Thus, both theoretically and experimentally, the identification of single events with discrete dichotomic spin variables is not justified.

The same applies to other arguments with photons purporting to show that, for example, Lorentz-invariance and local realism are incompatible. These are again based on the assumption of the dichotomic nature of photon, that it goes either through a polarizer (or detector) or not, with nothing in between. Indeed, we should talk, as we argued in this paper, of a localized light lump characterized by an angle of incidence θ with continuous polarization, for example.

4 CONCLUSIONS

In conclusion, I summarize the goals and achievements of the Quantum Theory of Single Events:

(1) It completes the standard probabilistic quantum theory by adding to it an underlying new dimension of deterministic single particle behavior. This is probably what de Broglie and Schrödinger intended to do, a goal which was arrested by Born's introduction of probability amplitudes .

(2) By distinguishing two wave functions, $\psi(x;a)$ and $\Psi(x)$, one depending on the hidden parameters a, the other not, it avoids quantum paradoxes and the denial of physical reality and other confusions arising from identifying these two different entities; in particular it eliminates the notion of the "collapse of the wave function," because $\psi(x,a)$ is already collapsed and $\Psi(x)$ never collapses.

(3) It gives a new procedure to obtain the classical limit of quantum theory, as well as a new meaning to the process of quantization. Classical theory is necessarily a single event theory: Each trajectory or each solution of field equations have the parameters of the initial conditions. Therefore we have a direct limit, in fact a derivation of classical mechanics, from the localized solutions of the wave equation. The general idea that quantum mechanics goes into classical mechanics in the limit $\hbar \to 0$ is very problematic and does not work in general. Furthermore we have to change also the interpretation from probability to determinism. Recent problems such as the quantum analog of the chaotic systems are best discussed in the framework of the quantum theory of single events. Moreover, since a "quantization" is obtained from singled events by averaging, the reverse process should also proceed via single events. Hence "quantization" could be better called the "wave-ization" of trajectories to wavelet motion.

(4) It points out to new directions of research by constructing detailed models of electrons and other particles, in particular those which would avoid infinite self energies, and would allow us to construct nonperturbative theories of interactions.

Quantum theory of single events is a more complete and accurate rendering of nature than both of its limits, the classical mechanics as one limit, and the usual quantum mechanics as the other limit.

REFERENCES

Quantum Theory of Single Events has been reviewed in:

A. O. Barut, "Quantum Theory of Single Events," in *Symposium on the Foundations of Modern Physics 1990*, P. Lahti and P. Mitelstaedt, eds. (World Scientific, Sinagapore, 1991), p. 31–46. "Brief history and recent developments in electron theory and quantum-electrodynamics," in *The Electron*, D. Hestenes and A. Weingartshofer, eds. (Kluwer, 1991).

Since then the following papers appeared on this topic:

A. O. Barut and S. Basri, "Path integrals and quantum interference," *Am. J. Phys.* **60**, 896 (1992).

A. O. Barut, "How to avoid 'quantum paradoxes'," *Found. Phys.* **22**, 137 (1992).

A. O. Barut, "There are two quantum theories–not one," in *Bell's Theorem and the Foundations of Modern Physics*, A. van der Merwe, F. Selleri, and G. Tarozzi, eds. (World Scientific, Singapore, 1992).

A. O. Barut and A. J. Bracken, "Particle-like solutions of the electromagnetic field," *Found. Phys.* **22**, 1267 (1992).

A. O. Barut, "Formulation of wave mechanics without the Planck's constant $\hbar$," *Phys. Lett. A* **17**, 1 (1992).

A. O. Barut, "Deterministic wave mechanics. A bridge between classical mechanics and probabilistic quantum theory," in *The Interpretation of Quantum Theory: Where we Stand*, L. Acardi, ed. (Acta Encyclopedia, Rome, 1993).

RESTORING LOCALITY WITH FASTER-THAN-LIGHT VELOCITIES[1]

Philippe H. Eberhard

Lawrence Berkeley Laboratory
University of California
Berkeley, California 94720

The idea of "locality" is a deep-rooted concept. It does not have to be abandoned even if "loophole-free" EPR experiments are performed and confirm the predictions of quantum theory. To satisfy locality, one can imagine that influences at a distance are exerted via mechanisms involving an ether and effects propagating in that ether at a velocity $V > c$. Such model of physical phenomena is not Lorentz invariant but, with V large enough, the model can be made to reproduce the results of all experiments where quantum mechanics and Lorentz invariance have been verified.

Key words: locality, superluminal, quantum-model, ether, EPR.

1. A HISTORICAL PRECEDENT

It is the contention of this paper that loophole-free EPR experiments [1] would not rule out all forms of "locality per say," i.e., would not necessarily imply that influences can be exerted at a distance instanteously without the mediation of an ether. The argument is inspired by a historical precedent, from Newton's times, when the principle of locality was threatened already. In Newton's theory of gravitation, celestial bodies exert attractive forces on each other, instantaneously at a distance and through vacuum, i.e., without an apparent material support to mediate that action at a distance. However, in a letter, Newton expressed himself in the following terms [2]:

> ...That gravity should be innate, inherent, and essential to matter, so that one body may act upon another at a distance through a *vacuum*, without the mediation of anything else, by and through which their action and force may be conveyed

> from one to another, is to me so great an absurdity that I believe no man who has in philosophical matters a competent faculty of thinking can ever fall into it. ...

This quote shows that, in Newton's times, the existence of "something" to mediate actions at a distance was not an idea that could be disposed of easily, even though the theory of gravity at the time seemed to point in another direction. Today, because predictions of quantum theory concerning the EPR paradox [3] have been shown to imply instantaneous actions at a distance [4], that locality principle is in jeopardy again, but, now as in Newton's time, it is natural that nevertheless one look for possibilities to hang on to that principle.

There was a remarkable development of gravitation theory, long after Newton's letter, when that theory was modified and did not involve instantaneous actions at a distance anymore. In "general relativity," indeed, gravitational effects propagate in gravitational waves at the finite velocity c. General relativity is compatible with experimental results of today, but it is also compatible with astronomical observations that justified Newton's gravitation theory, because the latter observations had measurement errors which made the difference between instantaneous and finite-speed actions at a distance unnoticeable.

In this paper, it is suggested that the same circumstances may apply again to the present interpretation of EPR experiments [5] against locality and that this can be true even if and when a loophole-free EPR experiment [1] is performed, if and when it demonstrates the existence of actions propagating faster than c. Then experiments that have confirmed Lorentz invariance and restricted relativity can be reinterpreted in order for the theory to cope with those superluminal velocities. To demonstrate this point, it is sufficient to construct one model that accounts for all experimental data to date and is local. Such a model exists [6]. It is not claimed to yield the correct description of phenomena in nature but it is an example to prove the argument.

2. THE LOCAL MODEL

2.1. Description of Reality

The model uses operators in Fock space. Let $\psi_1 \ldots \psi_n$ be any set of vectors in Fock's space. The ψ_k's are not necessarily normalized to 1. The model uses operators in Fock space of the same mathematical form as density matrices of quantum mechanics,

$$M = \sum_k \psi_k \psi_k^\dagger, \tag{1}$$

$$Tr\{M\} = 1. \tag{2}$$

At each point of space of coordinates x and at time t, there is an operator called the "quantum-state matrix," $\underline{Q}(x,t)$, of the form M, Eq. (1), and satisfying Eq. (2). It describes *reality* at the point of coordinates x at time t and contains all the information that is available there. $\underline{Q}(x,t)$ is a local quantity. It does not depend on the origin of coordinates of space and time.

As in quantum mechanics, an observable is associated with a set of projection operators. The probability p_j of observing outcome $\#j$ of a measurement performed at x and t is given by a trace

$$p_j = Tr\{\underline{\Pi}_j \underline{Q}(x,t)\} \tag{3}$$

where the $\underline{\Pi}_j$ are projection operators that depend on the kind of measurement performed but not on x and t. The projection operator $\underline{\Pi}_j$ is related to its counterpart $\widetilde{\Pi}_j$, in the Heisenberg representation of quantum mechanics, by the relation

$$\underline{\Pi}_j = e^{-iHt} e^{iPx} \widetilde{\Pi}_j e^{-iPx} e^{iHt}, \tag{4}$$

where P is the total momentum and H the Hamiltonian operators as they are defined in quantum field theory. Equation (4) makes $\underline{\Pi}_j$ an operator independent of translations of the origin of space and time coordinates, unlike $\widetilde{\Pi}_j$.

Let us define the operator

$$\widetilde{Q}(x,t) = e^{iHt} e^{-iPx} \underline{Q}(x,t) e^{iPx} e^{-iHt}. \tag{5}$$

$\widetilde{Q}(x,t)$ is a mathematical quantity that is useful to compare the predictions of the model with those of quantum mechanics. It cannot be used to describe a *local reality* at x and t because, unlike $\underline{Q}(x,t)$, it does depend on the origin of coordinates of space and time.

Equations (3) and (4) show that the probability p_j of outcome $\#j$ is the same in the model and in quantum mechanics if the density matrix in Heisenberg representation is [7]:

$$\rho = \widetilde{Q}(x,t), \tag{6}$$

x and t being the space coordinates and the time of the measurement.

2.2. Evolution of $\underset{\sim}{Q}(x,t)$

$\underset{\sim}{Q}(x,t)$ evolves in time as the result of two effects, the "Schrödinger evolution" and the "collapse phenomenon." Most of the time, $\underset{\sim}{Q}(x,t)$ evolves according to the equation

$$\frac{\partial \underset{\sim}{Q}(x,t)}{\partial t} = -i\left(H\underset{\sim}{Q}(x,t) - \underset{\sim}{Q}(x,t)H\right). \tag{7}$$

This is the "Schrödinger evolution." Since P and H commute, it follows that, as long as the evolution is of the Schrödinger type between two times t_0 and t, $\widetilde{Q}(x,t)$ is constant in time and, if Eq. (6) is satisfied for some x at some initial time t_0, it is also satisfied at time t. Then, for a measurement made at point x at time t, Eq. (3) yields the same predictions as quantum mechanics [7].

From time to time, $\underset{\sim}{Q}(x,t)$ evolves according to a "collapse" scenario. This happens when a measurement is performed, but it can be made to occur more often if the model is asked to reproduce the results of other models of quantum mechanics where there are spontaneous wave-function collapses without the presence of observers [8]. In any case there is a point m with space coordinates x_m where the collapse first occurs at a time t_m and there are projection operators $\underset{\sim}{\Pi}_{m,i}$. If the collapse is initiated by a measurement, m and t_m are the point and the time at which the measurement occurs and $\underset{\sim}{\Pi}_{m,i}$ is the projection operator associated with the measurement outcome. The collapse mechanism first affects $\underset{\sim}{Q}(x_m,t_m)$ at point m,

$$\underset{\sim}{Q}(x_m,t_m+\epsilon) = \left(\frac{1}{p_{m,i}}\right)\underset{\sim}{\Pi}_{m,i}\,\underset{\sim}{Q}(x_m,t_m-\epsilon)\underset{\sim}{\Pi}_{m,i}, \tag{8}$$

where

$$p_{m,i} = Tr\left\{\underset{\sim}{\Pi}_{m,i}\,\underset{\sim}{Q}(x_m,t_m-\epsilon)\right\}. \tag{9}$$

After the time t_m, there is a propagation of the collapse in space at a velocity V, which is a parameter yet of unknown value in the model, but which may be larger than c. At each point of space coordinates x, there is a time $t_c(x), (t_c(x) \geq t_m)$, at which a collapse occurs following the measurement at m,

$$t_c(x) = t_m + \frac{|x-x_m|}{V}. \tag{10}$$

Local (differential) equations can be written for $\underset{\sim}{Q}(x,t)$ and the propagation of the collapse phenomenon, [6], in such a way that, for all x's and in *extremely* good approximation,

$$\widetilde{Q}\left(x,t_c(x)+\epsilon\right) = \left(\frac{1}{q_{m,i}}\right)\widetilde{\Pi}_{m,i}\,\widetilde{Q}\left(x,t_c(x)-\epsilon\right)\widetilde{\Pi}_{m,i}, \tag{11}$$

where

$$q_{m,i} = Tr\left\{\underline{\Pi}_{m,i}\,\widetilde{Q}\left(x, t_c(x) - \epsilon\right)\right\}, \tag{12}$$

which are the same rules as those given for the evolution of density matrices during a measurement [7]. Therefore, if $\widetilde{Q}(x,t)$ satisfies Eq. (6) during the Schrödinger evolution that preceded $t_c(x), \widetilde{Q}(x,t)$ satisfies it also during the Schrödinger evolution that follows $t_c(x)$.

3. COMPARISON WITH QUANTUM MECHANICS

3.1. Quasi-Equivalence

If, in the model, one sets the parameter V at ∞, collapses everywhere occur at time $t_c(x) = t_m$, i.e., at the same time as in quantum mechanics. If, in addition, Eq. (6) is verified at all x's at an initial time t_0, Eq. (6) will be satisfied every time later. Then Eq. (3) will yield the same probabilities as quantum mechanics.

Actually the condition of Eq. (6) for all x's at the initial time t_0 can be removed.[2] In quantum mechanics, all experiments first require a "preparation" of the quantum system under study. This preparation consists of preliminary observations which determine the initial conditions well enough to make the predictions meaningful. In mathematical terms, this means that the density matrix ρ_0 before the preliminary observations is in principle unknown, but the preparation induces collapses of that density matrix ρ_0 into another, ρ, where the ambiguities caused by our uncertainties about ρ_0 can be neglected. The matrix ρ is the one involved in the measurements later on. All that is predicted by quantum mechanics are correlations between the observations made during the "preparation" and those made during the "measurements."

In the model with $V = \infty$, consider a preparation at time t_0 involving a measurement at a point of coordinates x_0 and a second measurement at time t at a point of coordinates x. Assume $\widetilde{Q}(x, t_0 - \epsilon)$ to be different from $\widetilde{Q}(x_0, t_0 - \epsilon)$. According to the procedure described above, the probability $p_{0,\ell}$ of outcome #ℓ in the "preparation" and the subsequent conditional probabilitity $p_j^{(\ell)}$ of outcome #j in the "measurement" will be

$$p_{0,\ell} = Tr\left\{\widetilde{\Pi}_{0,\ell}\,\widetilde{Q}(x_0, t_0 - \epsilon)\right\}, \tag{13}$$

$$p_i^{(\ell)} = \left(\frac{1}{q_{0,\ell}}\right) Tr\left\{\widetilde{\Pi}_j \widetilde{\Pi}_{0,\ell}\,\widetilde{Q}(x_0, t_0 - \epsilon)\widetilde{\Pi}_{0,\ell}\right\}, \tag{14}$$

where $\widetilde{\Pi}_{0,\ell}$ is the projection operator associated with outcome #ℓ during preparation and

$$q_{0,\ell} = Tr\left\{\widetilde{\Pi}_{0,\ell}\,\widetilde{Q}(x, t_0 - \epsilon)\right\}. \tag{15}$$

These probabilities $p_{0,\ell}$ and $p_i^{(\ell)}$ are the same as those computed in quantum mechanics, if, before preparation, the unknown density-matrix ρ_0 was assumed to be

$$\rho_0 = \sum_\ell \left(\frac{p_{0,\ell}}{q_{0,\ell}}\right)\ \widetilde{\Pi}_{0,\ell}\,\widetilde{Q}(x, t_0 - \epsilon)\widetilde{\Pi}_{0,\ell}. \tag{16}$$

Thus the same correlations $p_j^{(\ell)}$ between preparation and measurements will be predicted by the model and by quantum theory.

If V is not ∞ but measurements are spaced in time by intervals larger than $\frac{1}{V}$ times the distance between the points at which the subsequent measurements occur, the delay $\frac{|x-x_m|}{V}$ in Eq. (10) becomes irrelevant. Consider two measurements, first one in x_m at t_m, then one at x at time t. The quantum state matrix $\underline{Q}(x,t)$, thus $\widetilde{Q}(x,t)$, will have been subjected to the collapse of the type of Eq. (11) initiated by the measurement at x_m and t_m before the measurement at x and t takes place. Since $\widetilde{Q}(x,t)$ is constant between collapses, $\widetilde{Q}(x,t)$ will have the same value at time t as if the collapse had occurred at time t_m instead of $t_c(x)$. Therefore all predictions will be the same as if $V = \infty$, i.e. as in quantum theory.

3.2. Differences with Quantum Theory

From what is said above, it follows that the only circumstances where the model leads to different predictions than quantum mechanics are ones where there is one measurement at point x at time t after another at point x_m at time t_m, and they are spaced in time such that

$$|t - t_m| < \frac{|x - x_m|}{V}. \tag{17}$$

Then the probabilities p_j for the second measurement are computed using Eq. (3) with a matrix $\underline{Q}(x,t)$ that has not been subjected to the collapse generated by the first measurement. In the model, the probability $p_{i,j}$ of observing outcomes $\#i$ at the measurement at point (x_m, t_m) and $\#j$ at point (x,t) is

$$p_{i,j} = Tr\left\{\widetilde{\Pi}_{m,i}\,\widetilde{Q}(x_m, t_m)\right\}\ \times\ Tr\left\{\widetilde{\Pi}_{m,i}\,\widetilde{Q}(x,t)\right\}, \tag{18}$$

which, in general, will be different than the predictions of quantum mechanics.

These probabilities may be also different from those made by other local models where collapses propagate at a finite velocity V. However, in all those local models, Bell's inequalities must be holding when Ineq. (17) holds, while they do not always hold for the predictions of quantum theory. Therefore if V is not too large, there is a possibility to detect a failure of orthodox quantum theory in future experiments and our model could account for certain kinds of these violations. The cases where Ineq. (17) is satisfied have been studied in more detail in Ref. 6.

In any event, V is a free parameter of the model and can be assumed to be as large as necessary to make the difference with $V = \infty$ negligible in the data of all the experiments where quantum mechanics has been verified. Therefore the model is compatible with any *tested* prediction of quantum theory and the model cannot be ruled out by any past or future experiment where the predictions of quantum theory are verified. The experiment can only set a lower limit for V.

4. LORENTZ INVARIANCE

In relativistic quantum mechanics, if probabilities p_j have been computed using a given space-time rest frame, the same results will be obtained using the same computation rules in another rest frame. For relativistic quantum theory, there is no fundamental property that distinguishes one rest frame from another. In this sense, relativistic quantum mechanics is Lorentz invariant.

For the model referred to in this paper, if $V = \infty$, the computation of the probabilities p_j can also be carried out using the same rules in all rest frames. This is obvious since, in all rest frames, the predictions will be the same as in quantum mechanics. However the model does not only provide a procedure to compute the p_j's, it is also supposed to describe *reality* at each point of coordinates x at time t by the quantity $\underline{Q}(x,t)$. The collapse of $\underline{Q}(x,t)$ at the same time t_m for all x's is not Lorentz covariant. The description of reality by the model needs a special Lorentz rest frame where collapses occur at the same time for all x's. This rest frame can be considered to be the rest frame of an *ether* and, in this sense, the model is not Lorentz invariant. However, since the probabilities of observable quantities are invariant, there is no experiment that can tell what the ether rest frame is. As in quantum mechanics, any rest frame can be used to make the predictions.

If V is finite but larger than c, the model is also not Lorentz invariant and one can define the rest frame of the ether as the rest frame in which collapses propagate with the same velocity V in all directions. If, in the ether rest frame, measurements are spaced in time by an interval so large that Ineq. (17) is never satisfied, the probabilities p_j are the same as for $V = \infty$, therefore it is not possible to identify the ether rest frame. All rest frames can be used equally to compute the predictions.

If $V < \infty$ and condition (17) applies to two measurements, then the predictions for the second measurement are different from what they are for $V = \infty$, therefore from those of quantum mechanics. These differences exist only if condition (17) is fullfilled in the ether rest frame and, for $V > c$, condition (17) is not Lorentz invariant. One can study experimentally the space-time conditions under which the correlations between two observables measured at two different locations are different from those predicted by quantum mechanics, and determine what the ether rest frame is. This would be the best way to illustrate the violation of Lorentz invariance in nature if the predictions of the model are correct. However this rest frame cannot be identified if condition (17) is never fulfilled in a way that can be seen in the data of an experiment.

In conclusion, the model provides a frame work to test both quantum mechanics and Lorentz invariance. In addition, before any violation is found, the model can be used by anyone agreeing with Newton's above quoted statement and wanting to justify his belief in "locality."

REFERENCES

1. N. D. Mermin, "The EPR experiment—Thoughts about the 'loophole,'" *New Techniques and Ideas in Quantum Measurement Theory*, D. M. Greenberger, ed. (New York Academy of Science, New York, 1986), pp. 422-428.
2. I. Newton in a letter to R. Bentley, 1692/3, quoted in *Theories of the Universe*, M. K. Munitz, ed. (Free Press, New York, 1957), p. 217.
3. A. Einstein, B. Podolsky, and N. Rosen, *Phys. Rev.* **47** (1935) 777.
4. J. S. Bell, *Physics* **1** (1964) 195; followed by generalizations: J. F. Clauser, M. A. Horne, A. Shimony, and R. A. Holt, *Phys. Rev. Lett.* **23** (1969) 880; H. P. Stapp, *Phys. Rev. D* **3** (1971) 1303; J. F. Clauser and M. A. Horne, *Phys. Rev. D* **10** (1974) 526; J. S. Bell, CERN preprint Th 2053, reproduced in *Epistem. Lett.* (Assoc. F. Gonseth, CP 1081, CH-205 Bienne) **9** (1975) 11.
5. S. J. Freedman and J. F. Clauser, *Phys. Rev. Lett.* **28** (1972) 938; J. F. Clauser, *Phys. Rev. Lett.* **37** (1976) 1223; E. S. Fry and R. C. Thompson, *Phys. Rev. Lett.* **37** (1976) 465; A. Aspect, P. Grangier, and G. Roger, *Phys. Rev. Lett.* **47** (1981) 460; *ibid.* **49** (1982) 91; A. Aspect, J. Dalibard, and G. Roger, *Phys. Rev. Lett.* **49** (1982) 1804; and many others.

6. P. H. Eberhard, in *Quantum Theory and Pictures of Reality*, W. Schommers, ed. (Springer, Heidelberg, 1989), p. 169.
7. J. von Neuman, *Mathematical Foundations of the Quantum Mechanics* (Princeton University Press, Princeton, 1955).
8. G. C. Ghirardi, A. Rimini, and T. Weber, *Phys. Rev. D* **34** (1986) 470; G. C. Ghirardi, P. Pearle, and A. Rimini, *Phys. Rev. D* **42** (1990) 78.

NOTES

1. This work is supported in part by the Director, Office of Energy Research, Office of High Energy and Nuclear Physics, Division of High Energy Physics of the U. S. Department of Energy, under Contract DE-AC03-76SF00098.
2. This had not been noticed at the time Ref. 6 was written.

THE WAVE-PARTICLE DUALITY AND THE AHARONOV-BOHM EFFECT

Miguel Ferrero[1] and Emilio Santos[2]

1. *Universidad de Oviedo, Departamento de Física, 33007 Oviedo, Spain*
2. *Universidad de Cantabria, Departamento de Física Moderna, 39005 Santander, Spain*

We review some ideas put forward by Louis de Broglie in his seminal papers of 1923, where he set up the dual nature of matter starting from some formal relativistic considerations. In the light of its origin it can be said that the de Broglie waves were born as non-material waves associatted with the canonical momentum. Next we use the Aharonov-Bohm effect in order to show the correctness of this conclusion. As a consequence, the ontological status of the de Broglie waves should be the same as the status of the canonical momentum. Finaly, the recent experiments on interferometry of atoms are used to argue that we cannot avoid duality and we conclude that a consistent view of the de Broglie waves would imply that the world consists of particles submerged in a sea of waves.

Key words: wave-particle duality, local realism, vacuum fluctuations.

1. THE DE BROGLIE WAVE-PARTICLE DUALITY

Louis de Broglie wrote three seminal papers [1] in September-October 1923 in which he put forward his hypothesis about the wave-particle duality of matter. A summary of these three *Comptes Rendues* papers with an important note added was published shortly after in the *Philosophical Magazine* [2] and in fact his famous thesis [3] was the consequent development of the main ideas contained in these papers.

It seems that his first thoughts about the necessity of associating always the character of waves with that of the particles was generated in his early experimental work with his brother Maurice in the X-ray spectra and the photoelectric effect, and developed through the long discussions they had in relation with the interpretation of these experiments [4]. These first thoughts were soon stimulated by a paradox which, de Broglie confessed, for a long time had puzzled him. The paradox was the following. Consider a body, whose mass at rest is m_0, moving with respect to a fixed observer

Waves and Particles in Light and Matter, Edited by
A van der Merwe and A Garuccio, Plenum Press, New York, 1994

with velocity v=βc (β<1). The internal energy of this body is m_0c^2 and, as Planck's quantum hypothesis suggests, to this energy it can be associated a frecuency $\nu_0=m_0c^2/h$. For the fixed observer the relativistic energy is given by

$$E=\frac{m_0c^2}{\sqrt{1-\beta^2}} \tag{1}$$

so the corresponding associated frequency will be

$$\nu=\frac{1}{h}\frac{m_0c^2}{\sqrt{1-\beta^2}} \tag{2}$$

But, if the fixed observer looks at the "internal phenomenon" related to the frequency ν_0, he will see its frequency, owing to the relativistic dilation of time, lowered and equal to $\nu_1=\nu_0\sqrt{1-\beta^2}$. Now, the two frecuencies ν and ν_1 are widely different. What is the meaning of this difference? Which is the physical nature of ν? And, what is the physical relation between ν and ν_1? This was the kind of questions which worried him for a long time, that determined all trend of his research work and which, in some way, have not been fully answered even today.

It is well known that de solution given by de Broglie to these questions through the "harmony of phases theorem" implied that a moving body is accompanied always by a wave, a "phase wave," *being impossible to separate the motion of the body and the propagation of the wave.* ν would be then the frequency of this "phase wave," whose propagation would have the same direction as the body but with a different velocity, a superluminal and observer dependent velocity, given by V=c/β.

Another argument in favor of this solution was that this result was contanied in Lorentz time transformation. This was put forward by de Broglie as follows. Suppose that at time t=0 the reference frames of the fixed observer and the moving body coincide and that the last is moving, for simplicity, along the x-axis. Let τ be the local time of an observer moving with the body. Using the Planck quantum hypothesis, he will describe the "associated internal periodic phenomenon" at any time t by the function $\sin 2\pi\nu_0\tau$. Now, the fixed observer would describe the same periodic phenomenon by the function $\sin 2\pi\nu_0\,(t-t')/(\sqrt{1-\beta^2})$ with t ′ equal to x/V, where x=vt is the distance from the origin to the body at time t, and V the velocity of propagation of the "periodic phenomenon". This last function can be understood as representing a wave of frecuency $\nu_0/\sqrt{1-\beta^2}$, which indeed coincides with ν.

The next step he gave, almost at the same time (Oct.-Nov. 1923), was the generalization of the previous results. This was done in a very short note added to his *Phil. Mag.* paper. The essential part of his argument, as is well known, was the following. Hamilton's principle of least action for a *material point* can be expressed in space-time by the equation

$$\delta \int \sum_{1}^{4} J_{\mu} dx^{\mu} = 0, \tag{3}$$

where J_{μ} is the four-vector (E/c, p_x, p_y, p_z), E is the energy and $\boldsymbol{p}$ the *canonical momentum* of the material point. We want to emphasize that $\boldsymbol{p}$ is *not* the *kinematical* momentum or quantity of movement $\boldsymbol{g}=m\boldsymbol{v}$, and de Broglie says that explicitly in his thesis considering a particle in a magnetic field (pgs. 32, 33).

On the other hand, Fermat's principle for the propagation of waves can be written

$$\delta \int \sum_{1}^{4} O_{\mu} dx^{\mu} = 0, \tag{4}$$

where O_{μ} is the four-vector: (ν/c, n_x, n_y, n_z), and $\boldsymbol{n}$ is the wave vector, except for a factor of 2π [$\boldsymbol{n}=\frac{1}{\lambda}\boldsymbol{n}^0$, where $\boldsymbol{n}^0$ is a unit vector].

De Broglie points out that the quantum relation establishes the identity between the first components of this two four-vectors, that is,

$$J_1 = hO_1,$$

but there is no real reason why we should restrict the equality to only one component, this suggesting that the quantum relation should be writen as

$$\boldsymbol{J} = h\,\boldsymbol{O}$$

a relation from which he got immediately the identity of the two principles and his famous relation for the associate de Broglie wave length

$$\boldsymbol{p} = h\boldsymbol{n} = \frac{h}{\lambda}\boldsymbol{n}^0 \tag{5}$$

[or $\lambda = h/p$ if there is no magnetic field].

We have reviewed some ideas of de Broglie in order to show that his final conclusion, that is, that any moving body must be accompanied by a non-material wave of λ=h/p was, to some extent, the *formal* consequence of trying to reconcile relativity and the quantum postulate. To be more especific: Given the experimental evidence about the reality of light quanta available at that time –mainly the photoelectric and Compton effects– he was trying to understand the quantum postulate in the light of the special theory of relativity combining the corpuscular and the wave characters of light.

Which is the physical nature of the duality envisaged by de Broglie? Are the de Broglie waves real? If both the wave and the particle are real, which is the physical relation between material points and associated waves?

We all know that his first interpretation was that both the wave and the particle were real, duality meaning that a microscopic object is at the same time wave and particle, opening thus the way for the experiments made by Davisson and Germer and by Thompson. He concreted this physical interpretation of duality in his theory of the double solution and its simplified version called the pilot wave. But we also know that this is not the only possible interpretation to the duality dilemma. Heisenberg, Born, and Bohr worked out another interpretation in which a microscopic object is never at the same time wave and particle. It is either wave or particle depending on the whole experimental apparatus. But this second interpretation, in contrast to de Broglie's first idea, means that, to some extent, neither the wave nor the particle are real, because a physical property cannot exist until in has been observed, a solution whose consequences about the reality and objectivity of things is not acceptable for everybody.

In the following we will use the Aharonov-Bohm effect to show that the wave associated to the canonical momentum cannot be real, and we will see that the proposal of trying to make this waves "real waves" leads inevitably to the conclusion that the whole space should be filled with waves, giving rise to a more philosophically acceptable interpretation of the duality dilemma.

2. THE AHARONOV-BOHM EFFECT AND THE DE BROGLIE WAVES

Let us consider an Aharonov-Bohm [5] experimental device, which consists of a source of charged particles and a double-slit diffraction apparatus with a long solenoid placed behind the slits and perpendicular to the plane of the figure, so that a magnetic field can be created inside the solenoid, while the region external to the solenoid remains field-free (Fig. 1). The solenoid is located in a region that the particles cannot reach and moreover it may be surrounded by a cylindrical shielding impenetrable to the charged particles.

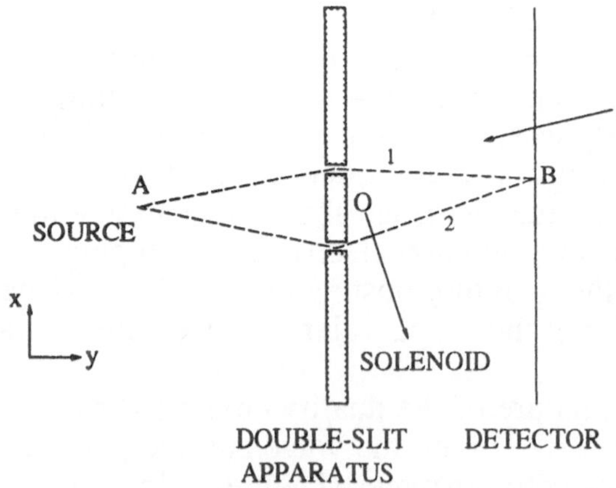

Fig. 1 Aharonov-Bohm experimental device.

It is a very well known result that the interference pattern is sensitive to the magnetic flux inside the cylinder even though the particles never pass through the region in which the magnetic field is different from zero.

The interference pattern in the screen depends on the phase difference of the waves coming by the paths 1 and 2. Calling $(\Delta S)_{B=0}$ the phase difference when there is no magnetic field, the difference $(\Delta S)_{B=B}$ when there is a field can be written [6]:

$$(\Delta S)_{B=B} = (\Delta S)_{B=0} + \frac{q}{hc}\Phi, \tag{6}$$

where $\Phi = \oint A \cdot dr$ is the flux inside the cylinder, q is the charge of the particles (electrons) and A the vector potential. This equation tells us how the motion of the particle is influenced by the field B, thus giving the new position of the interference pattern.

The phase difference for two specific points like A and B in the figure is given by

$$\Delta S = \frac{2\pi a}{\lambda} \tag{7}$$

where a is the difference of the geometrical path lengths.

Now, when the magnetic field is created inside the cylinder the interference pattern changes, which means that the phase difference has changed, as it is indicated in expression (6). For the particular point we are considering, B, the geometrical path lengths have not changed, so the only possibility left is that the change is due to a change in the de Broglie wavelength λ.

How may it have changed? Given that the particle goes always through a field-free region, that is where E=B=0, the Lorentz force on the particle

$$F = q(E + v \wedge B)$$

is always equal to zero. This means that $d(mv)/dt=0$ or that $g=mv$=const. So, if g is constant, the de Broglie wavelength, that has changed, cannot be associated with it and the only possibility left is, as we have said in the previous paragraph, the one chosen by de Broglie, that is, to associate the waves with the canonical momentum p

$$p = g + \frac{q}{c}A. \tag{8}$$

The correctness of this conclusions can be easily shown by the following calculation of expression (6).

The phase difference $(\Delta S)_{B=B}$ is given by the equation $(\Delta S)_{B=B} = \oint(dl/\lambda)$, where dl is the element of the path, and λ the de Broglie wave-length. Using the de Broglie relation and the previous formula (8) that we have written for p, we get

$$(\Delta S)_{B=B} = \oint \frac{pdl}{h} = \oint \frac{(g+qA/c)}{h} dl = \oint \frac{mv}{h} dl + \oint \frac{q}{ch} A dl =$$

$$= (\Delta S)_{B=0} + \frac{q}{ch}\Phi.$$

If this conclusion is correct, it means that the de Broglie wave has the same *ontological status* as the canonical momentum p; this being a *gauge dependent* quantity, we conclude that either the de Broglie wave *cannot be* real or we have to modify the criteria by which we decide whether a magnitud is a true physical quantity or a mathematical tool to do calculations.

3. A MODERN VIEW OF THE WAVE-PARTICLE DUALITY

Does this mean that we can avoid duality? We think not. As an example of the unavoidability of the de Broglie duality, we will comment on the recent interference experiment on atoms by Carnal and Mlynek [7].

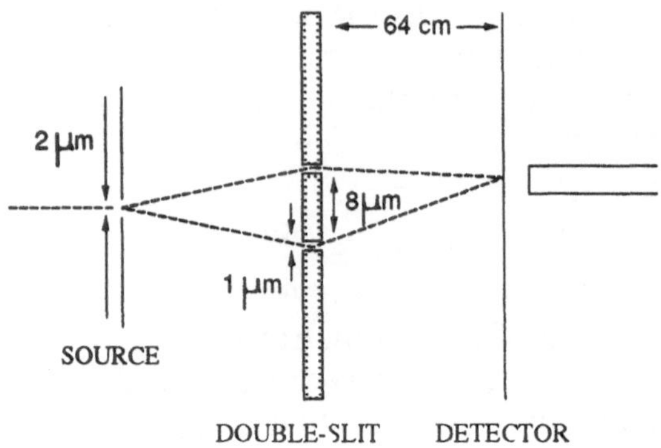

Fig. 2 Experiment of interference of atoms.

The apparatus consists of a source of He atoms at thermal velocity, with a de Broglie wavelength of approximately 1Å, which pass through a double slit of width 1μm in each slit and a mutual separation of 8μm, and a detector, situated far away (at 64 cm). Moving the detector up and down, the quoted authors got the typical interference pattern. A similar result was obtained also by Keith *et al.* [8] using Na atoms and a grating.

Looking at the results of this experiments it is not difficult to arrive at the conclusion that the de Broglie duality is unavoidable. To put it in another way: We need fields and particles. In this experiment the atom cannot pass at the same time for the two slits. This can be seen looking at the dimension of the experimental device: The distance between the slits is 5 orders of magnitude higher than the size of the atom. However, *there is something* that guides the motion of the particle because it knows if the other slit is open or not. Let us admit that "the something" that guides the particle is a kind of de Broglie wave. This posses the following problem: When a quantum entity reaches a beam spliter (e.g., a two-slit apparatus), the particle goes one way and "something" goes the other or both ways (see Fig. 3).

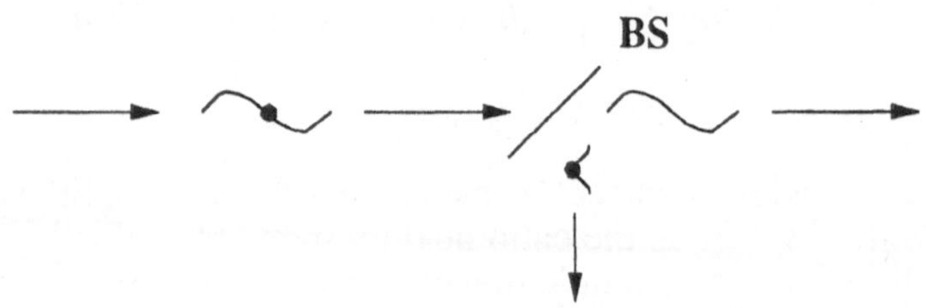

Fig. 3.

If we want to keep the de Broglie wave-particle duality, we must conclude that in every splitting a wave is produced; so, if we repeat this process many times, we are left with the following two possibilities:

(1) The *intensity* of the accompanying wave decreases with time in each splitting. The repetition of the process progresively would cause the wave to die out, so at the end we will have only particles *without the capacity* or producing new interferences. This is not true.

(2) The *total intensity* of the waves increases with time just because in each interaction with the beam splitter a new wave is created. This will lead us to the conclusion that the whole space is filled with waves. We can see this second solution in a different way, without the necessity of creating waves. The beam splitter do not create the waves because they were already there. The space is already filled with waves, and what we have is a particle submerged in a sea of waves; the particle acts on waves, the waves act on the particle, and the external objects (e.g., screens) act in both. Following this philosophy, our view of the de Broglie waves will tell us just that: *the world consist of particles submerged in a sea of waves.*

This "consistent" view of the duality dilemma is far from giving us a complete answer; it poses some problems that we should mention here. The first problem is: *How does this "complex thing" leads to Schrödinger's equation?* The answer is: We don't know. There have been several attempts at derivation, for example, through stochastic mechanics, but in our opinion they have not been truly successful. The second problem it poses is: *What is the nature of the waves?* Following the principle that we should not multiply the entities we already have, we conjecture that the waves are the "vacuum fluctuations" of the fundamental Bose fields: electromagnetic, nuclear, weak and strong, and gravitational.

The use of this "qualitative picture" in the case of the double slit experiment with atoms will give us the following picture:

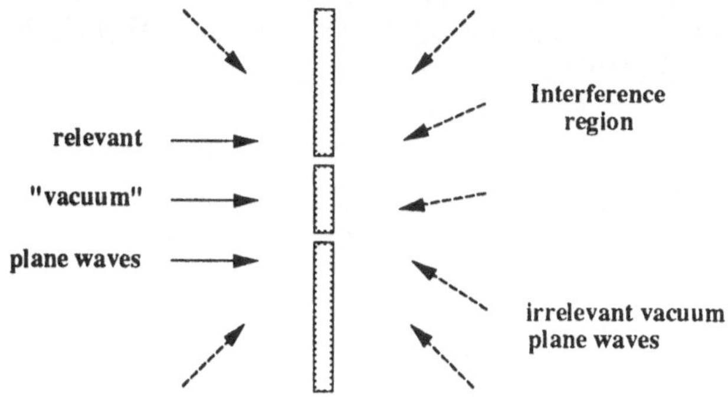

Fig. 4.

If the space is filled with waves, there should be waves coming over the screen from all sides. We don't know exactly what happens, but our conjecture is that the "vacuum field" is modified by the screen and that this modified "vacuum" guides the atom: Everything happens as if there were "relevant" and "irrelevant" vacuum waves, the relevant ones being precisely those which satisfy the de Broglie relation $\lambda = h/p$. Why is then that only the "vacuum" plane waves coming in front of the slits and having precisely the wave lenght $\lambda = h/mv$ are relevant? We have not yet an answer for this question. We conjecture that the random effect of all others waves averages to zero. This is all we can say for the moment.

In any case, as we have said at the begining, this is only a qualitative picture and we have not been able, for the time being, to derive a quantitative model of the quantum phenomena starting from this picture. This will be the subject of our future work.

Acknowledgement: We acknowledge financial support of DGICYT Projects No. PS-89-0171 and No. PB-90.0098 (Spain).

REFERENCES

1. Louis de Broglie, "Ondes et quanta," *C.R. Acad. Sc. (Paris)* **177**, 507-510; 548-550, 630-633 (1923).
2. Louis de Broglie, "A tentative theory of light quanta," *Phil. Mag.* **47**, 446-458 (1924).
3. Louis de Broglie, "Recherches sur la théorie des quanta," *Ann. Phys. (Paris)* **3**, 22-128 (1925).
4. See Louis de Broglie, "Autobiographie scientifique," in *Louis de Broglie, Physicien et Penseur* (Michel, Paris, 1953).
5. Y. Aharonov and D. Bohm, *Phys. Rev.* **115**, 485 (1959).
6. R. Feynman, *The Feynman Lectures on Physics* (Addison-Wesley, Reading, 1965).
7. O. Carnal and J.M. Mlynek, *Phys. Rev. Lett.* **66**, 286 (1991).
8. D.W. Keith, C.R. Ekstrom, Q.A. Turchette, D.E. Pritchard, *Phys. Rev. Lett.* **66**, 2693 (1991).

DE BROGLIE'S WAVE IN SPACE AND TIME

Augusto Garuccio

Dipartimento di Fisica dell'Università
INFN - Sezione di Bari
I-70126 Bari, Italy

The possible experiments which lead to conflicting predictions of quantum mechanics and the de Broglie pilot wave theory are presented and discussed. Recently, one of these has been performed by Wang, Zou, and Mandel, and the results are well in agreement with quantum mechanical prediction. The criticisms of this experimental set-up are discussed. In particular, Selleri's model, based on enhanced photon detection, is analyzed. A proposal of a new experiment to discriminate between the Selleri model and quantun mechanics is presented.

Key words: de Broglie wave, interference, down-conversion, photon pairs, henanced photon-detection.

In most textbooks on quantum mechanics the wave-particle duality is presented as a step of the process which, starting fron the papers on blackbody radiation of M. Planck in 1900, arrives in 1927 at the final formulation of quantum theory. This presentation, if it is correct from a chronological point of view, is reductive as concerning the debate which accompanied the birth of quantum mechanics and has been alive in the physics community up to the present.

The conception of the world and the approach to physical problems by researchers such as de Broglie, Einstein, and Schrödinger was completely different from the approach of Bohr, Heisenberg, Pauli; and completely different was the solutions that these people gave to interpretative problems of the new theory.

So the heuristic significance that de Broglie in his paper of 1924 gave to the wave-particle duality (waves and particles in real space interacting togehter) has been lost, and duality is now presented as an introduction to the complex Schrödinger wave.

Another misleading approach to de Broglie's work is the identification of his theory with the so called Bohm-Vigier approach to quantun mechanics. These authors, starting from the original idea of de Broglie, developed a causal, but nonlocal, theory, which reproduces completely all the results of quantum mechanics.

Waves and Particles in Light and Matter, Edited by
A. van der Merwe and A. Garuccio, Plenum Press, New York, 1994

John Bell, discussing the de Broglie's approach, wrote correctely: "In my opinion the pilot wave picture undoubtedly shows the best craftsmanship among the pictures we have considered ... However it would be wrong to leave the reader with the impression that, with pilot-wave picture, quantum theory simply emerges into the light of day, with the transparency of pure water. The very clarity of this picture puts in evidence the extraordinary "non-locality" of quantum theory."[1]

Indeed, there is a link between de Broglie theory (i.e., a causal realistic theory) and Bell's works on the locality of quantum mechanics. From Bell's inequality we know now that any local and causal theory is incompatible with quantum mechanics; more precisely, not all the results of quantum mechanics can be reproduced in a local realistic theory. Then - in principle - it is possible to find experiments that lead to conflicting testable predictions of quantum mechanics and of the de Broglie pilot wave theory. In particular, the problem concerning the reality of de Broglie waves has been discussed recently in several experimental proposals.

Essentially the question is related to the "correct" interpretation of the wave function Ψ – whether it must be interpreted as a simple tool for predicting mathematical probabilities or, on the contrary, whether it represent, as suggested by de Broglie, a real physical wave propagating in spacetime.

Let us state explicitely the basic assumptions within the de Broglie model:

(1) *A photon is composed of a localized particle and a real wave* Φ *propagating in space and time in accordance with d'Alembert's equation..*

Because of this assumption, when a photon impinges on a semitransmitting mirror, the associated wave is partly transmitted and partly reflected, while the particle is either transmitted or reflected. Obviously, the detection of the particle in one of two channels does not induce the collapse of the real wave in the other channel. For example, speaking about the theory of double solution, de Broglie wrote:

This solution involves the existence of two different waves, one of which is objective and represents a physical reality which, because it is intimately linked with the particle, enables its behaviour to be determined. The other is a subjective construction, which is based on the information which we posses about the objective wave and provides us with a probability representation for the particle. [2]

(2) *If in a region of space, n waves* $\Phi_1, \Phi_2, \ldots, \Phi_n$ *are present, the total wave is given by the sum*

$$\Phi = \Phi_1 + \Phi_2 + \ldots + \Phi_n$$

We stress that the previous assumption holds even when the waves come from different sources, as in the famous Pfleegor and Mandel[3] experiment. De Broglie and Andrade e Silva, commenting on the results of the experiment, wrote:

A photon coming from one laser or the other and arriving in the interference zone is guided, and this seems to us physically certain, by the superposition of the wave emitted by the two lasers ... The movements of the photon in the interference zone are actually guided by that superposition, and not by the single wave that carried it. [4]

(3) *"In an interference field, the probability of a photon's being present at a given point is ... proportional to the square of the amplitude of the wave present at that point."* [2]

THE EXPERIMENTAL PROPOSALS

In the recent years several experiments have been proposed in order to test the validity of de Broglie ideas. Most of these was based on coincidence detection of photon pairs in different branches of a particular experimental set up, in which conflicting results are possible according to quantum mechanics and the de Broglie theory, respectively.

A first proposal[5] was based on a modification of the famous Mandel and Pfleegor experiment[3], in which the authors have observed interference figures with the light originating in two independent laser beams under conditions where the light intensity was so low that the mean interval between the two photons was great compared with their transit time through the apparatus. Two laser beams, after passing through two attenuators, are superposed (Fig. 1) with the help of the semitrasmitting mirror *BS*. A small portion of the unattenuated beams is split off by means of the unsilvered glass M_1 and M_2 and passes through a pinhole to a phototube which activate a gate for the interference detector, for a 20 μs period, whenever the beat frequency falls below 50 KHz.

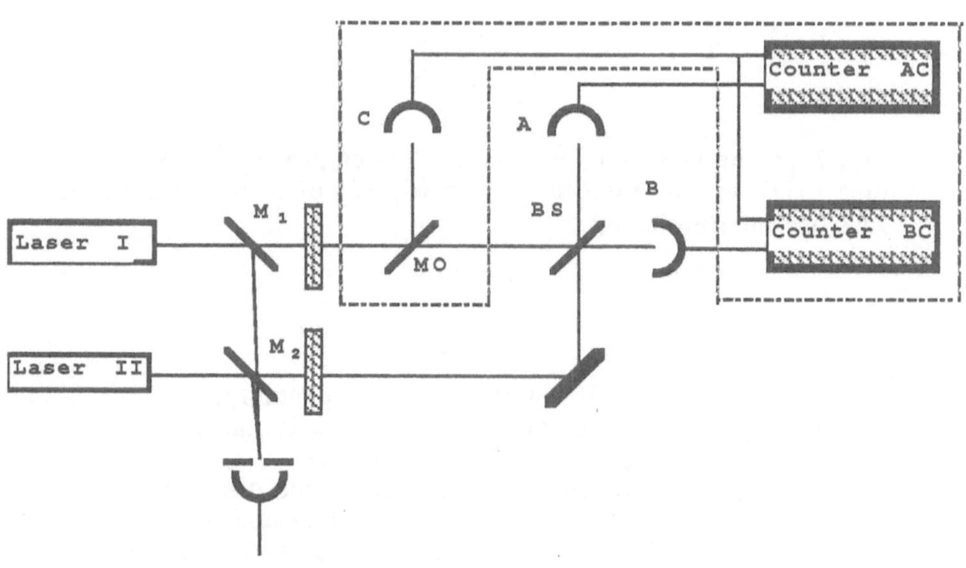

Fig. 1. The Garuccio, Popper, Vigier experimental set-up.

The proposal was to modify the initial apparatus, adding a semi-transparent mirror M_0 which splits the beam of the first laser into two parts. The reflected part is then observed in the photomultiplier C put in coincidence with the photomultipliers A and B.

We then cut down the intensity of both beams so that there can exist only one photon in each beam during the measurement. The de Broglie theory predicts that one should observe the Mandel-Pleegor interference pattern. Indeed, in this theory the wave function is just the classical Maxwell electromagnetic wave and the states (particle plus wave) evolve causally in time. This yields interference in the region R, due to the superposition of the photonless (empty) transmitted part of wave of laser I and the photon

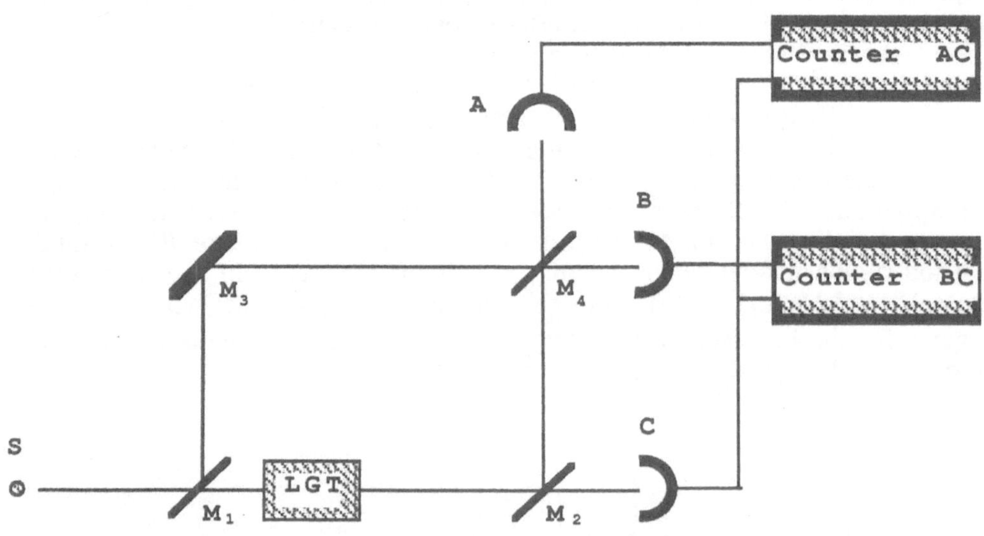

Fig. 2. In the Garuccio, Rapisarda, Vigier proposal a laser gain tube *LGT* was inserted in the arm M_1-M_2 of a modified Mach-Zhender interferometer.

carrying wave of laser II. It is not exactly so in quantum mechanics since the detection of the photon in C collapses instantaneously the wave of laser I, so that no interference should be observed in principle.

This proposal provoked a complex and heated discussion. Besides its experimental difficulty, it has been essentially contested on the basis of the laser nature of the source. In fact, the existence of a Poisson distribution of photons in the laser sources has been used to justify the point that no testable difference exists between the consequence of de Broglie theory and quantum mechanics[6]

In order to overcome the objections to the first proposal, a new experimental set-up was presented[7] based on a laser gain tube.

A single [very weakened] pulse source produces, one by one, independent wave packets, each containing a single photon. These successive one-photon wave packets impinge on a Mach-Zhender interferometer (Fig.2). The semitransmitting mirror (with a transmission coefficient 1/2) M_1 splits the wave packets into two identical parts, whereas one knows that the photon travels either in the reflected beam or in the transmitted beam: There is a 100% anticorrelation between reflected and transmitted photons, as has been verified in several experiments[8]. We then introduce along the path M_1-M_2 a laser gain tube (LGT), which multiplies (on the average) the photon number by two and preserve the phase of the impinging photon. The output pulse impinge on the semireflecting mirror M_2, and the transmitted part is detected in the photomultiplier C. The experiment is now designed to analyze the rate of observed coincidences between photomultipliers C, and A and B.

Two perfectly conflicting results are now theoretically possible:

- According to quantum mechanics no interference should be observed (i.e., the rate of coincidence C-A/C-B should equal 1). Indeed if two photons appear at C, no photon should appear in A or B, moreover if one photon appears at C and one appears in A or B then we know that no photons have taken the M_1-M_3-M_4 path and we are sure that no interference should appear. The passage of a photon in LGT path has collapsed the wave packet.
- According to de Broglie model the interference should appear with the maximum contrast. Indeed, even if the photon pass through the LGT, then a real physical de Broglie wave is moving along the path M_1-M_3-M_4 and interferes with the photon carrying wave travelling along the path M_1-M_2-M_4.

The objections to the preceding proposal was based on the statistical behaviour of a laser gain tube. When the stimulating wave intensity is equivalent to one photon, the rate of spontaneous (and incoherent) emission is equivalent to the rate of stimulated (coherent) emission and this make equivalent the predictions of the two theories.

Now, recent interference of two identical photons produced in parametric down conversion have opened the way to clarify the problem of reality of empty wave[9].

A parametric-down converter (pumped by U.V. laser light) produces pairs of linearly polarized photons; the two photons are generated simultaneously and, following different paths, form two beams, the signal and the idler beams. The beams go through a modified Mach-Zhender interferometer (Fig. 3) in which all the mirrors are semitransmitting and the optical lengths can be varied by a phase-shifter PS. We will consider the events in which the idler photon, after traversing BS_1 and BS_2, is detected by the photomultiplier D_2, and the signal photon is detected by the photomultiplier D_1 after traversing BS_2 and BS_3. The measured quantity is the joint-detection probability in D_1 and D_2 as a function of the optical length difference between the two paths BS_1-BS_4-BS_3, and BS_1-BS_2-BS_3.

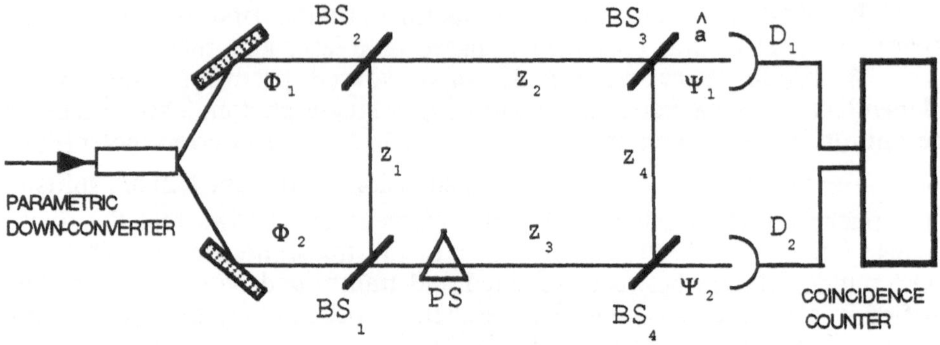

Fig. 3. The Croca, Garuccio, Lepore, Moreira experimental set-up. A parametric-down converter produced photon pairs; the two photons go through a modified Mach-Zhender interferometer in which all the mirrors are semitransmitting. The optical lengths can be varied by a phase-shifter *PS* of a modified Mach-Zhender interferometer. The coincidences between D_1 and D_2 is observed as function of phaseshift.

Then introducing the usual transmission and reflection coefficients t and r satisfying the relations

$$|r|^2 + |t|^2 = 1, \quad rt^* + r^*t = 0,$$

the outgoing wave can be written as

$$\psi_1 = t^2\Phi_1 + r^2t\Phi_2 + tr^2\Phi_2\, e^{i\delta}, \qquad \psi_2 = t^2\Phi_2\, e^{i\delta},$$

and the coincidence probability between D_1 and D_2 is

$$P(D_1,D_2) = \alpha_1\alpha_2|t|^2\left[|t|^4|\Phi_1|^2 + 2|r|^4|t|^2|\Phi_2|^2(1+\cos\delta)\right],$$

since Φ_1 and Φ_2 have random phases. $P(D_1,D_2)$ depends upon the phase difference between the two optical lengths of the interferometer. This effect is due to the fact that in the beamsplitter BS_3 there is an overlapping of the signal wave with the empty wave, generated by the idler photon going through BS_1 and BS_4.

In quantum optics, the joint probability is

$$P_{QM}(D_1,D_2) = C(|t|^2)^4$$

Figure 4 represents the ratio between the two joint detection probability as a function of phase difference between the two arms of interferometer.

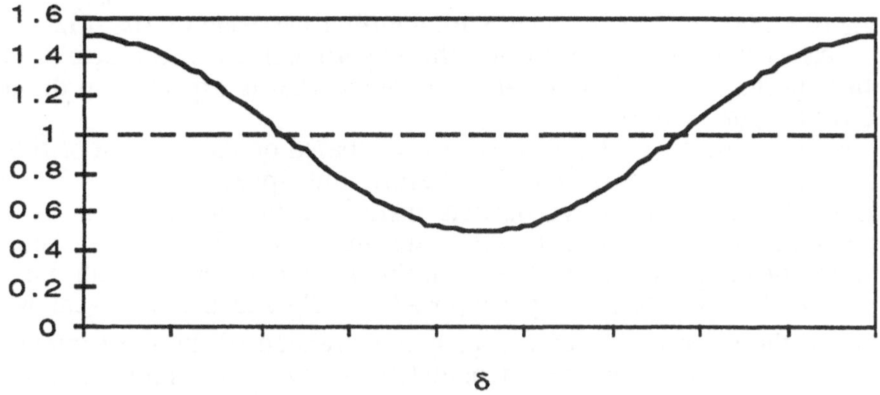

Fig. 4. Predicted results of de Broglie's model (continuous line) and quantum optics (broken line) as a function of phasehift δ.

This experiment has been performed by Wang, Zou, and Mandel[10]. The experimental set-up is lightly different from our proposal, and it is based on the use of three beamsplitter (Fig.5). The interferometer is the *PQRU* apparatus: I-M-R-U-D_1 represents the path of the detected idler photon, P-R-U-D_1 and Q-U-D_1 the path of empty waves, while S-P-Q-D_2 is the path of signal photon.

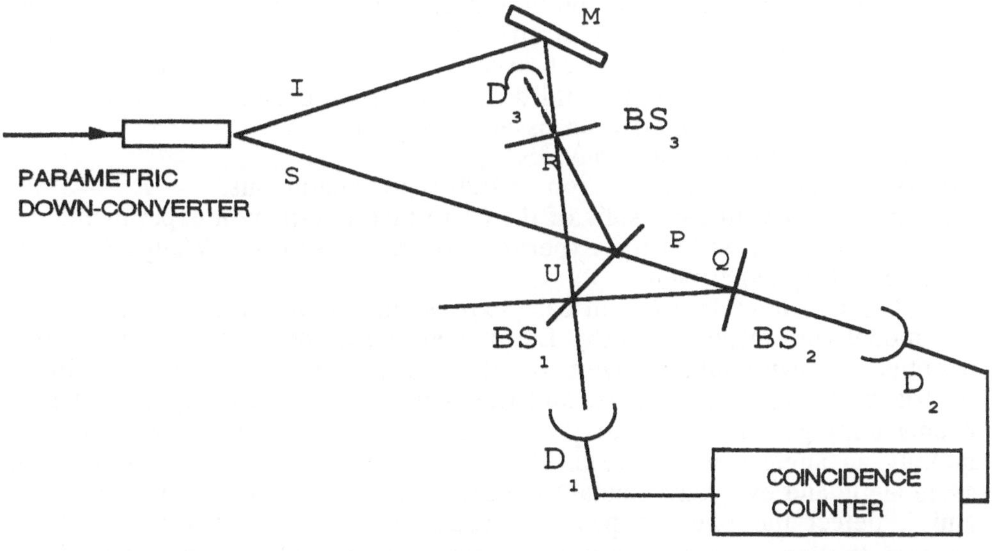

Fig. 5. The Wang, Zou, Mandel experimental set-up. The calibration of the apparatus is delicate, because it is necessary to assure the spatial superposition in BS_1 of the empty wave and of the full wave associated with the idler photon.

The crucial point of the experiment is the overlapping of the full signal wave with the empty wave on the beamsplitter R, and this is checked measuring the coincidence rate R_{13} between the detectors D_1 and D_3. From the measured coincidence rate R_{13} results a coherence length of photon packets of ≈ 200 μm, in agreement with the filter passband of 10^{12} Hz.

As is known, the results of the experiment are in agreement with quantum mechanical predictions and contradict what is expected on the basis of the de Broglie theory.

A first criticism to this experiment was based on the position of filters - before the photodetectors and not before the apparatus [11]. The real coherence length of the wave packets inside the interferometer is 10 time shorter than the coherence length determined by the filters; then the calibration of the photon paths based on this large coherence length does not assure that the packets are superimposed in BS_2 and that two real waves overlap in the same region of space. A new version of the experiment was performed by Wang, Zou, and Mandel[12], following the previous remarks and the results are essentially the same as before.

A different criticism of the results has been proposed by Selleri [13,14] and it is based on variable photon detection probability. Indeed, if the results of previous experiment are correct, it is necessary to deny one of three postulate of the de Broglie theory. The third postulate is the less essential in the theory, and it moreover is equivalent to the Born interpretation of ψ function of quantum mechanics(QM). Starting from a realistic and causal point of view, it is possible to develop variable probability detection models (VDPM) that split the set S of detected objects in a certain number of subsets S_i with probabilities P_i to be detected, so that the overall detection probability results from the average of different probabilities $P = \langle P_i \rangle_i$.These models agree with QM for a single photon detection probability; but, since the average of a product is in general different from the product of the averages, it is in the case of two-particle observations that one can expect a departure both from QM and de Broglie's third assumption on detection probability.

In particular, the model discussed by Selleri:

(a) reproduces single photon physics;

(b) explains the observed violation of Bell-type experiments;

(c) is consistent with the results of the performed two-photon experiments;

(d) is compatible, within the experimental errors, with the Wang, Zou, and Mandel (1991) experiment.

We will now describe an easy experimental configuration that can test this model with respect QM[15]. Let us consider a photon pair produced in a parametric down-converter (Fig. 6); the two photons have the same linear polarization, viz. along z axis, and travel in xy plane. The signal and idler beams impinge on two linear polarizers oriented along the same direction making an angle θ with z axis. The two polarizers can rotate through the same angle and everyday mantain the same direction. Two photomultipliers 1 and 2 detect the photons passed through the polarizers; the outputs of photomultipliers are collected in two counters C_1 and C_2 and in a coincidence circuit C_{12}. The measured quantity is the ratio R between the probability over the product of the probabilities of the single detection as a function of the angle θ of both polarizers.

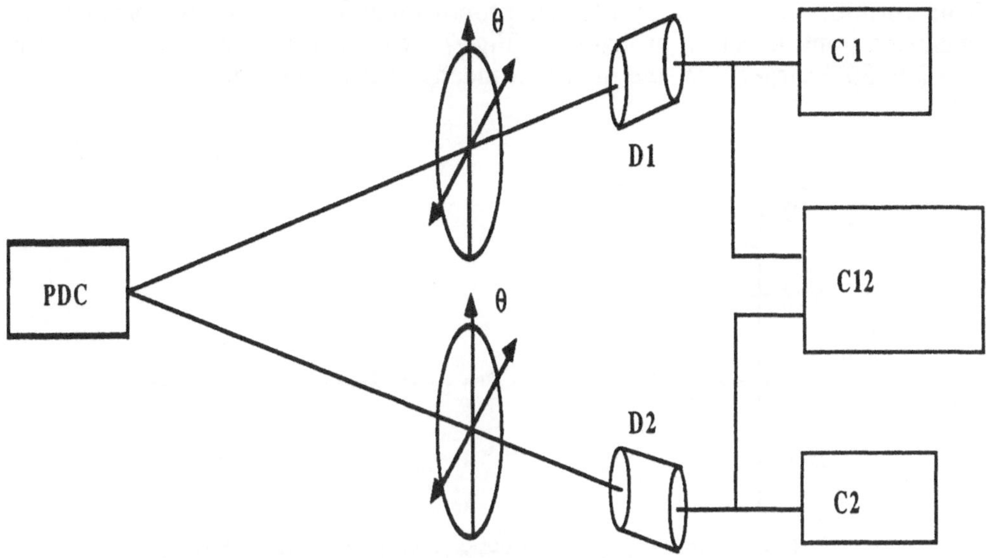

Fig. 6. The proposed experimental set-up for testing variable detection probability.

In the Selleri model the detection probability depends both on a new variable μ, which, for simplicity, is assumed equal to +1 or -1 and equiprobable, and on amplitude of wave function:

$$D(\mu) = \eta[1+ \mu(1-\eta)(1-2\,|\psi|^2\,],$$

where η is the (average) measured quantum efficiency of photomultiplier. It is easy to see , using the previous formula, averaging over μ and assuming that the Malus low holds for the transmission probability of the polarizer, that the probability of single photon detection is

$$P(1) = P(2) = \eta\,\cos^2\theta.$$

The joint probability detection of two photons results:

$$P(1,2) = \eta^2\cos^4\theta[1+ (1-\eta)^{\,2}\cos^2 2\theta].,$$

and the ratio between the joint probability detection and the product of single detection probability becomes

$$R= P(1,2)/P(1)P(2) = [1+ (1-\eta)^{\,2}\cos^2 2\theta].$$

This formula exibits an oscillation proportional to $\cos^2 2\theta$ and predicts an enhanced joint detection probability (Fig. 7), which in the case of $\theta = 0$ and $\eta = 0.1$ is 80% larger than the quantum mechanical prediction.

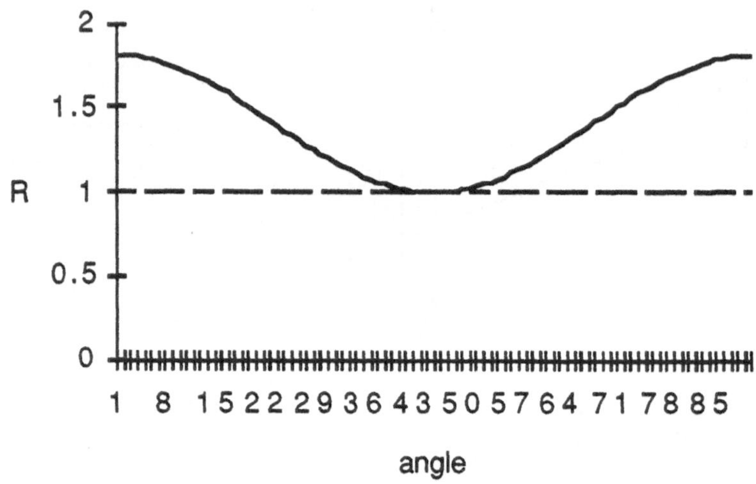

Fig. 7. Predicted results of Selleri's model (continuous line) and quantum optics (broken line) as function of angle of the polarizers.

The experiment is in principle easy to perform and can furnish information on the behavior of strong correlated photon pairs when coincidence measurements are performed. An experiment to measure the ratio between joint detection probability and single detection probability in parametric down-conversion has been performed by Friberg, Hong, and Mandel in 1985 without polarizers[16]; but, due to the non-totally perfect and unknown correlation between the two detected beams, the results are not conclusive.

REFERENCES

[1] J. Bell, *Proceedings of the Nobel Symposium: Possible Worlds in Art and Science* (Stockholm, August 11-15, 1986).
[2] L. de Broglie, *Ondes electromagnétique et photons* (Gauthier-Villars, Paris, 1968).
[3] R.L. Pfleegor and L. Mandel, *Phys Rev.* **159**, 1084 (1967).
[4] L. de Broglie and J. Andrade e Silva, *Phys. Rev.* **172**, 1284 (1968).
[5] A. Garuccio, K. Popper, and J.P. Vigier, *Phys. Lett.* **86A**, 397 (1981).

[6] W.M. de Muynk, *Epist. Lett.* **28**, 33 (1980). J. Andrade e Silva and A. Andrade e Silva, *Epist. Lett.* **29**, 39 (1980). O. Costa de Beauregard, *Epist. Lett.* **30**, 30 (1980).
[7] A. Garuccio, V.A. Rapisarda, and J.P. Vigier, *Phys. Lett.* **90A**, 17 (1982).
[8] P. Grangier, G. Roger, and A. Aspect. *Europhys. Lett.* **1**, 173 (1986).
[9] J.R. Croca, A. Garuccio, V.L. Lepore, and R.N. Moreira, *Found. Phys. Lett.* **3**, 557 (1990).
[10] L.J. Wang, X.Y. Zou, L. Mandel, *Phys. Rev. Lett.* **66,** 1111 (1991).
[11]J.R. Croca, A. Garuccio, V.L. Lepore, and R.N. Moreira,*Phys. Rev. Lett.* **68,** 3813 (1992).
[12]L.J. Wang, X.Y. Zou, and L. Mandel, L.*Phys. Rev. Lett.* **68,** 3814 (1992).
[13]F. Selleri, in *Bell's Theorem and the Foundations of Modern Physics",* A. van der Merwe, F. Selleri, and G. Tarozzi, eds. (Word Scientific, Singapore, 1992)
[14]F. Selleri, in *Wave-Particle Duality,* F. Selleri, ed.(Plenum, New York, 1992), Chap. 16.
[15] A. Garuccio, V*istas in Astronomy* **37** , 217 (1993).
[16] S. Friberg, C.H. Hong and L. Mandel, *Optics Comm.* **54**, 311 (1985).

INTERFEROMETRY WITH DE BROGLIE WAVES

Franz Hasselbach

Institut für Angewandte Physik
der Universität Tübingen
Auf der Morgenstelle 12
D-7400 Tübingen, Germany

A new type of miniature low-voltage UHV electron interferometer allows, due to its insensitivity to mechanical vibrations and alternating magnetic fields, experiments with de Broglie electron waves not feasible with conventional interferometers. A Wien filter is incorporated into the interferometer as a novel and unique optical component. It permits one to shift the coherent wave packets longitudinally by amounts that may be chosen at will by adjusting its excitation and enables us to measure the typical wavelike features of matter radiation: the coherence length and and the spectrum of electron waves (Fourier spectroscopy). Lost longitudinal coherence in the interference plane of the interferometer can be restored or optimized by this novel electron optical component. The extraordinary mechanical stability of the interferometer allowed us to place the whole battery-operated instrument on a turntable and to perform a Sagnac experiment with electron waves. Its mechanical stability makes it likely that a biprism interferometer for complex particles with inner degrees of freedom, namely ions with even shorter wavelengths, can be realized.

Key words; electron biprism interferometry, longitudinal shift of electron wave packets by a Wien filter, coherence length, Fourier spectroscopy of charged matter waves, Sagnac effect with electrons.

1. INTRODUCTION

De Broglie's brilliant matter-wave hypothesis [1] was proved experimentally as early as 1927 by Davisson and Germer [2] with the crystalline diffraction of low energetic electron waves. The first diffraction experiment using a macroscopic diffractor, an edge, was performed by Boersch [3,4] in

Waves and Particles in Light and Matter, Edited by
A. van der Merwe and A. Garuccio, Plenum Press, New York, 1994

an electron microscope in 1940. The development of an interferometer was pushed by the Marton [5] group in the United States, in the beginning of the fifties, and in Germany by Möllenstedt and coworkers. While Marton tried to split an electron wave with amplitude division by crystalline diffraction into two coherent partial waves, Möllenstedt and Düker [6] staked on wavefront division by an electron optical biprism consisting of a very fine (less than a μm in diameter) metallized filament stretched in the middle of two earthed electrodes. The "Möllenstedt biprism" has become the standard beam splitter in electron interferometry, whereas amplitude division has experienced its renaissance in Bonse and Hart's X-ray [7] interferometer and subsequently in Bonse, Rauch, and coworker's famous neutron interferometer [8]. Thomas Young's classical double-slit experiment — a pure *gedanken* experiment for electrons, which cannot be realized experimentally according to R. Feynman's remark in his famous textbook — has been performed by Faget and Fert [9] in 1957 and by Jönsson and Möllenstedt in 1959 [10], respectively.

In the present paper we focus our interest exclusively on new achievements obtained in the last few years with a new type of electron biprism interferometer [11]. Experiments inconceivable with conventional electron interferometers, which usually are slightly modified electron microscopes, could be performed. A rather up to date review on the state of the art in conventional electron interferometry and neutron interferometry may be found in a special volume *Physica B* [12].

2. THE NEW ELECTRON INTERFEROMETER

The extraordinary sensitivitiy of conventional electron interferometers to mechanical vibrations is due to the fact, that their low intrinsic mechanical resonance frequency nearly matches the freqency range of vibrations characteristic of the floor of the buildings. Their sensitivity to electromagnetic stray fields is caused by an inefficient magnetic shielding. These drawbacks have been overcome by a novel design [11] namely by a miniaturized self-centering construction principle of the interferometer. Mechanical alignment facilities are no longer required. This construction principle guarantees on the one hand a very high mechanical stiffness corresponding to a high resonance frequency of the instrument, with the result that, if the instrument vibrates, it vibrates as a whole. This is not impairing fringe visibility. The detrimental vibrations of the components of the instrument *relative to each other* cannot be excited by the low-frequency vibrations travelling along the floor of the building. On the other hand, the miniaturized construction permits one to shield the whole interferometer with a single

μ-metal cylinder very effectively, leading to an unsurpassed insensitivity to alternating magnetic fields. While for conventional electron interferometers a laboratory with extremely low vibrational and electrical disturbances is mandatory, with the new instrument electron interferometry is possible in virtually every environment.

Figure 1 shows a picture of the actual electron optical components on top and a schematic diagram of the electron optics in the middle. The path of the electron beams is given on the bottom. The total length of the interferometer from the cathode to the second quadrupole lens is 30 cm. It works in an ultra high vacuum environment with electrons of 150 eV to about 4 keV in energy.

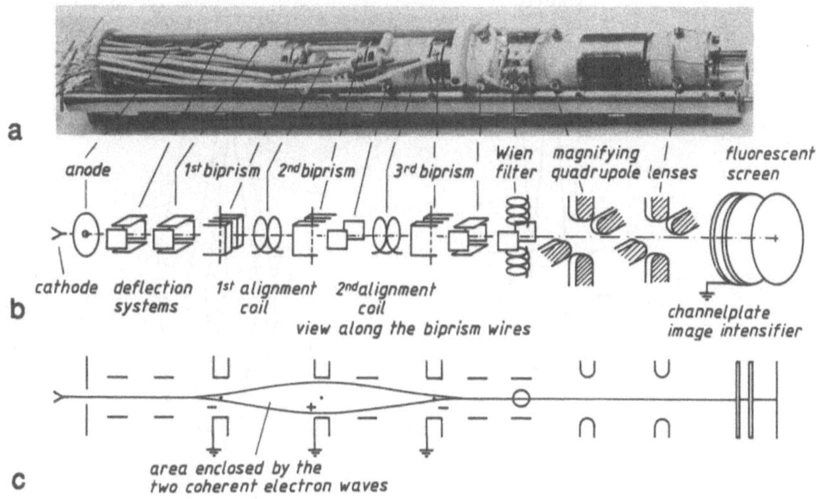

Fig. 1. (a) Technical realisation of the triple biprism interferometer; (b) electron optical setup; (c) beam path when an enclosed area is realized by a three biprism arrangement.

The virtual source of a diode field emission electron gun illuminates the the triple biprism component coherently. The triple biprism component allows one to operate the interferometer in different modes. For example, in order to observe electron biprism interferences a single, positively charged biprism is sufficient (Fig. 2a). The other two biprisms are switched off. The positively charged filament of the biprism bends the partial waves towards each other. The primary, narrowly-spaced interference fringes, downstream from the biprism, are magnified by two cascaded quadrupole lenses, intensified by a dual-stage channel plate image intensifier observed on a fluorescent screen and/or fed into a image processing system for quantitative evaluation.

When an enclosed area between the coherent beams is necessary, e.g., as in the case of the Sagnac experiment, a minimum of two biprisms must be used (Fig. 2b). The first biprism filament, then charged negatively, is bending apart the two coherent partial waves in order to enclose a larger area. The second one, positively charged bends them together again to form the interference fringes. The third biprism (see Fig. 1 at the bottom), charged slightly negative, may be used to bend the beams a little bit apart from each other in order to reduce their angle of superposition and in turn to enlarge the spacings between the interference fringes [13]. All experiments described below where done without this third biprism.

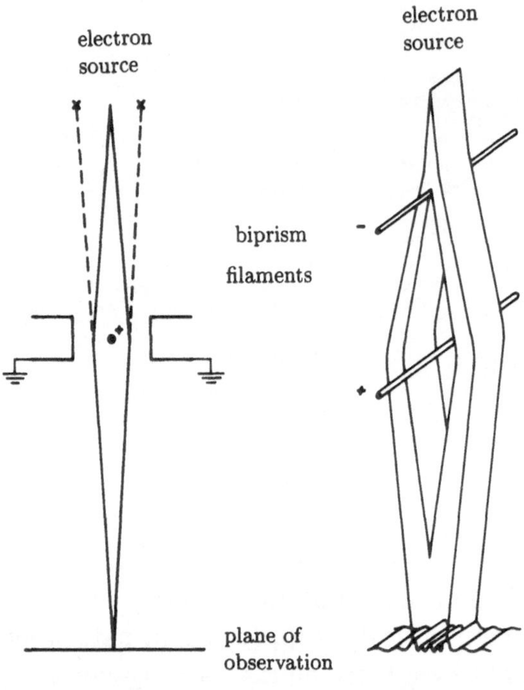

Fig. 2. Schematic setup for achieving wide separation of the coherent electron waves by means of a dual biprism arrangement.

The Wien filter a novel component in an electron interferometer is placed between the third biprism and the magnifying quadrupoles. Möllenstedt and Wohland [14] showed in 1979 that a Wien filter — originally invented as a spectrometer for charged particles — can be used to shift wave packets of charged particles longitudinally and to measure their coherence length. In addition to that it has proved to be indispensable in our low voltage instrument in order to reestablish temporal coherence in the interference plane.

3. EXPERIMENTS

3.1 Quantitative Analysis of de Broglie Electron Wave Packets

Wave packets of all kinds can be characterized coarsely by their coherence length and precisely by the spectrum of the plane waves that compose the wave packet. As we will see, both features can be measured in an electron interferometer equipped with a finely tunable Wien-filter.

It should be mentioned here that matter wave packets disperse in time due to the different velocities of the plane waves which they are composed of [15]. Invariants of this motion are the coherence length and the spectrum making up the packet. In the following, when we speak about wave packets, we always mean these time invariant features of the wave packets.

In order to measure *coherence lengths* of electron waves, the electron beam is split by an electron optical biprism into two coherent wave packets. Both packets are separated laterally by a small distance Δx when they enter the crossed electric and magnetic fields of the Wien filter (Fig. 3). The electric and magnetic fields in the Wien filter are chosen in such a way that there is no resulting deflection of the electron beam (compensated state of the Wien filter).

Inside the Wien filter the wave packets travel in regions of different electric potential, i.e., the group velocity of the beam (in Fig. 3) on the left hand-side is retarded and on the right-hand side is accelerated. Consequently the wave packets leave the Wien filter shifted longitudinally. The overlap of the wave packets in the interference plane is no more complete, the contrast of the interference fringes is reduced. Möllenstedt and Wohland [14] increased the excitation of the crossed electric and magnetic fields, while always staying in the compensated state of the Wien filter until the two wave packets arrive in the plane of interference one after the other, i.e., longitudinal coherence no more exists in the interference plane. They calculated the coherence length from the geometrical dimensions of their Wien filter, the lateral separation Δx of the beams inside of the Wien filter and the electric field strength for which fringe contrast just vanished.
We refined [16,17] this method in the following way: As a first step a Wien filter in which the E and B fields can be varied in very subtle steps had to be developed. The new measuring method works as follows: At first the electric field only is increased. The Wien filter then works as a deflection element and shifts the fringe field, e.g., by 15 fringes to the right on the fluorescent screen. Subsequently the magnetic field is adjusted to exactly compensate for this deflection. This state of the now compensated Wien filter corresponds to the following physical situation: The wave packet tra-

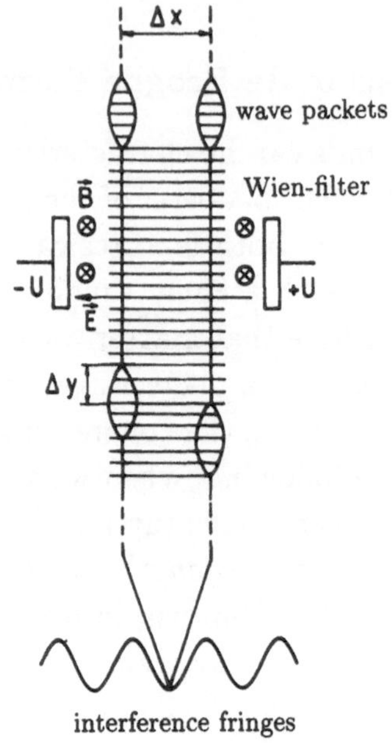

Fig. 3. Influence of a Wien filter in its compensated state on two spatially separated electron wave packets and on the phase of the waves. The wave packets are shifted longitudinally, the phase velocity is not affected (see text). Therefore, the positions of the horizontal lines, which symbolize the crests of the waves, are not shifted at all by the electromagnetic fields inside the Wien filter.

velling through the filter on the side where the potential is more negative is delayed compared to that on the other side. It leaves the filter with a longitudinal shift of exactly 15 wavelengths with respect to the other one. We observe now a reduced contrast of the fringes corresponding to this longitudinal shift.

In order to measure coherence lengths, this procedure is repeated while counting the number of fringes until a decrease in contrast from its maximum by, e.g., a factor of 1/e is observed. We define the coherence length by twice this number of fringes times the wavelength of the electrons. For field-emitted electrons of 4 keV of energy and a natural energy width of 0.36 eV, a coherence length L_c of 284 nm is measured (Table 1). Shorter coherence lengths were achieved artificially in our experiment by superimposing triangularly shaped voltages U of different peak-to-peak amplitudes to the extraction voltage of the field-emission gun of the interferometer. The experimental values of the coherence lengths as a function of the energy width ΔE of the electrons are given in table 1 [15]:

Table 1. Coherence lengths of electron beams of different relative energy spread ΔE.

U/V	0.0	0.5	1.0	1.5	1.9	2.5	3.0
L_c/nm	284	119.4	60.2	42.5	38.7	31.4	26.8
$\Delta E/eV$ (approx.)	0.36	0.68	1.36	1.93	2.10	2.61	3.12

An interesting wave-optical feature of a Wien filter is that, in its compensated state, its electron optical index of refraction equals 1. In analogy to light optics, Glaser (1933) defined the electron optical index of refraction n:

$$n(\mathbf{r}, \tfrac{\mathbf{v}}{v}) = \sqrt{\tfrac{\Phi(\mathbf{r})}{\Phi_0}} - \sqrt{\tfrac{e}{2mU_B}}\mathbf{A}(\mathbf{r})\tfrac{\mathbf{v}}{v}$$

containing an isotropic term, depending on the electric potential $\Phi(\mathbf{r})$ and an anisotropic term depending on the vector potential $\mathbf{A}(\mathbf{r})$. In this definition,

$$\Phi_0 = -U_B,$$

$$\Phi(\mathbf{r}) = \mathbf{E}\cdot\mathbf{r} - U_B = E\cdot x - U_B,$$

where U_B is the accelerating voltage of the electrons and $\mathbf{E}$ the electric field in the Wien filter, so that

$$\sqrt{\tfrac{\Phi(\mathbf{r})}{U_B}} = \sqrt{1 - \tfrac{E\cdot x}{U_B}} = 1 - \tfrac{E\cdot x}{2U_B} + O(x^2);$$

finally,

$$\mathbf{A} = (0, B\cdot x, 0).$$

This gauge (phase shifting field) [18] must be chosen such that the tangential component of $\mathbf{A}$ becomes continous at the entrance and exit planes of the Wien filter:

$$n = 1 - \tfrac{Ex}{2U_B} - \sqrt{\tfrac{e}{2mU_B}}Bx + O(x^2).$$

In the compensated state, the Wien condition is fullfilled:

$$E = -Bv, \qquad v = \sqrt{2U_B e/m},$$

i. e., in the electron optical index of refraction, the first-order terms cancel, so that

$$n = 1 + O(x^2)$$

n = 1 means that the phase of the electron matter wave is not affected at all when travelling through the electromagnetic fields of the compensated Wien filter (see Fig. 3). This is also observed experimentally: When the excitation of the Wien filter is increased while always staying in its compensated state, the interference fringes on the fluorescent screen remain stationary. Only their contrast decreases continously with increasing excitation due to the increasing relative longitudinal shift of the wave packets.

In order to measure the *power spectrum* of the plane waves composing a wave packet interferometrically, i.e., to realize *Fourier spectroscopy of charged particle waves*, the fringe contrast as a function of the longitudinal shift in the whole interference field (consisting of up to 20 000 fringes for field emitted electrons at an acceleration voltage of 2.5 kV) has to be recorded quantitatively. This recording has been done in groups, each consisiting of 15 fringes, by a CCD camera. The fringe contrast data is digitized, arranged to the complete interference field and Fourier-analyzed in a VAX computer [17]. Figure 4 shows the first spectrum of field emitted electrons obtained by Fourier spectroscopy. The resolution of better than 0.4 eV, as obtained in this first experiment, equals that of state of the art electron spectrometers which are routinely used in analytical electron microscopes dedicated to solve problems in material science. On the right-hand side a spectrum containing two peaks with a difference in energy of 30 eV is given. The cor-

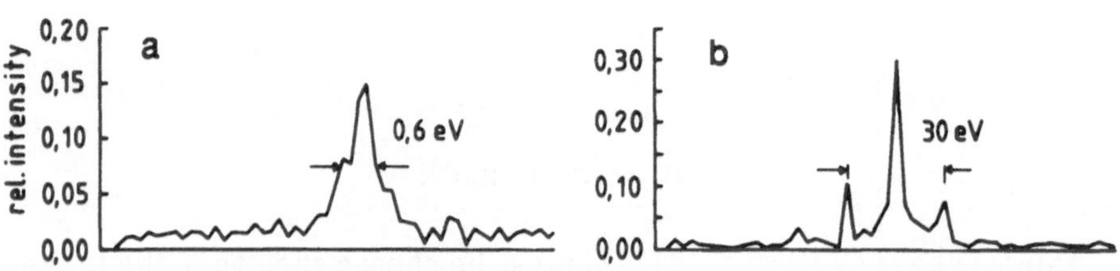

Fig. 4. Spectra obtained by Fourier-analysis. (a) Spectrum of a field emission cathode; accelerating voltage 2.5 kV (see text). (b) In order to simulate a spectrum containing two discrete lines, a square wave voltage of 30 V in amplitude was superimposed on the extraction voltage of the field-emitter. The two peaks are clearly visible. The central peak is a mathematical artifact.

responding pattern of interference fringes showed — in analogy to two beating waves of different wavelengths — the characteristic periodically vanishing and reappearing of contrast of the interference fringes as a function of the longitudinal shift of the coherence length packets.

3.2. Re-establishing and/or Optimizing Longitudinal Coherence in the Interference Plane. Which-Path Information

In the last chapter, we have seen that electron wave packets travelling in different electric potentials will acquire, due to their different group velocities, a longitudinal shift. Electron wave packets travel in different electric potentials in our interferometer, not only in the Wien filter, but also in every one of the electrostatic deflection systems, which are indispensable in the interferometer for fine alignment of the beams to the optical axis. Consider the horizontal deflection system behind the second biprism [Fig. 1(c)]. The electric potentials of the two condenser plates are +U and −U, respectively. Clearly the spatially separated wave packets travel in regions differing in their electric potentials. As a consequence of the relative delay of the two wave packets, we observe at least a reduced, if not a vanishing, fringe contrast. In our low-voltage interferometer all deflection elements cause such delays which usually add up to an amount larger than the coherence time. This means, even after correct alignment of the electron beams to the optical axis of the interferometer, that we usually observe no interference fringes at all [19]. The topmost micrograph in Fig. 5 was obtained without excitation of the Wien filter. Longitudinal coherence seems to be lost. The last, yet most important, alignment step of the interferometer is still missing: We have to restore longitudinal coherence by choosing the compensating delay of the Wien filter exactly opposite to the net relative delay caused by the other electron optical components combined (Fig. 5, middle). If we excite the Wien filter higher than its optimum value, contrast decreases and vanishes again (Fig. 5, bottom). This possibility of restoring "lost" coherence shows that in particle interferometry the quantum mechanical coherence is not as easily truly destroyed, as the experimental fringe visibility suggests, but it is often only "hidden" [20]. This experimentally evident robustness of the quantum mechanical coherence has been demonstrated recently in a conceptually similar experiment also for neutrons [21].

The transit time of the two wave packets on the two paths to the interference plane is exactly equal for both packets when optimum fringe contrast is observed. It differs by an amount equal or larger than the coherence time when fringe contrast vanishes. If we would measure the transit times of many electrons for the case of vanishing fringe contrast, we would

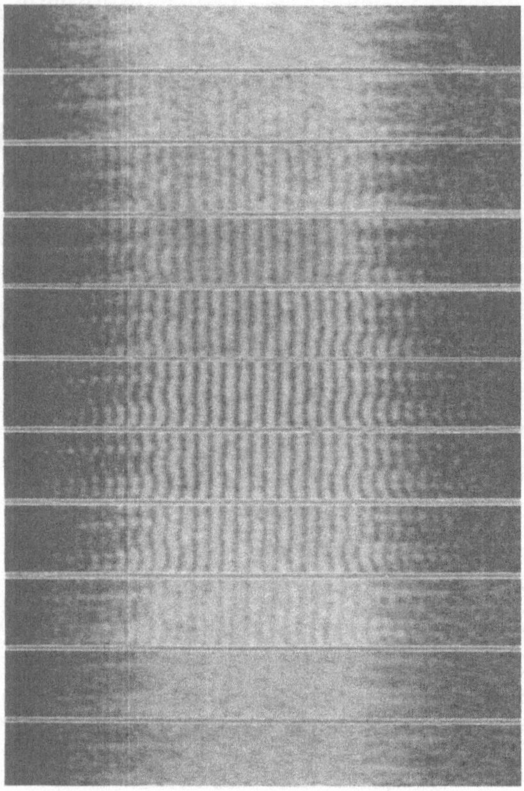

Fig. 5. Re-establishing of temporal coherence by means of a Wien filter. Without excitation of the Wien filter, the wave packets arrive in the plane of interference one after another (top). With increasing excitation of the Wien filter, the interference fringes appear. The excitation is increased until maximum contrast of the interference fringes is observed corresponding to a full overlap of the wave packets (middle). Only after this step our low voltage biprism interferometer is fully adjusted and ready for an experiment. By further increasing the excitation, the overlap decreases again and fringe contrast vanishes (bottom).

find that it is centered around two discrete values. The larger transit time corresponds to the path crossing the area with lower electric potential and the shorter one to that in the higher; that is, for a transit time difference just in excess of the coherence time, we know in principle the path of the electron. For transit time differences smaller than the coherence time, interference fringes are visible and the Heisenberg uncertainty principle $\Delta E \cdot \Delta t \sim h$

prevents us from measuring an exact value for the transit time. Which path information is lost.

3.3. Rotational Phaseshift of Electron Waves (Sagnac Effect)

The Sagnac [22] effect links classical physics, relativistic physics, and quantum mechanics and helped to clarify many interrelations. It has been proved for electromagnetic waves and neutrons [23], and recently for atoms [24] and for charged matter waves [25]. What is the Sagnac effect? The Sagnac effect is the relative phase shift $\phi_1 - \phi_2$ which two coherent waves experience in an interferometer wherein they circulate in opposite directions around a finite area A, when the whole interferometer is rotated with angular velocity Ω (Fig. 6).

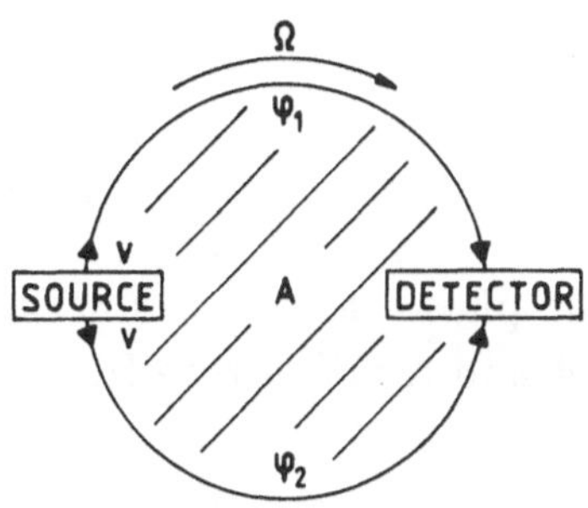

Fig. 6. Sagnac effect: An Ω dependent shift of the interference fringes is observed in the plane of the detector.

To observe the Sagnac effect of electron waves [19,25] we placed the whole interferometer on a turntable. The interferometer's insensitivity to mechanical vibrations and magnetic AC fields was a crucial prerequisite for this experiment. The rotational (Sagnac) phase shift $\Delta\phi$ is given by

$$\Delta\phi = \tfrac{4\pi}{hc^2} E A \Omega$$

where E denotes the total energy of the radiation (for electrons E corresponds to the relativistic mass), Ω the rotation rate, and A the area enclosed by the two coherent beams. We measured the (phase) shift of the fringes between clockwise and counterclockwise rotation of the interferometer with *exactly* the same rotation rate in order to avoid effects of centrifugal forces. Note that the Sagnac phase shift is proportional to the *total energy* of the particles or electromagnetic waves. As a consequence ultra-high sensitive rotation sensors could be realized with a Sagnac gyroscope working, e.g., with heavy ions [26,27].

Our results are given in Fig. 7. For rotation rates of about 0.5 rev/sec and an enclosed area of a few mm^2, phase differences of the order of $\pi/10$ (5% of a fringe) were expected and observed within error limits of about 30% caused by long term drifts of the electronic supplies and instabilities of the field-emitter [22].

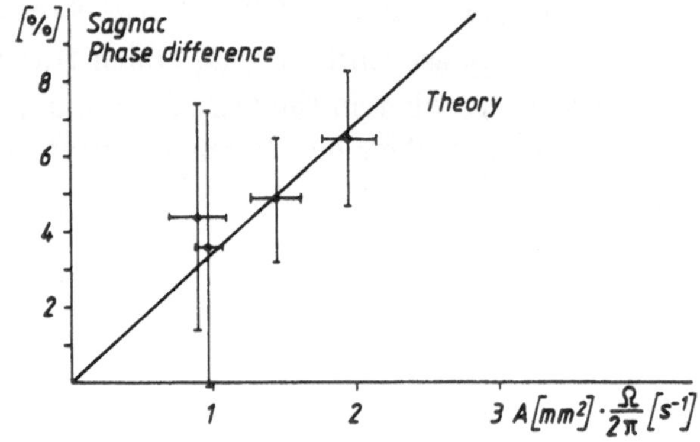

Fig. 7. The Sagnac phase shift as a function of the product of the enclosed area A and rotation rate Ω.

Formally the Sagnac effect is the mechanical analog of the Aharonov-Bohm [28] effect. This becomes evident when we compare the corresponding phase shifts:

Sagnac phase shift	Aharonov-Bohm phase shift
$\phi_1 - \phi_2 = \frac{2m}{\hbar} \int_A \mathbf{\Omega} \cdot d\sigma$	$\phi_1 - \phi_2 = \frac{e}{\hbar c} \oint \mathbf{A} \cdot ds$
	$= \frac{e}{\hbar c} \int_A \mathbf{B} \cdot d\sigma$

Herein the surface integrals are over the oriented enclosed area A. The following quantities correspond to each other:

$$\begin{array}{ccc} \mathbf{\Omega} \text{ Angular velocity} & \Longleftrightarrow & \mathbf{B} \text{ magnetic field} \\ 2m & \Longleftrightarrow & e/c \end{array}$$

In their recent paper [29] Hendriks and Nienhius demonstrate this by showing that a rotation has the same effect on the Schrödinger equation as an electromagnetic field described by a vector potential $\mathbf{A}$ or by the corresponding magnetic field. Due to the full formal mathematical analogy, both effects are manifestations of a geometric- or Berry phase shift [30].

4. CONCLUSION AND FUTURE APPLICATIONS OF THE NEW INTERFEROMETER

The miniaturized electron interferometer has proven to be orders of magnitude more insensitive to mechanical vibrations and electromagnetic stray fields than conventional instruments. Electron interferometry has become possible in practically every environment. The de Broglie wave character of electrons has been demonstrated by measuring coherence lengths and power spectra of plane wave components that make up the wave packets. It has been shown that in matter wave interferometry coherence is often "hidden" only and can be restored by a Wien filter. Last but not least the Sagnac effect for charged particle waves has been demonstrated. An interferometer working with complex particles with inner degrees of freedom, namely ions with their even smaller wavelengths and correspondingly higher sensitivity to vibrations, seems within reach with the new instrument.

ACKNOWLEDGEMENT

I would like to thank Prof. Möllenstedt for his interest in these experiments, my coworkers Dr. M. Nicklaus, H. Gauch, A. Schäfer, H. Wachendorfer for helpful discussions and D. Kehrer for carefully producing the whole interferometer out of a block of titanium and another one of machinable glass ceramics. The Sagnac research project was supported by the Deutsche Forschungsgemeinschaft under grant No. Ha-1063/2-1,2,3.

REFERENCES

[1] L. de Broglie, "Ondes et quanta", *C. R. Acad. Sci.* **177**, 507-510(1923); "A tentative theory of light quanta", *Phil. Mag.* **47**, 446(1924); "Recherches sur la théorie des quanta", *Ann. Phys. (Paris)* **3**, 22-128(1925).
[2] C.J. Davisson, L.H. Germer, "Diffraction of electrons by a crystal of nickel", *Phys. Rev.* **30**, 705(1927).
[3] H. Boersch, "Fresnelsche Elektronenbeugung", *Naturwiss.* **28**, 709(1940).
[4] H. Boersch, "Fresnelsche Beugung im Elektronenmikroskop", *Phys. Z.* **44**, 202-211(1943).
[5] L. Marton, "Electron Interferometer", *Phys. Rev.* **85**, 1057-1058(1952).
[6] G. Möllenstedt and H. Düker, "Fresnelscher Interferenzversuch mit einem Biprisma für Elektronenwellen", *Naturwiss.* **42**, 41(1954); G. Möllenstedt and H. Düker, "Beobachtungen und Messungen an Biprisma-Interferenzen mit Elektronenwellen", *Z. Phys.* **145**, 377-397(1956).
[7] U. Bonse and M. Hart, "An X-ray interferometer", *Appl. Phys. Lett.* **6**, 155(1965).

[8] H. Rauch, W. Treimer and U. Bonse, "Test of a Single Crystal Neutron Interferometer", *Phys. Lett.* **47A**, 369-371(1974).
[9] J. Faget, C. Fert, "Diffraction et interferénces en optique electronique", *Cahiers de Physique* **83**, 285(1975).
[10] G. Möllenstedt and C. Jönsson, "Elektronen-Mehrfachinterferenzen an regelmässig hergestellten Feinspalten", *Z. Phys.* **155**, 472-474(1959).
[11] F. Hasselbach, "A ruggedized miniature UHV electron biprism interferometer for new fundamental experiments and applications", *Z. Phys. B* **71**, 443-449(1988); F. Hasselbach, "Ein kleines UHV Biprisma-Elektronen-Interferometer für 2 keV-Elektronen", 19. Tagung der Deutschen Gesellschaft für Elektronenmikroskopie, Tübingen 1979, Abstract 7L1, p. 90.
[12] "Matter wave interferometry", *Proceedings of the International Workshop on Matter Wave Interferometry in the Light of Schrödingers Wave Mechanics*, (Vienna, Austria 14–16 September 1987), G. Badurek, H. Rauch, A. Zeilinger, eds.; *Physica B*, **151**.
[13] W. Bayh, "Messung der kontinuierlichen Phasenschiebung von Elektronenwellen im kraftfeldfreien Raum durch das magnetische Vektorpotential einer Wolfram-Wendel", *Z. Phys.* **169**, 492-510(1962).
[14] G. Möllenstedt and G. Wohland, "Direct interferometric measurement of the coherence length of an electron wave packet using a Wien filter", in *Electron Microscopy 1980, (Proceedings Seventh European Congress on Electron Microscopy Foundation, Leiden, 1980)* P. Bredoro and G. Boom eds., Vol.1, pp.28-29.
[15] A.G.Klein, G.I. Opat and W.A. Hamilton, "Longitudinal coherence in neutron interferometry", *Phys. Rev. Lett.* **50**, 563-565(1983); H. Kaiser, S.A. Werner and E.A. George, "Direct measurement of the longitudinal coherence length of a thermal neutron beam", *Phys. Rev. Lett.* **50**, 560-563(1983); G. Comsa, Comment on "Direct measurement of the longitudinal coherence of a thermal neutron beam", *Phys Rev. Lett.* **51**, 1105-1106(1983).
[16] I. Daberkow, H. Gauch and F. Hasselbach, "Measurement of the longitudinal coherence of electrons from a fieldemission source", talk at the Joint Meeting on Electron Microscopy, Antwerp 1983; Programm and Abstract book, p. 100.
[17] F. Hasselbach and A. Schäfer, "Interferometric (Fourier-spectroscopic) measurement of electron energy distributions", *Proceedings 12th International Congress for Electron Microscopy, Seattle 1990*, L.D. Peachey and D.B. Williams, eds., (San Francisco Press, Box 6800, San Francisco), Vol. 2, (1990) p. 110-111.
[18] G. Wohland, "Messung der Kohärenzlänge von Elektronen im Elektroneninterferometer mit Wien-Filter", PhD Thesis, Universität Tübingen, 1981.

[19] F. Hasselbach and M. Nicklaus, "An electron optical Sagnac experiment", *Physica B* **151**, 230-234(1988).
[20] M. Nicklaus and F. Hasselbach, "Wien filter: a wave packet shifting device for restoring longitudinal coherence in charged matter wave interferometers", *Phys. Rev. A* **48**, 152(1993).
[21] R. Clothier, H. Kaiser, S.A. Werner, H. Rauch and H. Wöllwitsch, "Neutron phase echo", *Phys. Rev. A.* **44**, 5357(1991).
[22] G. Sagnac, "L'éther lumineux démotré par l'effet du vent relatif d'éther dans un intertféromètre en rotation uniforme", *C. R. Acad. Sci.* **157**, 708-710(1913); G. Sagnac, "Sur la preuve de la réalité de l'éther lumineux par l'expérience de l'interféromètre tournant", *C. R. Acad. Sci.* **157**,1410-1413(1913); E.J. Post, "Sagnac effect", *Rev. Mod. Phys.* **39**, 475-493(1967).
[23] S.A. Werner, J.-L. Staudemann and R. Collela, "Effect of earth's rotation on quantum mechanical phase of the neutron, *Phys.Rev. Lett.* **42**, 1103(1979).
[24] F. Riehle, Th. Kisters, A. Witte, J. Helmcke and Ch.J. Bordé, "Optical Ramsey spectroscopy in a rotating frame: Sagnac effect in a matter wave interferometer", *Phys. Rev. Lett.* **67**, 177(1991).
[25] F. Hasselbach and M. Nicklaus, "Phase shift of electron waves in a rotating frame of reference", *Proceedings 12th International Congress for Electron Microscopy, Seattle 1990*, Vol. 1, p. 212-213; F. Hasselbach and M. Nicklaus, "A Sagnac experiment with electrons: Observation of the rotational phase shift of electron waves in vacuum", *Phys. Rev. A* **48**, 143(1993).
[26] F. Hasselbach, "Einrichtung zur Messung von Drehbewegungen und Drehbeschleunigungen", German patent DE 3504278 C12, 8 February 1985.
[27] J.F. Clauser, "Ultra-high sensitivity accelerometers and gyroscopes using neutral atom matter-wave interferometry", *Physica B* **151**, 262-272(1988).
[28] See, e.g., M. Peshkin and A. Tonomura, *The Aharonov-Bohm Effect*, (Lecture Notes in Physics **340**) (Springer, New York, 1989).
[29] B.H.W.Hendriks and G. Nienhuis, "Sagnac effect as viewed by a corotating observer", *Quantum Opt.* **2**, 13-21(1990).
[30] M.V. Berry, "Quantal phase factors accompanying adiabatic changes", *Proc. Roy. Soc. London A* **392**, 45-57(1984); H.J. Bernstein and A.V. Phillips, "Fiber bundles and quantum mechanics", *Sci. Am.* **245** 95-109(1981).

[illegible]

QUANTUM MECHANICS OF ULTRACOLD NEUTRONS

V.K. Ignatovich

Laboratory of Neutron Physics
Joint Institute for Nuclear Research
141980 Dubna, Moscow region, Russia

A short review of the physics of ultracold neutrons (UCN) and their application to investigation of quantum mechanical problems is presented. The meaning of the UCN anomaly is explained. A possibility to understand this anomaly with the help of a nonspreading wave packet is discussed. The interpretation of the quantum mechanics based on L. de Broglie's ideas is proposed. The wave function is supposed to be a classical field of the particle with coordinates and momentum being defined simultaneously. The motion of the particle is determined by the interaction of its field with the environment. An experimental evidence showing that the field of a free neutron may be described by a nonspreading wave packet similar to the singular de Broglie solution of the inhomogeneous Schrödinger equation is pointed out. The width of this wave packet is estimated. An additional experiment for measurement of this width is considered. A new description of a spinor particle is found.

Key words: quantum mechanics, wave packet, absorption, ultracold neutrons.

1. INTRODUCTION

The ultracold neutrons (UCN) are very slow neutrons with velocities of about 5 m/s. They can be stored in closed vessels for time periods as long as 1000 s. Their properties make them a very suitable instrument for different applications in quantum mechanics. At first look these properties seem to be very paradoxical:

1. Aside from neutrinos, neutrons are the most penetrating particles, *but UCN can be trapped in a bottle and stored there.*

2. Almost all material walls repel them, *but almost every nucleus of the wall will attract them.*

4. The gravitational, the weak, the electromagnetic and the strong interactions are very different, *but they meet on equal footing in the physics of UCN.*

5. The magnetic field cannot penetrate into superconductors, *but it does not prevent the neutron with its magnetic moment enter superconductors.*

These paradoxes are well resolved, and we shall not dwell upon them too long in this paper (see, for example, the books [1]-[3]). The main topic of this paper is the discussion of some insight into the fundamental problems of quantum mechanics that can be gained through the experience with UCN. With some precautions we can say that the experimental results related to UCN anomaly may be interpreted as an evidence in favor of the de Broglie's wave packet for the neutron wave function. With this result in mind, one can go further and discuss the interpretation of the neutron ψ function as a classical field. The particle itself can be thought of as a source of the field, and its motion is determined by interaction of this field with the environment. At present we can only try to put down all the equations related to this problem but cannot solve them even in the simplest case of one dimensional reflection from a semitransparent mirror. Nevertheless such an approach seems to be fruitful since it shows the source of uncertainties and probabilities entering into physics through quantum mechanics.

In the next section of this paper, we discuss what the UCN are, how they are obtained and what interactions they have with the environment. In the third section we consider some experiments performed with these particles and point out the main UCN problem. In the fourth section the fundamental problems of quantum mechanics related to UCN and other particles are discussed. Finally, a new description of a spinor particle is proposed in the appendix.

2. WHAT IS AN ULTRACOLD NEUTRON (UCN)

A neutron is called ultracold (see the books [1]-[3]) if its energy E (see Fig. 1) is less than $E_{\text{lim}} \equiv u \approx 10^{-7}$ eV and, respectively the velocity ≈ 5 m/s. These values are actually determined by the neutron — matter interaction. This interaction varies from substance to substance, and the values of E and v also vary, but the magnitudes 10^{-7} eV and ≈ 5 m/s are typical. The main feature of UCN is the ability of total reflection from many substances at any angle of incidence. The neutrons of higher energies can

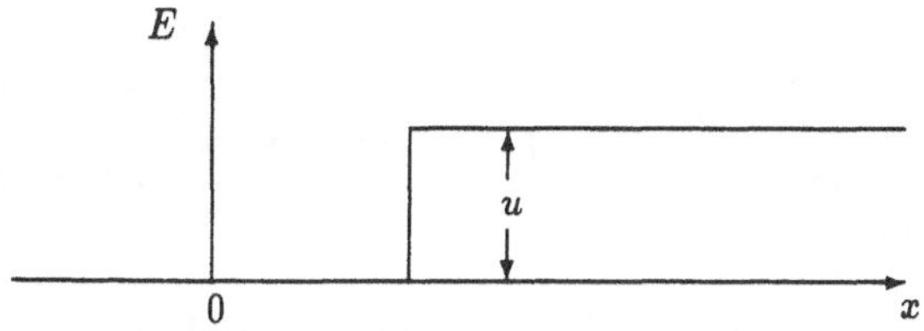

Fig. 1: The optical potential u seen by the UCN at reflecting walls.

be totally reflected only in a limited range of angles of incidence, and this range decreases with increasing energy.

The UCN usually are extracted from the low-energy tail of Maxwellian distribution of neutrons issuing from the moderator at neutron sources and then are guided along neutron guides (generally vacuumed metallic tubes) into an experimental hall. Sometimes they are obtained from a beam of neutrons with higher energy. In this case neutrons are slowed down, for instance mechanically, by reflection from a running away mirror [4] or with the help of an external gravitational or magnetic fields.

It is possible also to extract neutrons from a neutron cloud with the help of a rapidily moving bottle [5, 6] (Fig. 2), that suddenly stops inside the cloud and then slowly moves away to the experimental hall with the UCN captured in it.

In the UCN energy domain all the interactions meet on an equal footing. Indeed, let us consider all four known interactions and their role in UCN physics:

Strong: Strong interaction is attractive (negative) and enters only through the neutron-nucleus elastic scattering amplitude $b \approx 10^{-12}$ cm, which is positive for almost all chemical elements. A neutron that penetrates a substance undergoes coherent multiple scattering with amplitude b on the substance nuclei and as a result finds itself to be refracted or specularly reflected from the interface as if there were a positive potential that repels the neutron from the substance (Fig. 1):

$$u = (\hbar^2/2m)4\pi N_0 b \approx 10^{-7}\ \mathrm{eV},$$

where N_0 is atomic density and m is the neutron mass. This potential is very small but is nevertheless sufficient to store UCN in bottles and guide them along neutron guides.

Electromagnetic: Electromagnetic field enters through the interaction $-\mu\mathbf{B}$ of the magnetic field $\mathbf{B}$ with the neutron magnetic moment μ. The field $B \approx 3$ T gives the potential $\mu B \approx u$. It means that it is possible

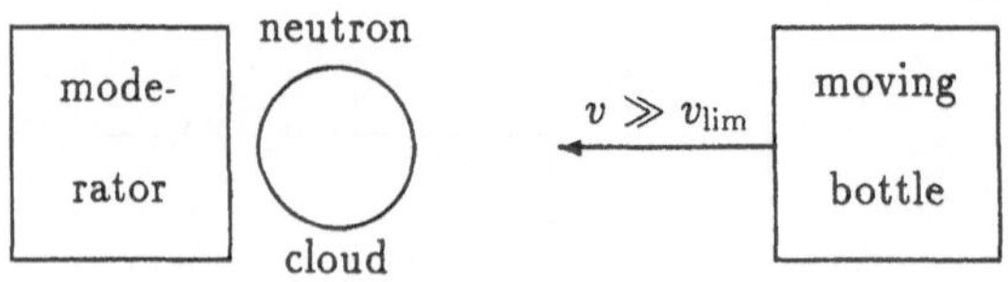

Fig. 2: A fast moving closed vessel is suddenly stopped inside a UCN cloud, capturing some of the neutrons which happened to be at this location at that very moment.

to store neutrons in a bottle with one or all walls being magnetic [7].

Gravity: The gravity enters through the earth gravitational potential mgz. It becomes equal to 10^{-7} eV when the neutron has risen up by a meter. It means that a reflecting floor of a bottle and the gravity create together a potential well (Fig. 3), and there are the discrete bound states for neutron in this well. The lowest ground state has the energy of the order of 10^{-12} eV. It seems to mean that the neutrons with vertical velocities lower than 1 cm/s cannot exist on the earth, and their vertical range of motion cannot be less then 0.1 mm. But it is not so [8], since with an inhomogeneous magnetic field it is possible to compensate the gravity field and to make the ground state energy for one of the spin components as low as one wishes. The action of gravity can be used to store the neutrons in an open saucepan with side walls of enough height to prevent neutrons from jumping over them. Such a bottle is used indeed in real experiments [9].

Weak: The weak interaction is responsible for the β decay of the neutron. Its lifetime $T_\beta \approx 900$ s may be of the same order of magnitude as the loss time of the neutron in the bottle, determined by the capture and inelastic scattering in the walls.

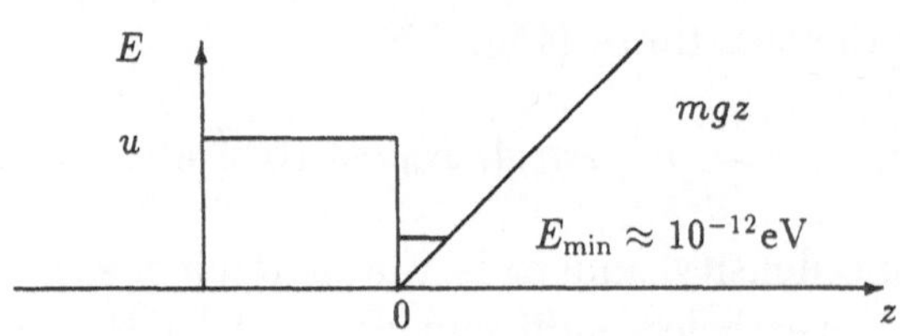

Fig. 3: The gravity field mgz together with the optical potential u of the horizontal plane create a potential well within which UCN can have bound states.

3. SOME EXPERIMENTS PERFORMED WITH UCN, AND THE UCN PROBLEM

The experiments with UCN were started to improve the precision of the neutron lifetime T_β measurement and to diminish the upper limit of its electric dipole moment (EDM). Comparing with the neutrons of higher energy, the UCN can be observed for considerably longer periods of time and it gives the possibility to measure very small quantities of energy, supplied to them by external actions. At present they are permitted to reach the best precision [9, 10] of 0.3% in measurement of T_β and the lowest upper limit [11] 10^{-25} cm of EDM length. The experiments, closely related to quantum mechanics, demonstrate the wave nature of the neutron. They are:

1. reflection from foils (see, e.g. [12]),
2. transmission through foils (see, e.g. [13]),
3. diffraction on gratings (see, e.g. [14]).

In Fig. 4 a curve similar to one, obtained in the experiment [14], is showm. It demonstrates the splitting of a resonance peak in transmission of UCN through five foils — three barriers system (see the insert in the Fig. 4). This example is enough to demonstrate the wave nature of the UCN, and we shall not dwell any more on this subject (for review see [15, 16]), because no contradiction with quantum mechanics was found here.

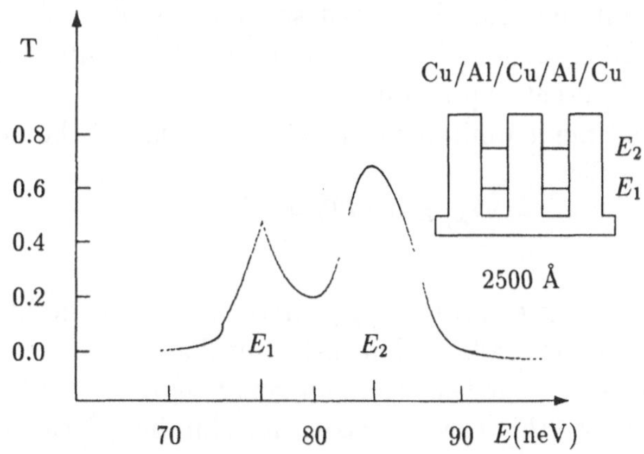

Fig. 4: The UCN transmission curve through three barriers exhibits two peaks corresponding to the quasibound energy levels E_1 and E_2 of the system of five layers shown in the insert.

Now we formulate the main UCN problem (anomaly). This problem is connected with the storage experiment, the scheme of which is shown on Fig. 5. Here s_1 and s_2 denote two shutters that can be opened and closed.

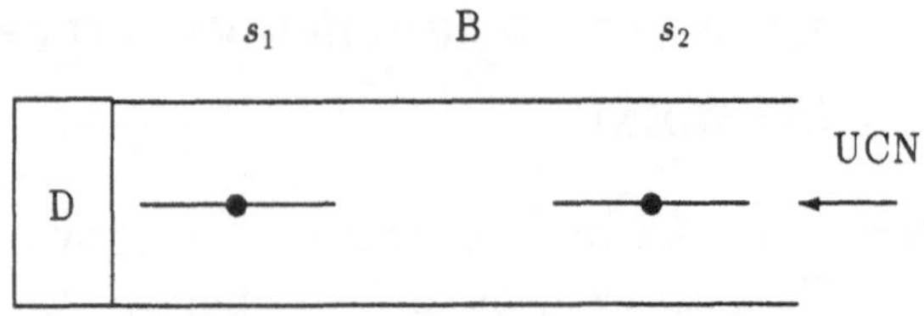

Fig. 5: UCN storage experiment scheme: One opens the shutter s_1, fills the bottle, then closes it, waits some time $t_{\rm exp}$ called the exposition time, opens the shutter s_2 and finally counts the number of UCN left in the bottle.

When s_1 is closed and s_2 is opened, neutrons fill the bottle. When the s_2 is closed the UCN are trapped in the bottle.

After waiting some "exposition time," $t_{\rm exp}$, the shutter s_1 is opened and the number of UCN left in the bottle are counted by the detector.

The storage curve similar to the one found in [17], is plotted in Fig. 6. It shows in logariphmic scale the number of neutrons in the bottle that survived after some exposition time $t_{\rm exp}$. This curve can be represented by a function

$$N(t) = N_0 \exp(-t/\tau(t)),$$

where $\tau(t)$ is called "storage time." For monoline spectrum of stored neutrons τ is constant, but for the broad spectrum it depends on exposition time, since it is different for different energies. The numbers on the picture denote the τ at different expositions.

The storage time is limited by the own life time of the neutron T_β and by losses in the walls:

$$1/\tau = 1/T_\beta + 1/T_l.$$

The loss time T_l is equal to t_f/μ, where t_f is a free-flight time between two collisions with the walls and μ is the probability of losses at a single collision with the wall. The last value for small velocity v of the neutron and for normal incidence (this case is chosen for simplicity) can be represented as

$$\mu = 2(v/v_{\rm lim})\eta, \tag{1}$$

where $v_{\rm lim}$ is the limiting velocity of UCN: $v_{\rm lim} \sim \sqrt{u}$ and η is the reduced loss coefficient (see any of cited textbooks on UCN) equal to the ratio: $\eta = b''/b'$ of imaginary and real parts of the amplitude b. The imaginary part b'' can be represented as $b'' = \sigma_l/2\lambda$, where λ is the neutron wave length (for UCN it is of the order of 100 nm) and σ_l is the loss cross section.

Let us enumerate the factors responsible for losses (T_l).

1. Absorption cross section σ_a (own and impurities)
3. Inelastic scattering cross section $\sigma_{\text{inel}}^{\text{coh,inc}}$ (own, impurities and vacuum gas)
4. Geometry: dimension, surface roughnesses, gaps.
5. Gravity: change of trajectories, variation of spectrum along the height.

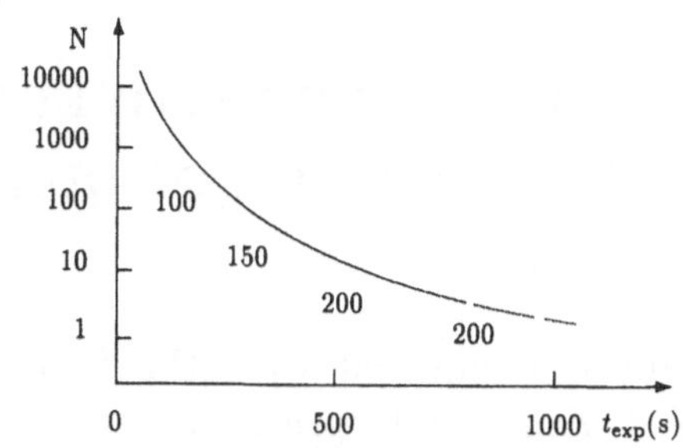

Fig. 6: The storage curve: The different numbers show the storage time for different expositions.

Taking into account all these factors, we can say that at present for the coldest, purest bottle (Beryllium walls covered with oxygen) with the smallest self-absorption, we observe [18] *an additional loss cross section of 1 barn ($\eta \approx 3 \times 10^{-5}$) of unknown nature, which is two order of magnitude higher than the theoretical one.* And this is the main UCN problem or UCN anomaly. To explain the anomaly, it is necessary to go out of the enumerated factors.

One of the attempts is connected with a suggestion that the neutron has two different states $|n>$ and $|\overline{n}>$, characterized by some quantum number, hiddeness, that is conserved in almost all interactions and corresponds to an operator Y:

$$Y|n>=|n>, \qquad Y|\overline{n}>=-|\overline{n}>.$$

The neutron is supposed to be created in nuclear decay always in the state $|n>$. But in the universe there exists some interaction that violates the hiddeness, so after the time t we get the mixture

$$\Psi(t)=\sqrt{1-(\gamma t)^2}|n>+\gamma t|\overline{n}>,$$

where γ is the speed of Y violation. The anomaly can be explained if in in the state $|\overline{n}>$ the neutron interact with substance differently than in the state $|n>$ and becomes absorbed at every collision with the wall.

This scheme looks attractive, but the question is how to check it. It depends on the properties of the state $|\overline{n}>$. If, for instance, its magnetic moment of it is different from $|n>$, the rate of transition γ and the storage time should depend on magnetic field. The experiment [19] disproved such a hypothesis.

If the mass and magnetic moment of the two states are the same, the effect of the Y violation should be more pronounced the larger the dimensions of the bottle are and the longer the neutron's free flight time is. So we should expect that the loss coefficient is proportional to an average dimension L of the bottle: $\eta \propto L$. No experiment to seek such an effect has yet been performed.

4. DE BROGLIE'S IDEAS REDISCOVERED WITH UCN, AND THE INTERPRETATION OF QUANTUM MECHANICS

The other attempt is connected with the fundamental problems of quantum mechanics and with its interpretation. It is very close to the ideas of "double solutions" by L. de Broglie [20], of reality of the wave function by A.A.Tyapkin [21] and to the idea of an "empty" wave by F.Selleri [22].

4.1. An Attempt to Explain the UCN Anomaly

To explain the UCN anomaly, it was supposed that the free neutron, moving with a velocity v, has its own wave function

$$\psi(s,\mathbf{v},\mathbf{r},t) = c\exp(i\mathbf{v}\mathbf{r} - i\omega t)\frac{\exp(-s|\mathbf{r}-\mathbf{v}t|)}{|\mathbf{r}-\mathbf{v}t|}, \tag{2}$$

that does not spread in time. Here s is some parameter, which should provide the observed loss coefficient, v is the neutron velocity and c is the normalization constant defined by the equality:

$$\int d^3r\psi(s,\mathbf{v},\mathbf{r},t) = \frac{|4\pi c|^2}{(2\pi)^3}\int\limits_0^\infty \frac{1}{[(\mathbf{p}-\mathbf{v})^2+s^2]^2}d^3p = 1. \tag{3}$$

The Fourier representation of this wave packet

$$\psi(s,\mathbf{v},\mathbf{r},t) = \frac{4\pi c}{(2\pi)^3}\int \frac{\exp[i\mathbf{p}\mathbf{r} - ip^2t/2 + i((\mathbf{p}-\mathbf{v})^2+s^2)t/2]}{(\mathbf{p}-\mathbf{v})^2+s^2}d^3p,$$

shows that the spectrum is centered around $p^2 = v^2$ but extends to infinity.

A part of the spectrum of this function is always higher than u. So, the neutron escapes through the wall with the probability

$$\mu = \frac{|4\pi c|^2}{(2\pi)^3}\int\limits_0^\infty \frac{1-|R(p)|^2}{[(\mathbf{p}-\mathbf{v})^2+s^2]^2}d^3p. \tag{4}$$

It is postulated that it is the neutron as a whole and not a part of it that escapes through the wall. This postulate looks very like the projection hypothesis in the theory of measurement.

To estimate the parameter s entering the relations (2,4), we substitute $|R(p)| = 1$ for $p_\perp^2 < u$, where u is the wall potential, and $|R(p)| = 0$ for $p_\perp^2 > u$. Then, for small v and perpendicular incidence of the incoming wave, the expression (4) can be represented in the form

$$\mu = \frac{\int_0^\infty d^3p\theta[(p_\perp + v)^2 > u]/(p^2+s^2)^2}{\int_0^\infty d^3p/(p^2+s^2)^2} \approx 2s/\pi v_l, \tag{5}$$

where the θ function is equal to unity when the inequality in its argument is satisfied and to zero in the opposite case.

Comparing with (1), we get

$$s = \pi\eta v \approx v \times 10^{-4}.$$

Thus the storage experiments give a possibility to estimate the parameter s of the singular de Broglie solution.

This approach may be criticized, since the wave function (2) satisfies the inhomogeneous equation

$$(i\partial/\partial t + \Delta/2)\psi(\mathbf{r},t) = -C(t)\delta(\mathbf{r} - \mathbf{r}(t)). \tag{6}$$

So the challenge is to construct a wave packet which satisfied the normal equation

$$(i\partial/\partial t + \Delta/2)\Psi(\mathbf{r},t) = 0.$$

To achieve this, the de Broglie's approach may be used.

We start from the equation

$$(i\partial/\partial t + \Delta/2)\psi(r,t) = 0,$$

find its solution

$$\psi(r,t) = j_0(sr)\exp(-is^2t/2),$$

where

$$j_0(x) = \sin(x)/x$$

is the spherical Bessel function. Then we go to the moving reference frame via transformation of variables

$$x = x_0 + vt,$$

transform derivatives

$$\partial_{x0} \to \partial_x, \qquad \partial_{t0} \to \partial_t + v\partial_x,$$

represent the wave function ψ in the form

$$\psi(\mathbf{r}_0, t) = \exp(-i\mathbf{vr} + iv^2t/2)\Psi(\mathbf{r}, t),$$

and we get the normal homogeneous equation for the wave function Ψ and its solution

$$\Psi(\mathbf{r}, t) = \exp(i\mathbf{vr} - i\omega t)j_0(s|\mathbf{r} - \mathbf{v}t|), \tag{7}$$

with

$$\omega = (v^2 + s^2)/2,$$

where the first term describes the kinetic energy of the particle and the second gives the energy that is necessary to prevent the wave packet from spreading.

The Fourier representation of the function (7) has the form

$$\exp(i\mathbf{vr} - i\omega t)j_0(s|\mathbf{r} - \mathbf{v}t|) = \int \exp(i\mathbf{pr} - ip^2t/2)\delta((\mathbf{p} - \mathbf{v})^2 - s^2)\, d^3p/2\pi s.$$

So its spectrum in the momentum space is like a sphere of radius s (Fig. 7) centered at the end of the velocity $\mathbf{v}$ (in units $m/\hbar = 1$).

But, to describe small losses it is necessary to have small s, implying that the spectrum of such a packet is very narrow. Therefore the neutron with velocity v lower than $v_{\text{lim}} - s$ (the majority of UCN have just such velocities) has no components with the energy higher than u and has no extra losses. So the parameter s is of no help, and we should conclude that such a wave packet cannot explain the UCN anomaly.

4.2. The Classical Interpretation of Quantum Mechanics of a Free Particle

The attempts to explain the UCN anomaly taking an insight into the fundamental problems lead to an idea that the wave function may be considered to be a classical field. The action of this field on the motion of the

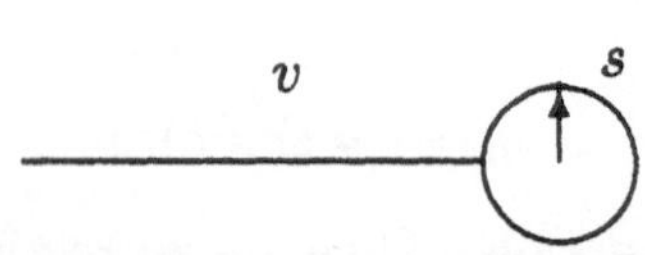

Fig. 7: The spectrum of the nonsingular de Broglie's wavepacket is represented by a sphere of radius s centered at the extremity of vector $\mathbf{v}$.

particle can be even greater than the action of other known fields. For instance, let us consider the neutron with the magnetic moment. This moment creates the magnetic field around the neutron. We know, that a magnetic field does not penetrate superconductors, so the question is whether the neutron can penetrate superconductors or not? The experiment shows that it can. It is a common practice to study the behavior of a superconductor in an external magnetic field measuring depolarization of neutrons transmitted through it [24].

The theory also shows that it can. The magnetic properties of the neutron are used to determine the neutron-atom scattering amplitude, and this amplitude makes a contribution into the reflecting potential of the substance. If the energy of the neutron is higher than the potential, the neutron can enter the superconductor neglecting the classical law prohibiting penetration of magnetic field inside it. It is necessary to remind here that the potential resulted from the wave properties of the neutron — from its multiple coherent scattering in the medium. So the motion of the particle is determined mainly by its wave function and not by a property such as the magnetic moment.

Now we can enumerate the properties of the wave function as a classical field:

1. Since we are accustomed to have only real classical fields, the wave function can be considered to be composed of two real fields oscillating in time and shifted in phase by $\pi/2$.

2. It should have a source - the particle with well defined position and momentum.

3. It has a hidden parameter s. If s depends on the momentum v of the particle and increases with it, we have a very natural transition from wave optics to geometrical optics, or, in other words, from wave properties to corpuscular ones.

4. If described by (7), it is not normalized.

5. It does not violate any uncertainty relations, since, for the field (7), $<(\Delta x)^2>=\infty$ and, for the field (2), $<(\Delta p)^2>=\infty$.

6. It does not contradict the probabilistic interpretation, since the probability to detect a particle is proportional to interaction with it, and the last can be taken proportional to $|\psi|^2$.

7. In the Bohm-de Broglie's approach,

$$\Psi(\mathbf{r},t)=R(\mathbf{r},t)\exp(iS/\hbar),$$

we have

$$\partial\rho/\partial t + \nabla(\rho\nabla S/m) = 0, \qquad \rho = R^2,$$

$$\partial S/\partial t + (\nabla S)^2/2m + V + Q = 0,$$

with

$$Q = (\Delta R)/R = \begin{cases} +s^2 & \text{for } (2), \\ -s^2 & \text{for } (7), \end{cases}$$

which means that the quantum potential is a constant and can be interpreted as an interaction of the particle with the vacuum.

Such an interpretation is very close to ideas of A.A.Tyapkin [21] and F.Selleri [23], because it supposes the reality of the wave function. The empty wave, which changes the probabilities, should have some interaction energy with the medium, and so it can be imagined as a field. In this way it is easy to understand the meaning of interference. The interference can be observed even with classical particles. Indeed, let us consider a screen with two holes and an electron, which moves through one of the holes, and we know which one exactly. The motion of the electron depends on whether the other hole is open or not, since it depends on the interaction of its field with the screen, and the last is different for the two cases. So we see that the other hole interfere with the motion through the first one, and this is the phenomena called "interference."

Of course, with the electrostatic field it is not possible to obtain an oscillating pattern, since there is no such a parameter as the wavelength, but the above example shows that there is no principal difference between the wave function and an electromagnetic field.

4.3. A Thought Experiment to Measure s

The intrinsic parameter s of a particle, defines the form of its wave function. In order to measure s it is necessary to compare the wave function at two different points. However, experimentally we measure $|\psi(\mathbf{r})|^2$ only at a single point. In order to compare the wave packet in two different points it is necessary to use some sort of interference, where we have two packets prepared from the single one, shift them a little with respect to each other and then superpose them in one point. Of course, when we are preparing two packets from a single one we have to use some sort of splitters, for instance a semitransparent mirror. These splitters do not split the particle but only its field.

Let us consider the reflection and transmission of two semitransparent barriers (Fig. 8) separated by a distance L, measured by time-of-flight (TOF) method.

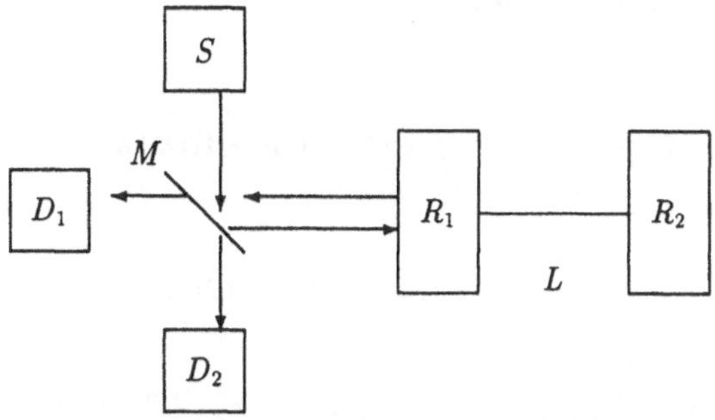

Fig. 8: Illustration of a Time-Of-Flight spectrum measurement experiment involving neutrons reflected from two semi-transparent barriers.

The reflection amplitude of two barriers with the same reflection R and transmission T amplitudes can be represented by

$$\begin{aligned} R_{12} &= R + T^2 R e^{2ikL} + T^2 R^3 e^{4ikL} + T^2 R^5 e^{6ikL} + \dots \\ &= R + T^2 R \exp(2ikL)/[1 - R^2 \exp(2ikL)] \\ &= R[1 - (R^2 - T^2)\exp(2ikL)]/[1 - R^2 \exp(2ikL)], \end{aligned}$$

where k is a wave number of the neutron. If $b'' = 0$ we have always $R^2 - T^2 = \exp(-2i\varphi)$, where the phase φ is real. And the transmission amplitude is

$$T_{12} = T \exp(ikL) T/[1 - R^2 \exp(2ikL)].$$

If we had only one barrier, the idealized TOF spectra would looks like the one in Fig. 9, where $\tau = S/v$, $I_1 = \xi |R|^2 I_0$, I_0 is intensity of the source, and ξ is the detector efficiency.

The question is how the TOF spectrum in the case of two barriers will look? If it looks like that one shown in Fig. 10, where $t_0 = 2L/v$, it will correspond to the algorithm of calculation, i.e. to multiple reflections from the barriers, but then the total intensity reflected by two barriers will be always greater than I_1. But we know, that due to interference, the reflection from two barriers for some distance L can be lower than I_1. How can it happen, and how does the TOF spectrum looks in that case?

In order to understand this point, we are going to directly calculate the reflection of the simplest (Gaussian) wave packet with the width s from the simplest two barrier system (two δ functions) and find out what conditions should be imposed on the experiment to have the possibility to extract the parameter s.

Let us consider the simplest case when the barriers are described by a potential $u = 2p\delta(x)$. The reflection and transmission amplitudes for it are

$$R(k_x) = -p/(p - ik_x), \qquad T(k_x) = -ik/(p - ik_x).$$

Using the integral representation

$$1/(p-ik)=\int_0^\infty \exp(-\alpha p+i\alpha k)\,d\alpha,$$

we can obtain the wave, reflected from one barrier, in the form

$$\begin{aligned}\Psi_R(\mathbf{r},t) &= -\frac{p}{2\pi s}\int_0^\infty d\alpha\, e^{-\alpha p+i(\alpha-x)k_x+i\mathbf{k}_\perp\mathbf{r}_\perp-ik^2t/2}\,d^3k\delta[(\mathbf{k}-\mathbf{v})^2-s^2]\\ &= -p\int d\alpha e^{-\alpha p}e^{iv(|x|+\alpha)-i\omega t}j_0(s||x|+\alpha-vt|),\end{aligned}$$

where $|x|$ is the flight path from the source (at $t=0$) to the barrier and then to the detector.

We see that the reflected wave is composed of the waves similar to the incident one, but delayed. It means that the reflected pulse is broadened, so we can say that TOF spectrum consists of a pulse with a width $<t>=t_1+t_2+t_3$, where $t_{1,2}=l_{1,2}/v$, $l_1=1/s$, $l_2=1/p$, and t_3 is the experimental resolution limited by the incident pulse duration, spectral width and the spectrometer resolution.

If we can change p (suppose it is a magnetic barrier), we can find $t_{\lim}=1/sv+t_3$ for $p\to\infty$. It is seen that the reflection from one barrier does not permit one to distinguish that part of pulse width which is due to s. The wave transmitted through one barrier can be calculated similarly.

Now consider two barriers. A possible realization of such an experiment can be found in [14, 25]. In the case of [14] we have a three-film system, where one of the films after measurement of the resonance dip in reflection, can be destroyed in situ for the next measurement of the reflection from a single barrier.

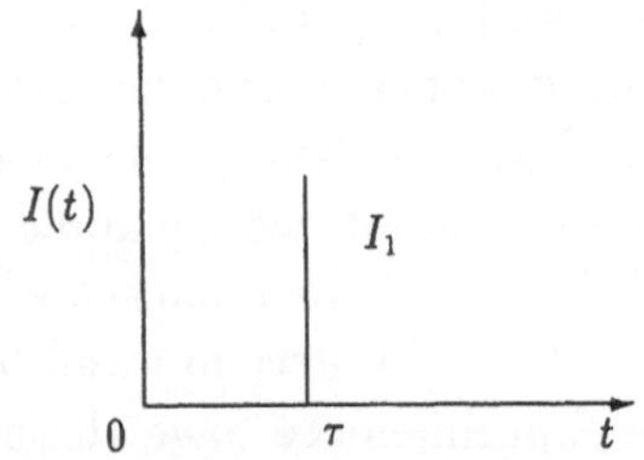

Fig. 9: An ideal Time-of-Flight spectrum of monoenergetic neutrons reflected from one of the barriers featured in Fig. 8.

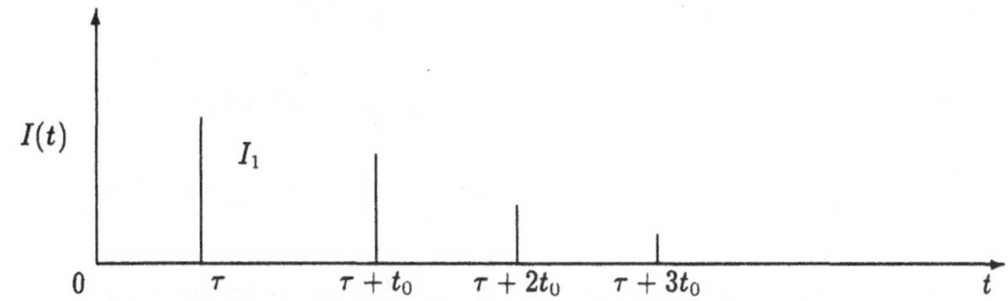

Fig. 10: Expected idealized Time-of-Flight spectrum of neutrons in the experiment shown in Fig. 8 from the classical point of view.

In the case [25] the measurements can be repeatable, since the two-barrier system can be composed of magnetic films with different hysteresis properties.

The reflection of two barriers gives again the combination of packets similar to the incident one but with an additional delay determined by L namely:

$$\Psi_R(\mathbf{r},t) = -p\{\int d\alpha e^{-\alpha p} e^{iv(|x|+\alpha)-i\omega t} j_0(s||x|+\alpha-vt|)$$
$$+\sum_{n=0}^{\infty} f(n,\alpha,p) e^{iv(x+\alpha+2Ln)-i\omega t} j_0(s|x+\alpha+2L(n+1)-vt|)\}, \quad (8)$$

where

$$f(n,\alpha,p) = (\alpha p)^{2n}\left(\frac{1}{(2n)!} - \frac{2\alpha p}{(2n+1)!} + \frac{(\alpha p)^2}{(2n+2)!}\right).$$

The terms under the sum sign in (8) show the wave packets with different number of reflections between barriers and the different delay.

For simplicity we limit ourselves only with single reflection from the both barriers, i.e. we suppose that R is small. Then

$$R_{12}) \approx R[1+\exp(2ikL)]. \quad (9)$$

It means that we use only the term with $n = 0$ in the sum of (8). It is correct if $p \ll v$.

To perform some analytical calculations to see the effect of the s we substitute for j_0 in (8) a gaussian function

$$F(s|\mathbf{x}|) = (s/\sqrt{\pi})^{3/2}\exp(-s^2\mathbf{x}^2/2). \quad (10)$$

Such a wave function satisfies the non-linear Schrödinger equation (known as Bialynicki-Birula and Mycielski equation) and was used in [27].

After reflection the wave function takes the form

$$\Psi(\mathbf{x},t) = -p\int \exp[i\mathbf{v}(\mathbf{x}_r - \mathbf{x}' + \mathbf{e}_\perp\alpha) - i\omega t]\exp(-p\alpha)\times$$

$$\{F(s|\mathbf{x}_r - \mathbf{x}' + \mathbf{e}_\perp\alpha - \mathbf{v}t|) + F(s|\mathbf{x}_r - \mathbf{x}' + \mathbf{e}_\perp(\alpha + 2L) - \mathbf{v}t|)\exp(2iLv_\perp)\}\,d\alpha. \tag{11}$$

where $\mathbf{x}_r$ is the reflected radius vector $\mathbf{x}$ of the detection point: $\mathbf{x}_r = (\mathbf{x}_{||}, -x_\perp)$, $\mathbf{x}'$ is the radius vector of the particle at time moment $t = 0$, $\mathbf{e}$ is a unit vector and the index $\perp$ denotes the vector components, perpendicular to barriers.

For p, $v_\perp \gg s$ it can be approximated (we neglect the retardation $t_2 = 1/pv$) by

$$\Psi(\mathbf{x},t) = R(v_\perp)\exp[i\mathbf{v}(\mathbf{x}_r - \mathbf{x}') - i\omega t]\times$$

$$\{F(s|\mathbf{x}_r - \mathbf{x}' - \mathbf{v}t|) + F(s|\mathbf{x}_r - \mathbf{x}' + 2L\mathbf{e}_\perp - \mathbf{v}t|)\exp(2iLv_\perp)\}. \tag{12}$$

The TOF distribution of intensity of a particle measured by the detector is defined by the relation $I(t) = C << |\Psi(t)|^2 >>$ or

$$I(t) = C\int d^3x \int d^3x' \int d^3v\ |\Psi(\mathbf{x}-\mathbf{x}',t)|^2\Omega(\mathbf{x})\Omega_0(\mathbf{x}')S(\mathbf{v}), \tag{13}$$

where C is a normalization constant. This expression means that the intensity is considered to be averaged over the distribution of detection points, $\Omega(\mathbf{x})$, over the distribution of origins of the wave-packet, $\Omega_0(\mathbf{x}')$, and over the spectrum, $S(\mathbf{v})$.

For simplicity we suppose, that

$$\begin{aligned} \Omega(\mathbf{x}) &= \exp(-|\mathbf{x}-\mathbf{x}_0|^2/2\sigma^2)/(\sigma\sqrt{2\pi})^3, \\ \Omega_0(\mathbf{x}') &= \exp(-|\mathbf{x}'-\mathbf{x}_1|^2/2\zeta^2)/(\zeta\sqrt{2\pi})^3, \\ S(\mathbf{v}) &= \exp(-|\mathbf{v}-\mathbf{v}_0|^2/2\Delta^2)/(\Delta\sqrt{2\pi})^3, \end{aligned} \tag{14}$$

where $\mathbf{x}_0$, $\mathbf{x}_1$, $\mathbf{v}_0$ are the centers and σ, ζ, Δ are the dimensions of the respective volumes. If ζ is determined by a duration t_p of the pulsed source, then $\zeta \approx v_0 t_p$. In the following we shall put $\mathbf{x}_1 = 0$.

With such functions all the integrals we meet are reduced to the form:

$$\int \exp\left[-\frac{1}{2}\sum_{i=1}^{n} a_i^2(x-x_i)^2 - ixu\right]dx =$$

$$\sqrt{\frac{2\pi}{A}}\exp\left\{-\frac{1}{2A}\left[\sum_{i>j=1}^{n} a_i^2a_j^2(x_i-x_j)^2 + u^2 + 2iu\sum_{j=1}^{n} a_j^2x_j\right]\right\}, \tag{15}$$

where $A = \sum_{i=1}^{n} a_i^2$.

After substitution of (12) into (13) we get an expression of the form

$$I(t) = C(X^2 + X'^2 + 2AXX' \cos\alpha), \tag{16}$$

where

$$X = \exp(-\bar{s}^2 v^2 (\tau + \tau_0)^2)/2, \qquad X' = \exp(-\bar{s}^2 v^2 (\tau - \tau_0)^2)/2,$$
$$A = \exp[-(s^2 - \bar{s}^2 + 2\overline{\Delta}^2)L^2], \qquad \alpha = 2Lv[1 - 2\bar{s}^2\Delta^2\tau(\tau + \tau_1)].$$

Here were introduced the variable $\tau = t - \tau_1$, where $\tau_1 = |\mathbf{x}_0 + \mathbf{e}_\perp L|/v$ denote the flight time between source — midpoint between barriers — detector, and the parameter $\tau_0 = L/v$, denoting the flight time between the barriers. The parameters $\bar{s}^2$ and $\overline{\Delta}^2$ are the s^2 and Δ^2 renormalized by other uncertainties:

$$\bar{s}^2 = \frac{s_1^2}{1 + 2s_1^2\Delta^2 t^2} \qquad \overline{\Delta}^2 = \frac{\Delta^2}{1 + 2\Delta^2 t^2 s_1^2} \qquad s_1^2 = \frac{s^2}{1 + 2s^2(\zeta^2 + \xi^2)}.$$

The expression (16) consists of two parts. The first one gives two packets coming separately with time delay $t_0 = 2L/v$, as shown in fig. 10. They have the width $1/sv$ and represent well the spectrum in fig. 10 only if $2Ls \gg 1$, because in that case the peaks do not overlap and the third interference term is always exponentially small. But if $Ls \leq 1$, then the two peaks are overlapping and the interference term becomes essential. In particular, when $2Lv = (2n + 1)\pi$ the peaks extinguish each other and the maximal interference takes place at the time $t = \tau_1$.

Now we see, that the expression (16) contains such parameters as s^2, Δ^2 and $\bar{s}^2$, $\overline{\Delta}^2$. Measuring the profile of the reflected pulse and dependence of the interference contrast on distance L one can extract all these parameters and find the value s.

The width s is directly connected with the coherence length of the neutron. In stationary conditions, as was shown in [28], it is difficult to separate the coherence length restricted by the preparation of the beam from the intrinsic one. So the nonstationary experiment is required. Here we considered the direct TOF method, but other methods, for example that one with oscillating external fields (see [29]) also can be useful.

4.4. The Equations of Motion of the Particle

Now, what does it means, that the wave function is a field, and how this field influence the motion of the particle? To say this means to introduce all the equations determining the field and coordinates of the particle. It seems to us that the field can be determined from boundary conditions and Eq. (17),

$$(i\partial/\partial t + \Delta/2 - u(\mathbf{r}))\psi(\mathbf{r}, t) = -C(t)\delta(\mathbf{r} - \mathbf{r}(t)). \tag{17}$$

where $\mathbf{r}(t)$ are the coordinates of the particle. The dependence of these coordinates on time is determined from the equations

$$md^2\mathbf{r}/dt^2 = -\nabla V(\mathbf{r},t), \tag{18}$$

where the force (for the case of UCN it gives a reasonable result) can be put down tentatively as

$$\nabla V(\mathbf{r},t) \propto \mathbf{n}|\Psi(\mathbf{r},t)|^2 u(\mathbf{r}), \tag{19}$$

u being the usual potential, $\mathbf{n}$ being the vector normal to the interface. The solution of (17), entering (19), can contain also the part, satisfying the homogeneous Schrödinger equation.

Of course, one cannot solve these equations even in the simplest case, but it seems that, due to quantum mechanics it is not necessary. Quantum mechanics gives us a very good route to avoid this, but we should pay for that by dealing only with probabilities.

APPENDIX: A WAVE PACKET FOR A SPINOR PARTICLE

Let us consider the Shrödinger equation

$$[i\partial/\partial t + \Delta/2 - \sigma B]\Psi(\mathbf{r},t) = 0.$$

In the same way as in sec. 4.1, the particle at rest can be described with the wave function

$$\psi = \exp(-is^2t/2)j_0(\hat{s}r)\xi,$$

where $\hat{s} = \sqrt{s^2 - 2\sigma B}$ and ξ is a spinor. In the moving reference frame, we get

$$\Psi = \exp[i\mathbf{v}(\mathbf{r} - \mathbf{v}t) - i(s^2 - v^2)t/2]j_0(\hat{s}|\mathbf{r} - \mathbf{v}t|)\xi.$$

We see, that there is no precession at all. We have a "fat" and a "thin" component.

There is another possibility:

$$\Psi = \exp[i\mathbf{v}\mathbf{r} - i(s^2 + v^2 + 2\sigma B)t/2]j_0(s|\mathbf{r} - \mathbf{v}t|)\xi,$$

and also:

$$\Psi = exp[i\hat{\mathbf{v}}\mathbf{r} - i(s^2 + v^2)t/2]j_0(s|\mathbf{r} - \hat{\mathbf{v}}t|)\xi,$$

where $|\hat{\mathbf{v}}| = \sqrt{v^2 - 2\sigma B}$. Are all these representations equivalent?

This approach is applicable to other particles too. For instance, the different speeds of different components of the photon may be due to a different averaging of the medium, and, as a result, the different velocities of its fat and thin components.

Acknowledgment

The author is very grateful to Prof F.Selleri and A.Garuccio for their invitation to Trani conference, to Prof. A.A. Tyapkin for his interest and support of this work, and to Igor Carron for his invaluable contribution.

REFERENCES

[1] A.Steyerl, in *Neutron Physics* (Springer Tracts of Modern Physics, vol. 80), Part 2: *Very Low Energy Neutrons*, p. 57.
[2] V.K.Ignatovich, *The Physics of Ultracold Neutrons* (Clarendon Press, Oxford, 1990).
[3] R.Golub, D.Richardson, and S.K.Lamoreaux, *Ultra-Cold Neutrons* (Adam Hilger, Bristol, 1991).
[4] A.Steyerl and S.S.Malik, *Nucl. Instr.&Meth.* **A284**, 200 (1989).
[5] A.D.Stoika, A.V.Strelkov, and V.M.Shvetsov, *JINR communication*, JINR P3-92-116, Dubna, 1992 (in Russian).
[6] Yu.N.Pokotilovskii, *Nucl. Instr.& Meth.* **A314**, 561 (1992).
[7] Yu.G.Abov et al., *Sov. J. Nucl. Phys.* **38**, 70 (1983).
[8] V.I.Lushikov and A.I.Frank, *JETP Lett.* **28**, 40 (1978).
[9] V.P.Alfimenkov, V.P.Gudkov et. al., *Preprint*: LNPI-1629, Leningrad, 1990.
[10] K.Scheckenbach and V.Mampe, *J.Phys. G., Nucl. Phys.* **18**, 1 (1992).
[11] I.S.Altarev et al., *JETP Lett.* **44**, 360 (1986).
[12] H.Sheckenhofer and A.Steyerl, *Phys. Rev. Lett.* **39**, 1310 (1977).
[13] K.A.Steinhauser et al., *Phys. Rev. Lett.* **44**, 1306 (1980).
[14] A.Steyerl et al., *Z. Phys.* **B41**, 283 (1981).
[15] K.Eder et al., *Physica* **B172**, 329 (1991).
[16] R.Gähler and A.Zeilinger, *Am. J. Phys.* **59**, 316 (1991).
[17] R.Golub et al., *Rutherford Appleton Laboratory Report* RL-91-001, 1981.
[18] V.P.Alfimenkov et al., *JETP Lett.* **55**, 92 (1992).
[19] Yu.Yu.Kosvintsev et al., *At. Energy* **53**, 191 (1982).
[20] L. de Broglie, *Non-Linear Wave Mechanics: A Causal Interpretation.* (Elsevier, Amsterdam, 1960).
[21] A.A.Tyapkin, JINR E4-3687, 1968; in *L'interpretazione Materialistica Della Meccanika Quantistica, Fisica e Filosofia in URSS.* (Feltrinelli, Editore Milano), p. 413.
[22] F.Selleri, *Quantum Paradoxes and Physical Reality*, A.van der Merwe, ed. (Kluwer Academic, Dordrecht, 1990).
[23] F.Selleri, *Lett. Nuovo Cimento* **1**, 908 (1969).
[24] R.J.Popular and G.Collin, *Phys. Rev.* **B38**, 768 (1988).

[25] M.I.Novopoltsev and Yu.N.Pokotilovskii, *Sov. J. Tech. Phys.* **52**, 1243 (1982).
[27] A.Steyerl and S.S.Malik, *Ann. Phys.* (*New York*) **217**, 222 (1992).
[28] H.J.Bernstein and F.E.Low, *Phys. Rev. Lett.* **59**, 951 (1987).
[29] R.Golub and S.K.Lamoreaux, *Phys. Lett.* A**162**, 122 (1992).

THE PHYSICAL INTERPRETATION OF SPECIAL RELATIVITY

S. J. Prokhovnik

School of Mathematics
University of New South Wales
Kensington, NSW 2033, Australia

It is suggested that the astronomical evidence of the last sixty years presents us with a new cosmological imperative—the existence of a unique and observable fundamental reference frame for light propagation. It is shown that the existence of such a frame offers an intelligible physical interpretation of special relativity in terms of the anisotropy effects associated with motion relative to the fundamental frame. It is further suggested that this "realist" view of special relativity offers the possibility of a realist interpretation of quantum mechanics; in particular, it resolves a number of problems arising in John Bell's discussion of such an interpretation.

Key words: cosmology, special relativity, quantum mechanics, Bell.

1. THE NEW COSMOLOGICAL IMPERATIVE

Our "Milky Way" galaxy consists of about one hundred-thousand million stars, stretches across a distance of one hundred thousand light years, it appears to have a hot, active nucleus, and rotates once every 250 million years. Yet, as far as we can see in all directions, it is only one of many such galaxies, unevenly separated from one—another by an average of about a million light-years, but usually associated in clusters or even superclusters which may be separated by far greater distances.

Intergalactic (and interstellar) space is by no means "empty"; it is populated by a low density of ions, atoms and simple molecules, and is permeated by a near-uniform blackbody microwave radiation of temperature ~2.7K—this apart from the various forms of radiation which propagate between the stars and between the galaxies.

From spectroscopic information, the universe appears to be much

Waves and Particles in Light and Matter, Edited by
A. van der Merwe and A. Garuccio, Plenum Press, New York, 1994

the same in all directions, no matter how distant the region of observation—it has no apparent edge or bounds. The spectroscopic evidence also presents another regularity: The more distant the galactic source (using a luminosity criterion), the greater the red-shift in the spectrum of the light received. Indeed, the discoverer of this result proposed a linear law: For any particular spectrum line λ,

$$\Delta\lambda/\lambda \propto \text{luminosity distance}. \tag{1}$$

And by analogy with Doppler effect observations, it is suggested that

$$\Delta\lambda/\lambda = z = w/c, \tag{2}$$

where w is the recession speed of the source and c is the speed of light. Combining (1) and (2), we obtain Hubble's law:

$$w = Hr, \tag{3}$$

where r is the luminosity distance of a galactic source, and H is called the Hubble constant whose value is still uncertain but may be taken as $H \sim 18$ Km s^{-1} per million light years. It is seen that H has the dimension of the reciprocal of a time T, and corresponding to the above estimate of H, T is equal to about 18×10^9 years.

The enormous progress in astronomical observation since 1930, when Hubble first proposed his linear law, has supported the validity of this law; and independent evidence, suggesting that the universe was far denser and hotter in the past, supports the interpretation that "the universe is expanding," or rather that galaxies (and their clusters) are receding from one another in a systematic way.

From the evidence outlined above—a new picture of our universe which has been emerging only since about 1930—we can think of the cosmos in terms of a model based on three generally-accepted assumptions, viz:

I. The universe can be considered as an ensemble of fundamental particles, these being the galaxies and clusters thereof; associated with each such particle we can imagine the presence of an associated "fundamental observer".

II. Hubble's law,

$$w = Hr = r/T,$$

applies precisely as between any pair of fundamental particles/observers.

III. A cosmological principle applies to our universe at large, such that the universe would appear much the same from the standpoint of any fundamental particle/observer, and that the laws of nature apply equally everywhere; this implies that our model universe is homogeneous and isotropic from the viewpoint of any fundamental observer—every point of the universe can be considered as its centre!

The assumption II also offers a universal measure of time, available to any fundamental observer with a telescope/spectroscope, known as "cosmic time"; it is given by

$$t \propto r/w$$

for any pair of "typical" galaxies. This measure of time dates back to a defined zero time, when the universe was very hot and dense and ostensibly started expanding, and its present value is closely related to T, depending on how we view the rate of expansion of the cosmic system. The nature of this universal expansion is expressed—assumed—by a "scale factor," usually denoted by $R(t)$, that is as a function of cosmic time.

The three basic assumptions of modern cosmology are expressed mathematically by the Robertson-Walker metric:

$$dr^2 = dt^2 - \frac{R^2(t)}{c^2(1 + \frac{1}{4}k\underline{r}^2)}(d\underline{x}^2 + d\underline{y}^2 + d\underline{z}^2) \qquad (4)$$

in terms of the cosmic time t and the comoving coordinates $\underline{x}, \underline{y}, \underline{z}$ (with $\underline{r}^2 = \underline{x}^2 + \underline{y}^2 + \underline{z}^2$), corresponding to the coordinates $(\underline{x}, \underline{y}, \underline{z}, t)$ of a fundamental particle considered in respect to any such particle taken to be at the *origin* of the frame. *There is no unique centre in accordance with the cosmological principle*, and the metric is invariant in respect to all frames with a fundamental particle origin. The symbol k denotes the 3-space curvature parameter which, following a Gaussian simplification, may take the values of 0, 1, or -1, depending on whether the cosmic 3-space is assumed to be Euclidean, elliptic, or hyperbolic.

Thus the metric defines a fundamental reference frame associated with the ensemble of fundamental particles, and each fundamental particle may be considered as at the the origin of a sub-frame of the fundamental frame, with all such frames being observationally equivalent. We note, finally, that the fundamental frame, defined by the metric (4), is observationally definable in two ways:

(1) only for an observer 'stationary' with respect to this frame, that is a fundamental observer, will the Hubble Law appear to operate in the same way in all directions; for a 'moving' observer, that is one moving relative to the fundamental frame, the Hubble Law will appear directionally distorted, and the distortion may be employed to deduce the speed and direction of the observer's movement relative to the fundamental frame;

(2) following the discovery by Penzias and Wilson (1965) of a universal microwave background (of temperature 2.7K), we can now actually

measure our speed and its direction relative to this radiation background and hence relative to the fundamental frame associated with the essential matter and radiation constituting our universe.

We are interested, in particular, in the kinematics of cosmic light-paths which must satisfy the null-geodesic of the metric (4). For this purpose it is convenient to express this in terms of spherical polar (comoving) coordinates, and consider the radial light-path of a light-signal emitted, at cosmic time t_0, from a fundamental particle source, F_0, which we may take as the (convenient) origin of a (conveniently-employed) fundamental subframe. Further, for an Einstein-de Sitter model of the universe—the currently-accepted "standard" model for which,

$$R(t) \propto t^{2/3}, \quad k = 0,$$

the null-geodesic, for the radial light-path, reduces to

$$0 = c^2 dt^2 - R^2(t) d\underline{r}^2. \tag{5}$$

However, the comoving-coordinate meaning of $\underline{r}$ remains anomalous, unless we can relate it to a real (expanding) distance of a fundamental particle from the chosen origin. We can achieve this by noting that $R(t)d\underline{r}$ must have the dimension of such a distance (or length) in order that the metric and its null-geodesic be dimensionally homogeneous.

To this end, Paparodopoulos (1988) suggested putting

$$R(t) = (t/t_0)^{2/3},$$

which is dimensionless, and considering $\underline{r}$ as the distance (of a fundamental particle) from the chosen origin at time t_0. For our problem we can, of course, take t_0 as the time of emission of the light signal from F_0, treated as the origin of the subframe employed.

We can now define the *varying* distance r of a fundamental particle from F_0 by

$$r = \underline{r}R(t) = \underline{r}(t/t_0)^{2/3}, \tag{6}$$

noting that, when $t = t_0$, then $r = \underline{r}$, in accordance with our definition of $\underline{r}$. Hence also

$$\dot{r} = \underline{r}\dot{R}(t) \equiv w, \tag{7}$$

the speed of recession of a fundamental particle at $(\underline{r}, r)$ relative to F_0.

It is convenient to define s as another distance from F_0, this being the distance from F_0 which the light signal has travelled when it passes the fundamental observer at $(\underline{r}, r)$. Clearly,

$$s = r = \underline{r}R(t). \tag{8}$$

Though formally equal, s and r have different meanings; thus

$$\dot{s} = \underline{r}\dot{R}(t) + \dot{\underline{r}}R(t), \tag{9}$$

since the light signal's $\underline{r}$ coordinate also varies as it passes a succession of fundamental particles. Hence, invoking (7) and (5), the Robertson-Walker metric constraint, we obtain

$$\dot{s}(r,t) = c + w(r,t), \tag{10}$$

noting that we have employed (5) in the form

$$\dot{\underline{r}}R(t) = +c,$$

since for our light-path, $\dot{\underline{r}}$ is positive as the light signal moves *away* from F_0.

The result (10) signifies that when the light signal reaches *a* (any) fundamental particle receding from the origin at speed w, it will pass that particle at speed c. Many other kinematic results are derivable from the null-geodesic using the meaning of r and s as defined above, but the result (10) is the most important one for our present purposes: it states that our (model) universe manifests a unique fundamental frame for light propagation, such that the speed of light relative to any fundamental particle (that is, to any point in this frame) is precisely c in all directions, irrespective of the distance or speed of the source of the light. The result is, of course, in complete accordance with the cosmological principle. We note, also, that the result (10) will hold for any (model) universe whose 3-space is Euclidean (that is, $k = 0$), irrespective of the nature of the scale factor, $R(t)$. The result does not follow readily for models whose 3-space is not Euclidean, since for such models the *form* of $R(b)$ also varies with time.

The result (10), is also consistent with all physical observation and experience. It is in accordance with the observation that "the speed of light is independent of the velocity of its source, and with the maxim that "light does not overtake light"—thus light from all sources, whether cosmic or local, reaches us at the same speed. The observation that the path of light is affected by the presence of matter (e.g., the sun through its gravitational field) is a pointer to the notion that the behaviour of radiation is closely dependent on the presence of matter on the cosmological scale also—on the changing distribution of this matter.

The suggestion that the propagation of light on the cosmological scale takes place according to the result (10), $\dot{s} = c + w$, was first proposed by McCrea (1962) as an hypothesis. However, the result is now no longer an hypothesis; it is a direct consequence of our basic assumptions of cosmology consistent with astronomical observation, and it implies that, in an expanding universe a propagating light ray is being "stretched" (and hence its energy diluted but not lost) by the expansion to the same degree as the stretching of the distance between any pair of fundamental particles. As first noted by McCrea, the cosmological red-shift can then be considered as an observable manifestation of the stretching effect on a light ray's wavelength.

Thus our cosmological assumptions, as expressed by the Robertson-Walker metric imply the existence of a unique (expanding), astronomically-observable frame relative to which fundamental particles/observers are "stationary," and light propagation (all radiation) is constant and isotropic.

We note that this frame is not akin to an aether or absolute space; it is the frame associated with the (expanding) distribution of matter in our unique universe and, as such, is the arena for all large-scale manifestations and interactions of matter and energy. It may be shown (Prokhovnik, 1985) that the fundamental frame is an inertial frame, and hence the source frame for the inertial property of all reference frames in uniform motion relative to it. The existence of such a frame is an anathema to most conventional physicists; it appears to violate a cornerstone of special relativity and raises "unspeakable" questions about the speed of light.

Yet, the notion of such a frame presents no threat to special relativity, nor to any other fundamental tenets of physics. It offers, on the contrary, an enhanced understanding of special relativity and, perhaps also, of quantum mechanics.

2. A COSMOLOGICAL VIEW OF SPECIAL RELATIVITY

Following our notion of the fundamental reference frame, we can now consider special relativity as a consequence of a single "principle," that is, *relative to the fundamental particle at any locality, the speed of light is precisely isotropic.*

We note, first, that at any locality of the universe, involving (say) only a few light years in any direction, the expansion factor can be considered as negligibly small, remembering that even the highest current estimates of the Hubble constant suggest that

$$H < 3 cm\, s^{-1} \text{per light year.}$$

Also over a "short" period of time the scale factor $R(t)$ varies negligibly. For example, for $R(t) \propto t^{2/3}$, the scale factor increases by only one part in 10^4 over a period of 10,000 years. So that even for such a period we can take $R(t) \approx 1$, whereby the Robertson-Walker metric reduces to

$$d\sigma^2 \approx c^2 dt^2 - dx^2 - dy^2 - dz^2,$$

implying the (almost) Lorentz equivalence of inertial frames within such a region. Within a region as small as the solar system such equivalence would be close enough to exact. However to deduce a physical meaning for the effects and assumptions of special relativity, we require a more detailed argument involving bodies/observers and their reference frames in uniform motion relative to the local fundamental observer. For such observers the Hubble law would appear directionally distorted, and the temperature of the black-body background radiation would appear not to be the same in all directions, that is anisotropic.

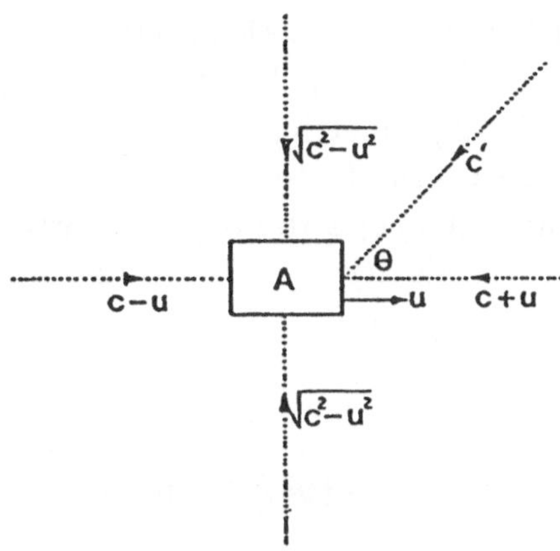

Fig. 1. Speed of light relative to a body moving with velocity $\boldsymbol{u}$ with respect to the fundamental frame I.

More important, but not necessarily observable, the speed of light would *not* be isotropic relative to such bodies/observers and their associated reference frames. It is easily seen that relative to such a body (and etc.), moving at velocity $\boldsymbol{u}$ relative to the fundamental frame I, the speed of light approaching it will be as shown in Fig. 1, where

$$c' = (c^2 - u^2 \sin^2 \theta)^{1/2} + u \cos \theta \tag{11}$$

for the direction making an angle θ with the direction of $\boldsymbol{u}$. We may call the result (11) the *primary anisotropy effect* due to motion relative to I.

A second consequence of such movement results from the retarded potential effect on the fields (gravitational and/or electromagnetic) associated with the moving body and its constituent particles. Following Einstein's view of gravitational fields and in the context of the existence of a frame such as I, the variation of the potential at any point of a gravitational field, due to the movement of its source, will involve a time-lag depending on the distance of the point from the source, exactly as is well-known for the electromagnetic field of a moving body. It follows (Prokhovnik, 1985) that the usual inverse-square law for such a field is then modified by a factor involving the body's speed u relative to I, so that for a moving source of mass M,

$$F = \frac{-GmM \ (1 - u^2/c^2)}{r^2 \ \left[1 - (u^2/c^2)\sin^2\theta\right]^{3/2}} \tag{12}$$

for any direction making an angle θ with the direction of $\boldsymbol{u}$, and where m is the rest mass of a test particle, stationary in I.[1]

The result (12) implies that the Coulomb and gravitational fields (and any other field of this type), linking a moving system of particles, will be correspondingly asymmetric, so that the system must contract in the direction of motion by the factor $(1 - u^2/c^2)^{1/2}$ in order to maintain its internal equilibrium. Hence, a rod of rest length ℓ inclined at an angle θ to the direction of its velocity $\boldsymbol{u}$, relative to I, will assume a length ℓ' (according to I observers), where

$$\ell' = \frac{\ell(1 - u^2/c^2)^{1/2}}{\left[1 - (u^2/c^2)\sin^2\theta\right]^{1/2}}. \tag{13}$$

The time-dilation effect now follows as a direct consequence of the interaction of the primary anisotropy and contraction effects. Following Builder (1958), consider a light-clock consisting of a rod of rest length ℓ with a mirror at each end to reflect a beam of light to and fro along the length of the rod. Let the unit of time be taken as the interval between successive light reflections on one of the mirrors which is connected to a photocounter. When the rod is stationary in I, the unit of time $\tilde{t}$ is given by

$$\tilde{t} = 2\ell/c.$$

However, when such a clock moves with velocity $\boldsymbol{u}$ relative to I, the speed of light relative to the moving rod, will, in general, be different for the two directions depending on the orientation of the rod-clock relative to the direction of $\boldsymbol{u}$. For an angle θ to the direction of $\boldsymbol{u}$, the two light speeds c_1 and c_2 are given by (11), so that

$$c_1 = (c^2 - u^2\sin^2\theta)^{1/2} + u\cos\theta, \quad c_2 = (c^2 - u^2\sin^2\theta)^{1/2} - u\cos\theta.$$

Hence, the unit of time $\tilde{t}'$ for the moving rod-clock is given by

$$\tilde{t}' = (\ell'/c_1) + (\ell'/c_2) = (2\ell/c)(1 - u^2/c^2)^{-1/2} = \beta\tilde{t}, \tag{14}$$

invoking (13) to relate ℓ' and ℓ. It is seen that the time-dilation result is independent of the orientation of the light clock, and that it must also apply to all phenomena involving electromagnetic impulses and energy exchanges. It may be shown that other types of clocks (though not pendulum clocks), will be similarly affected on account of the anisotropy consequences (11) and (13). Hence Builder contended that this effect will manifest itself not only in physical clocks, but in all natural phenomena, both physical and biological, which have an electromagnetic basis.

It is well known that the contraction effect (13) is sufficient to conceal the light speed anisotropy to an observer moving uniformly relative to I; hence the null-result of all Michelson-Morley type experiments. It follows that, in consequence of the time-dilation effect (14), such an observer will further find that his *measure* of the speed of light, with respect to his comoving reference frame, is precisely c in all directions—as it is for (fundamental) observers stationary in I. Thus Einstein's light principle becomes physically intelligible in our context.

It was in accordance with this principle that Einstein specified light signal measurement conventions (which assume light isotropy) for synchronising clocks and estimating the space and time coordinates of distant event, etc. However, it now becomes clear that if observers, associated with a "moving" reference frame, treat their inertial system[2] as "stationary" and so synchronise the clocks according to Einstein, then these clocks are bound to appear non-synchronous to an observer in a different frame, and vice-versa. In effect, the employment of Einstein's conventions produces (Prokhovnik, 1985) a synchronism discrepancy effect given by

$$(\beta u d/c^2)\cos\theta, \tag{15}$$

where d is the I measure of the length interval separating the (moving) Einstein-synchronous clocks in question, and θ is, as usual, the angle made by this interval with the direction of $\boldsymbol{u}$. The result is, of course, equivalent to the relativity of simultaneity factor deduced by Einstein from the Lorentz transformation, and since it is a function of $\boldsymbol{u}$ it leads to the relativity of simultaneity in respect to *any* pair of inertial frames in relative motion and hence having different velocities with respect to I. It is seen that these results demystify Einstein's light principle, time dilation and the "relativity of simultaneity."

The anisotropy consequences (11)–(15), which affect moving bodies and the observations of moving observers, provide a complete physical interpretation, free of any ambiguity, of special relativity. Their interaction is expressed by the Lorentz transformation. They show how and why the light principle operates in respect to all inertial frames, they explain why any local experiment designed to detect an absolute velocity is bound to yield a null-effect (cf. Prokhovnik, 1979). The existence of a fundamental reference frame provides a physical basis for these absolute anisotropy effects, and their interaction produces the *local* observational equivalence of all inertial frames in respect to the laws of nature as expressed by the Lorentz transformation. This latter result, a manifestation of the principle of relativity, is by no means fortuitous: it is a consequence of a widely-operating, action-reaction principle proclaimed by Newton, where in this case the anisotropy reactions to uniform motion, relative to I, nullify precisely the observation by a comoving observer of such motion. Only by astronomical observation can we discern the existence of the fundamental frame and our movement relative to it.

The interpretation, as above, of special relativity completes Lorentz's programme for such an interpretation. Lorentz's developing concept of an aether converged towards the notion that it had only the single light-propagation property of our fundamental frame. He was aware of the Heaviside retarded-potential effect which must produce the length-contraction result, but remained to the end rather confused about the nature of time dilation and about the relativity of simultaneity result which is a direct intelligible consequence of Einstein's measurement conventions. It remained for Builder (1958) to disclose the source of these two separate time results and hence present a fully-integrated "neo-Lorentzian" interpretation of Einstein's special theory.

3. THE BELL CONNECTON

In discussing the relativistic extensions of quantum field theory and the null-results implied by special relativity, Bell (1987) complains ("it is lamented") "that a preferred frame of reference is involved behind the phenomena." Bell does not reject outright the notion of such a frame; and, indeed, in his Paper 9 he shows that the results of special relativity are easier to understand and hence to teach in the context of such a preferred frame than otherwise. He considers that it is only a question of philosophical taste whether one prefers Lorentz's approach (which deduces the basis of the null-effects of uniform motion relative to a unique rest-frame), or Einstein's approach which treats these null effects as a starting point for his theory.

However, with the growth of modern cosmology the choice between these approaches now actually favours Lorentz, though Einstein's *theory* remains intact and enhanced thereby. Yet, taste, tradition and authority still weigh heavily in favour of Einstein for most physicists—they are prepared to ignore the challenge (and opportunity!) of the astronomical evidence of a cosmologically-based preferred reference frame.

This compelling evidence actually resolves the problems raised by Bell (1987) for his beables quantum field theory (BQFT) in his Paper 19.

He notes that "the formulation of BQFT relies heavily on a particular division of spacetime into space and time. *Could this be avoided?*" It does not need to be avoided if we admit the existence of an *expanding* fundamental frame from which we can define a universal measure of (cosmic) time. Note that cosmic time offers us a criterion for absolute simultaneity, and also for comparing the rates at which clocks are operating. Seeing that the material bodies of our universe are (practically) all in some sort of proper motion relative to the fundamental frame—in translational, orbital and/or rotational motion—we might say that "(practically) all clocks run slow!"

Note also, that in our context, special relativity is essentially *a local theory*: It applies, at any locality, to systems in uniform motion relative to the fundamental frame. This then satisfies Bell's requirements—no longer contradictory—for both a null result for the Michelson-Morley experiment and for the possibility of detecting motion relative to the cosmos, so that "in the case of a more or less isolated object ... motion relative to the world as a whole (may) be deemed more or less irrelevant." (Bell, 1987, p. 179.)

Two other points should be mentioned. The notion of non-local causality, discussed by Bell, requires a criterion of simultaneity which has some absolute significance; it is seen that the cosmologically-based notion of a universal measure of (cosmic) time resolves this problem also.

Next, our interpretation of special relativity rests on a single assumption about a cosmologically-based fundamental reference frame; but I do not deny that the underlying properties of the space-time, associated with this frame, may well depend on a "physical vacuum" or "Higgs field" as proposed, for example, by Dirac (1951), Winterberg (1990), Selleri (1990), and Bell (1987). For my own purposes, I do not need to involve any mention of an "aether" which (historically) has undertones of ad hoc properties introduced to explain phenomena. I require only a single assumption, imposed by astronomical observation and cosmological theory, of a unique reference frame in respect to which, at any point, light travels at a constant speed for all directions. This assumption is *sufficient* to derive and illuminate the basic consequences and assumption of special relativity.

This approach to special relativity, which we note is preferred by Bell on pedagogical grounds, is of course contrary to the way conventional physicists and text-books present the topic; they consider that the introduction of a preferred frame is unnecessary and violates the spirit of Einstein's approach. After all, why bother with a preferred frame; the theory is derivable by taking any inertial frame as the "stationary" frame (as did Einstein); as such, it is self-consistent and its results are valid. A preferred frame does not offer any new result or prediction! However, the employment of a preferred frame makes the theory physically-intelligible without introducing esoteric notions of time. For example, it shows how absolute effects (e.g., time dilation) arise naturally from a relativity theory, and how the light principle can apply to all inertial frames as a result of these absolute effects. In this way our approach dissolves ambiguities and paradoxes and makes the theory easier to understand and teach. And, of major importance, it links the theory with other fields, for example with cosmology and quantum theory.

Note that the existence of a preferred cosmologically-based reference frame is now—unlike in 1905—manifestly observable, as was admitted even by Einstein (1956) in the last edition of his book on *The Meaning of Relativity*. Yet this is tacitly ignored by physicists, who are nevertheless prepared to embrace notions of "missing" matter, unobservable matter and "confined" tachyons in order to satisfy their theories!

The climate in modern theoretical physics is for the rejection of "realist," physically-intelligible, interpretations of relativity and quantum mechanics in favour of a reliance on mathematical elegance and the predictive properties of these theories. However, the absence of a realist interpretation does not mean an absence of attempts at esoteric interpretations. In relativity this has led to the rejection of an objective notion of time, and its replacement by subjective and obscurantist meanings which appeal to sensationalist and mystical views of the world. In quantum mechanics it has led to a rejection of the very existence of an objective world, and to a "collapse" of wave functions giving rise to the continual multiplication of universes!

REFERENCES

Bell, J. S. (1987), *Speakable and Unspeakable in Quantum Mechanics* (Cambridge University Press, Cambridge).
Builder, G. (1958), *Aust. J. Phys.* **11**, 279.
Dirac, P. A. M. (1951), *Nature* **168**, 906.
Einstein, A. (1956), *The Meaning of Relativity* (Methuen, London).
McCrea, W. H. (1962), *Proc. Math. Soc. Univ. Southampton* **5**, 15.
Paparadopoulos, A. E. (1990), *Proceedings, Physical Interpretation of Relativity Conference* (London), p. 180.
Penzias, A. A. and Wilson, R. W. (1965), *Astrophys. J.* **142**, 419.
Prokhovnik, S. J. (1979), *Found. Phys.* **9**, 883.
Prokhovnik, S. J. (1985), *Light in Einstein's Universe* (Reidel, Dordrecht).
Selleri, F. (1990), *Quantum Paradoxes and Physical Reality*, A. van der Merwe, ed. (Kluwer, Dordrecht).
Winterberg, F. (1990), *Z. Naturforsch.* **45a**, 1102.

NOTES

1. The effect on a *moving* test particle involves only a component of the local field effect. Thus in respect to a comoving particle the effect is diminished by the factor $[1-(u^2/c^2)\sin^2\theta]^{1/2}$, depending on the angle θ made by $\boldsymbol{u}$ with the direction of the field effect at the locality of the particle.
2. Given that the fundamental frame I is an inertial system, in the sense as described above, then any frame in uniform motion relative to I at any locality must also be an inertial system.

QUANTUM NEUTRON OPTICS

Helmut Rauch

Atominstitut der Österreichischen Universitäten
Schüttelstrasse 115, A-1020 Wien, Austria

The invention of neutron interferometry in 1974 stimulated many experiments related to the wave-particle dualism of quantum mechanics. Widely separated coherent beams can be produced within a perfect crystal interferometer, and they can be influenced by nuclear, magnetic and gravitational interaction. The verification of the 4π symmetry of spinor wave functions and of the spin superposition law at a macroscopic scale and the observation of gravitational effects including the Sagnac effect have been widely debated in literature. The coupling of the neutron magnetic moment to resonator coils permitted the coherent energy exchange between the neutron quantum system and the macroscopic resonator. This phenomenon provided the basis for the observation of the magnetic Josephson effect with an energy sensitivity of 10^{-19}eV. Partial beam path detection experiments are in close connection with the development of quantum mechanical measurement theory. The very high sensitivity of neutron interferometry may be used in future for new fundamental, solid-state and nuclear-physics application. A striking spectral modulation effect has been observed by means of a proper post-selection procedure under conditions where the spatial shift of the wave trains greatly exceeds the coherence length of the neutron beams traversing an interferometer. It is shown that Schrödinger-cat-like states are created by the superposition of two coherent states generated in the interferometer. These entangled states exhibit under certain circumstances characteristic squeezing phenomena indicating a highly non-classical behavior. Analogies with light optical experiments are discussed.

Key words: Quantum mechanics, neutron interferometry, squeezed states, postselection procedures.

1. INTRODUCTION

Different kinds of neutron interferometers have been tested in the

Waves and Particles in Light and Matter, Edited by
A. van der Merwe and A. Garuccio, Plenum Press, New York, 1994

past. The slit interferometer is based on wavefront division and provides long beam paths but only a very small beam separation [1,2]. The perfect crystal interferometer [3,4] is based on amplitude division. The interferometer based on grating diffraction is a more recent development with its main application for very slow neutrons [5]. A schematical comparison is shown in Fig. 1. The perfect crystal interferometer provides highest intensity and highest flexibility for beam handling and is now most frequently used due to its wide beam separation and its universal availability for fundamental-, nuclear- and solid-state physics research.

The perfect crystal interferometer represents a macroscopic quantum device with characteristic dimensions of several centimeters. The basis for this kind of neutron interferometry is provided by the undisturbed arrangement of atoms in a monolithic perfect silicon crystal [3,6]. An incident beam is split coherently at the first crystal plate, reflected at the middle plate and coherently superposed at the third plate (Fig. 1b). From general symmetry considerations follows immediately that the wave functions in both beam paths, which compose the beam in the forward direction behind the interferometer, are equal ($\psi_0^I = \psi_0^{II}$), because they are transmitted-reflected-reflected (TRR) and reflected-reflected-transmitted (RRT), respectively. The de Broglie wavelength of the neutrons diffracted from such crystals is about 1.8 Å and their energy is about 0.025 eV. The theoretical treatment of the diffraction process from the perfect crystal is based on the dynamical diffraction theory, which can also be found in the literature for the neutron case [7,10]. Inside the perfect crystal two wave fields are excited when the incident beam fulfills the Bragg condition, one of them having its nodes at the position of the atoms and the other in between them. Therefore, their wave vectors are slightly different ($k_1 - k_2 \simeq 10^{-5} k_0$) and due to mutual interference processes, a rather complicated interference pattern is built up, which changes substantially over a characteristic length Δ_0—the so-called *Pendellösung* length—which is of the order of 50μm for an ordinary silicon reflection. To preserve the interference properties over the length of the interferometer, the dimensions of the monolithic system have to be accurate on a scale comparable to this quantity. The whole interferometer crystal has to be placed on a stable goniometer table under conditions avoiding temperature gradients and vibrations.

A phase shift between the two coherent beams can be produced by nuclear, magnetic or gravitational interactions. In the first case, the phase shift is most easily calculated using the index of refraction [11,12]:

$$n = \frac{k_{in}}{k_0} = 1 - \frac{\lambda^2 N}{2\pi}\sqrt{b_c^2 - \left(\frac{\sigma_r}{2\lambda}\right)} + i\frac{\sigma_r N \lambda}{4\pi}, \tag{1.1}$$

which simplifies for weakly absorbing materials ($\sigma_r \to 0$) to

$$n = 1 - \lambda^2 \frac{N b_c}{2\pi}, \tag{1.2}$$

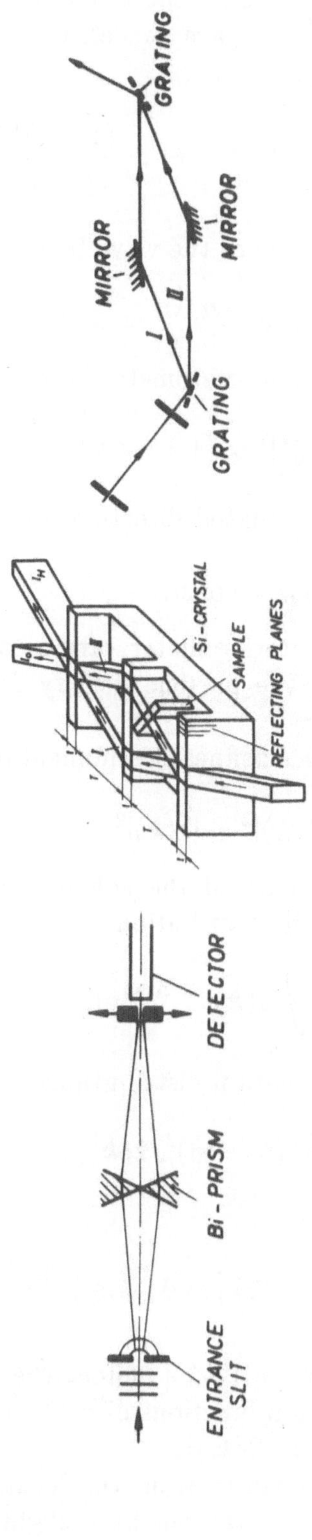

Fig. 1. Scheme of a slit-, a perfect-crystal and a grating interferometer.

where b_c is the coherent scattering length and N is the particle density of the phase shifting material. The different k vector inside the phase shifter causes a spatial shift $\boldsymbol{\Delta}$ of the wave packet which depends on the orientation of the sample surface $\hat{s}$:

$$\boldsymbol{\Delta} = \frac{(\boldsymbol{k} - \boldsymbol{k}_0)}{k} D_0 \quad \text{with} \quad \boldsymbol{k} - \boldsymbol{k}_0 = \frac{(1-n)k\hat{s}}{(\hat{\boldsymbol{k}} \cdot \hat{\boldsymbol{s}})}. \tag{1.3}$$

As in ordinary light optics the change of the wave function is obtained as follows:

$$\psi \to \psi_0 e^{i\boldsymbol{\Delta}\cdot\boldsymbol{k}} = \psi_0 e^{-iNb_c\lambda D} = \psi_0 e^{i\chi}. \tag{1.4}$$

Therefore, the intensity behind the interferometer becomes

$$I_0 \propto |\psi_0^I + \psi_0^{II}|^2 \propto (1 + \cos\chi). \tag{1.5}$$

The intensity of the beam in the deviated direction follows from particle conservation:

$$I_0 + I_H = \text{const.} \tag{1.6}$$

Thus, the intensities behind the interferometer vary as a function of the thickness D of the phase shifter, the particle density N or the neutron wavelength λ.

Standard quantum mechanics defines the momentum distribution of the beam by

$$g(\boldsymbol{k}) = |\psi(\boldsymbol{k})|^2 = |a(\boldsymbol{k})|^2, \tag{1.7}$$

and, one therefore gets the real part of the coherence function as the Fourier-transform of the momentum distribution

$$|\Gamma(\boldsymbol{\Delta})| \propto |\int g(\boldsymbol{k}) e^{i\boldsymbol{k}\cdot\boldsymbol{\Delta}} d^3\boldsymbol{k}|, \tag{1.8}$$

which simplifies for Gaussian momentum distributions

$$g(\boldsymbol{k}) \propto \exp\left[-(\boldsymbol{k} - \boldsymbol{k}_0)^2/2\delta\boldsymbol{k}^2\right], \tag{1.9}$$

with characteristic widths δk_i, to

$$|\Gamma(\boldsymbol{\Delta}_0)| = \prod_{i=x,y,z} \exp\left[-(\Delta_i \delta k_i)^2/2\right]. \tag{1.10}$$

The mean square distance related to $|\Gamma(\boldsymbol{\Delta})|$ defines the coherence length Δ_c^i which is for Gaussian distribution functions directly related to the minimum uncertainty relation $\left(\Delta_c^i = 1/(2\delta k_i)\right)$.

Any experimental device deviates from the idealized assumptions made by the theory: the perfect crystal can have slight deviations from its perfectness, and its dimensions may vary slightly; the phase shifter contributes to imperfections by variations in its thickness and inhomogeneities; and even the neutron beam itself contributes to a deviation from the idealized situation because of its momentum spread δk. Therefore, the experimental interference patterns have to be descibed by a generalized relation

$$I \propto A + B\cos(\chi + \phi_0), \tag{1.11}$$

where A, B and ϕ_0 are characteristic parameters of a certain set-up. It should be mentioned, however, that the idealized behaviour described by Eq. (1.5) can nearly be approached by a well balanced set-up [13]. The reduction of the contrast at high order results from the longitudinal coherence length which is determined by the momentum spread of the neutron beam $(\Delta_x = (2\delta k_x)^{-1}$. This causes a change in the amplitude factor of Eq. (1.7) as $(B \to B\exp\left[-\Delta_i\delta k_i)^2/2\right]$. The wavelength dependence of χ in Eq. (1.4) disappears in a special sample position where the surface of the sample is oriented parallel to the reflecting planes and the path length through the interferometer becomes $D_0/\sin\Theta_B$ and, therefore, the phase shift $\chi = -2d_{hkl}Nb_cD_0$ becomes independent of the wavelength. In this case the damping at high interference orders due to the wavelength spread does not appear as in the standard position. Related results of a recent experiment where the interference pattern in the 256th interference order have been measured in the dispersive and the nondispersive sample position are shown in Fig. 2 [14]. The much higher visibility of the interferences in the nondispersive sample arrangement is visible and is caused by the much smaller momentum spread perpendicular to the reflecting planes.

Various post-selection measurements in neutron interferometry have shown that interference fringes can be restored even in cases when the overall beam does not exhibit any interference fringes due to spatial phase shifts larger than the coherence length of the interfering beams [15–18]. This indicates that the simple picture which predicts interference only when wave packets spatially overlap is untrue. Interference actually occurs no matter how large the optical path difference may be. From classical optics it has been known for many years that the coherence properties manifest themselves in a spatial intensity variation for phase shifts smaller than the coherence length and in a spectral intensity variation for large phase shifts [19–23]. This phenomenon becomes more apparent for less monochromatic beams and can cause overall spectral shifts [24,25] and even squeezing phenomena [26,27].

The related phenomena for matter waves have been discussed recently [28,29] and will be elucidated in more detail in this paper. The experimental verification has been performed with a perfect crystal interferometer. Figure 1 depicts the general scheme of the measurements. Due to the rather low intensity of any neutron source one deals with selfinter-

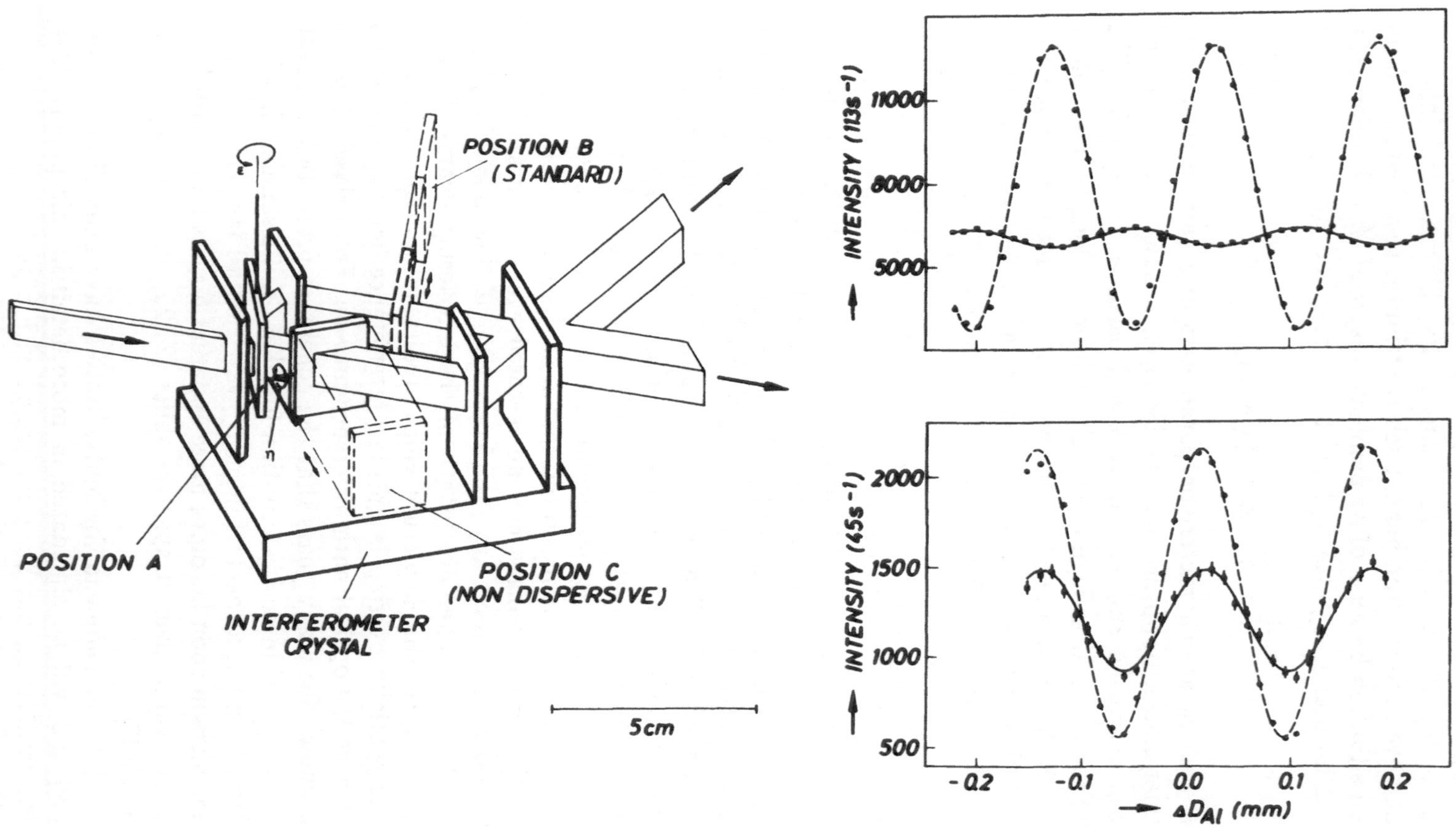

Fig. 2. Interference pattern observed at high order ($m = 256$) with a dispersively (above) and a nondispersively arranged sample [14]. (Dashed lines correspond to measurements at low order.)

Table 1. Properties of the neutrons

Particle properties	
mass:	$m_0 = 1.6749543(86).10^{-24}$ gm
spin:	$s = 1/2\ h$
magnetic moment:	$\mu = -1.91304308(54)\ \mu_K$
half live:	$T_{1/2} = 641(8)$ sec
electric charge:	$q < 2.2.10^{-20}$ e
electric dipole moment:	$d < 4.8.10^{-25}$ e.cm
confinement radius:	$R = 0.7$ fm
quark structure:	$n = u - d - d$
Wave properties	
Compton wavelength:	$\lambda_c = h/mc = 1.32.10^{-13}$ cm
de Broglie wavelength:	$\lambda_B = h/mv \simeq 1.10^{-8}$ cm*
coherence length:	$\Delta_c = \lambda^2\Delta\lambda \simeq 1.10^{-6}$ cm*
packet length:	$\Delta_p = v.\Delta t \simeq 1.10^0$ cm*
decay length:	$\Delta_d = v.T_{1/2} \simeq 2.10^8$ cm*
phase difference:	$0 \leq \chi \leq 2\pi$

* values belong to thermal neutrons ($\lambda_B = 1.8\ A, v = 2\ 200$m/s).

ference phenomena only. All the results of interferometric measurements obtained till now can be explained well in terms of the wave picture of quantum mechanics and the complementarity principle of standard quantum mechanics. Nevertheless, one should bear in mind that the neutron also carries well defined particle properties, which have to be transferred through the interferometer. These properties are summarized in Table 1 together with a formulation in the wave picture. Both particle and wave properties are well established and, therefore, neutrons seem to be a proper tool for testing quantum mechanics with massive particles, where the wave-particle dualism becomes obvious.

All neutron interferometric experiments pertain to the case of self-interference where during a certain time interval only one neutron is inside the interferometer, if at all. Usually, at that time, the next neutron has not yet been born and is still contained in the uranium nuclei of the reactor fuel. Although there is no interaction between different neutrons, they have a certain common history within predetermined limits which are defined, e.g., by the neutron moderation process, by their movement along the neutron guide tubes, by the monochromator crystal and by the special interferometer set-up. Therefore, any real interferometer pattern contains single particle and ensemble properties together. In the following chapters typical experiments performed mainly by our group within the last 15 years will be presented. Time has come to switch from classical to quantum neutron optics.

2. STOCHASTIC VERSUS DETERMINISTIC ABSORPTION

A certain beam attenuation can be achieved either by a semi-transparent material or by a proper chopper system. The transmission probability in the first case is defined by the absorption cross section $\sigma_a$of the material $[a = I/I_0 = \exp(\sigma_a N D)]$ and the change of the wave function is obtained directly from the complex index of refraction [Eq. (1.1)]:

$$\psi \rightarrow \psi_0 e^{i(n-1)kD} = \psi_0 e^{i\chi} e^{-\sigma_a ND/2} = e^{ix}\sqrt{a}\,\psi. \tag{2.1}$$

Therefore, the beam modulation behind the interferometer is obtained in the following form

$$I_0 \propto |\psi_0^I + \psi_0^{II}|^2 \propto \left[(1+a) + 2\sqrt{a}\cos\chi\right] \tag{2.2}$$

On the other hand, the transmission probability of a chopper wheel or another shutter system is given by the open to closed ratio, $a = t_{\mathbf{open}}/(t_{\mathbf{open}} + t_{\mathbf{closed}})$, and one obtains after straightforward calculations

$$\begin{aligned} I &\propto \left[(1-a)|\psi_0^{II}|^2 + a|\psi_0^I + \psi_0^{II}|^2\right] \\ &\propto \left[(1+a) + 2a\cos\chi\right], \end{aligned} \tag{2.3}$$

i.e., the contrast of the interference pattern is proportional to $\sqrt{a}$, in the first case, and proportional to a in the second case, although the same number of neutrons are observed in both cases. The absorption represents a measuring process in both cases because a compound nucleus is produced with an excitation energy of several MeV, which is usually deexcited by capture gamma rays. These can easily be detected by different means.

Figure 3 shows the dependence of the normalized contrast of the measured interference pattern on the transmission probability [30–32]. The different contrast becomes especially obvious for low transmission probabilities where the interfering part of the interference pattern is distinctly larger than the transmission probability through the semi-transparent absorber sheet. The discrepancy diverges for $\underline{a} \rightarrow 0$ but it has been shown that in this regime the variations of the transmission due to variations of the thickness or of the density of the absorber plate have to be taken into account which shifts the points below the $\sqrt{a}$ curve [33]. This can most easily be understood if the variation of the beam attenuation due to variations of the thickness or density fluctuations is included $a = \bar{a} + \Delta a$, which yields after averaging

$$\overline{\sqrt{a}} < \sqrt{\bar{a}} \tag{2.4}$$

indicating that the points fall below the $\sqrt{\bar{a}}$ curve.

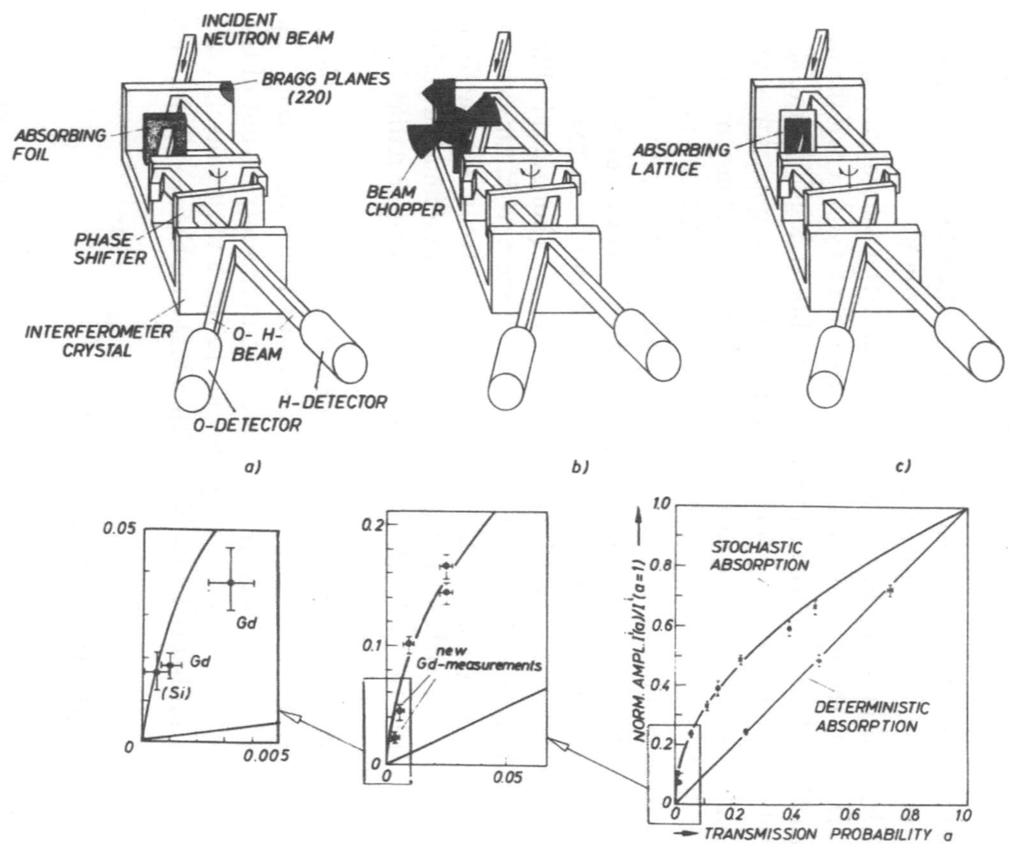

Fig. 3. Sketch of the experimental arrangement for absorber measurements (above). (a) stochastic absorption, (b) deterministic absorption, (c) attenuation by a transmission grating. Reduction of the contrast as a function of beam attenuation for different absorption methods (below) [31,32].

The region between the linear and the square root behaviour can be reached by very narrow chopper slits or by narrow transmission lattices, where one starts to loose information of through which individual slit the neutron went. This is exactly the region which shows the transition between a deterministic and a stochastic view and, therefore, it can be formulated by a Bell-like [34,35] inequality

$$\sqrt{a} > X > a \tag{2.5}$$

The stochastic limit corresponds to the quantum limit when one does not know anymore through which individual slit the neutron went. Which situation exists depends how the slit widths ℓ compares to the coherence lengths $(\Delta_i \sim (2\Delta k_i)^{-1})$ in the related direction. In case that the slit widths become smaller than the coherence lengths, the wave function behind the slits show distinct diffraction peaks which correspond to new quantum states $(n \neq 0)$, which now do not overlap with the undisturbed reference beam. The creation of the new quantum states means that those labeled neutrons carry information about the chosen beam path and, therefore, do not contribute to the interference amplitude [36]. A related experiment has been carried out by rotating an absorption lattice around the beam axis where one changes from $\ell \ll \Delta_x$ (vertical slits) to $\ell \gg \Delta_y$ (horizontal slits), Fig. 4, because the coherence length parallel to the reflecting lattice vector is much larger than in any other directions. Thus, the attenuation factor $\underline{a}$ has to be generalized including not only nuclear absorption and scattering processes but also lattice diffraction effects if they remove neutrons from the original phase space.

A very similar situation exists if a very fast chopper produces beam bursts (packet lengths) shorter than the coherence time $\Delta t_c = \Delta/v$. In this case, diffraction in time occurs which also removes neutrons out of the original phase space. This limit is very difficult to reach with a mechanical chopper but it can probably be tackled with a high frequency spin flipper.

3. NEUTRON JOSEPHSON EFFECT

This phenomenon is based on the dipole coupling of the magnetic moment $\boldsymbol{\mu}$ of the neutron to a magnetic field $\boldsymbol{B}$ $(H = -\boldsymbol{\mu} \cdot \boldsymbol{B})$, which causes the famous 4π-symmetry of spinor wave functions, as measured in early neutron interferometer experiments [37,38]. The change of the wave function reads as:

$$\psi \rightarrow \psi_0 e^{-i(Ht/\hbar)} = \psi_0 e^{-i(\mu B t/\hbar)} = \psi_0 e^{-i\alpha \cdot \sigma/2} = \psi(\alpha) \tag{3.1}$$

where $\boldsymbol{\alpha}$ represents a formal description of the Larmor rotation angle around the field $B(\alpha = (2\mu/\hbar)\int B dt \simeq (2\mu/\hbar v)\int B ds)$. This enabled

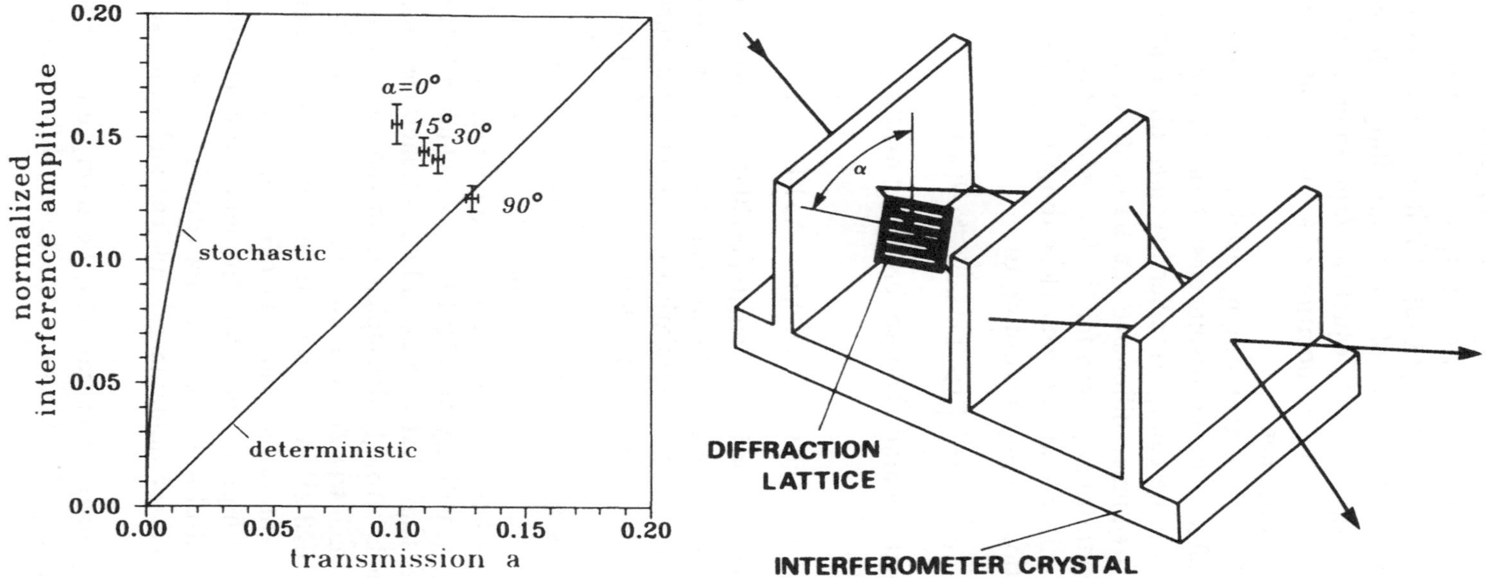

Fig. 4. Lattice absorber in the interferometer approaching the quantum limit [31,36].

also the realization of the spin superposition experiments where spin-up ($|\uparrow>$) and spin-down states ($|\downarrow>$) are superposed producing a final state perpendicular to both initial states [39,40]. It is interesting to mention that in the case of spin reversal by means of a resonance flipper the spin term is accompanied by an energy exchange equal the Zeeman energy $\hbar\omega_r = 2\mu B$. This provided the basis for the observation of a new quantum beat effect; the magnetic Josephson analog.

A double coil arrangement can be used for the observation of a new quantum beat effect. If the frequencies of the two coils are chosen to be slightly different, the energy transfer becomes different too ($\Delta E = \hbar(\omega_{r1} - \omega_{r2})$). The frequency difference can be made very small, if high quality frequency generators are used for the field generation. The flipping efficiencies for both coils are always very close to unity (better than 0.99). Now, the wave functions change according to

$$\psi \rightarrow e^{i(\omega-\omega_{r1})t} \mid\downarrow> + e^{i\chi} e^{i(\omega-\omega_{r1})t} \mid\downarrow> . \tag{3.2}$$

Therefore, the intensity behind the interferometer exhibit a typical quantum beat effect, given by

$$I \propto 1 + \cos\left[\chi + (\omega_{r1} - \omega_{r2})t\right] . \tag{3.3}$$

Thus, the intensity behind the interferometer oscillates between the forward and deviated beam without any apparent change inside the interferometer [41]. The time constant of this modulation can reach a macroscopic scale which is correlated to an uncertainty relation $\Delta E \Delta t \leq \frac{\hbar}{2}$. Figure 5 shows the result of an experiment, where the periodicity of the intensity modulation, $T = 2\pi/(\omega_{r1} - \omega_{r2})$, amounts to $T = (47.90 \pm 0.15)$ sec caused by a frequency difference of about 0.02 Hz. This corresponds to a mean difference ΔE of the energy transfer between the two beams, $\Delta E = 8.6 \times 10^{-17}$ eV, and to an energy sensitivity of 2.7×10^{-19} eV, which is better by many orders of magnitudes than that of other advanced spectroscopic methods. This high resolution is strongly decoupled from the monochromaticity of the neutron beam, which was $\Delta E_B \simeq 5.5 \times 10^{-4}$ eV around a mean energy of the beam $E_B = 0.023$ eV in this case. It should be mentioned, that the result can also be interpreted as being the effect of a slowly varying phase $\Delta(t)$ between the two flipper fields, but the more physical description is based on the argument of a different energy transfer. The extremely high resolution may be used for fundamental, nuclear and solid- state physics applications.

The quantum beat effect can also be interpreted as a magnetic Josephson effect analog. In this case, the phase difference is driven by the magnetic energy

$$\frac{\delta}{\delta t}(\Delta_2 - \Delta_1) = \omega_{r2} - \omega_{r1} = \frac{1}{\hbar} 2\mu B_0, \tag{3.4}$$

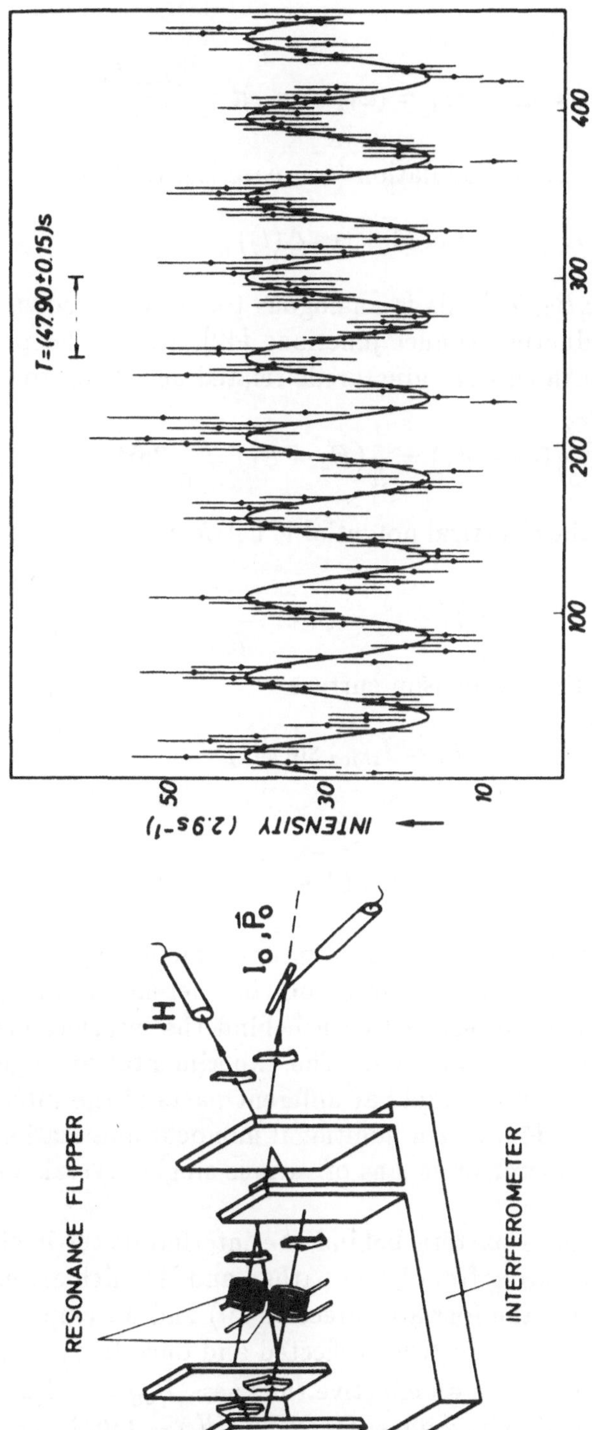

Fig. 5. Magnetic Josephson effect observed when the frequencies of the two flipper coils differ by about 0.02 Hz around 71.89979 kHz [41].

and, therefore,

$$\Delta(t) = \Delta_2 - \Delta_1 = (\omega_{r2} - \omega_{r1})t = \frac{2\mu\Delta B_0}{\hbar} \cdot t \tag{3.5}$$

This yields the observed modulation [compare Eq. (3.2)]

$$I \propto (1 + \cos\Delta(t)), \tag{3.6}$$

where $\Delta(t) = 2\mu\Delta B_0/\hbar$. This is analogous to the well-known Josephson effect in superconducting tunnel junctions [42], where the phase of the Cooper- pairs in both superconductors is related according to

$$\frac{\delta}{\delta t}(\phi_2 - \phi_1) = \frac{1}{\hbar}(E_2 - E_1) = \frac{1}{\hbar}\, 2\mathrm{eV}, \tag{3.7}$$

which is driven by the electrical potential V between both superconductors. This gives

$$\phi(t) = \phi_2 - \phi_1 = \frac{2\mathrm{eV}}{\hbar} t \tag{3.8}$$

and a superconducting Josephson current

$$I_S = I_{s\mathrm{Max}} \sin\phi(t) \tag{3.9}$$

4. POSTSELECTION OF STATES

In the course of several neutron interferometer experiments [15–18] it has been established that smoothed out interference properties at high interference order can be restored even behind the interferometer when a proper spectral filtering is applied. The experimental arrangement with an indication of the wave packets at different parts of the interference experiment is shown in Fig. 6. An additional monochromatization is applied behind the interferometer by means of various single crystals brought into Bragg-position.

The interference pattern behind the interferometer is given by the wave functions originating from beam paths I and II which are equal in amplitude and phase for the forward direction (0) and an empty interferometer because both beams are twice reflected and once transmitted (TRR$\hat{=}$ RRT). A phase shifter with an effective thickness $D_{\mathrm{eff}} = D/(\hat{\boldsymbol{k}} \cdot \hat{\boldsymbol{s}})$ changes the related wave function by a phase factor $\exp[i(n-1)kD_{\mathrm{eff}}] = \exp(i\boldsymbol{\Delta}\cdot\boldsymbol{k})$ and one gets for the individual plane wave components an intensity

$$\begin{aligned} I_0(\boldsymbol{r},\boldsymbol{k}) &= |\psi_0^I(\boldsymbol{r},\boldsymbol{k}) + \psi_0^{II}(\boldsymbol{r}+\Delta,\boldsymbol{k}|^2 \\ &\propto |a(\boldsymbol{k})|^2\, |1 + \cos(\boldsymbol{\Delta}(\boldsymbol{k})\cdot\boldsymbol{k})\,|^2 \end{aligned} \tag{4.1}$$

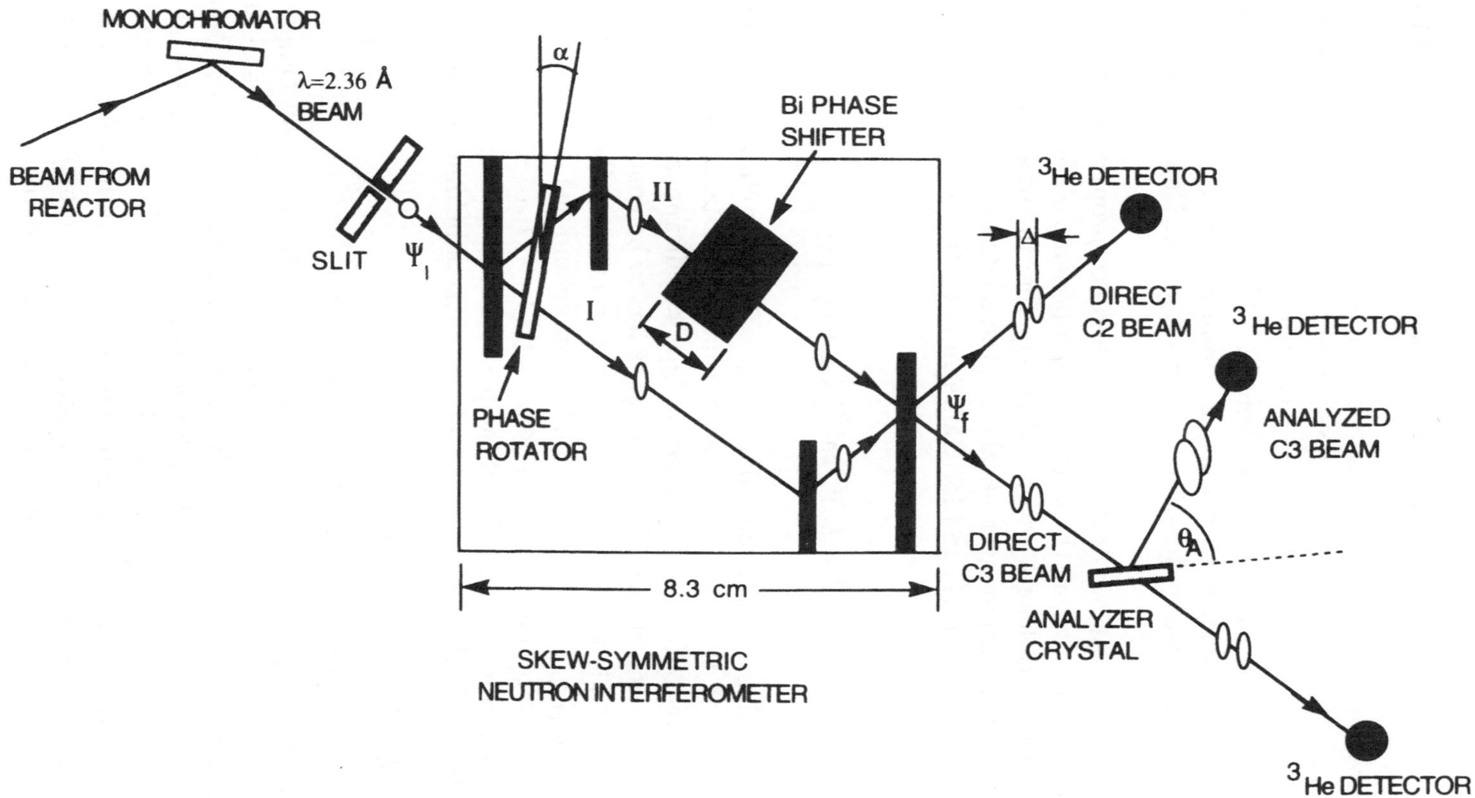

Fig. 6. Scheme of the experimental arrangement with a skew-symmetrically cut perfect crystal interferometer and a post-selection analyzer crystal [18,28].

and, for the overall beam,

$$I_0(\boldsymbol{\Delta}_0) \propto 1 + |\Gamma(\boldsymbol{\Delta}_0)| \cos(\boldsymbol{\Delta}_0 \cdot \boldsymbol{k}_0), \tag{4.2}$$

where $\boldsymbol{\Delta}_0$ represents the spatial phase shift for the $\boldsymbol{k}_0$ component of the packet. Equation (4.2) describes the interference fringes when $\boldsymbol{\Delta}_0$ is varied. The formula also shows that the interference fringes disappear for spatial phase shifts larger than the coherence lengths $[\Delta_i \geq \Delta_c^i = 1/(2\delta k_i)]$ (see Eq. (1.10)). This behavior is shown in Fig. 7 and has been verified experimentally by several investigations for Gaussian and non-Gaussian neutron beams [43–45].

In our experiment, we deal with the coherence properties along the interferometer axis (x), where the tangential components of the momentum vectors (and coherence length) does not change due to Bragg diffraction. According to basic quantum mechanical laws, the related momentum distribution follows from Eq. (4.1) and for Gaussian packets it can be rewritten in the form ($k = k_x, k_y, k_z$)

$$I_0(k) \propto \exp\left[-(k-k_0)^2/2\delta k^2\right]\left\{1 + \cos\left[\chi_0 \frac{k_0}{k}\right]\right\}, \tag{4.3}$$

where the mean phase shift is introduced ($\chi_0 = k_0\Delta_0 = Nb_c\lambda_0 D_{\text{eff}}$). The surprising feature is that $I_0(k)$ becomes oscillatory for large phase shifts where the interference fringes described by Eq. (4.2) disappear (see Fig. 7). This indicates that interference in phase space has to be considered [46,47] rather than the simple wave function overlap criterion described by the coherence function [Eq. (1.10)]. The second beam behind the interferometer (H) just shows the complementary modulation ($I_H = I_{\text{total}} - I_0$).

The amplitude function [48] of the packets arising from beam paths I and II determines the spatial shape of the packets behind the interferometer:

$$\begin{aligned} I_0(x) &= |\psi(x) + \psi(x+\Delta)|^2 \\ &\longrightarrow \exp\left[-x^2/2\delta x^2\right] + \exp\left[-(x+\Delta_0)^2/2\delta x^2\right], \end{aligned} \tag{4.4}$$

for $\Delta \gg \delta x$, which separates for large phase shifts into two peaks (Fig. 7). For Gaussian packets, δx corresponds to the coherence length Δ_c and fulfills the minimum uncertainty relation $\delta x \delta k = 1/2$. For an appropriately large displacement ($\Delta \gg \Delta_c$), the related state can be interpreted as a superposition state of two macroscopically distinguishable states, that is, a stationary Schrödinger-cat-like state [49,50], but here first for massive particles. These states—separated in ordinary space and oscillating in momentum space—seem to be notoriously fragile and sensitive to dephasing effects [51,54].

Measurements of the wavelength spectrum were made with a narrow mosaic silicon crystal which reflects in the parallel position a very narrow

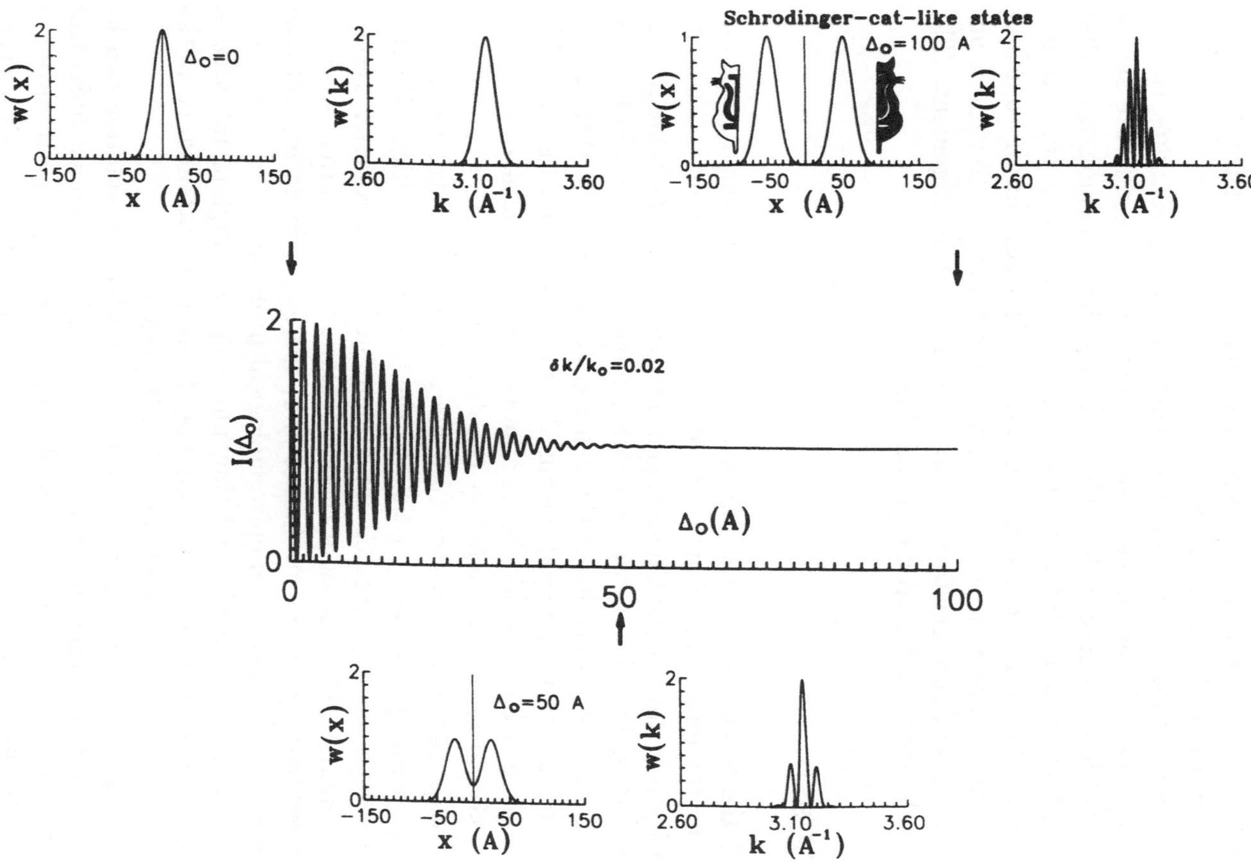

Fig. 7. Interference pattern as a function of the relative phase shift and related wave packets and momentum spectra behind the interferometer [19].

band of neutrons only ($\delta k'/\delta k \approx 0.05$) causing an enhanced visibility at large phase shifts (Fig. 8). This feature shows that an interference pattern can be restored even behind the interferometer by means of a proper post-selection procedure. In this case the overall beam does not show interference fringes anymore and the wave packets originating from the two different beam paths do not overlap.

The momentum distribution has been measured by scanning the analyzer crystal through the Bragg-position. The related results are shown in Fig. 9 for different phase shifts. These results clearly demonstrate that the predicted spectral modulation [Eq. (4.3)] appears when the interference fringes of the overall beam disappear. The modulation is somehow smeared out due to averaging processes across the beam due to various imperfections, unavoidably existing in any experimental arrangement. The contrast of the empty interferometer was 60%.

5. DISCUSSION

All the results of the neutron interferometric experiments are well described by the formalism of quantum mechanics. According to the complementarity principle of the Copenhagen interpretation, the wave picture has to be used to describe the observed phenomena. The question how the well-defined particle properties of the neutron are transferred through the interferometer is not meaningful within this interpretation, but from the physical point of view it should be an allowed one. Therefore, other interpretations should also be included in the discussion of such experiments. The particle picture can be preserved if pilot waves are postulated or if a quantum potential guides the particle to the predicted position. Related calculations have been performed for a simplified interferometer system [55]. Unfortunately, the results of these calculations are identical with the results of ordinary quantum mechanics and, therefore, to decide between both points of view remains an epistemological problem.

The newly discovered persisting phase space coupling in cases of large spatial shifts of the wave packets may bring some attendance to the action of plane wave components outside the packet.

The results clearly demonstrate that a spectral modulation can be observed in neutron interference experiments at high interference order and that interference has to be treated in phase space rather than in ordinary space. It looks like that the plane wave components of the wave packets, i.e. narrow band width components, interact over a much larger distance than the size of the packets. This interaction guides neutrons of certain momentum bands to the 0 or H beam, respectively. These phenomena throw a new light on the discussion on Schrödinger-cat-like situations in quantum mechanics and, therefore, on the discussion about the EPR ex-

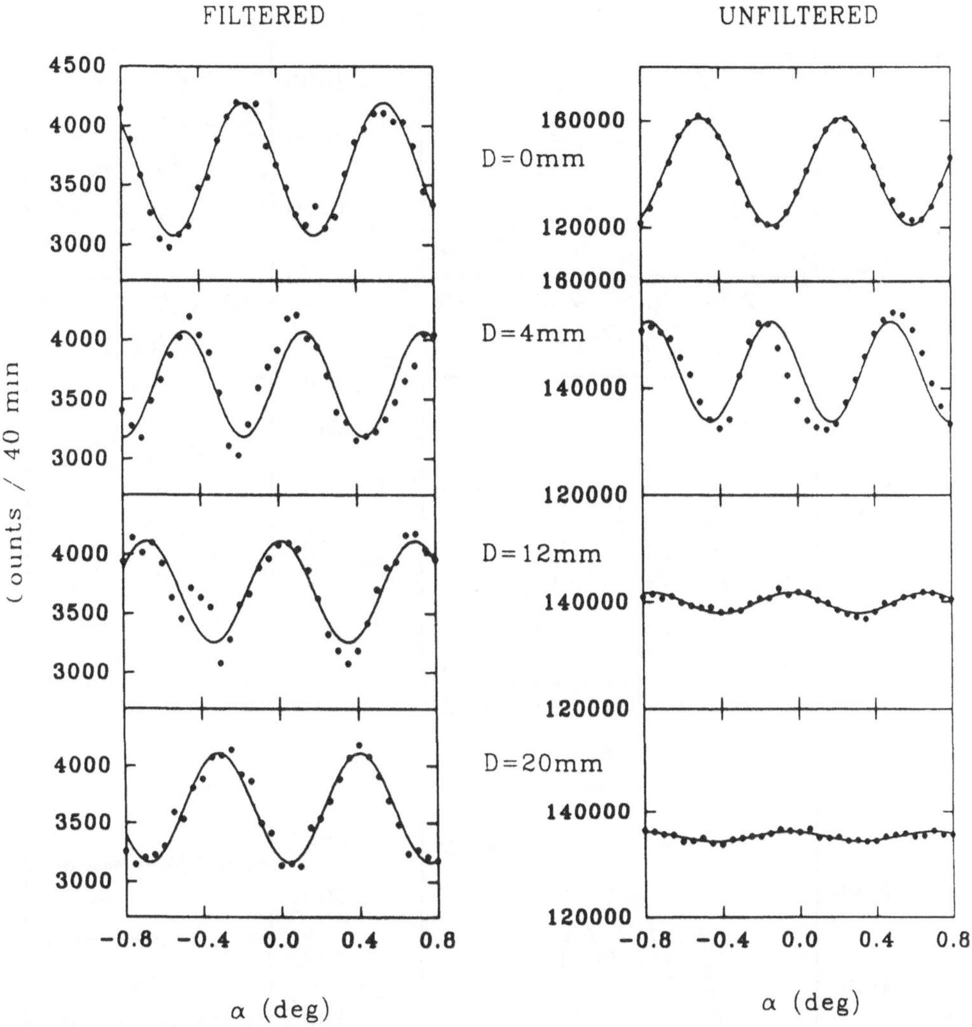

Fig. 8. Interference pattern of the overall beam ($\delta k/k_0 = 0.012$) and the beam reflected from a nearly perfect crystal analyzer in the antiparallel position ($\delta k'/k_0 = 0.0003$) [18].

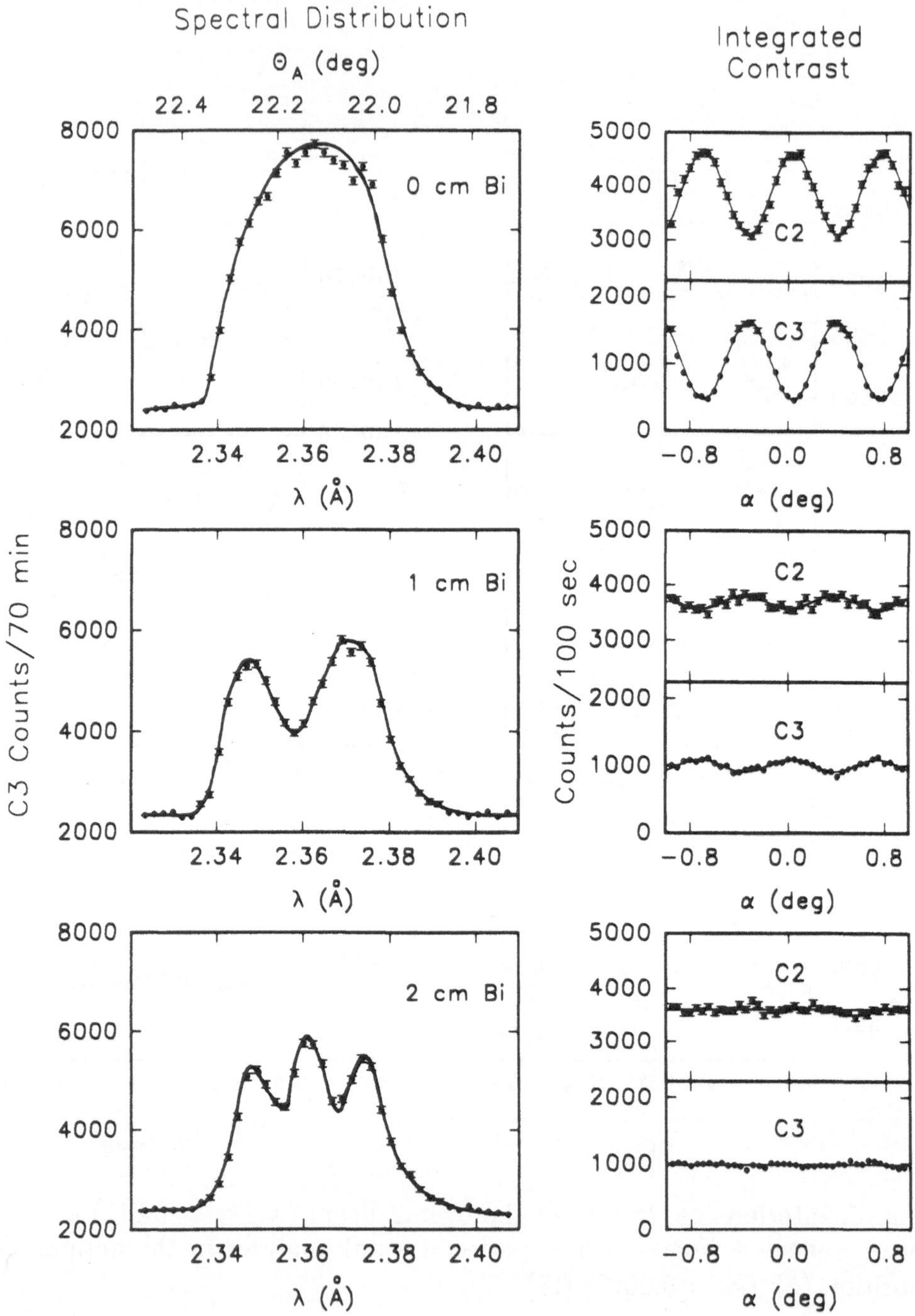

Fig. 9. Measured spectral modulation of the outgoing beam and the residual interference pattern for different bismuth phase shifter thicknesses. The wavelength resolution of the analyzer was 0.002 Å [18].

periments too [28,56–58]. Spatially separated packets remain entangled in phase space and nonlocality appears as a result of this entanglement. The analogy with optical experiments performed in the time-frequency domain is striking [23]. An analogue situation exists in neutron spin-echo systems where multiple spin rotation plays an equivalent role as high order interferences discussed here [59].

Each peak in the momentum distribution corresponds to a different number of phase shifts experienced by the neutrons of that wavelength band during its passage through the interferometer. In that sense, the minimum quantum unity of the incident wave packet becomes a new quantity representing different quantum states with distinguishable properties. This kind of labeling shows that constructive interference is restricted to that wavelength band only; a situation similar to that where new states have been created due to lattice diffraction inside the interferometer [36] (see Fig. 4).

The new quantum states created behind the interferometer can be analyzed with regard to their uncertainty properties. Analogies between a coherent state behavior and a free but coherently coupled particle motion inside the interferometer have been addressed previously [32]. In such cases, the dynamically conjugate variables x and p minimize the uncertainty product with identical uncertainties $(\Delta x)^2 = (\Delta k)^2 = 1/2$ (in dimensionless units). Using $I_0(k)$ and $I_0(x)$ [Eqs. (4.3) and (4.4)] as distribution function we get, in our case, for $\Delta \gg \delta x$,

$$\langle(\Delta x)^2\rangle = \langle x^2\rangle - \langle x\rangle^2 \longrightarrow (\delta x)^2 + \left[\frac{\Delta}{2}\right]^2 \tag{5.1}$$

and, for $\delta k/k_0 \ll 1$,

$$\begin{aligned}\langle(\Delta k)^2\rangle &= \langle k^2\rangle - \langle k\rangle^2 \\ &= (\delta k)^2\left\{1 - \left[\frac{\Delta_0}{2\delta x}\right]^2 \frac{e^{-(\Delta_0/2\delta x)^2}\cos(\Delta_0 k_0) + e^{-(\Delta_0/2\delta x)^2}}{\left[1 + e^{-(\Delta_0/2\delta x)^2/2}\cos(\Delta_0 k_0)\right]^2}\right\}\end{aligned} \tag{5.2}$$

These relations are shown in Fig. 10 indicating that for $(\Delta k)^2$ a value below the coherent state value can be achieved, which in quantum optic terminology means squeezing [35,60,61]. One emphasizes that a single coherent state does not exhibit squeezing, but a state created by superposition of two coherent states can exhibit a considerable amount of squeezing. Thus, highly nonclassical states can be made by the power of the quantum mechanical superposition principle.

We have always tried to perform unbiased experiments and do not wish to interfere with any epistemological interpretation of quantum mechanics. Perhaps in the future new proposals for experiments will be formulated which permit a unique decision between different interpretations.

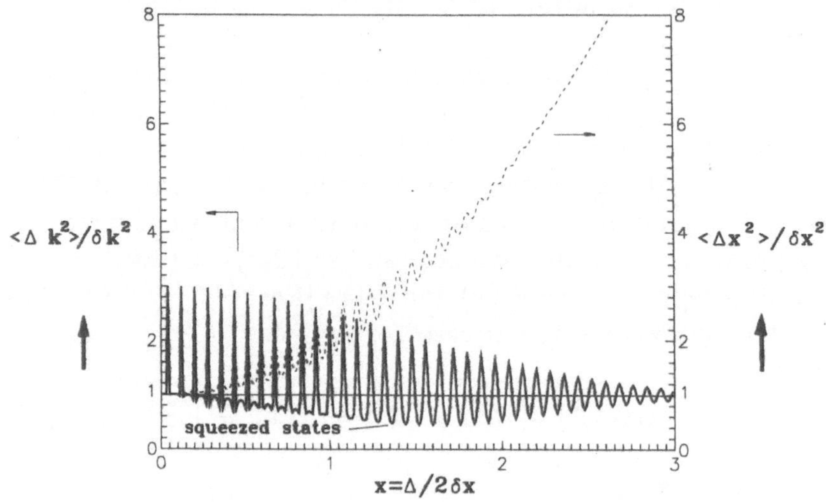

Fig. 10. Spatial and momentum uncertainties of the outgoing beams with the indication of squeezing in the momentum domain [18].

As an experimentalist, one appreciates the pioneering work of the founders of quantum mechanics, who created this basic theory with so little experimental evidence. Now we have much more direct evidences, even on a macroscopic scale but, nevertheless, one notices that the interpretation of quantum mechanics goes beyond human intuition in certain cases. Only few aspects of the experiments discussed before should be mentioned again: How can every neutron have information about which beam to join behind the interferometer, when a slightly different energy exchange occurs in both beams inside the interferometer and the time constant of the beat effect is by many orders of magnitudes larger than the time of flight through the system? How can the interference pattern be influenced in a pulsed beam when the mean occupation number of a single pulse is in the order of 10^{-4} only? [17] How can the wave packets influence each other when they are shifted more than their dimensions? These are the speakable and unspeakable questions of quantum mechanics.

Some recent experiments of our group which are related to fundamental physics problems have been discussed in this article. Several review articles can give a broader scope about the status of neutron interferometry [62–66].

ACKNOWLEDGEMENT

Most of the experimental results discussed in detail have been obtained by our Dortmund-Grenoble-Vienna interferometer group working at the high flux reactor in Grenoble, and some recent ones stem from our cooperation with the Columbia-Missouri group working at the MURR-reactor. The cooperation within these groups and especially the cooperation with colleagues from our Institute which are cited in the references are gratefully acknowledged.

REFERENCES

1. H.Maier-Leibnitz, and T.Springer, *Z.Phy.* **167** (1962).
2. R.Gaehler, J.Kalus, and W.Mampe, *J.Phys.E* **13** (1980) 546.
3. H.Rauch, W.Treimer, and U.Bonse, *Phys.Lett.A* **47** (1974) 369.
4. W.Bauspiess, U.Bonse, H.Rauch, and W.Treimer, *Z. Phys.* **271** (1974) 177.
5. A.I.Ioffe, V.S.Zabiyankan, and G.M.Drabkin, *Phys.Lett.* **111** (1985) 373.
6. U.Bonse, and M.Hart, *Appl.Phys.Lett.* **6** (1965) 155.
7. H.Rauch, and D.Petrascheck, *Neutron Diffraction*, H.Dachs, ed. (Springer, Berlin, 1978), Chap.9.

8. V.F.Sears, *Can.J.Phys.* **56** (1978) 1261.
9. W.Bauspiess, U.Bonse, and W.Graeff, *J.Appl.Cryst.* **9** (1976) 68.
10. D.Petrascheck, *Acta Phys.Austr.* **45** (1976) 217.
11. M.L.Goldberger, and F.Seitz, *Phys.Rev.* **71** (1947) 294.
12. V.F.Sears, *Phys.Rep.* **82** (1982) 1.
13. U.Bonse, and H.Rauch, eds, *Neutron Interferometry* (Clarendon Press, Oxford, 1979).
14. H.Rauch, E.Seidl, D.Tuppinger, D.Petrascheck, and R.Scherm, *Z.Phys. B* **69** (1987) 69.
15. S.A.Werner, R.Clothier, H.Kaiser, H.Rauch, and H.Wölwitsch, *Phys. Rev. Lett.* **67** (1991) 683.
16. H.Kaiser, R.Clothier, S.A.Werner, H.Rauch, and H.Wölwitsch, *Phys. Rev.A* **45** (1992) 31.
17. H.Rauch, H.Wölwitsch, R.Clothier, H.Kaiser, and S.A.Werner, *Phys. Rev.A* **46** (1992) 49.
18. D.L.Jacobson, S.A.Werner, and H.Rauch, *Phys.Rev. A*, in print.
19. L.Mandel, *J.Opt.Soc.Am.* **52** (1962) 1335.
20. L.Mandel, and E.Wolf, *Rev.Mod.Phys.* **37** (1965) 231.
21. F.Heineger, A.Herden, and T.Tschudi, *Opt.Comm.* **48** (1983) 237.
22. D.F.V.James, and E.Wolf, *Phys.Lett.A* **157** (1991) 6.
23. X.Y.Zou, T.P.Grayson, and L.Mandel, *Phys.Rev.Lett.* **69** (1992) 3041.
24. E.Wolf, *Phys.Rev.Lett.* **63** (1989) 2220.
25. D.Faktis, and G.M.Morris, *Opt.Lett.* **13** (1988) 4.
26. W.Schleich, M.Pernigo, and Fam Le Kien, *Phys.Rev. A***44** (1991) 2172.
27. J.Janski, and A.V.Vinogradov, *Phys.Rev.Lett.* **64** (1990) 2771.
28. H.Rauch, *Phys.Lett. A***173** (1993) 240.
29. H.Rauch, *Proceedings Quantum Measurement and Control*, E.Ezawa and Y.Murayama, eds. (North Holland, Amsterdam, 1993), p.223.
30. H.Rauch, and J.Summhammer, *Phys.Lett. 104***A** (1984) 44.
31. J.Summhammer, H.Rauch, and D.Tuppinger, *Phys.Rev. A***36** (1987) 4447.
32. H.Rauch, J.Summhammer, M.Zawisky, and E.Jericha, *Phys.Rev. A***42** (1990) 3726.
33. M.Namiki, and S.Pascazio, *Phys.Lett.* **147** (1990) 430.
34. J.Bell, *Physics* **1** (1965) 195.
35. D.Home, and F.Selleri, *Nuovo Cimento* **14** (1991) 1.
36. H.Rauch, and J.Summhammer, *Phys.Rev.* **46** (1992) 7284.
37. H.Rauch, A.Zeilinger, G.Badurek, A.Wilfing, W.Bauspiess, and U. Bonse, *Phys.Lett. A***54** (1975) 425.
38. S.A.Werner, R.Colella, A.W.Overhauser, and C.F.Eagen, *Phys.Rev. Lett.* **35** (1975) 1053.
39. J.Summhammer, G.Badurek, H.Rauch, U.Kischko, and A.Zeilinger, *Phys.Rev. A***27** (1983) 2523.

40. G.Badurek, H.Rauch, and J.Summhammer, *Phys.Rev.Lett.* **51** (1983) 1015.
41. G.Badurek, H.Rauch, and D.Tuppinger, *Phys.Rev.* *A***34** (1986) 2600.
42. B.D.Josephson, *Rev.Mod.Phys.* **46** (1974) 251.
43. H.Rauch, in *Neutron Interferometry*, U.Bonse and H.Rauch, eds. (Clarendon Press, Oxford, 1979), p.161.
44. H.Kaiser, S.A.Werner, and E.A.George, *Phys. Rev.Lett.* **50** (1983) 563.
45. R.Clothier, H.Kaiser, S.A.Werner, H.Rauch, and H.Wölwitsch, *Phys. Rev.* *A*44 (1991) 5357.
46. W.Schleich, and J.A.Wheeler, *Nature* **326** (1987) 574.
47. W.Schleich, D.F.Walls, and J.A.Wheeler, *Phys.Rev.* *A***38** (1988) 1177.
48. J.-M.Levy-Leblond, and F.Balibar, *Quantics* (North-Holland, Amsterdam, 1990).
49. A.Legett, *Proceedings, Foundations of Quantum Mechanics*, S.Kamefuchi et al., eds. Physical Society and Japan, Tokyo, 1984, p.74.
50. B.Yurke, W.Schleich, and D.F.Walls, *Phys.Rev.* *A***42** (1990) 1703.
51. D.F.Walls, and G.J.Milburn, *Phys.Rev.* *A***31** (1985) 2403.
52. R.J.Glauber, *New Techniques and Ideas in Quantum Measurement Theory*, D.M.Greenberger, ed., *Proc. N.Y. Acad. Sci.*, 1986, p.336.
53. M.Namiki, and S.Pascazio, *Phys.Rev.* *A***44** (1991) 39.
54. H.Zurek, *Physics Today,* Oct. 1991, p.36.
55. C.Dewdney, *Phys.Lett.* *A***109** (1985) 377.
56. A.Einstein, B.Podolsky, and N.Rosen, *Phys.Rev.* **47** (1935) 777.
57. D.M.Greenberger, M.A.Horne, and A.Zeilinger, in *Bell's Theorem, Quantum Theory and Conceptions of the Universe*, M.Kafatos, ed. (Kluwer Academic, Dordrecht, 1989), p.69.
58. N.D.Mermin, *Phys.Rev.Lett.* **65** (1990) 1838.
59. G.Badurek, and H.Weinfurter, *Phys.Rev.Lett.*, in print.
60. D.F.Walls, *Nature* **306** (1983) 141.
61. S.L.Braunstein, and R.I.McLachlan, *Phys.Rev.* **A35** (1987) 1659.
62. A.G.Klein, and S.A.Werner, *Rep.Prog.Phys.* **46** (1983) 259.
63. D.Greenberger, *Rev.Mod.Phys.* **55** (1983) 875.
64. H.Rauch, *Contemp.Phys.* **27** (1986) 345.
65. S.A.Werner, and A.G.Klein, in *Methods of Experimental Physics* **23A** (Academic, New York, 1986), p. 259.
66. V.F.Sears, *Neutron Optics*, (Oxford University Press, Oxford, 1989).

SOME COMMENTS ON THE DE BROGLIE-BOHM PICTURE BY AN ADMIRING SPECTATOR

Euan J. Squires

Department of Mathematical Science
University of Durham
Durham City, DH1 3LE, United Kingdom

The pilot-wave model of quantum theory, as introduced by Bohm in 1952, is derived and reviewed. The possibility of obtaining the model through a Lagrangian formulation is considered, and it is shown how it might be applied in quantum cosmology. Finally, problems of Lorentz invariance and locality are discussed.

Key words: quantum theory, pilot wave, locality.

1. INTRODUCTION

The title of this talk reflects the fact that there are many people in this meeting who are experts on the "pilot-wave interpretation" - and that I am not of their company. Many years ago John Bell told me that I should learn about the causal model. I tried, but I met a serious barrier. As an undergraduate I had been taught about the Hamilton-Jacobi equation and had completely failed to understand what it was all about. Hence, when I saw that Bohm's work seemed to depend on this equation, I gave up.

It was much later before I realised that this obstacle need not be surmounted and that there was a way around - an easier way in for simple minds like mine. Since then I have become increasingly impressed by the model, by the insight it gives into the wonders of quantum theory, by the way it - *naturally* - evades all the irrelevant "no-go" theorems, and by the possibility that it might actually be true, i.e., be a correct representation of what *is*. In this talk I shall give various possible ways of looking at the model, perhaps of changing it, of using it in new circumstances, etc. Throughout I shall be concerned with the version of the pilot-wave model introduced by Bohm[1] in 1952. This differs significantly from the earlier version of de Broglie.

Waves and Particles in Light and Matter, Edited by
A. van der Merwe and A. Garuccio, Plenum Press, New York, 1994

2. DERIVATION

First, since there may be some people here who are not aware of how simple the model is, I shall begin with an elementary derivation. We first need to note that all measurements are ultimately measurements of *position*, e.g., of some macroscopic object. Then, recalling that the predictions of quantum theory refer to probabilities, we ask what is the requirement on particle trajectories such that we always obtain the correct position probabilities. Clearly we have to postulate that the distribution of positions at one time agrees with that given by quantum theory (this is an initial condition, to which we shall return below) and that the particles move so as to preserve this agreement for all subsequent times.

To be as simple as possible here, we consider one particle moving in one dimension. Then, if $\rho(x,t)$ is the distribution at time t, the latter requirement leads immediately to the equation

$$v\frac{\partial\rho}{\partial x}+\frac{\partial\rho}{\partial t}+\rho\frac{\partial v}{\partial x}=0, \tag{1}$$

where $v(x,t)$ is the required velocity. The solution of this equation can be written in the form

$$v=-\frac{1}{\rho}\int_{x_0}^{x}\left(\frac{\partial\rho}{\partial t}\right)dx. \tag{2}$$

If we now insert the quantum distribution,

$$\rho=|\psi|^2, \tag{3}$$

into Eq.(2) and use the Schrödinger equation,

$$i\hbar\frac{\partial\psi}{\partial t}=H\psi, \tag{4}$$

with

$$H=-\frac{\hbar^2}{2m}\frac{\partial^2}{\partial x^2}+V(x), \tag{5}$$

we find

$$v=\frac{1}{m}Re(\frac{p\psi}{\psi})+\frac{constant}{\rho}, \tag{6}$$

where p is the momentum operator. The usual neglect of the second term can be justified on the grounds of simplicity, or by requiring that the velocity should be bounded, or, more generally, by the requirement of rotational invariance.

The result can be written in a different form if we put

$$\psi = Re^{\imath S/\hbar}, \tag{7}$$

where R and S are real. Then Eq.(6), without the final term, becomes

$$v = \frac{1}{m}\frac{\partial S}{\partial x}. \tag{8}$$

The model is completely defined by these equations. The generalisation to many particles and three-dimensional space is immediate: We replace, for example, Eq.(6) with

$$\dot{\mathbf{x}}_\imath \equiv \mathbf{v}_\imath = Re\frac{\mathbf{p}_\imath \Psi(\mathbf{x}_1, \mathbf{x}_2, ...)}{m_\imath \Psi(\mathbf{x}_1, \mathbf{x}_2, ...)}. \tag{9}$$

It is important to emphasise that *this is it!*. Nothing more is required, even to describe (orbital) angular momentum. This equation *automatically* provides the contextuality of measurement which ensures that the model avoids the restrictions of the Kochen-Specker type "no-go" theorems. No contrivance is needed.

If we wish to obtain the analogue of Newton's second law of motion, we consider

$$\ddot{\mathbf{x}}_\imath = \frac{\partial}{\partial t}\dot{\mathbf{x}}_\imath + \sum_\jmath (\dot{\mathbf{x}}_\jmath . \nabla_\jmath)\dot{\mathbf{x}}_\imath. \tag{10}$$

Then, using Eq.(9), we find, after some rearrangement,

$$m_\imath \dot{\mathbf{x}}_\imath = -Re\nabla_\imath \left(\frac{i\hbar\dot{\psi}}{\psi}\right) + Re\sum_\jmath (\dot{\mathbf{x}}_\jmath . \nabla_\jmath)\left(\frac{\mathbf{p}_\imath \psi}{\psi}\right) \tag{11}$$

$$= -\nabla_\imath \Big[Re\left(\frac{H\psi}{\psi}\right) - \sum_\jmath \frac{m_\jmath \dot{x}_\jmath^2}{2}\Big]. \tag{12}$$

Hence we can write

$$\ddot{\mathbf{x}}_\imath = -\nabla(V + Q), \tag{13}$$

where Q is the *quantum potential*, given by

$$Q = Re\sum_\jmath \frac{[p_\jmath^2 - m_\jmath^2 v_\jmath^2]\psi}{2m_\jmath \psi}. \tag{14}$$

As with Eq.(9), there is a certain naturalness about this equation - we almost could have guessed it! We note however a very surprising feature. If the wave function is an eigenstate of energy, taken to be real apart from the energy factor, $\exp(iEt)$, then $V + Q = 0$, i.e., the quantum potential exactly cancels the classical potential, and everything is eternally at rest. We return to this later.

At this stage the model looks very like classical mechanics. The only change caused by quantum theory is the introduction of new forces, and, of course, the history of classical mechanics contains other examples of the discovery of new forces. The potential Q is non-local, since it depends on the positions of all the particles. However, again such non-locality is neither surprising nor alarming - the potentials of Newtonian gravity and classical electrostatics are similarly non-local. In these cases the non-locality is removed in a proper relativistic theory.

Another difference with classical mechanics is that the equation of motion for the trajectories is first order. In other words the analogue of Newton's law has to be solved with a particular initial condition for the velocities. More seriously there is even an initial condition on the positions, namely, that they have to be distributed according to the quantum probability law. This is a strange condition, first, because it implies that the wave function plays a dual role: It determines the quantum force and also acts as a probability. Second, because, if we think in terms of the cosmological context, there is only *one* initial condition, i.e., one point in a space of 3.10^N dimensions, where N is the number of particles in the universe; so it is hard to see how probability can arise. (Here we are taking a rather simplistic view of regarding the "hidden variables" as just particle positions. This is probably adequate to explain the point.)

To some extent these questions can be answered if we consider how an experiment to test a prediction of quantum theory is actually done. It is necessary to suppose that a set of identical experiments is performed, i.e., that the wave function of the universe has the form

$$\Psi = \prod_i \psi(\mathbf{x}_i)\Phi(remaining\ variables), \tag{15}$$

where the product is taken from 1 to n, n being the (large) number of identical experiments. (Actually this will be an approximate expression for the wave function, valid over a certain range of the other variables.) Now suppose that we choose the (boundary condition) point in configuration space at this particular time, and select it totally at random, with a weight given by $|\Psi|^2$. It is then inevitable that in the vast majority of cases the distribution of the points $\mathbf{x}_i$ will be that given by $|\psi|^2$. Thus the agreement between the quantum distribution and that of the particle trajectories is automatic. One argument that might be made against this conclusion, namely that the selection of the initial condition should be made at an earlier time, does not seem to be valid, because the trajectories are determined by the condition that this particular distribution is an equilibrium distribution, i.e., if it holds at one time it holds at all times. Thus what is typical at one time will be typical at others. Detailed discussion along the lines of this paragraph are given by Dürr, Goldstein, and Zanghi[2].

Of course, in making these type of arguments, we are using the quantum wave function not just to provide a new force, but also to define a probability measure; it plays a dual role in the theory. We are also ignoring the fact that the initial condition of the universe cannot really be selected at random. Indeed it has been suggested[3] that it has to be special to an accuracy of 1 part in $10^{10^{123}}$.

3. DERIVATION FROM A LAGRANGIAN[1]

Equations (4) and (9) show a pleasing symmetry, since both are of first order in time derivatives, i.e., they are equations for $\dot{\psi}$ and $\dot{\mathbf{x}}$, respectively. Their solutions, which can be written in the form

$$\psi_t = \psi_t(\psi_0) \tag{16}$$

and

$$\mathbf{x}_t = \mathbf{x}_t(\psi_0, \mathbf{x}_0), \tag{17}$$

do not, however, show this symmetry. Whereas the wave function clearly affects the trajectory, the latter has no effect on the wave function. This is an unusual situation in physics, where *action* is normally accompanied by *reaction*. Indeed, this fact was used by Einstein as one of the motivations for general relativity[4]: "...it is contrary to the mode of thinking in science to conceive of a thing (the space-time continuum) which acts itself, but which cannot be acted upon." Thus it is interesting to consider the possibility of modifying the equations so as to introduce such a reaction. Other possible reasons why we might want to do this are that it could play a role in making the distribution of particles equal to $|\psi|^2$, e.g., by somehow "collapsing" the wave function around the trajectories; it could remove the "empty" waves, so that we do not have to contemplate the possibility of all the other worlds (in an Everett sense) actually existing but being somehow less real than the wave which has the particles; it would also make the theory different in its predictions to orthodox quantum theory, so there would be the possibility of experimental test. (Note, however, that I do not regard the fact that the Bohm model agrees with quantum theory, i.e., as far as we know, with experiment, as a *fault* - in spite of rather bizarre suggestions of this nature that have sometimes been made!)

A natural way of introducing the effect we are seeking here is to derive the model from a Lagrangian. There is no difficulty in doing this for the Schrödinger equation. For the case of a single particle, we define the action to be

$$A_S = \int dt d\mathbf{x} \left(\frac{i\hbar}{2}[\psi^* \partial_t \psi - \psi \partial_t \psi^*] - \frac{\hbar^2}{2m} \nabla\psi^* . \nabla\psi - V\psi\psi^* \right). \qquad (18)$$

Then variation with respect to ψ^* will give the required equation.

Unfortunately, however, it is not possible to obtain the Bohm equation for $\dot{\mathbf{x}}$ by a similar method (at least not without introducing a complex "position" $\mathbf{z} = \mathbf{x} + i\mathbf{y}$, which would be hard to interpret). Instead, we regard this equation as a constraint which we obtain from the action

$$A_{\boldsymbol{\lambda}} = \int dt \boldsymbol{\lambda} . \left[\dot{\mathbf{q}} + \frac{i\hbar}{2m} \int d\mathbf{x} \left(\frac{\nabla\psi}{\psi} - \frac{\nabla\psi^*}{\psi^*} \right) \delta(\mathbf{x} - \mathbf{q}) \right], \qquad (19)$$

where $\boldsymbol{\lambda}$ is a Lagrange multiplier. We can also include the standard action for a non-relativistic particle:

$$A_N = \int dt \left(\frac{m\dot{q}^2}{2} - V(\mathbf{q}) \right). \qquad (20)$$

Thus we put

$$A = A_S + A_{\boldsymbol{\lambda}} + kA_N, \qquad (21)$$

where k is an arbitrary parameter. Note that there is no need to include an arbitrary parameter in the second term, since it can be absorbed into the $\boldsymbol{\lambda}$.

Variation with respect to $\boldsymbol{\lambda}$, ψ and $\mathbf{q}$, respectively, yields the equations

$$\dot{\mathbf{q}} = -\frac{1}{m} Re \left(\frac{i\hbar\nabla\psi}{\psi} \right)_{\mathbf{x}=\mathbf{q}}, \qquad (22)$$

$$i\hbar\dot{\psi} = H\psi - \frac{i\hbar}{2m\psi^*} \boldsymbol{\lambda} . \nabla\delta(\mathbf{x} - \mathbf{q}), \qquad (23)$$

$$\dot{\boldsymbol{\lambda}} = \frac{i\hbar}{2m} \nabla \left[\boldsymbol{\lambda} . \left(\frac{\nabla\psi}{\psi} - \frac{\nabla\psi^*}{\psi^*} \right) \right]_{\mathbf{x}=\mathbf{q}} - k(m\ddot{\mathbf{q}} + \nabla V). \qquad (24)$$

Work is in progress on these equations, but unfortunately I do not have any solutions available. They have the expected form. The last equation gives a non-zero value to $\boldsymbol{\lambda}$ due to the difference between the particle's actual motion and that given by Newton's second law. This value of $\boldsymbol{\lambda}$ then scales the extra term in the modified Schrödinger equation. It is interesting to note that this extra term is similar to a "source-like" term, e.g., as in electromagnetism, and suggests that the particles might somehow be regarded as the source of the wave function which produces the quantum force. This is indeed a novel way of thinking about quantum theory. The existence of the free parameter k means, of course, that we can make the modification to the usual Bohm model, i.e., to orthodox quantum theory, as small as we please. There are other ambiguities; for example, the Bohm condition, Eq.(22), is obtained by requiring that the distribution of trajectories remains constant relative to a wave function evolving according to the unmodified Schrödinger equation; maybe it should also be modified to satisfy the same condition for the new equation.

4. A DERIVATION FROM A COLLAPSE MODEL

One motivation for the above approach was that we might obtain modifications to the Schrödinger equation that would cause it to collapse. An alternative method would be to begin with an explicit collapse model and see if we can obtain anything like the Bohm trajectory. In fact[5] there is a sense in which such trajectories are implicit in the collapse models. To see how this comes about, we consider first the simplest such model, namely, that in which there are just two states. The modified Schrödinger equation is[6]

$$i\hbar|\dot{\psi}>= H|\psi> +i[w(t)\eta - \lambda\eta^2]|\psi>, \tag{25}$$

where η is a Hermitian operator with two eigenvalues a and b, and corresponding eigenstates $|a>$ and $|b>$. The function w is formally the derivative of some random process given by $B(t)$. This random function however is correlated with the state of the system in such a way that, at time t, the probability of a given value of B is given by

$$P(B,t) = N\left[|\alpha|^2 \exp\left(-\frac{(B-2\lambda at)^2}{2\lambda t}\right) + |\beta|^2 \exp\left(-\frac{(B-2\lambda bt)^2}{2\lambda t}\right)\right], \tag{26}$$

where N is a normalisation constant and where the state at $t=0$ is taken to be

$$|\psi>_0= \alpha|a> +\beta|b> . \tag{27}$$

Then, after a time large compared to λ^{-1}, B is approximately either $2\lambda at$ or $2\lambda bt$ and, correspondingly, the state is almost entirely in one or the other eigenstate, with probabilities given by the usual quantum rule[6].

We now note that the probability given in Eq.(26) is the result of a random walk with one of two possible drifts, which must be chosen at the outset, and for repeated experiments chosen with the probabilities $|\alpha|^2$ and $|\beta|^2$, respectively. It is the choice of this drift which determines the final result, independent of the initial wave function or any residual stochasticity. We could for example regard this "choice" as a hidden variable. Further we can obtain its effect on the wave function by inserting the choice directly into the evolution of the state vector, i.e., by replacing Eq.(25) with

$$i\hbar|\dot{\psi}>= H|\psi> -i\lambda(\eta - q)^2|\psi>, \tag{28}$$

with q chosen to be either a or b. If, for simplicity, we ignore the effect of the Hamiltonian evolution, i.e., drop the H term in this equation, then it readily solves to give

$$|\psi>= \exp[-\lambda(\eta - q)^2]|\psi>_0, \tag{29}$$

which tends to one of the eigenstates according to the value of q.

At this stage we note that we can drop the restriction to two eigenvalues and can instead consider η to be the position operator. Then we have a collapse equation that collapses to (approximate) eigenvalues of position. The requirement of Gallilean invariance demands that the "hidden variable" q should be a function of time and, since the condition that it is selected at random with probability weight given by the quantum probability rule should hold at all times, it is clear that the trajectory has to be similar to the de Broglie-Bohm trajectory described above. Indeed it is identical, except for the fact that the wave function no longer satisfies the simple Schrödinger equation. The resulting modification will again lead to results that differ in some respects from those of ordinary quantum theory.

One objection that could be made to this model is that it appears to allow faster-than-light signalling. This follows from a theorem due to Gisin[7], which implies that any non-linear, deterministic modification to the Schrödinger equation allows such signalling. The *stochastic* collapse models do not have this problem, essentially because they have a linear evolution equation for the density matrix. I am not clear as to the proper reaction to this objection, particularly since the stochastic and non-stochastic models appear to give predictions that are indistinguishable from each other. Also, it should be remenbered that we are working with a non-relativistic theory, for which there is nothing exceptional about faster-than-light signals. It is indeed a curiosity of orthodox non-relativistic quantum theory that, in spite of its subtle non-locality, it does not allow such signals.

5. THE BOHM MODEL IN QUANTUM COSMOLOGY

If wave functions are a part of reality, as they have to be in the models we are considering, then, in fact, there is only one wavefunction, namely, the wave function of the universe, $|\Psi>$. All others are approximate, effective wavefunctions, derived from $|\Psi>$, and of use in particular restricted situations. Thus a theory which claims to be fundamental must be able to deal with the wavefunction of the universe, i.e., it must be applicable to quantum cosmology. Quite apart from the fact that this involves a major extrapolation from the micro-world to which quantum theory is known to apply, there are new problems, both conceptual and technical, which arise when we try to include gravity in quantum theory (see, for example, Isham[8]).

We shall use the canonical quantisation procedure (see Halliwell[9] for a review and references), for which it is first necessary to write classical general relativity in Hamiltonian form. To this end we introduce a foliation of four-dimensional spacetime, in which the sheets are spacelike, labelled by a time parameter t and having a three-metric which we denote by h_{ij}. The embedding of these surfaces in the four-dimensional space, with metric $g_{\mu\nu}$, is described by the equation

$$ds^2 = g_{\mu\nu}dx^\mu dx^\nu = -(N^2 - N_i N_j)dt^2 + 2N_i dx^i dt + h_{ij}dx^i dx^j. \qquad (30)$$

The action for general relativity can then be written in the form

$$S = \int d^3\mathbf{x}dt \left[\dot{h}_{ij}\pi^{ij} - NH - N^i H_i\right], \qquad (31)$$

where π^{ij} is the "momentum" conjugate to h_{ij}. The quantities N and N^i only appear here as factors in the Hamiltonian, so they play the role of Lagrange multipliers. For example, variation of the action with respect to N yields the constraint

$$H \equiv G_{ijkl}\pi^{ij}\pi^{kl} - (R - 2\Lambda) + H_m = 0, \qquad (32)$$

where

$$G_{ijkl} = \frac{1}{2}h^{-\frac{1}{2}}(h_{ik}h_{jl} + h_{il}h_{jk} - h_{ij}h_{kl}) \qquad (33)$$

and H_m is a matter Hamiltonian.

In quantum theory these constraints become conditions on the wave function. Thus Eq.(32) becomes the so called Wheeler-deWitt equation

$$H|\Psi>= 0, \qquad (34)$$

which, together with the other constraints, implies that the wave function of the universe is independent of time:

$$i\hbar\partial_t|\Psi>=H|\Psi>=0. \tag{35}$$

This is a surprising result, not easily reconciled with the changing world we experience. There have been many attempts in the literature to reintroduce time, and the possibility of change, into this theory, but none are without problems. For a critical review we refer to Isham[10].

It was emphasised many years ago by John Bell[11] that cosmologists have an original insight into the measurement problem of quantum theory, since for them the notion of an outside observer has a rather different status. Not being able to rely on a nebulous separation of the world into system and apparatus, they have an urgent need for a proper solution to the problem, and the Bohm model is an obvious candidate. What is more, when we use the Bohm model in this situation, time is automatically reintroduced. For example, we have the analogue of Eq.(9):

$$\dot{h}_{ij} = Re\frac{G_{ijkl}\pi^{kl}\Psi}{\Psi}, \tag{36}$$

which shows that the metric evolves with time as expected.

The only exception to this occurs when $|\Psi>$ is purely real, or has a phase which is independent of the observables. Then we have a universe in which nothing ever changes. (A discussion of whether a similar disastrous result would hold in other interpretations of quantum theory is given in [12]. Now a real wave function might be considered to be unlikely (although in some cases it could be a natural choice, e.g., the Hartle-Hawking wave function is real), but nevertheless there is a significant problem here. We observe what appears to be an approximately classical macroscopic universe, i.e., one in which the classical potentials are much greater than the quantum potentials. Since for a real wavefunction the two are identically equal, such a universe appears to require that the wave function is very carefully selected.

These issues are addressed very clearly in articles by Kowalski-Glikman and Vink[13] and by Vink[14]. The metric is restricted to that of so-called mini-superspace with

$$ds^2 = -N^2dt^2 + a(t)^2d\Omega_3^2, \tag{37}$$

which leads to a Wheeler-deWitt equation of the form

$$\left[\frac{1}{2}a^{-3}(a\partial_a)^2 + a^3V_G(a) - \frac{1}{2}a^{-3}\partial_\phi^2 + a^3V_M(\phi)\right]|\Psi>=0, \tag{38}$$

where operator-ordering ambiguities have been ignored and a (uniform) scalar field has been introduced. If, for simplicity, we neglect the kinetic energy of this field, then we have a second-order ordinary differential equation with two linearly-independent solutions, both of which may be taken to be real. Thus the only non-trivial complexity comes from the phase difference between the two arbitrary constants multiplying these functions in the general solution. It is this phase that gives rise to the time dependence of the scale factor $a(t)$. Some possible solutions of the above equation are studied by Vink[14], whilst in [15] it is suggested how in this simple model the quantum potential might be used to cancel exactly an (unwanted) cosmological constant.

Finally here we refer to the interesting work of Valentini[16], who has applied the Bohm model to quantum cosmology, but with a preferred frame of reference, so that the wave function is not constant. There is clearly scope for more work in this general area; in particular some of the ideas mentioned at the end of Sec.2 and in Sec.3 might be relevant in the cosmological context.

6. LORENTZ INVARIANCE AND LOCALITY

As we noted earlier the problems of Lorentz invariance and locality are already evident in Eq.(9). The velocity $\mathbf{v}_i$ on the right is a function of time, as are the positions $\mathbf{x}_1, \mathbf{x}_2$, etc., on the left. But which times do we use? In non-relativistic physics it makes sense to use a common time t for all these variables, and this is the procedure that has been adopted in calculations of Bohm trajectories. Clearly. however, it is not Lorentz invariant since it implies a particular definition of simultaneous times at spatially separate points.

An obvious suggestion for obtaining a relativistic-invariant form of the Bohm model is to use retarded times in equations like (9). Thus, for a two-particle system, the dynamics would be described by the coupled equations (cf. Eq.8)

$$\dot{\mathbf{x}}_1(t) = \frac{1}{m_1}\nabla_1 S\left(\mathbf{x}_1(t), \mathbf{x}_2(t_r)\right) \tag{39}$$

and

$$\dot{\mathbf{x}}_2(t) = \frac{1}{m_2}\nabla_2 S\left(\mathbf{x}_1(t_r), \mathbf{x}_2(t)\right), \tag{40}$$

where the retarded time t_r is defined by

$$t_r = t - \frac{|\mathbf{x}_1 - \mathbf{x}_2|}{c}. \tag{41}$$

Here there is no non-locality or violation of Lorentz invariance; signals are sent along the forward light cone, and each particle is affected by the position of the other at an earlier time.

Inevitably such a theory will violate quantum theory. This follows from Bell's theorem. (An analogous demonstration that any *Lorentz-invariant* hidden-variable theory must violate quantum theory is given in [17] and [18].) It is not clear, however, whether this theory violates currently available experiments. The key question here is whether the time for a light signal to pass between the two ends of the apparatus in the Aspect *et al.* experiments[19] is greater than the time taken for the "measurement" of a photon spin to be completed. It is not completely obvious how the latter should be calculated in the Bohm model, but is hard to see how it can be significantly less than the signal time, which is about 10^{-7} sec. in the actual experiments. A reliable experimental test of whether a relativistically-invariant version of the Bohm model can work almost certainly requires EPR-like correlations to be tested over distances of the order of a kilometre. (The same thing seems to be true of collapse models of measurement[20].)

7. SUMMARY

My own opinion is that the Bohm model is the most economical, *satisfactory*, version of quantum theory. It is therefore important to study different applications of the model, in particular to cosmology, and to consider various possible ways of deriving it and/or modifying it.

If we wish not to believe it, then the following are the best arguments that might be used.

(i) It very clearly violates both locality and Lorentz invariance. To a large extent, of course, this is because it exposes properties that are intrinsic to quantum theory. Only in some version of the many-worlds interpretation is there any hope of avoiding these unpleasant features[21], unless of course quantum theory proves to be incorrect when tested over large distances.

(ii) There may be difficulties in understanding why the "initial distribution" of the hidden variables should conform to quantum theory, or indeed to understand how probability ever arises.

(iii) It is not clear that the model explains why we see an approximately classical macroscopic world. But is this asking too much of a theory?

(iv) We might want to argue that a purely deterministic world, in which time is redundant, is inadequate, and that the richer possibilities of non-deterministic quantum theory, perhaps coupled with some effects which are explicitly non-symmetric in the direction of time, are too valuable to be discarded. But this is the start of another, much harder, story!

I am happy to be able to present this talk at a meeting held in honour of Louis de Broglie, one of the great figures of twentieth-century physics. Conversations with many people, in particular David Hind, Lucien Hardy, and Jan Govaerts, are gratefully acknowledged.

REFERENCES

1. D.Bohm, *Phys. Rev.* **85** (1952) 166 and 180.
2. D.Dürr, S.Goldstein, and N.Zanghi, *J.Stat.Phys.* **67** (1992) 843.
3. R.Penrose, *The Emperor's New Mind* (Oxford University press, Oxford, 1989).
4. A.Einstein, *The Meaning of Relativity* (Methuen, London, 1922).
5. E.J.Squires, *Phys. Lett.* **A157** (1991) 453.
6. G.C.Ghirardi and P.Pearle, "State-vector reduction I", ICTP preprint IC/91/22 (1991).
7. N.Gisin, *Helv.Phys.Acta* **62** (1989) 363.
8. C.J.Isham, Imperial College preprint TP/90-91/14 *Conceptual and Geometrical Problems in Quantum Gravity* (lectures at the Schladming winter school, 1991), Imperial College preprint TP/90-91/14.
9. J.J.Halliwell, *Introductory Lectures on Quantum Cosmology* (Proceedings of the Jerusalem Winter School on Quantum Cosmology, T.Piran, ed. (World Scientific, Singapore, 1990).
10. C.J.Isham, *Canonical Quantum Gravity and the Problem of Time* (lectures at the NATO Institute, "Recent Problems in Mathematical Physics", Salamanca, 1992), Imperial College preprint TP/91-92/25.
11. J.S.Bell, "Quantum mechanics for cosmologists," in *Quantum Gravity 2*, C.J.Isham, R.Penrose, and D.W.Sciama, eds. (Oxford University Press, Oxford, 1981).
12. E.J.Squires, *Found. Phys. Lett.* **5** (1992) 71.
13. J.Kowalski-Glikman and J.C Vink, *J.Class.& Quantum Grav.* **7** (1990) 901.
14. J.C.Vink, "Quantum potential interpretation of the wave function of the universe", Centre for High Energy Astrophysics, Amsterdam, preprint CHEAF-91-7 (1991).
15. E.J.Squires, *Phys. Lett.* **A162** (1992) 35.
16. A.Valentini, "Nonlocal hidden varibles and quantum gravity," SISSA preprint 105/92/A.
17. L.Hardy, *Phys. Rev. Lett.* **68** (1992) 2981.
18. L.Hardy and E.J.Squires, *Phys. Lett.* **A168** (1992) 169.
19. A.Aspect, P.Grangier, and G.Roger, *Phys. Rev. Lett.* **49** (1982) 91.

20. E.J.Squires, "Has quantum non-locality been experimentally verified?," talk at the 1991 Cesena conference, to be published in *Found. Phys. Lett.* **6** (5) (1993).

21. E.J.Squires, "How to test for Cartesian dualism by quantum experiments," talk at Joensuu conference; *Symposium on the Foundations of Modern Physics*, 1990, P.Lahti and P.Mittelstaedt, eds. (World Scientific, Singapore, 1991).

NOTES

1. The work reported in this section has been done jointly with Jan Govaerts.

THE RELATIONSHIP BETWEEN THE DIRAC VELOCITY OPERATOR AND THE DE BROGLIE POSTULATE

A. M. Awobode

Department of Physics
University of Ibadan
Ibadan, Nigeria

It will be demonstrated here that a consequence of the Dirac theory, that is, the eigenvalues of the velocity operator, can be deduced from arguments based on the existence and properties of the de Broglie waves. Furthermore, the method of establishing the proof described here suggests that the result has a general validity.

Key words: group, phase, and particle velocities.

1. INTRODUCTION

Both the Dirac electron theory and the de Broglie wave postulate are evidently successful attempts to unify relativistic and quantum principles [1,2]. It is therefore reasonable to expect that a relationship should exist betweeen the theory of relativistic particles given by Dirac [3] and the de Broglie postulate, quite apart from the observation that the Dirac theory is a Schrödinger-type wave equation. The Dirac wave function is a bispinor, whereas the de Broglie postulate refers to a plane wave. However, a form of the Schrödinger wave function has already been derived from arguments based on the Dirac Hamiltonian and the Heisenberg equation [4]. The eigenvalues of the velocity operator, where the operator is also obtained via the Heisenberg equation of motion, are identical with the particle velocity of the de Broglie postulate.

2. BRIEF REVIEW OF THE DE BROGLIE POSTULATE

For a brief review of the de Broglie postulate, emphasising its essential, relativistic nature, let us consider a relativistic particle of velocity of u with respect to a stationary observer. In its rest frame of reference, a periodic phenomenon of frequency $\nu = mc^2/h$ is attributed to the particle of rest mass m, in accordance with the quantum postulate.

With respect to the stationary observer, the frequency is

$$\nu_i = \left(mc^2/h\right)\left(1-\beta^2\right)^{1/2}, \tag{1}$$

where h is Planck's constant, c is the velocity of light in empty space, and $\beta = u/c$. Also with respect to the stationary observer, the frequency of the same periodic phenomenon is

$$\nu_j = \left(mc^2/h\right)\left[1-\beta^2\right]^{-1/2}. \tag{2}$$

De Broglie noted correctly that v_i and v_j are not equal and was led to postulate the existence of a plane wave fixed to the particle and travelling in the same direction as the particle at a phase velocity $w = c^2/u$ with respect to the observer. The wave has a frequency

$$\nu_j = \left(mc^2/h\right)\left(1-\beta^2\right)^{-1/2} \tag{3}$$

and always appears to the stationary observer to be in phase with the periodic phenomena.

The wave is assumed to be in constant phase agreement with the periodic phenomena, and for this condition to be satisfied it is required that the particle be located within the wave [3]. This is a valid requirement, which if not satisfied will imply that the wave could be detected at a remote location earlier than the particle—an implication which obviously is not supported by observations.

Moreover, in the proof of the phase agreement between the periodic phenomenon and the travelling wave, it is implicitly presupposed that the wavefront of the matter wave coincides with the position of the particle [6]. It will be explicitly demonstrated here that, when the phase of the wave is equal to that of the periodic phenomenon, as required, the particle is located with the wave, and from this we obtain the eigenvalues of the Dirac velocity operator.

3. THE MATHEMATICAL DESCRIPTION OF A TRAVELLING WAVE

The second-order wave equation is satisfied by the function

$$G(x,t) = g(x-vt) + g(x+vt), \tag{4}$$

where the first and second terms are functions representing waves propagating in the forward and backward directions, respectively. The trigonometric functions *sine* and *cosine* satisfy the periodicity required by the general solution and may therefore be used to describe waves in propagation.

A travelling wave is described by the function [7]

$$\phi = A \sin 2\pi f(t - x/v), \tag{5}$$

where A is the maximum amplitude and the argument $f(t - x/v)$ of the sinusoidal function always has a constant value zero, where f is the frequency, t is the time elapsed from $t = 0$, x is the displacement from the origin, and v is the phase velocity of the wave.

The phase of a plane wave of frequency v_j and velocity w is given by

$$\nu_j(t - x_j/w) = \left(mc^2/h\right)\left(1 - \beta^2\right)^{-1/2}(t - x_j/w). \tag{6}$$

Following the general expression for waves, we note that $x_j = wt$ is the position of the wavefront with respect to the stationary observer at time t. The phase of the periodic phenomenon fixed to the particle at time t is given by

$$\nu t = \left(mc^2/h\right)\left(1 - \beta^2\right)^{1/2}(x_i/\beta c), \tag{7}$$

since the particle has moved a distance $x_i = \beta ct$ in the time $t = x_i/\beta c$ with respect to the stationary observer.

For the phases to be equal to one another at all t, we must have

$$\nu_i t = \nu_j(t - \beta x_j/c). \tag{8}$$

Substitution for ν_i and ν_j and t from Eqs. (1) and (2) gives

$$\left(1 - \beta^2\right)^{1/2} x_i/\beta_c = \left(x_i/\beta c - \beta x_j/c\right), \tag{9}$$

which implies that $x_i = x_j$ and therefore the particle is located within the wave, as required.

Equating (or) setting $x_i = x_j$, from $x_i = \beta ct = ut$ and $x_j = wt = (c^2/u)t$ above, we obtain

$$u = c^2/u, \tag{10}$$

and thus $u^2 - c^2 = 0$, the solutions of which are $u = \pm c$, which are the eigenvalues of the velocity operator $u = c\alpha$ in the Dirac theory.

The Dirac velocity operator $c\alpha$ is consistently derived from the Heisenberg equation of motion $x = [x, H] = c\alpha$, using the Dirac Hamiltonian $H = c\alpha p + \beta mc^2$, which correctly describes the properties of electrons and other spin $-1/2$ particles; and therefore the eigenvalues of the Dirac velocity operator $u = c\alpha$ is applicable to these particles. The result described above has also been shown to follow from the phase velocity w when it is made to be consistent with relativistic principles [8], and experiments have been suggested in order to test the assertion.

It is necessary to emphasize that the final conclusion that the eigenvalues $u = \pm c$ of the Dirac velocity operator $u = c\alpha$ can be derived from the de Broglie postulate is not inconsistent with Eq. (1), as it may initially appear. Equation (1) represents the phase of a travelling wave, which may be described by a sine or cosine function as sin $\nu(t - x/w)$, where

$$\nu = \left(mc^2/h\right)\left(1 - u^2/c^2\right)^{-1/2}. \tag{11}$$

As $u \to c$, ν tends to a finite quantity k, because $m \to 0$ [9]. At extreme relativistic or ultra-relativistic energies, it is appropriate to make the approximations $m = 0, m \to 0$ as $u \to 0$. Therefore, the result is quite consistent with relativistic quantum mechanics and its method of calculating observable quantities [10,11].

To dispel doubt concerning the equality established between the eigenvalues of the Dirac velocity operator and the particle/group velocity of the de Broglie postulate, we shall derive the result in a different way. De Broglie demonstrated that, for matter waves of phase velocity $w = c^2/u$, the group velocity v_g is equal to the particle velocity $u(v_g = u)$, by using the formula

$$(1/v_g) = d(\nu/w)d\nu = d(1/\lambda)/d\nu, \tag{12}$$

where ν is the frequency, λ is the wavelength, and $w > u$. The general equation relating the phase velocity w, the group velocity v_g, and the wavelength λ for all wave is

$$\lambda(dw/d\lambda) - w = -v_g. \tag{13}$$

This equation is easily derived from Eq. (12) together with the well-known relation $w = \nu\lambda$, and is found to be true for all waves.

The solution of the above equation is

$$w = v_g(1 + \lambda s), \tag{14}$$

where s is an integration constant. We can also express Eq. (14) above as

$$w/(\lambda v_g) = 1/\lambda + s. \tag{15}$$

Multiplying both sides of Eq. (15) by Planck's constant, we obtain

$$(h/\lambda)(w/v_g) = h/\lambda + hs. \tag{16}$$

Now, using the de Broglie-Einstein equations $p = h/\lambda$ and $E = h\nu = pw$, we obtain

$$E/v_g - p = hs. \tag{17}$$

From this it is easy to see that, when $p = 0, \lambda = \infty, E = 0$ and hence $hs = 0$, which implies that $s = 0$, since h is a natural constant.

It therefore follows from Eq. (14) that, if $s = 0$, then

$$w = v_g. \tag{18}$$

That is, in addition to de Broglie's demonstration that the *group velocity* v_g *is equal to the particle velocity* u, we have shown that, assuming the validity of the de Broglie-Einstein relations, the *wave phase velocity* w *is also equal to the group velocity* v_g. Upon substituting the de Broglie expresions for w and v_g, we obtain

$$c^2/u = u, \quad \text{i.e., } c^2 - u^2 = 0, \tag{19}$$

and therefore $u = \pm c$, as shown previously.

Moreover, it has been shown that arguments based on energy relations also lead to the conclusion that, at velocities other than $u = \pm c$, the wave is forbidden by the principle of relativity to carry any energy, in contrast to expectations arising from quantum theory. It is shown that, when $u < c, w > c$, E_w, the wave energy is real but incompatible with relativity; however, when $u > c, w < c$, the particle velocity exceeds c and the wave/particle energy becomes imaginary [12].

Another implication of the result obtained above in Eq. (18) is that, for matter waves in empty space, the phase velocity is independent of the wavelength. Under such a condition, free space is a non-dispersive medium, as is to be expected, and de Broglie waves should therefore propagate without deviation.

4. CONCLUDING REMARKS AND OBSERVATIONS

The result described above and the method used in obtaining it are important for several reasons:

(1) In the first place, it is necessary to recognize and describe the formal link between the Dirac theory and de Broglie's postulate in order to emphasize the relativistic nature of the latter. The postulate has been found to be in close agreement at very high energies, and it therefore becomes possible, on this basis, to further understand the nature of the wave-particle dualism and the relationship between the wave and particle for material corpuscles.

(2) Although, it is perhaps possible to guess and arrive at the above conclusion on reasoning by analogy, inferences based on mere qualitative analogies are however logically inadequate and are often misleading. The result described above has been validly deduced from the fundamental, relativistic and self-evident requirement of phase invariance and for this reason has been placed on a more secure theoretical basis. Moreover, because the result that $u = \pm c$ has been validly derived from the wave postulate, it follows that this may not be peculiar to spin $-1/2$ particles described by the Dirac theory but could be characteristic of all particles at relativistic energies.

(3) From the Dirac theory, it is possible to obtain the eigenvalues of the velocity operator by using the Heisenberg equation of motion and the commutator of the position operator with the Dirac Hamiltonian. This at first appeared to be a defect because of the associated trembling motion, Zitterbewegung, obtained as a solution of the Heisenberg equation of motion, and therefore further developments were directed towards redefining the Hamiltonian and the position operators into forms which, however, do not transform relativistically. By deducing the result as described above in an elegant and direct way based on fundamental and general principles, the use of the familiar position operators is further justified and the result is thereby given a greater degree of validity.

(4) It must also be noted that in order to fully comprehend the nature of a phenomenon, all its physical characteristics must be known. For example, the nature of the de Broglie waves will only become completely clear when all the essential parameters have been experimentally determined. The result described above has implications for the wave velocity of the de Broglie waves, since as previously described, $w = c^2/u$ and $u = c$ implies that $w = c$. This is consistent with views that have been expressed in the approximate treatment of relativistic electrons. It needs to be pointed out as well that the adoption of $w = c^2/u$ as the accurate phase velocity of de Broglie waves is not compatible with the $v_g = u$, where u is the particle velocity and v_g is the group velocity.

5. SUMMARY

(1) If we accept the Einstein-de Broglie relations $E = h\nu$ and $p = h/\lambda$ together with the well-known definitions of phase and group velocities, then it can be shown that, analogous to de Broglie's demonstration, that the *group velocity* v_g is equal to the *particle velocity* u, the *phase velocity* w is also equal to the *group velocity* v_g.

(2) If we accept de Broglie's assumption that the travelling matter wave associated with a particle is always in phase with the attributed internal periodic phenomenon, then it can be shown that the *wave and*

particle are located together, as required by the de Broglie Postulate.

From these results follow the implication that the particle velocity has the values $u = \pm c$, which are also recognized as the eigenvalues of the Dirac velocity operator. Thus, mathematically and logically, the relationship between the Dirac velocity operator and the de Broglie postulate has been demonstrated and made explicit.

APPENDIX: DERIVATON OF THE "DISPERSIVE" EQUATION

To derive the equation

$$v_g = w - \lambda(dw/d\lambda), \tag{20}$$

where v_g is the group velocity, w is the phase velocity, and λ is the wavelength, we first consider the definition of the group velocity and the relationship between the parameters.

The group velocity v_g is defined as

$$v_g = d\omega/dk, \tag{21}$$

where ω, the angular frequency, is related to the frequency ν by the expression $\omega = 2\pi\nu$ and the wave vector $k = 2\pi/\lambda$. Equation (21) given above can be shown to be equivalent to the one employed by de Broglie if its reciprocal is taken and the expressions for the frequency and wave length substituted to get

$$(1/v_g) = d(1/\lambda)/d\nu = d(\nu/w)/d\nu. \tag{22}$$

The phase velocity w is related to the angular frequency and the wave vector as follows

$$w = \omega/k, \quad \text{i.e., } \omega = kw, \tag{23}$$

and therefore $w = \nu\lambda$.

Differentiating ω in Eq. (23) above with respect to k, we find

$$(d\omega/dk) = w + k(dw/dk),$$

which implies that

$$v_g = w + k(dw/dk), \tag{24}$$

which can be expressed in terms of the wavelength if we recall the expression for the wave vector and differentiate k with respect to λ:

$$dk/d\lambda = -2\pi/\lambda^2 = -k/\lambda \text{ or } dk/k = -d\lambda/\lambda, \tag{25}$$

and so

$$v_g = w - \lambda(dw/d\lambda). \tag{26}$$

REFERENCES

1. L. de Broglie, *C. R. Acad. Sc. (Paris)* **177**, 507–510 (1923); *B. Phil. Mag.* **47**, 446–458 (1924).
2. G. Lochak, *Wave-Particle Dualism*, S. Diner, D. Fargue, G. Lochak, and F. Selleri (Reidel, Dordrecht, 1983).
3. P. A. M. Dirac, *Proc. Roy. Soc. A* **117**, 610 (1928). G. L. Trigg, *Quantum Mechanics* (Van Nostrand, New York, 1964).
4. A. M. Awobode, *Found. Phys. Lett.* **3**(2), 167 (1990).
5. L. de Broglie, in *Old and New Questions in Physics, Cosmology, Philosophy, and Theoretical Biology: Essays in Honor of Wolfgang Yourgrau*, Alwyn van der Merwe, ed. (Plenum, New York, 1983).
6. L. de Broglie and G. Ludwig, *Wave Mechanics* (Pergamon, New York, 1968), pp. 73–93.
7. David Halliday and Robert Resnick, *Physics* (Parts I and II, combined edition) (Wiley, New York, 1966), pp. 466–470.
8. A. M. Awobode "On the determination of the phase velocity of relativistic matter waves," Daresbury Laboratory Preprint DL/SCI/P751T (1991).
9. W. G. V. Rosser, *An Introduction to the Theory of Relativity* (Butterworths, London, 1964), p. 223.
10. M. E. Rose, *Relativistic Electron Theory* (Wiley, New York, 1961), p. 246. B. J. Bjorken and S. D. Drell, *Relativistic Quantum Mechanics* (McGraw-Hill, New York 1964), p. 115.
11. N. F. Mott and H. S. W. Massey, *The Theory of Atomic Collisions*, 3rd edn. (Clarendon Press, Oxford, 1965), p. 232.
12. A. M. Awobode, "Propagation of electron phase waves," to be published. It has also been suggested here that massless particles are possible carriers of the wave energy.

OPTICS AND INTERFEROMETRY WITH ATOMS

V. I. Balykin

Fakultät für Physik, Universität Konstanz
D-78434 Konstanz, Germany

Atom optics and interferometry based on laser manipulation of atoms and a use of microfabricated structures are reviewed.

Key words: laser cooling, microfabricated structures, atom optics, interferometry.

1. INTRODUCTION

The established area of matter-wave optics, electron and neutron optics, is enriched by new type of optics; *atom optics.* The term of *atom optics* is due to the natural analogy with *light optics* or the optics of photons. Atom optics, in analogy to neutron and electron optics, deals with the realization of as a traditional elements, such as lenses, mirrors, beam splitters and atom interferometers, as well as a new "dissipative" elements such as a slower and a cooler, which have no analogy in another types of optics. The main fundamental interest in optics with atoms is due to fact, that new types of particles can be used (1) to explore the idea of wave-particle duality with "non-elemental" particles, (2) to test the fundamental laws of physics (to interferometers), (3) to reach, by using a combination of "traditional" atom optics elements and "dissipative" elements, extremely low temperature and simultaneously high atom density which could follow to the Bose-Einstein condensation of trapped sample of atoms and another collective quantum phenomena. The practical interest lies in opportunities to create atom microprobe with high resolution due to a very short de Broglie wavelength and a minimal damage of an investigated object due to low atom energy.

2. TYPES OF MASSIVE-PARTICLES OPTICS

According to the de Broglie theory, wave like properties are associated with any particles of matter, and the de Broglie wavelength is defined by the fundamental relation

$$\lambda_{Br} = h/p = h/mv, \quad (1)$$

where h is Planck's constant and p, m, and v are the momentum, mass, and velocity of particle, respectively. The wave properties of massive particles were verified in experiments on the diffraction of electrons and used in the first light optics analog for particles - *electron optics* (Grivet, 1972).

Electron optics is based on (a) the wave properties of electrons and (b) the electromagnetic interaction between moving electronic charge and electrical and magnetic fields of appropriate configuration (Grivet, 1972). The most familiar application of electron optics is electron microscopy (Ruska, 1980).

Another light optics analog is *neutron optics* based again on (a) the wave properties of ultracold neutrons and (b) the interaction between neutrons and atomic nuclei, which can be described by means of what is known as the optical potential (Sears, 1989). As distinct from electron optics, we deal here with more massive particles whose wave properties manifest themselves at low temperatures (ultracold neutrons ; Shapiro, 1976). The effect of gravitation and low intensity of ultracold neutron sources make experiment in neutron optics more complex than in electron optics. Nevertheless neutron interferometers (Bonze and Rauch, 1979) and microscopes (Shutz et al. 1980; Frank, 1987, 1991) have already been successfully realized.

The next natural object are neutral atoms or molecules. The wave properties of atoms and molecules and varies types of their interaction with matter and electromagnetic fields (from static to optical) make it possible to implement *atom* and *molecular optics*. It is precisely the great variety of methods for exerting action on atoms (or molecules) possessing a static electrical or magnetic moment, a quadruple moment, optical resonance transitions (or a high-frequency dipole moment) that form the basis for several possible ways to realize atomic (molecular) optics. Let us consider them briefly.

3. METHODS OF REALIZATION OF ATOM OPTICS

The known methods to implement atom optics (atomic - optical effects) can be classed in the following three categories:

(a) methods based on the interaction between atoms and matter;
(b) methods based on the interaction between atoms having a magnetic or electrical dipole moment and a static electrical or magnetic field of a suitable configuration;
(c) methods based on the resonance (or quasi -resonance) interaction between atoms and a laser field.

The first experiment on atom optics realized by method (a) and (b) were successfully conducted almost a century ago. The advent of tunable laser allowed the possibility to be demonstrated of atom optics based on the atom-light interaction. Let us consider briefly the different methods of realization of atom optics.

4. ATOM AND MATTER INTERACTION

In his classical monograph, Ramsey (1956) considered the mirror reflection and diffraction of molecular beams on the surfaces of a solid. According to Ramsey (1956), for mirror reflection to occur, it is necessary that the following two condition be satisfied.

(a) The projection of the height of surface irregularities on the direction of molecular beam must be shorter than the de Broglie wavelength. If δ the average height of surface irregularities and ϕ_0 the grazing angle of incident beam, the above requirement may be expressed as

$$\delta \sin \phi_0 < \lambda_{Br} \tag{2}$$

(b) The average residence time of the particle on the surface must be short. In this case, the state of reflected particles will be the same as that of incident particles.

The roughness of most thoroughly mechanically polished surfaces of the order 10^{-5} cm, whereas the de Broglie wavelength of hydrogen at 300 K amounts to 10^{-8} cm. Therefore, according to (1) and (2), the condition for the reflection has the form $\phi_0 < 10^{-3}$ rad. It was more than 50 years ago that experimentalists managed to observe a 5 % reflection of hydrogen beam from polished bronze mirror at the grazing angle of $\phi_0 = 10^{-3}$ rad (Krauer and Stern, 1929). Cleaved crystal surfaces are much more smooth. The thermal vibrations of the crystal lattice limit the roughness of the surface to the 10^{-8} cm. In that case a beam of He atoms should undergo reflection at grazing angles less than 20 - 30 rad. This was confirmed in the experiments (Estermann and Stern,1930) with He atoms and LiF crystal. The temperature dependence of the grazing angle marking the onset of simple reflection of atoms bears witness of the fact that thermal vibrations have an effect on the surface roughness of crystal.

Experiments on the simple reflection of atoms at the surface of condensed medium continue to draw investigator's attention. Recall the experiments on the reflection of ^{4}He atoms grazing the surface of liquid ^{4}He (Nayak et al. 1983) and thermal Cs atoms grazing a polished glass surface (Anderson et al. 1986).

The first experiment aimed at observing the diffraction of atoms by a cleaved crystal surface acting as a two- dimensional, plane grating were conducted by Stern (1929), and the results of detailed research into this phenomena were presented in (Frish and Stern, 1933).

The diffraction of atoms by a fabricated periodic structure (a slotted membrane) with a much more grating period was observed in the work reported in (Keith et al. 1988).

Atomic interferometry based on the microfabricated structures was realized in two elegant experiments: the atomic Young's two slit interferometer (Carnal and Mlynek, 1991) and the atomic Michelson interferometer (Keith et al. 1991).

The MIT group`s interferometer consist of three diffraction gratings with period of 400 nm spaced 0.65 m apart. They used a beam of sodium atoms (de Broglie wavelength of 0.16 A). The first two grating separate and redirect the atom beam forming the interference pattern in the atomic flux at the third grating which act as a mask to sample interference pattern. The third grating has exactly the same spacing as expected fringe pattern arriving at its place, so the grating transmits the waves when its slits coincide with the fringes and blocks the waves when they do not. There are variety of experimental complications in atomic interferometer. One of the most serious is a vibrational isolation and grating alignment. The three grating must be stationary relative to each other to with 1/4 period during the time the final grating integrates the intensity at given position. The motion of the grating due to acceleration during the time of flight of the atoms in the interferometer must be also less 1/4 period. The problem were solved by using a combination of passive isolation and active feedback. The interference fringes were recorded as a transverse position of the second grating by measuring the count rate by a hot wire placed behind the third grating. The transverse splitting of atomic beams at the central grating was about 30 μm.

The group of Constanz University realized another type of atomic interferometer: the Young's double slit configuration. A 2 μm entrance slit is forming a spatially coherent atomic beam which further illuminates the microfabricated gold structure consisting of two 1 μm wide slits that are separated by 8 μm. The periodicity in the interference pattern was observed for two de Broglie wavelength 0.56 A and 1.03 A. The fringe visibility was achieved up to 60%.

The group of Tokyo University has realized the two slit interferometer with laser cooled metastable neon atoms. Laser cooling permits considerably decrease the atomic velocity and accordingly the period of interference pattern: the observed period was up to 100 μm.

Although zone plates have been already recognized as a focusing element in 19th century, they have found very little application in classical optics. In recent years microfabricated Fresnel zone plates have been applied in X- ray microscopy (Schmahl and Rudolph, 1984) and in the focusing of slow neutron beam (Kearney et al. 1980). At Constanz University, the Fresnel zone plate was at the first time used also as a focusing device for atoms (Carnal et al. 1991). The Fresnel zone plate was 210 μm in diameter, with an innermost zone diameter of 18 μm. For λ_{dB} = 1.96 Å the focal lens of Fresnel zone plate is 0.45 m. In a focusing experiment atoms of metastable helium with an atomic de Broglie wavelength λ_B = 0.5 - 2.5 Å are passed through either a single or a double slit with dimensions in the 10 μm range and they are served as the atomic source. The images of single slit and double slit were observed.

One of the latest achievement in the focusing of atom by their interaction with matter is the reflection of hydrogen beams at a liquid helium- vacuum interface (Berkhout et al. 1989). The mirror consists of a fused- quartz substrate polished to optical precision and coated with a liquid ^{4}He film to obtain high reflectivity. Reflection of He atoms at a liquid- vacuum interface has been observed earlier (see Nayak V. U. et al. 1983); however, the high reflection was not achieved except at grazing incidence. Berkhout et al (1989) demonstrated focusing of a high divergent atomic beam at normal incidence. The measured reflectivity was 80% and limited by a static surface roughness due to substrate and dynamic surface roughness of the helium film.

5. ATOM AND STATIC ELECTRIC AND MAGNETIC FIELDS INTERECTION

Some elements of the optics of atoms and molecules, based on the interaction between spatially nonuniform static magnetic or electrical fields and the magnetic or electrical dipole moment of the particles, have long been known and used successfully in experimental physics. A review on the early experiments in this field was presented by Ramsey (1956).

In the presence of a magnetic or electric field, the quantum state of atom or molecule are shifted, the shift depending on the initial quantum state of the particle and the field strength (the Zeeman and Stark effects). In the adiabatic approximation (the field varies not very rapidly in time and space, the particles move slow enough), the internal state of particles follows the field strength variation, in other words the particles remain in one and the same quantum sublevel whose energy W depends on the field strength.

In the adiabatic approximation, the motion of the center of mass of a neutral particle with a mass M obeys the Schrödinger equation for the wave function ψ (r, t):

$$ih[\, \partial\psi_i (\mathbf{r}, t) / \partial t \,] = \{ - h^2/ 2M) \nabla^2 + W_i (r)\} \psi_i (\mathbf{r}, t) \qquad (3)$$

where W_i (**r**) is the internal energy of the particle in the quantum state *i* at the point **r** which depends on the electrical field strength E(**r**) or the magnetic field strength H(**r**).

(a) Magnetic Interaction: In the simple case of a constant magnetic moment μ , the effective potential energy W of an atom or a molecule in an external magnetic field of strength H is given by

$$W = - \boldsymbol{\mu} H = - \mu_{eff} H \qquad (4)$$

where μ_{eff} is the projection of $\boldsymbol{\mu}$ on the direction of **H**. It follows from the relationship between force and potential energy that the force acting on the atom or molecule is

$$\mathbf{F} = - \nabla W = - (\partial W / \partial H) \nabla H = \mu_{eff} \nabla H. \qquad (5)$$

A particle in nonuniform magnetic field is acted upon by a force directed along the field strength gradient.

The authors of (Friedburg and Paul, 1950, 1951; Korsynskii and Fogel, 1951; Vanthier, 1949) proposed to use nonuniform magnetic fields to focus molecular beams issuing at different angle from the source. Such configuration of the magnetic field was used by Frieburg and Paul (1950, 1951) for focusing of atoms. The method was extended by Benewitz and Paul (1954) to atoms whose magnetic moments were depend on the strength of the external magnetic field.

The focusing properties of magnetic lens depends on the magnetic sublevel of atoms. That was successfully used by Ramsey and coworkers to create the hydrogen maser (Goldenberg et al. 1960; 1962). The hydrogen atoms in the state F = 1, M = 0 were focused in a small hole in the wall of the storage cell and accumulated there, while the atom in the lower atom state F = 0 are defocused.

The effect of splitting of wave packet in inhomogeneous magnetic field (the Stern Gerlach effect)was successfully used to build a longitudinal Stern Gerlach interferometer (Miniatura 1992). The principle of this interferometer is the following. First a polarization of the beam is achieved by passing it through magnetic field with transverse gradient. Then a mixer produces a coherent superposition of magnetic sublevels. The sublevel are split by a longitudinal field gradient. As sublevel are orthogonal, they are have to be remixed by second mixer in order to interfere. Finally, an analyzer filters a particular polarization of the beam and population of filtered state contain interference term arising from the different potentials experienced by the sublevels in the longitudinal magnetic field.

(b) Electrical Interaction: Since the energy of atom or molecule in an electrical field depends on the strength of the latter, it then can be presumed, by analogy with (4) and (5), than the atom or molecule possesses an effective dipole moment given by

$$\mu_{eff} = - (\partial W/\partial E). \tag{6}$$

The force acting on an atom or a molecule in a non uniform electrical field is defined by the following expression similar to (5):

$$F = \mu_{eff} \nabla E = \mu_{eff} (\partial E/\partial z) \tag{7}$$

where the direction of the field strength gradient is taken to be the z- axis.

Paul and co- workers (Benewitz et al. 1955) created focusing electrical fields for a beam of polar molecules. The electrical focusing of a beam of molecules in a certain (excited) quantum state was used by Townes in developing the NH_3 maser (Gordon et al. 1955). The hexapolar electrical field configuration possesses not only focusing properties, but also selectivity with respect to the quantum state of the molecule, for the quantity μ_{eff} depends on its quantum numbers J, K, and M. This latter properties was successfully used in experiment on molecular dynamics with a beam of molecules in a specified quantum state, including the experiments on the orientation of molecules (Bernstein, 1982).

When speaking of the optics of atomic or molecular beams, we almost always mean their focusing, for it is exactly this effect that has found practical application. But one can also speak electrical or magnetic mirrors and grating for slow moving neutral atoms and molecules (Opat et al. 1992).

6. ATOMS AND LIGHT FIELD INTERACTION

Atoms or molecules having no static magnetic or electrical dipole moment cannot change their mechanical trajectory in a static magnetic or electrical field. However, new possibilities are being opened up for such particles, based on the induction in them of a high frequency (optical) electrical dipole moment in a quasiresonant or resonant laser light field. Before the advent of the laser, it was only possible to induce microwave transitions in atoms and molecules, which allowed one to alter efficiently their quantum state, hence the character of motion in external spatially non uniform electric or magnetic field (Ramsey, 1956).

An atom in quasiresonant laser field acquires a high frequency polarizability and, if the intensity of laser field is spatially nonuniform, the atom is acted upon by *the gradient (the dipole)* force (Askarian, 1962;1972).

In the optical region of the spectrum, the recoil effect resulting from atom light interaction is significant. This effect predicted by Einstein (Einstein, 1909, 1917) as far as back as 1909, was experimentally corroborated by the slight deflection of a beam of sodium atoms scattering the resonant radiation of D line of Na (Frish, 1933).

An intense laser radiation tuned to the resonance with some allowed dipole transition in an atom can make it re-emit millions of photons, and so such a radiation can exert a substantial effect on the atomic velocity and mechanical trajectory. Hänsch and Schawlow (1975) proposed to use the resonance force due to spontaneous re-emission of photons for cooling neutral atoms (Fig. 3b), and Wineland and Dehmelt (1975) for cooling ions in an electromagnetic trap.

The gradient and spontaneous forces are at the root of a great many of experiments on controlling the motion of atoms by means of light, which permit to create a large variety as *dissipative* and *nondissipative* optical elements, which was already considered in the reviews (Ashkin, 1980; Letokhov and Minogin, 1981; Dehmelt, 1983; Stenholm, 1986; Wineland and Itano, 1987; Chu, 1991; Phillips et al, 1991; Cohen- Tannoudji, 1991), special issue of scientific journals (Meystre and Stenholm, eds., 1985; Chu and Wieman, eds., 1991;), monographs (Minogin and Letokhov, 1978; Kazantsev et al, 1991) and conference reports (Moi et al. eds., 1991). The problem of atom optic have been recently discussed in (Balykin and Letokhov, 1989; Pritchard, 1991;, Ekstrom et al. 1992) and a special issue (Mlynek, Balykin and Meystre, eds., 1992).

Laser- induced optics of atom can be separated in two parts: (1) atom optics with dissipative elements which is based on the use a spontaneous light pressure fprce and which have no analog in classical optics and (2) atom optics without dissipation based on the use a gradient force.

Let us very briefly consider each of this type of atom optics.

6. 1. ATOM OPTICS WITH DISSIPATIVE ELEMENTS

The most interesting field of application of atom optics with dissipative elements is a formation of atom sources which are both slow and bright. The atoms are exposed to laser radiation which is tuned to the atomic resonance. The atoms accumulate momentum from the light due to spontaneous scattering of the incident laser photons. At first stage of preparation of atomic source, the atoms are slowed down by a counter-propagating laser beam (of decreasing de Broglie wavelength); next (or simultaneously with a slowing down) the atoms can be collimated, compressed, or focused by different configuration of laser field (considerable increasing of the phase density of atoms). At the final stage, the atom can be also trapped by magnetic and (or) optical fields. The works (Kasevich and Chu, 1991; F. Shimizu, 1992) are an example of the use such dissipative optics to build an atom source for the atom interferometers.

6. 2. ATOM OPTICS WITH NON- DISSIPATIVE ELEMENTS

6. 2. 1. Beam splitter

***(a) Diffraction from optical standing wave*:** Diffraction of atoms by light has received considerable interest with regard to its practical importance for atomic interferometer (coherent beam splitter) and its intrinsic features. Pritchard and co workers have demonstrated the diffraction of sodium atoms at normal incidence to a transmission grating consisting of an optical standing wave (Gould et al. 1986; Martin et al. 1988). Hainal and Opat (1989) proposed to combine a reflection and diffraction of atoms by using a standing evanescent wave. In this case the required optical wavefield can be produced by totally internally reflecting a laser beam at the surface of a refractive medium and the retro-reflecting the light back along its original path. The evanescent field decreases exponentially in the direction to the surface and is modulated sinusoidally along the surface.

The result of diffraction of atom on the standing wave can be expected by considering the incident atomic beam as a plane de Broglie wave incident upon a periodic structure formed by standing evanescent wave. The diffraction angles of the reflected atomic beams are given by grating equation:

$$d(\sin\Theta_{rm} - \sin\Theta_i) = m\lambda_B, \tag{8}$$

where Θ_{rm} and Θ_i are the reflected and incident angles of the atomic beams, λB is the de Broglie wavelength corresponding the atomic beam, $d=\lambda/2$ the period of the standing wave, and m is the diffraction order.

Another consideration of diffraction of atoms by light is the photon picture, in which a diffraction is considered as a result of absorption and reemission of a photon, leading to exchange of the atomic momentum. The projection of this momentum along the laser axis is quantized in integral multiplies of 2✻k. If atoms are sufficiently detuned, so that spontaneous emission is negligible, the situation is analogous to lossless diffraction grating in optics. The momentum distribution in far field consists of several sharply defined diffraction peaks.

In the case of diffraction by a standing evanescent wave, the exponential profile of the evanescent wave in the direction perpendicular to the interface has contributions from waves of all directions. This permits the atom to acquire, through absorption- stimulated emission processes of photons of these waves, a momentum in a direction perpendicular to interface for a *specular reflection.* An atom, entering an evanescent wave, can absorb also a photon from either of two counter propagating waves along an interface vacuum- dielectric. The atom can then reemit that photon by a stimulated process back into the same wave. In this case there is a zero net change of atomic momentum parallel to interface. The atom can also reemit photon in to opposite evanescent wave. In this case, the atom momentum parallel to the interface will be changed by 2hk. The absorption and emission of photon pairs changes the momentum in the direction of standing wave, but not the kinetic energy of atoms due to energy conservation. This means that the modules of the total atomic momentum remains unchanged but the momentum normal to interface must be changed from that in specular reflection. This is a photon explanation of appearance of first *diffraction* order in reflection of atoms by standing evanescent wave. The change of parallel to interface component of atomic momentum on the value m*2hk gives m´th order of atomic diffraction.

At small glancing angle the diffraction angle can be considerably larger (a factor of 100) than one in the diffraction of atoms at normal incident to a transmission grating consisting of a standing wave. For atoms with a thermal velocity 10^5 cm/s and a glancing angle 10^{-3} rad, a diffracted by standing evanescent wave beams may be separated by angles of order 10^{-2} -10^{-3} rad.

(b) Optical Stern- Gerlach effect: The transverse Optical Stern- Gerlach effect was predicted in (Kazantsev, 1974): A beam of two-level atom interacting with an optical gradient can, under certain circumstances, be coherently split into two beams. This effect is related to the nonadiabatic coupling of two dressed states of atoms moving in an inhomogeneous light field. The momentum transfer to the atoms is determined by the amplitude of the electrical field and the time interaction of the atoms with the laser field and can sufficiently large for use in atom interferometer. The coherent splitting of a beam of metastable He atoms through the Optical- Stern effect was investigated in (Sleator, 1992). The coherent splitting was observed in the case of exact resonance of the laser with the atomic transition and the splitting angle was about 800 μrad.

(c) Atom diffraction on a traveling wave: In a plane traveling wave (when the spontaneous emission may be neglected)the atomic wave packet of a two level atom is split into packets differing in velocity approximately by the recoil velocity. The probability for an atom being in one of the packet is determined by the time of interaction and laser parameters The difference in atomic moment in ground and excited state leads to splitting of the atomic trajectory (Pusep et al. 77).

6. 2. 2. Atom Lens

The principal element in any sort of optics is a lens. It is therefore essential to create laser field configurations capable of focusing neutral atomic beams. There are at present two possibilities for focusing an atomic beam by laser light: by using the gradient force or by using the radiation pressure force.

The lenses used in light and electron optics (Grivet, 1972) satisfy the following condition: A divergent concentric beam is transformed by means of the lens into a convergent concentric beam. This requirement means that the force effecting the focusing of atoms in a beam should be proportional to the atomic displacement. This criterion must be satisfied for the force to produce the true "image": If this criteria is not met , the image is blurred, i. e., there are aberrations.

From point of view of the optical image theory: (*wave- optical approach*, Papoulis, 1968), an atomic beam should be considered in a form of de Broglie wave. In the optical image theory, the ideal objective lens is a transparency having the phase transmission function

$$T(x, y) = \exp[-ik(x^2 + y^2)/2f], \tag{9}$$

where $k = 2\pi/\lambda$ and f is the focal length of the lens. A light beam passing through such a transparency undergoes an additional phase change of $k(x^2 + y^2)/2f$.

The task is to find such a potential field U(x, y) that would make the phase change of a wave function (de Broglie wave) to satisfy Eq. (9), in which k is equal $2\pi/\lambda_B$ and λ_B is the de Broglie wavelength. It is known that, if the de Broglie wavelength is small in comparison with the characteristic size conditioning a given problem, the characteristic of the system are close to classical. In the quasi-classical approximation, the atomic wave function is defined by the expression (Landay and Livshits, 1985)

$$\psi = (C/p(z)^{1/2})\exp[(i/h)\int p(z)dz], \tag{10}$$

where C is a constant and $p(z) = [2M (E - U(z))]^{1/2}$ is the atomic momentum, M is the atomic mass, E the atomic total energy, and U(z) the atomic potential energy. In a quasi- resonant laser field the potential energy is given by (Gordon and Ashkin, 1980)

$$U(x, y) = (h\Omega/2) \ln(1 + s(x, y)) \tag{11}$$

where s(x, y) is the atomic transition saturation parameter and Ω is a detuning of laser frequency from atomic transition.

To calculate the atomic beam density distribution in the focal plane the Kirchoff's diffraction theory can be used . The effect of aberrations can be evaluated by considering the de Broglie wave front distortion caused by the objective lens.

The focusing of an atomic beam by means of the gradient force was demonstrated first at Bell Lab (Bjorkholm et al. 1978; Bjorkholm et al. 1980). In their scheme, the atomic lens was created by dye laser which was superimposed upon an atomic beam of sodium. The atomic beam propagated along and inside a narrow near Gaussian laser beam. The laser frequency was tuned below the atomic transition frequency, so that the gradient force was directed toward the laser beam axis. The minimum spot size was about 27 μm.

The resolution can be considerably improved by using the same idea of the gradient force, but now with a different laser field configuration and a different atom-field interaction geometry as was proposed in (Balykin and Letokhov, 1987b). A new atomic objective lens is a focused TEM^*_{01} laser beam tuned above resonance. The atomic beam propagates along the lens axis, where the intensity of light and therefore the rate of spontaneous emission and momentum diffusion, is minimal. This solves the diffusion problem. This configuration was analyzed in the thin lens approximation in (Balykin and Letokhov, 1987a). Later the path-integral techniques (Gallatin and Gould, 1991) and methods developed for particle optics (McClelland and Scheinfeind, 1991) were used to achieve a more general result. The atomic spot radius was in image plate may be as small as 10 Å.

6. 2. 3. Mirror for Atoms

The gradient force is directed along the gradient of the laser field and the sign of this force is determined by the sign of the detuning. At a positive detuning the gradient force expels an atom from the laser field. These properties of the gradient force can be used to reflect atoms by a laser field. There are a several configurations of laser field which can be used for the reflection of atoms. Cook and Hill (1982) suggested the use of an evanescent light wave as a mirror for atom. When a plane traveling light wave is totally internally reflected at the surface of a dielectric in a vacuum, a thin wave is generated at the surface.

The atom in the evanescent wave will experience a radiation force due to momentum transfer from the wave. For a two level atom, the radiation force has a component parallel to the surface (the light pressure force) and a component normal to the surface (the gradient force). The maximum value of gradient force is reached at the detuning of the order of the Rabi frequency. At such detuning the light pressure force is considerably smaller then the gradient force and it could be neglected in the interaction of atom with evanescent wave.

It can be shown (Cook and Hill, 1982; Ol´shanii et al. 1992) from a consideration of the motion of atoms under the action of the gradient force (and when the amplitude of the light is changed adiabatically slow in comparison with relaxation of the internal atomic motion) that the angle of reflection of the atom is equal to the angle of incidence. This means that the atoms are specular reflected. The first observation of a specular reflection of sodium atoms has been done in (Balykin et al. 1987d). In the experiment a beam of sodium atoms was used. The maximum angle of reflection observed in this experiment was about 0.4°. An important parameter of any mirror is its reflectivity. At small incident angle, the reflectivity of atomic mirror was observed close to 100%.

In another experiment (Kasevich et al. 1990) the specular reflection of atoms was observed by "dropping " a sample of laser cooled atoms on the evanescent wave. The sample of cold atoms was prepared in the following way. Sodium atoms were loaded into opto- magnetic trap by slowing down a thermal atomic beam with counter propagating laser beam. The trap was formed by three mutually orthogonal pairs of counter propagating, circular polarized laser beams intersecting in zero field region of a spherical quadrupole magnetic field. After loading of the atoms during 0.5 s, the field coils were turned off, leaving a small residual magnetic field in order to cool the atom further in the optical molasses. The final stage of cooling had been done by a gradual extinction of the light and the final temperature of the atoms (the cloud of 3 mm diameter) was $\approx 25\ \mu K$. The atoms were dropped from a height of 2 cm. Two bounces were registered at the initial trapping region. The main losses for atoms was the ballistic expansion of the sample of atoms due to the initial spatial and velocity extents.

One of the remarkable properties of an atomic mirror is its ability to reflect atoms in a specific quantum state. The quantum state selectivity is due to the relationship between gradient force and the detuning. For positive detuning, the gradient force repels the atom from the surface and thus the specular reflection takes place. For negative detuning, the force attracts an atom to the surface, and diffusive reflection is observed. The quantum state selective reflection was studied with sodium atoms (Balykin et al. 1988).

A mirror is an essential element of many types of interferometers. For applications of atom mirrors in atom interferometer the reflection of atom must be "coherent". In (Seifert et al. 1993) "coherent reflection" of well collimated beam of metastable argon atom by using "open" electronic transition was demonstrated.

6. 2. 4. Optical Atom Interferometer

These type of atom interferometers are based on the optical beam splitters and recombiners. The beam splitter is produced by exciting atom in a coherent superposition of internal states of the atom using laser field. The different internal states obtain the different momenta from laser light and the corresponding wavepackets drift apart. After a certain time of free evolution they are recombined by using another interaction of atoms with a laser field.

Different schemes of excitation of atoms can be used. For atoms with long-lived atom state is used an one-photon excitation (Riehle, 1991; Sterr, 1992). For atoms with a short excited state lifetime two-photon Raman excitation are used (Kasevich and Chu, 1991). The wave packets in optical atom interferometers are usually overlapping in space, and the interference pattern can be detected by counting atom in the excited state. These type interferometers were already successfully used for precise measurements of the Sagnac effect (Riehle, 1991) and the gravitational acceleration of atom (Kasevich and Chu, 1991).

6.2.5. Atom Cavity

The developed technique of laser cooling and trapping of atoms permits one to store nowadays atoms for a long time and with a high densities (see special issue of scientific journals: Meystre and Stenholm, eds., 1985; Chu and Wieman, eds., 1991; Mlynek, Balykin and Meystre, eds., 1992). One interest of studying trapped of atoms is the possibility of observing quantum statistical effects, for example the Bose-Einstein condensation of atoms. The Bose-Einstein condensation of spin polarized hydrogen atoms was predicted at high densities and low temperatues (Greytak and Kleppner, 1984). Several publications also analyzed a possibility Bose-Einstein condensation in other atoms by applying laser cooling techniques (Vigue, 1986; Bagnato et al. 1987). An alternative approach to the observation of collective quantum effects is the population of *highly excited quantum states* by on the average more than one atom per state. A laser resonator is well-known example of such system: It is quite easy now to reach a population of laser cavity mode by more than one photons.

It was proposed to create an atom cavity (Balykin and Letokhov, 1989e), that is similar to optical cavities with the material mirrors replaced by light-induced mirrors and instead of photons the cavity is filled by atoms bouncing between these mirrors. The atom cavity can be formed even by one mirror: Gravity plays the role of the second mirror and bends the atomic trajectory so that atom always bounce on the first mirror. In two-mirror cavity gravity plays a small role in the motion of atoms, and the calculation of mode structure can be done using the theory developed for optical resonators. In a one-mirror cavity (a gravitational cavity) gravity plays an essential role, and a direct application of the formalism developed for optical resonator is difficult. In (Wallis et al. 1992) both classical and quantum mechanical calculations of the atomic dynamics in the one-mirror cavity are given.

7. CONCLUSIONS

Optics and interferometry with atoms has made enormous progress during the last several years due to the latest development in technology and laser physics. This made the development of atom interferometer with high sensitivity for measurement of acceleration and rotational possible. Atom interferometers can be used as sensitive accelerometers in a variety of precision experiments for fundamental tests of physics, such as a searches for net charge of atom, "fifth" force measurements and measurements of gravity-inertial effects (Audretsch and Lämmerzahi, 1991). In contrast to neutron and electron interferometers, an interal energy structure of atoms allows of new type interferometer, in which the internal degree of freedom can be used. Two approaches were realized to build different elements of atom optics: the first, based on the microfabricated structures, and the second in which different configurations of a laser field serve as atom optical elements. The practical interest in atom optics lies in the opportunities to create atom microprobe with atom-size resolution and minimum damage of investigated objects.

ACKNOWLEDGEMENTS

The author wants to thank J. Mlynek and L. Meerts for many fruitful discussions. This work has been supported by the Deutsche Forschungsgemeinschaft.

REFERENCES

Anderson A., Haroche S., Hinds E. A., Jhe W., Meschede D., and Moi L. (1986), *Phys. Rev.* **34**, 3513.

Audretsch J. and Lämmerzahi C. (1991), *Appl. Phys.* **B54**, 351.

Ashkin A. (1980), *Science* **210**, 1081.

Askarian G. A. (1962), *Sov. Phys. JETP* **15**, 1088.

Askarian G. A. (1973), *Usp. Fiz. Nauk.* **110**, 115 (in Russian).

Bagnato V., Pritchard D. E., and Kleppner D. (1987), *Phys. Rev.* **A35**, 4354.

Balykin V. I., and Letokhov V. S. (1987a), *Opt. Commun.* **64**, 151.

Balykin V. I., Letokhov V. S., Sidorov A. I., and Ovchinnikov Yu. B. (1987b) *Pisma Zh. Eksp. Teor. Fiz.* **45**, 282 (*Sov. Phys. JETP Lett.* **45**, 353).

Balykin V. I., Letokhov V. S., Ovchinnikov Yu. B. and Sidorov A. I. (1988), *Phys. Rev. Lett.* **60**, 2137 (Errata **61**, 902).

Balykin V. I., and Letokhov V. S. (1989a), *Appl. Phys.* **B48**, 517.

Balykin V. I., and Letokhov V. S. (1989b), *Phys. Today* **4**, 23.

Balykin V. I., Letokhov V. S., Ovchinnikov Yu. B., and Sidorov A. I. (1989c), *Pisma Zh. Eksp. Teor. Fiz.* **49**, 383.

Benewitz G., and Paul W. (1954) *Z. Phys.* **139**, 489.

Benewitz G., Paul W., and Schlier C. (1955) *Z. Phys.* **141**, 6.

Bernstein R. B. (1982), *Chemical Dynamics via Molecular Beam and Laser Techniques* (Clarendon Press, Oxford).

Berkhout J. J., Luiten O. J., Setija I. D., Hijmans T. W., Mizusaki T., and Walraven J. T. M. (1989), *Phys. Rev. Lett.* **63**, 1689.

Bjorkholm J. E., Freeman R. E., Ashkin A. A., and Pearson D. B. (1980), *Opt. Lett.* **5**, 111.

Bonch-Bruyevich A. M., and Khodovoi A. A. (1968), *Sov. Phys. Usp.* **10**, 637.

Bonze U., and Rauch H., eds. (1979), *Neutron Interferometry* (Clarendon Press, Oxford).

Born M., and Wolf E. (1984), *Principles of Optics* (Pergamon Press, Oxford).

Carnal O., and Mlynek J. (1991), *Phys. Rev. Lett.* **66**, 2689.

Carnal O., Sigel M., Sleator T., Takuma H., and Mlynek J. (1991), *Phys. Rev. Lett.* **67**, 3231.

Chu S. (1991), *Science* **253**, 861.

Chu S., and Wieman C., eds. (1989), *J. Opt. Soc. Am.* **B6**, Special Issue No. B11, *Laser Cooling and Trapping of Atoms.*

Cohen-Tannoudji C., and Phillips W. D. (1990), *Phys. Today* **10**, 33.

Cook R. J., and Hill R. K. (1982), *Opt. Commin* **43**, 258.

Dehmelt H. (1983), in *Advances in Laser Spectroscopy*, Arecchi F. T., Strumia F., and Walter H., eds. (Plenum, New York), p. 153.

Einstein A. (1909), *Phys. Z.* **10**, 185; ibid. (1917) **18**, 121.

Ekstrom Ch. R., Keith D. W., and Prichard D. E. (1992), *Appl. Phys.* **B54**, 369.

Estermann L., and Stern O. (1930); *Z. Phys.* **61**, 95.

Frank A. I. (1987), *Usp. Fiz. Nauk.* **151**, 229 (*Sov. Phys. Uspekhi* **30**, 110).

Frank A. I. (1991), *Usp. Fiz. Nauk.* **161**, 95 (*Sov. Phys. Uspekhi* **34**, 980).

Friedburg H., and Paul W. *Naturwiss.* **37**, 20 (1950).

Friedburg H., and Paul W. *Naturwiss.* **38**, 159 (1951).

Frish O. R., and Stern O. (1933), *Z. Phys.* **84**, 430.

Frish O. R. (1933), *Z. Phys.* **86**, 42.

Gallatin G., and Gould P. J. (1991), *Opt. Soc. Am.* **B8**, 502.

Goldenberg H. M., Kleppner D., and Ramsey N. F. (1960), *Phys. Rev. Lett.* **8**, 361.

Goldenberg H. M., Kleppner D., and Ramsey N. F. (1962), *Phys. Rev. Lett.* **126**, 603.

Gordon J. P., Zeiger H. J., and Townes C. (1955), *Phys. Rev.* **99**, 1264.

Gordon, J. P., and Ashkin A. (1980) *Phys. Rev.* **A21**, 1606.

Gould P. I., Ruff G. A., and Pritchard D. E. (1986), *Phys. Rev. Lett.* **56**, 827.

Greytak T. J., and Kleppner D., in *New Trends in Atomic Physics* (Les Houches, Session XXXVIII, 1982, G. Grynberg and R. Stora, eds. (North-Holland, Amsterdam, 1984), p. 1125.

Grivet, P. (1972), *Electron Optics* (Oxford University Press, Oxford).

Hajnal J. V. and Opat G. I. (1989), *Opt. Comm.* **71**, 119.

Kasevich M., Weiss D., and Chu S. (1990), *Opt. Lett.* **15**, 607.

Kazantsev A. P. (1974), *Zh. Eksp. Teor. Fiz.* **66**, 1599 (*Sov. Phys. JETP* **39**, 784).

Kazantsev A. P. (1974), *Zh. Eksp. Teor. Fiz.* **66**, 1599 (*Sov. Phys. JETP* **39**, 784).

Kazantsev A. P., Surdutovich G. I., and Yakovlev V. P. (1991), *Mechanical Action of Lighton Atom* (World Scientific, Singapore).

Kearney P. D., Klein A. G., Opat G. I., and Gähler R. (1980), *Nature* **287**, 313.

Keith D. W., Schattenburg M. L., Smith H. I., and Pritchard D. E. (1988), *Phys. Rev. Lett.* **61**, 1580.

Keith D. W., Ekstrom C. R., Turchette Q. A., and Pritchard D. E. (1991), *Phys. Rev. Lett.* **66**, 2693.

Knauer F., and Stern O. (1929), *Z. Phys.* **53**, 799.

Korsynksii M. I., and Fogel Ya. M. (1951), *Zh. Eksp. Teor. Fiz.* **21**, 25; ibid., (1951) **21**, 38.

Landay L. D., and Livshits E. M. (1985) *Quantum Mechanics* (Addison-Wesley, Reading, Mass.).

Letokhov V. S., and Minogin V. G. (1981), *Phys. Rep.* **73**, 1.

Martin P. J., Oldaker B. G., Miklich A. H., and Pritchard D. E. (1988) *Phys. Rev. Lett.* **60**, 516.

McClelland J. J., and Scheinfein M. P. (1991), *J. Opt. Soc. Am.* **B8**, 1974.

Meystre P., and Stenholm S., eds. (1985), *J. Opt. Soc. Am.* **B2**, Special Issue N. 11, *Mechanical Effects of Light.*

Minogin V. G., and Letokhov V. S. (1987), *Laser Radiation Pressure on Atoms* (Gordon & Breach, New York).

Miniatura Ch., Robert J., Le Boiteux S., Reinhardt J., and Baudon J. (1992), *Appl. Phys.* **B54**, 347.

Mlynek J., Balykin V. I., and Meystre P., eds. (1992), *Appl. Phys.* **B54**, Special Issue, No 5, *Optics and Interferometry with Atoms.*

Moi L., Gozzini S., Gabbanini C., Arlmondo E., and Strumia F., eds. (1991), *Light Induced Kinetic Effects* (ETS Editrice, Piza).

Nayak V. U., Edwards D. O., and Masuhara N. (1983), *Phys. Rev. Lett.* **50**, 990.

Olśhani M. A., Letokhov V. S., and Minogin V. G. (1992), to be published.

Opat G. I., Wark S. J., and Cimmino A. (1992), *Appl. Phys.* **B54**, 396.

Papoulis M. (1968), *System and Transforms with Application in Optics* (McGraw-Hill, New York)

Pritchard D. W. (1991), *Atom Optic*, Zorn J. C. and Lewis R. R., eds., *XII International Conference on Atomic Physics* (American Institute Physics, Ann Arbor), P. 165.

Puset A. Yu., Doktorov A. B., and Burnstein A. L. (1977), *Zh. Eksp. Teor. Fiz.* **72**, 98.

Ramsey N. F. (1956), *Molecular Beams* (Clarendon, Oxford).

Ruska E. (1980), *The Early Development of Electron Lenses and Electron Miscroscopy* (Hirzel-Verlag, Stuttgart).

Riehle F., Kisters Th., Witte A., Helmke J., and Borde Ch. (1991), **67**, 177.

Schmahl G. andRudolph D. eds. (1984), *X-Ray Microscopy* (Springer Series in Optical Sciences **43** Springer, New York).

Sears V. F. (1989), *Neutron Optics* (New York).

Seifert W., Heine C., Ovchinnikov Y., Balykin V. I., and Mlynek J. (1993), in preparation.

Shapiro F. L. (1976), *Neutron Research* (Nauka, Moscow).

Shimizu F., Shimizu K., and Takuma H. (1992), *Phys. Rev. A* **46**, R17.

Shutz G., Steyerl A., and Mampe W. (1980), *Phys. Rev. Lett.* **44**, 1400.

Sleator T., Pfau T., Balykin V., and Mlynek J. (1992), *Appl. Phys.* **B54**, 375.

Stenholm S. (1986), *Rev. Mod. Phys.* **58**, 699.

Stern O. (1929), *Naturwiss.* **17**, 391.

Sterr U., Sengstock K. Müller J. H., Bettermann D., and Ertmer W. (1992), **B54**, 341.

Vigué J. (1986), *Phys. Rev.* **A34**, 4476.

Wallis H., Dallibard J., and Cohen-Tannoudji C. (1992), *Appl. Phys.* **B54**, 407.

Wineland D. J. and Dehmelt T. W. (1975), *Bull. Amer. Phys. Soc.* **20**, 637.

Wineland D. J. and Itano W. M. (1987) *Phys. Today* **6**, 34.

LOUIS DE BROGLIE'S WAVE-PARTICLE DUALISM: HISTORICAL AND PHILOSOPHICAL REMARKS

Hans-Peter Boehm

Technical University Dresden
Faculty of Philosophy
D-01062 Dresden, Germany

The realist's criticisms of the Copenhagen Interpretation (CI) try to reestablish an ontological understanding of the wave particle dualism by classical spacetime notions. Investigating some of the original papers it can be shown that the founders of quantum theory were very careful with respect to that classical pictorial way of thinking. Especially the history of the debate on the foundation of QM make it clear that up to now there is no alternative to the CI.

Key words: quantum mechanics, Copenhagen interpretation, wave-particle dualism, philosophy of science.

1. INTRODUCTION

In a recent paper F. Selleri criticizing the C.I. made the following statement:

"What happened was rather that a purely philosophical fashion swept across the world of physics and conquered minds to the idea that we should not try to explain the physical phenomena by causality and space-time pictures." [1]

However, careful investigation shows that:

- It was not a philosophical fashion which forced the majority of physicists to accept the CI, but experimental evidence.
- Physicists never gave up to think in space-time pictures, but objective reality in the quantum domain has not to be identical with theoretical pictures.
- Nevertheless, the phenomenons are described by classical space-time pictures because another way of thinking seems to be impossible. But the

phenomenons can't be ascribed a unique space-time model at the microscopical level.
- Revision of our theoretical concepts with which we try to understand the objective reality means not that we deny the existence of the latter ones.

If we compare the original works of Einstein, de Broglie, Schrödinger and Born we can observe what I call an epistemological shift. The classical way of identifying the concepts of theory with the objects of investigation led to contradictions.

But otherwise these concepts could not be mere patterns of descriptions because of there partial empirical adequacy. The way out is Bohr's complementarity. This principle maintains the applicability of causal spacetime pictures for the qualitative causal understanding of phenomenons but also marks their limits with respect to the quantum objects, what ever they are. All what we are doing in science (not only in physics) is to reflect subjectively what we assume to exist objectively. (This assumption is of course a philosophical decision but the very precondition for natural sciences). It would be rather strange if the world would be exactly the way we reconstruct it, especially with respect to that part of the world which is not immediately observable by our human senses. What Selleri calls a philosophical fashion seems to me to be a consequence of empirical facts and philosophical reasoning.

2. EINSTEIN ON WAVE-PARTICLE DUALITY

All his live Einstein was extremely cautious with respect to the nature of light. He never found an answer which convinced him. In 1951 nearly at the end of his life he wrote to his friend Besso: "All these 50 years of pondering have not brought me closer to answering the question 'What are light-quanta?'" [2]

In his 1905 work he stated an analogy between light in the Wien-region (high radiation density) an the entropy of an ideal gas. This led him to the concept of light as energy quanta within that region [3]. In his own investigations on energy fluctuations in 1908/09 he could show that for black body radiation the fluctuation formula contains a "wave" term and a "particle" term. But both for different energy regions [4].

Early in 1917 Einstein was ready to ascribe the light quantum a momentum. But he never said that light is a particle. What he did is to associate to the energy quantum a momentum quantum to explain the directedness of the energy transfer in emission and absorption processes [5].

From an classical point of view these are of course the ingredients of a particle but if we formulate carefully we can only say, that in emission and absorption processes the behaviour of light is in analogy to the classical particle picture. So what we are doing is to match a model with an special phenomenon. The same holds for Einstein's "ghost waves" he joked about in 1924. This ghost wave seemed to guide the light quantum, but he never spoke about an ontological dualism like the late de Broglie. It is not a pure accident that Einstein always used the notion ψ function, never wave function.

Nonetheless the late Einstein's philosophical ideal was of course the "real single event" (der reale Einzelfall) to be described in classical spacetime notions. But this realistic attitude is neither an argument nor a proof.

3. DE BROGLIE'S 1924 THESIS

Louis de Broglie's starting point was the hypothesis that, for a microscopic massive 'particle',both the relations $E = m_0c^2$ and $E = h\nu_0$ should be valid. That is, the 'particle' is ascribed a frequency. Consequently there must be a wave like phenomenon associated with the 'particle' [6].

Relativistic reasoning led de Broglie to two frequencies depending on the observer's point of view. To reconcile both oscillatory phenomenons, he proved his famous theorem on phase correlation:

Relative to an observer at rest, there appears a 'wave' associated with any moving 'particle' that is constantly in phase with a 'wave' of frequency

$$\nu = \frac{m_o c^2}{h} / \sqrt{1-\beta^2}\,.$$

The phase velocity of the latter wave is $V = c/\beta$, which is always greater than c [7]. Consequently de Broglie stated in his thesis that this 'wave' can't be an energy carrying 'wave'. It represents a distribution of phases in space and is therefore called a "phase wave".

Now de Broglie made an *epistemological shift* (unreflected of course) which became rationalized in Born's interpretation two years later. He found that the superposition of "phase waves" within a narrow interval of the relevant frequency leads to a beat phenomenon, whose energy propagation velocity -the group velocity- equals the velocity of the moving 'particles'.

(As was shown by Schrödinger in his second communication to wave mechanics, the phase correlation theorem is not a peculiarity of the relativistic approach; it is a necessity of classical optics for the wave representation of a bundle of rays.) Suddenly we have an ensemble of "phase waves" and a single 'particle'. The nature of this "waves" was more strange than before. L. de Broglie could proof that the analytical background of his results is the equivalence of the mechanical principle of least action and the optical principle of Fermat within the region where $m_0c^2=h\nu_0$ holds. Applied to the hydrogen atom de Broglie derived a resonance condition equivalent to the Bohr-Sommerfield quantization.

The way de Broglie interpreted his results was that the rays of the superposition of the "phase waves" represent the possible trajectories of the moving 'particle'. The dualism which de Broglie conceptualized in his 1924 work is an epistemological one. Only some years later (1927 was the turning point) de Broglie started to fall back and try to re-establish classical realism [8].

4. SCHRÖDINGER ON HIS WAVE FUNCTION (1926)

Schrödingers first - rather enthusiastic - idea (in his second 1926 communication) was the wave-mechanical reformulation of the Hamiltonian picture of classical mechanics to get a "natural" explanation of the quantum discontinuities [9].

However, soon it became clear to him that his wave function couldn't be interpreted as representing a 3-dimensional "wave phenomenon" in the general case of many quantum objects. As is well known, only for the one-electron system the wave function is three- dimensional. However, the relevant "wave packet" is not stable.

The equivalence of matrix mechanics and wave mechanics [10] shown by Schrödinger himself, gave a strong indication that there must be a relation between the two concepts ('particles' vs. 'waves') at the roots of the two theories.

In his last two works of 1926 Schrödinger realized that physical meaning has ascribed to $\psi\psi^*$ and not to the function ψ itself. First he tried to interpret it as the electrical charge density. However, this turns out to be wrong, because of the normalization condition. So he arrived at the formulation that $\psi\psi^*$ is a "kind of weight function": "The wave mechanical configuration of the system is the superposition of many - in the strict sense of all - kinematically possible point mechanical configurations." [11]

This is in full analogy to de Broglie's result of 1924. He also made what I called an epistemological shift from one 'wave' function to an ensemble representing the possible trajectories of a 'particle'. With his notion of a "weight function" he was very near to Born's statistical interpretation.

5. BORN'S STATISTICAL INTERPRETATION: THE SHIFT FROM REALISM TO EMPIRICAL ADEQUACY

Born, investigating linear motions of quantum systems in collision processes, succeeded in connecting, the empirically so powerful, particle-picture with Schrödingers, mathematically elegant, wave-mechanical method. The result was his well-known statistical interpretation [12].

It was immediately clear to him that this interpretation challenged the fundamentals of traditional physics. There was no room for a deterministic theory , and the substantial notions like 'wave' and 'particle' loosed their ontological meaning. With Born's statistical interpretation quantum mechanics became a phenomenology of microphysics.

Bohr's principle of complementarity, which is the epistemological core of the Copenhagen Interpretation, is exactly the philosophical reflection of Born's interpretation. The notions 'wave' and 'particle' are patterns of description of phenomenons caused by microscopic objects in well-defined experimental contexts. A mutual reduction of both concepts is not possible because the empirical facts are mutual exclusive. If this is the last answer microphysics can give (and there are strong indications for this), then there appear serious philosophical problems to be discussed in our last point.

6. PHILOSOPHICAL REMARKS

What Bohr, Born, and the adherents of the Copenhagen Interpretation brought about was (and is) an epistemological shift from the pictorial realism of classical physics to the weaker conception of empirical adequacy . This concept means that the relevant phenomenons are described in an adequate way by models (patterns) based on classical, but mutually exclusive notions. Because of the empirical adequacy there must be a *partial* correspondence between the properties of the models and the properties of the real systems. (The stress is on partial.)

All efforts to construct a hidden variable theory which is in accordance with the classical concepts run into contradictions or pathologies as shown by Bell's theorem and its experimental tests [13]. From my point of view, we come always back to Bohr.

The only acceptable consequence for me is the critical assessment of the concept of classical realism as the universal mode of physical description. Philosophical conceptions are like physical theories results of *our* thinking. But our thinking is conditioned by human cognitive abilities, which are limited to the mesocosmic world by natural evolution [14].

Because the human race is still alive, our cognitive abilities as well as our rational notions seems to be empirically adequate relative to that part of the real world we are fitted to. We can transcend this level by rational construction but only with respect to an genetically invariant base.

The psychologist J. Piaget empirically investigated this base with respect to the evolution of the mathematical thinking [15]. I think a similar research should be possible concerning the most fundamental physical concepts. Thus, in a certain sense, the epistemological foundation of quantum mechanics becomes a problem of psychology.

From my point of view the CI is exactly a consequence of our evolutionary adapted structure of thinking. Rational reasoning cannot surpass that structure. Consequently there is a kind of objective limitation with respect to our abilities of acquiring scientific knowledge. From a physical point of view this answer seems to be less satisfying. It indicates that we cannot get a unique causal picture of physical processes in the quantum domain at all. The proof of this statement is up to further experiment. In a sense we are faced with a situation Selleri calls "experimental philosophy".

REFERENCES

[1] F. Selleri, *Philosophia naturalis* **28,** 17-34 (1991).
[2] A. Einstein, letter to Besso, cited according to: A. Pais, *Rev.Mod.Phys.* **51**,163-914 (1979).
[3] A. Einstein, *Ann.Phys.* Leipzig **7,** 132-148 (1905).
[4] A. Einstein, *Phys.Z.* **10**,817 (1909).
[5] A. Einstein, in A. Pais, Ref.[2], p.888.
[6] L. de Broglie, Ph.D.thesis (1924); *Ann. Phys. (Paris)(*10) **III,** 22-128 (1925)
[7] compare the excellent paper by M. Mugur-Schächter, *Found.Phys.Lett.* **2**,261-286 (1989).
[8] L.de Broglie, *J.Phys.Rad.***5**,225 (1927).
[9] E. Schrödinger, *Ann.Phys. (Leipzig)* **80,** 437 (1926).
[10] E. Schrödinger, *Ann.Phys. (Leipzig)* **79,** 734 (1926).
[11] E. Schrödinger, *Ann.Phys. (Leipzig)* **81,** 109 (1926).
[12] M. Born, *Z.Phys.* **38**,803 (1926).
[13] D. Home and F.Selleri, *La Revista del Nuovo Cimento* **14**,(1991).
[14] G. Vollmer, *"Was können wir wissen?"* (2 Volumes) (Stuttgart,1985).
[15] J. Piaget, *Genetic Epistemology* (New York, 1970).

COMPATIBLE STATISTICAL INTERPRETATION OF INTERFERENCE IN DOUBLE-SLIT INTERFEROMETER

Mirjana Božić

Institute of Physics
P. O. Box 57, Beograd, Yugoslavia

De Broglian probabilities, associated with two characteristic sets of trajectories in the double-slit interferometer, are evaluated and graphically presented. The change of de Broglian probabilities with attenuation coefficient a shows a remarkable consistency with the underlying physical picture.

Key words: de Broglian probabilities, compatibility, wave and particle interference, quanton.

1. INTRODUCTION

The standard interpretation of quantum mechanics, founded by Bohr [1] and von Neumann [2], forbids us to make physical pictures of quantum objects and of events in the quantum domain. This interpretation gives importance only to the results of quantum measurements. Such approach is justified by the fact that all quantum measurements perturb the system which is studied as well as the evolution of its states.

Alternative interpretations of quantum mechanics [3–9] tend to follow great traditions of theoretical physics, in which the agreement of experimental results with the theoretical predictions derived from intuitive assumptions, physical pictures and verified laws, was considered to be satisfactory.

The differences in the standard and alternative interpretations are most evident if one compares their explanation of interference phenomena. For example, according to the standard interpretation, particle trajectories and interference phenomena are not coexistent, they are complementary[1,11]. In the interpretations which were developed from de Broglie's wave mechanics [5–9] it is allowed to speak about particle's trajectories in

Waves and Particles in Light and Matter, Edited by
A. van der Merwe and A. Garuccio, Plenum Press, New York, 1994

the interference phenomena. It is considered that interference and trajectories are coexistent, contemporary.

De Broglie could not accept the affirmation that trajectories do not coexist with interference because any attempt to detect trajectories destroys the interference. He argued that this conclusion was wrong and believed that particle trajectories exist and that their form is determined by the really existing wave field. Then, he concluded that the probabilistic scheme of the standard QM should be enriched by a new kind of probabilities, which he called hidden probabilities [4,11]. This enlarged probabilistic interpretation which we propose to call *compatible statistical interpretation* of QM contains three kinds of probabilities: two kinds, which de Broglie called present and predicted probabilities come from the standard interpretation [4,11]. The third one, hidden probabilities (or de Broglian, as we propose to call them) is the new one.

According to this classification, probabilities $|\Psi(\boldsymbol{r}, t)|^2 d^3\boldsymbol{r}$ are *present,* since they may be determined without any preparation of the physical system. Present probabilities correspond to a classical probabilistic set of observable events. The probabilities $|c_m|^2$, where c_m are the coefficients in the development of the function $\Psi(\boldsymbol{r}, t)$ in terms of eigenfunctions $\varphi_m(\boldsymbol{r}, t)$ of particular variable Q,

$$\Psi(\boldsymbol{r}, t) = \sum_m c_m \varphi_m(\boldsymbol{r}, t), \tag{1}$$

are called *predicted* probabilities. Predicted probabilities concern events which are not observable by simple inspection because for their measurement a special preparation (spectral splitting) is necessary [4,11].

Hidden probabilities should satisfy classical probability axioms. They are associated with events which are expected to happen objectively, although we might be unable to precisely follow their evolution [4,9,11]. But those events and their statistics should help us in explaining and understanding the statistics of observable events.

In the interferometers, hidden probabilities are associated with different sets of possible particle trajectories from the source to the detectors behind the interferometer. In the Young double-slit interferometer there are two characteristic sets of trajectories which end in the same point on the screen. To one set belong the trajectories which pass through the slit B, and to the other belong those which pass through the slit C. To each detector at the exit of the Mach-Zehnder interferometer lead per two trajectories shown in Fig. 1. With each set of trajectories we associate the probabilities of the third kind, de Broglian probabilities, by reasoning in the following way.

The wave function at the point $\boldsymbol{r}$ on the the screen of the Young double-slit interferometer is a superposition of wave functions $\varphi_B(\boldsymbol{r}, t)$ and

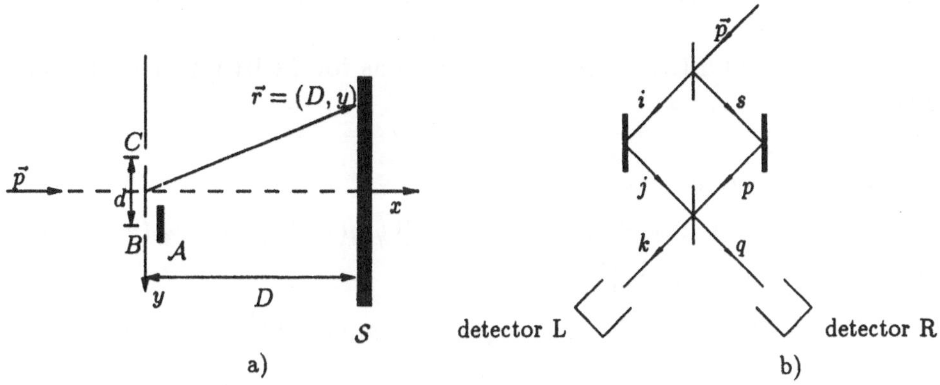

Fig. 1. (a) Double-slit interferometer; (b) Mach-Zehnder or neutron interferometer.

$\varphi_C(\boldsymbol{r}, t)$ which spread from slits B and C, respectively:

$$\Psi(\boldsymbol{r}, t) = \varphi_B(\boldsymbol{r}, t) + \varphi_C(\boldsymbol{r}, t) \tag{2}$$

Selleri and Tarozzi [12,13] concluded that probability density

$$P(\boldsymbol{r}, t) = \mid \Psi(\boldsymbol{r}, t) \mid^2 = \mid \varphi_B(\boldsymbol{r}, t) \mid^2 + \mid \varphi_C(\boldsymbol{r}, t) \mid^2 + 2\mathrm{Re}\varphi_B^*(\boldsymbol{r}, t)\varphi_C(\boldsymbol{r}, t) \tag{3}$$

is equal to the sum

$$P(\boldsymbol{r}, t) = P(E_B(\boldsymbol{r}, t)) + P(E_C(\boldsymbol{r}, t)) \tag{4}$$

of probability densities of two elementary events:

$E_B(\boldsymbol{r}, t)$ - quanton at $\boldsymbol{r}$, at moment t, whose particle passed through B when B and C are open;

$E_C(\boldsymbol{r}, t)$ - quanton at $\boldsymbol{r}$, at moment t whose particle passed through C when B and C are open.

To the Tarozzi-Selleri relation (4) was added [14,15,16] the following relation among de Broglian probability densities $P(E_B(\boldsymbol{r}, t))$ and $P(E_C(\boldsymbol{r}, t))$:

$$\frac{P(E_B(\boldsymbol{r}, t))}{P(E_C(\boldsymbol{r}, t))} = \frac{\mid \varphi_B(\boldsymbol{r}, t) \mid^2}{\mid \varphi_C(\boldsymbol{r}, t) \mid^2} \, . \tag{5}$$

This relation represents the generalization of the analoguous relation, which was written [14] for a Mach-Zehnder (neutron) type interferometer. In the latter case, this relation seems quite natural if one takes into account the experiments in neutron interferometer with stochastic and deterministic absorbers [17,18].

From (4) and (5) follow the expressions for de Broglian probability densities:

$$P(E_B(\mathbf{r},t)) = |\varphi_B(\mathbf{r},t)|^2 \cdot \{1 + 2\mathrm{Re}\varphi_B^*(\mathbf{r},t)\varphi_C(\mathbf{r},t)/[|\varphi_B(\mathbf{r},t)|^2 + |\varphi_C(\mathbf{r},t)|^2]\},$$
$$P(E_C(\mathbf{r},t)) = |\varphi_C(\mathbf{r},t)|^2 \cdot \{1 + 2\mathrm{Re}\varphi_B^*(\mathbf{r},t)\varphi_C(\mathbf{r},t)/[|\varphi_B(\mathbf{r},t)|^2 + |\varphi_C(\mathbf{r},t)|^2]\}. \quad (6)$$

2. DE BROGLIAN PROBABILITIES FOR WAVE PACKETS IN THE DOUBLE-SLIT INTERFEROMETER WITH ONE ATTENUATOR

The best way to see the meaning of those expressions is to evaluate and plot the dependence of $P(E_B(\mathbf{r},t))$ and $P(E_C(\mathbf{r},t))$ on $\mathbf{r} = (D, y)$ for the appropriately chosen wavefunction.

As is usual, we shall represent free quantons of the wave length $\lambda = \frac{2\pi}{k} = \frac{h}{p}$ which approach the slits by two-dimensional wave packets[19]

$$\Psi(x, y, t) = \frac{1}{\delta\sqrt{\pi}} \exp\left(-\frac{x^2 + y^2}{2\delta^2}\right) \exp(ikx) \exp(-it\hbar k^2/2m), \quad (7)$$

where the factor $1/\delta\sqrt{\pi}$ is introduced to normalize the Schrödinger function. This implies that the state of the quanton immediatly passing the slit can be considered to be represented by the superposition of two wave packets of the form (7), one centered at $y = d/2$ and the other at $y = -d/2$. Thus we have for the initial wave function immediately behind the slits

$$\Psi(x, y, 0) = \varphi_B(x, y, 0) + \varphi_C(x, y, 0),$$

with

$$\varphi_B(x, y, 0) = \frac{1}{\delta\sqrt{2\pi}} \exp\left[-\frac{x^2 + (y - d/2)^2}{2\delta^2}\right] \exp(ikx),$$
$$\varphi_C(x, y, 0) = \frac{1}{\delta\sqrt{2\pi}} \exp\left[-\frac{x^2 + (y + d/2)^2}{2\delta^2}\right] \exp(ikx). \quad (8)$$

In order to find the time variation of this wave function over screen $\mathcal{S}$ it is sufficient to calculate time variations of $\varphi_B(x, y, 0)$ and $\varphi_C(x, y, 0)$. Their time variation is known [19] and reads

$$\varphi_{\substack{B\\C}} = \frac{1}{\sqrt{\pi}\delta\{1+iht/2\pi m\delta^2\}} \exp\left[-\frac{(x-pt/m)^2+(y\mp d/2)^2}{2\delta^2\{1+(ht/2\pi m\delta^2)^2\}}\right]$$
$$\times \exp\left[\frac{i}{1+(ht/2\pi m\delta^2)^2}\left\{\frac{2\pi px}{h}-\frac{\pi p^2 t}{hm}+\frac{ht}{4\pi m\delta^4}(x^2+(y\mp d/2)^2)\right\}\right]. \tag{9}$$

If one takes into account that experimental parameters satisfy

$$\frac{h}{2\pi p}\frac{X}{\delta^2} \gg 1, \tag{10}$$

the value of functions φ_B and φ_C for $t = Dm/p$ takes the much simpler form

$$\varphi_{\substack{B\\C}}(D,\, y,\, Dm/p) = \exp\left\{\frac{\left(y\mp\frac{d}{2}\right)^2}{2\delta^2(\lambda D/2\pi\delta^2)}\right\}\exp\left\{i\frac{\pi}{\lambda}\left[D+\frac{\left(y\mp\frac{d}{2}\right)^2}{D}\right]\right\}. \tag{11}$$

If we put in front of the slit B the attenuator A having the coefficient of attenuation a we should substitute everywhere the wave function $\varphi_B(\boldsymbol{r},\, t)$ by the wave function

$$\varphi_B^a(\boldsymbol{r},\, t) = a\,\varphi_B(\boldsymbol{r},\, t). \tag{12}$$

Finally, de Broglian probability densities $P(E_B(R,\, y,\, Dm/p))$ and $P(E_C(D,\, y,\, Dm/p))$ take the form:

$$\begin{aligned} P(E_B(D,y,Dm/p)) &= a^2 \mid \varphi_B(D,y,Dm/p)\mid^2 \\ &\cdot\left[1+\frac{2a\mid\varphi_B(D,y,Dm/p)\mid\cdot\mid\varphi_C(D,y,Dm/p)\mid}{\mid\varphi_B(D,y,Dm/p)\mid^2+\mid\varphi_C(D,y,Dm/p)\mid^2}\cos\frac{kyd}{D}\right], \\ P(E_C(D,y,Dm/p)) &= \mid\varphi_C(D,y,Dm/p)\mid^2 \\ &\cdot\left[1+\frac{2a\mid\varphi_B(D,y,Dm/p)\mid\cdot\mid\varphi_C(D,y,Dm/p)\mid}{\mid\varphi_B(D,y,Dm/p)\mid^2+\mid\varphi_C(D,y,Dm/p)\mid^2}\cos\frac{kyd}{D}\right]. \end{aligned} \tag{13}$$

In Fig. 2 we have plotted de Broglian probability densities $P(E_B(D,\, y,\, Dm/p))$ and $P(E_C(D,\, y,\, Dm/p))$ and the total probability density $P(D,\, y,\, Dm/p)$ for five values of the attenuation coefficient. The other characteristic quantities D, d, and λ are the ones from the double-slit experiment with neutrons done by Zeilinger et al.[20].

From Fig. 2 we see that for $a = 0$ to the screen S arrive only particles which have passed through the upper slit, and the interference pattern is identical with the single-slit interference pattern. On increasing a, particles which passed through the slit B start to arrive at the screen. But,

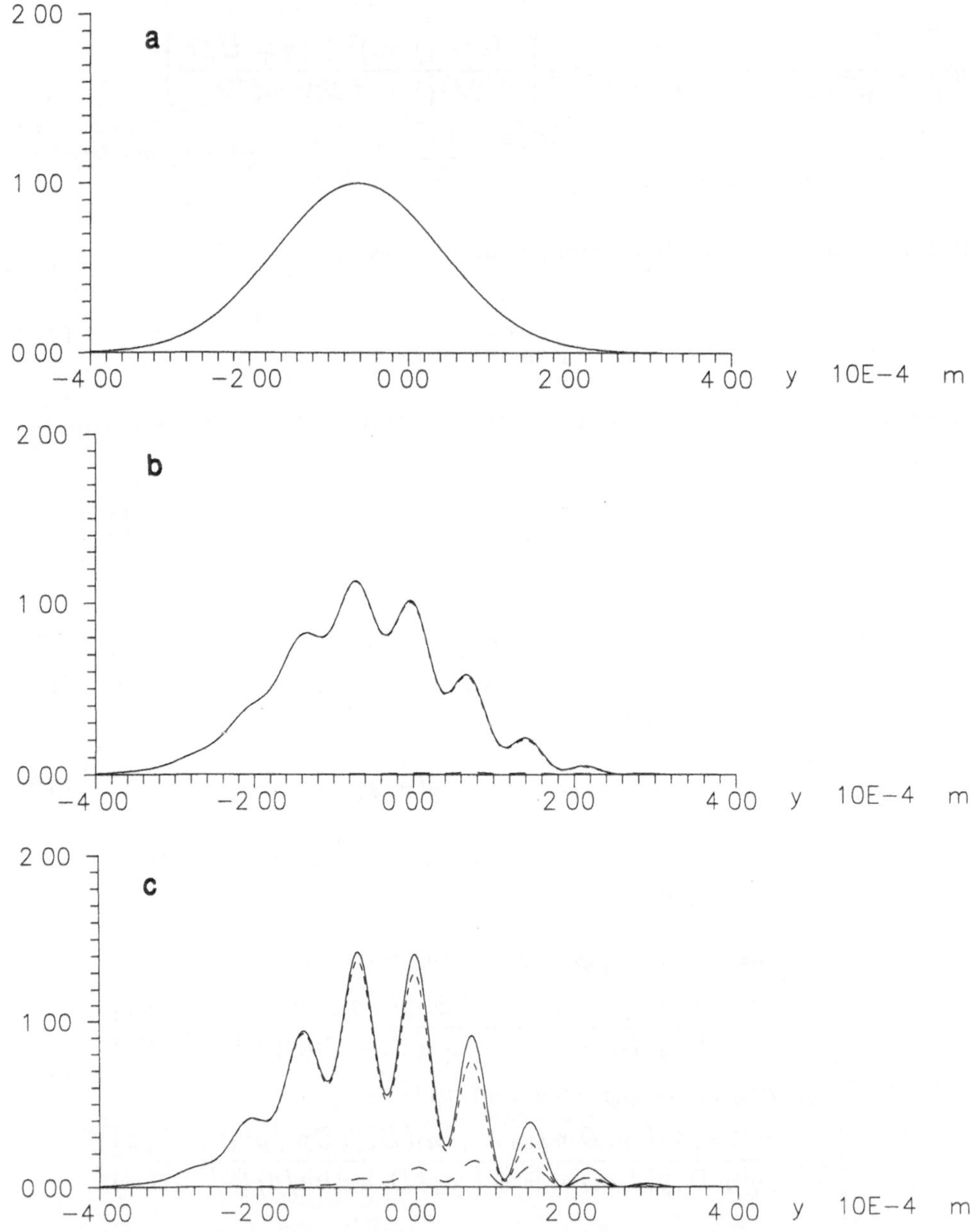

Fig. 2. Graphs of the de Broglian probability densities $P(E_B(D, y, Dm/p))$ (– – – –) and $P(E_C(D, y, Dm/p))$ (- - - -) and of the total probability density $P(D, y, Dm/p)$ (full line) for five values of the attenuation coefficient a: (a) $a = 0$, (b) $a = 0.1$, (c) $a = 0.3$, (d) $a = 0.7$, (e) $a = 1$. The values of other parameters are: $\lambda = 18.5 \cdot 10^{-10} m$, $D = 5m$, $\delta = 10^{-5} m$, $d = 1.26 \cdot 10^{-4} m$.

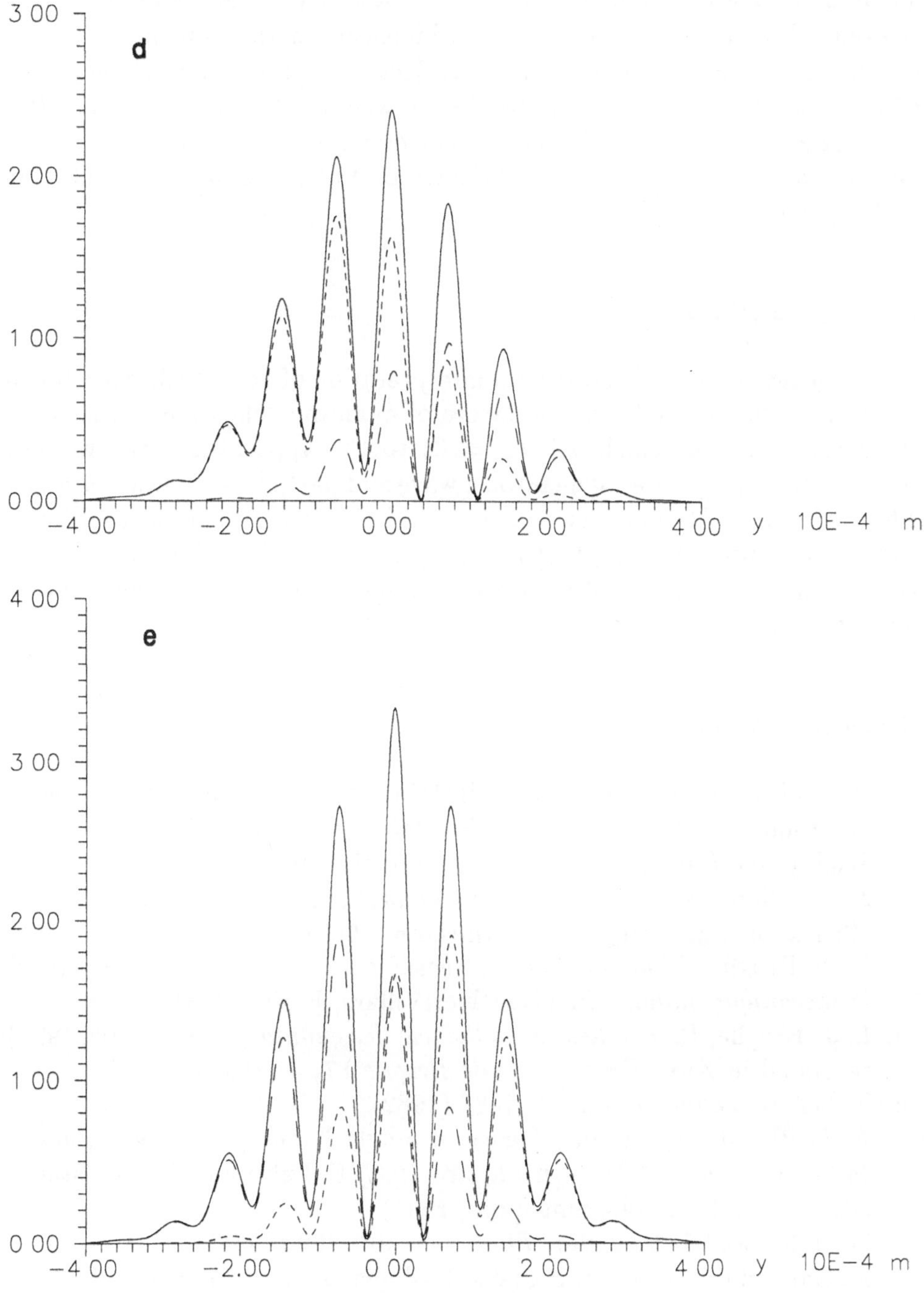

Fig. 2 (Continued)

the wave which accompanies a particle and passes through both slits influences the trajectory of particle which passed through B as well as of the one which passed through C. On increasing a the influence of waves which spread from both slits on the trajectories from B and C increases, and the assymetry of $P(D, y, Dm/p)$ as well as of $P(E_B(D, y, Dm/p))$ and $P(E_C(D, y, Dm/p))$ decreases. For $a = 1$ the interferometer is symmetric, and consequently the asymmetry of $P(D, y, Dm/p)$ as well as of de Broglian probabilities disappears.

3. CONCLUSION

In the compatible statistical interpretation of double slit experiment there is no discontinuity in the nature of quantons when one passes from the apparatus with single slit ($a = 0$) to the apparatus with two slits ($a \neq 0$). Both properties of quantons, wave and particle, are always present. The trajectory of the particle which passed through B is influenced by its wave which spreads through B and through C. De Broglian probabilities associated with two sets of trajectories describe consistently this physical picture.

REFERENCES

1. N. Bohr, "Discussion with Einstein on epistemological problems in atomic physics," in *Albert Einstein: Philosopher-Scientist*, P. A. Schilpp, ed. (Open Court, La Salle, Illinois, 1949), p. 200.
2. J. von Neumann, *Mathematical Foundations of Quantum Mechanics* (Princeton University Press, Princeton, 1955).
3. L. de Broglie, *Etude critique des bases de l' interpretation actuelle de la mecanique ondulatoire* (Gauthier-Villars, Paris, 1963).
4. L. de Broglie, *Recherches sur la théorie des quanta* (Thèse, Paris, 1924); reprinted in *Ann. Fond. Louis de Broglie* **17**, 1 (1992).
5. F. Selleri, *Found. Phys.* **12**, 1087 (1982).
6. A. O. Barut, "Quantum Theory of Single Events," in *Symposium on the Foundations of Modern Physics 1990,* P. Lahti and P. Mitelstaedt, eds. (World Scientific, Singapore, 1991).
7. D. Bohm and J. P. Vigier, *Phys. Rev.* **96**, 208 (1954).
8. A. Garuccio, V. Rapisarda, and J. P. Vigier, *Phys. Lett.* **90A**, 17 (1982).
9. N. C. Petroni, *Phys. Lett. A* **14**, 370 (1989); **160**, 107 (1991).
10. N. O. Scully, B. G. Englert, and H. Walther, *Nature* **351**, 6322 (1991).
11. G. Lochak, "De Broglie's initial conception of de Broglie waves," in *The Wave-Particle Dualism,* S. Diner, D. Fargue, G. Lochak, and F. Selleri, eds. (Reidel, Dordrecht, 1984).

12. F. Selleri and G. Tarozzi, *Nuovo Cimento* **43**, 31 (1978).
13. G. Tarozzi, in *The Wave-Particle Dualism*, S. Diner, D. Fargue, G. Lochak, and F. Selleri, eds. (Reidel, Dordrecht, 1984).
14. M. Božić and Z. Marić, *Phys. Lett. A* **158**, 33 (1991).
15. M. Božić, Z. Marić, and J. P. Vigier, *Found. Phys.* **22**, 1325 (1992).
16. M. Božić and Z. Marić, "Probability and interference," to be published in *"Courants, amers, écueils en microphysique,"* G. Lochak. ed.
17. J. Summhammer, H. Rauch, and D. Tuppinger, *Physica B* **151**, 103 (1988).
18. H. Rauch and J. Summhammer, *Phys. Lett. A* **104**, 44 (1984).
19. S. I. Tomonaga, *Quantum Mechanics, Vol. II* (North-Holland, Amsterdam, 1966).
20. A. Zeilinger, M. A. Horne, and C. O. Schull, in *Proc. Int. Symp. Foundations of Quantum Mechanics*, S. Kamifuchi et al., eds. (Physical Society of Japan, Tokyo, Japan, 1983), pp. 389-393.

WAVE FUNCTION STRUCTURE AND TRANSACTIONAL INTERPRETATION

Leonardo Chiatti

Department of Physics, "La Sapienza" University
P. le Aldo Moro 00185
Rome, Italy

The concept of "transaction," introduced by Cramer in his realistic nonlocal interpretation of quantum mechanics (QM), is herein reformulated in the language of the Feynman graphs technique.

Key words: nonlocal realism, transactional interpretation, Feynman graphs.

1. INTRODUCTION

The concept of "transaction" was introduced, in Wheeler-Feynman electrodynamics [1,2], by Cramer [3,4] in order to indicate the situation where two point events A and B (B is placed in the future light cone of A) are connected by a stationary wave which is the superposition of a retarded Ψ field and of an advanced Ψ^* field. Under particular conditions the retarded field emitted by B is cancelled by the retarded field emitted by A; meanwhile the advanced field emitted by A is cancelled by the advanced field emitted by B, and therefore only the field that connects A and B maintains itself. That field is obviously defined either by the initial conditions in A and the final ones in B.

The wave Ψ, called "offer," takes an amount $E > 0$ of energy in A and deposits it in B. The wave Ψ^*, called "confirmation," takes an amount of $-E$ of energy in B and deposits it in A, which is the same. The two waves are in fact the same field viewed in two different time directions, and therefore is not surprising that their effect is identical: An amount (quantum) $E > 0$ of energy is transmitted from A to B. The same thing happens for every other observable, and therefore the quantum of each observable is always traveling from the past to the future.

Waves and Particles in Light and Matter, Edited by
A. van der Merwe and A. Garuccio, Plenum Press, New York, 1994

Cramer generalized the concept of transaction for the Ψ fields which are solutions of quantum relativistic wave equations [5], therefore supplying a realistic nonlocal interpretation of QM. Others realistic nonlocal time symmetric interpretation of QM are also been proposed (see [9–30]).

In this paper a reformulation of the transaction concept in terms of the transition amplitude is attempted in order to obtain a clearer connection with the Feynman graphs' technique.

2. THE CONCEPT OF TRANSACTION

Let Γ be an oriented path on M^4. For each point event $x \in \Gamma$ there exists a portion $I_x(\Gamma)$ of Γ entering in x and a portion $O_x(\Gamma)$ of Γ coming from x. Let P_x, F_x be the past and future light cone of x. If:

$$\forall x \in \Gamma, \quad I_x(\Gamma) \subset P_x, \quad O_x(\Gamma) \subset F_x,$$

then Γ will be called a λ^+ path. If:

$$\forall x \in \Gamma, \quad I_x(\Gamma) \subset F_x, \quad O_x(\Gamma) \subset P_x,$$

then Γ will be called a λ^--path. Hereinafter each path will be assumed to be either a λ^+-path or a λ^--path, except where otherwise expressed.

We call "particle" a function $\omega : (2^\Lambda - \phi) \to C, \Lambda \subset \{\Gamma\}$, such that $\forall \Gamma, \Gamma_1, \Gamma_2 \in \Lambda$:

$$|\omega(\Gamma)| = 1,$$

$$\Gamma = \Gamma_1 + \Gamma_2 \to \omega(\Gamma) = \omega(\Gamma_1)\omega(\Gamma_2),$$

$$\omega(\Gamma_1 \cup \Gamma_2) = \omega(\Gamma_1) + \omega(\Gamma_2),$$

$$\omega(\Gamma^*) = (\omega(\Gamma))^*,$$

where $\Gamma_1 + \Gamma_2$ is the path attained on concatenating Γ_1, Γ_2, and Γ^* is the path attained on inverting the verse of Γ. If $\omega(\Gamma)$ exists, two point events x, y belonging to Γ are said to be "connectable" by that particle.

If more than one particle is defined on Γ, we say the copies x_i, y_i of x, y point events to be connectable by the i-labeled particle. If a point event u exists which is connectable to $M + N$ point events $x_i \in P_u, y_j \in F_u$ by, respectively, M i-labeled and N j-labeled particles, we say the x_i, y_j point events to be connectable by $M + N$ particles interacting in u. As a particular case the indexes i, j coincide ($M = N$) and no interaction actually occurs in u. Let $z(a, b)$ be a λ^+-path having the extremities a, b; the single particle transition amplitude between two point events is easily defined, in the usual Feynman-Hibbs paths method [31], as:

$$\sum_{z(x_i,y_i)} \omega(z(x_i,y_i)) = A(x_i,y_i),$$
$$\sum_{z^*(x_i,y_i)} \omega(z^*(x_i,y_i)) = A^*(x_i,y_i).$$

If no interaction occurs, the total transition amplitude is attained performing a productory on i of these expressions, otherwise it will be expressed as

$$\sum_u \{f\Pi_{i,j}A(x_i,u)A(u,y_j)\},$$
$$\sum_u \{f\Pi_{i,j}A^*(x_i,u)A^*(u,y_j)\},$$

where the factor f depends on the interaction occuring in u. This is the essence of the Feynman graphs' technique, seen as an extension of the paths' method [32].

We remark that in interacting case each particle is free everywhere except in u, the interaction being completely localized there. Therefore, also in the interacting case, a free particle wave function $\psi_i(x_i)$ can be defined in each point event x_i. Similarly, it will be possible to define a free particle wavefunction $\phi_j(y_j)$ in each y_j point event. This conclusion holds even if the $x_i(y_j)$ point events are placed in the past (future) infinity or if they are not simultaneous between themselves. Therefore, we say a "transaction" occurs if a complex number different by zero is defined on each x_i, y_j: The number $\psi_i(x_i)$ is called the "offer," the number $\phi_j(y_j)$ is called the "confirmation"; they will propagate in the spacetime obeying to well-known rules of the paths method. The offer propagates towards the future; the motion along $z(x_i,w)$ transforms ψ_i in a contribution $\omega(z(x_i,w))\psi_i$ to a new wave function χ_i; the value of χ_i in w is the sum of all the contributions of several $z(x_i,w)$.

The confirmation propagates towards the past; the motion along $z^*(w,y_j)$ transforms ϕ_j^* in a contribution $\omega(z^*(w,y_j))\phi_j^*$ to a new wave function φ_j^*; the value of φ_j^* in w is the sum of all the contributions of several $z^*(w,y_j)$. Let us suppose all the $x_i(y_j)$ events occur at the same time $t_A(t_B)$; in such a case is possible to define two global wave functions:

$$\psi(\{x_i\},t_A) = \Pi_i\psi_i(x_i) \rightarrow \chi(\{y_j\},t_B),$$
$$\phi^*(\{y_i\},t_B) = \Pi_j\phi_j^*(y_j) \rightarrow \varphi^*(\{x_i\},t_A);$$

otherwise it will need to multiply each single ψ_i, ϕ_j for the related amplitude factor between t and t_A or t_B before to insert it in these formulas.

Generally speaking, χ and φ cannot be factorized in single particle wave functions except respectively for $t = t_A, t_B$, where we have assumed such a factorization to be possible. These functions are global properties of the transaction process as a whole, like the total torsional angle of a tape is a global, nowhere localized characteristic of that tape.

Nevertheless, the particles participating in the transaction process perform a real propagation of these fields in the spacetime, through the mechanism explained above; in this sense we can say χ and φ are physical nonlocal fields. We observe that:

$$\chi(\{y_j\}, t_B) \not\equiv \phi(\{y_j\}, t_B),$$
$$\varphi(\{x_j\}, t_A) \not\equiv \psi(\{x_i\}, t_A).$$

In fact, ψ and ϕ (that is, the true initial and final states) will be components of, respectively, φ and χ, having the Fourier coefficients:

$$\int \phi^*(\{y_j\}, t_B)\ \chi(\{y_j\}, t_B)\, dy_j = \alpha,$$
$$\int \psi^*(\{x_i\}, t_A)\ \varphi(\{x_i\}, t_A)\, dx_i = \alpha^*.$$

Thus, the initial state is attained by conjugation of a $1/\alpha^*$ fraction of the ψ^* component of the advanced physical field φ^*. The final state is attained by the conjugation of a $1/\alpha$ fraction of the ϕ component of the retarded physical field χ. There are $|\alpha|$ several and equivalent ways to choose each of these fractions (the α, α^* phase being the same for all the possible choices); therefore there are $|\alpha||\alpha| = \alpha\ \alpha^*$ several and equivalent ways to perform a $\psi \rightarrow \phi$ transaction. Thus, $\alpha\ \alpha^*$ is the statistic weight of that transaction; what does it mean? If we know the final as well as the initial state, the transaction is, of course, completely determined. If one of these two states is unknown, we can measure its appearance probability in a series of transactions in which the other state is fixed; the appearance probability of the $\psi \rightarrow \phi$ transaction is proportional to $\alpha\ \alpha^*$. In this picture the quantum indetermination arises in a practical experiment because of the ignorance of final conditions.

If $\psi, \phi = 0$, no transaction occurs; even in such a case, however, the particles which are able to perform that transaction, actually exist on the spacetime (if the boundary conditions remain unchanged). They propagate the $\psi, \phi = 0$ fields with the same mechanism seen above (physical vacuum). Therefore, a sharp distinction must be made between the quanta (quantized amounts of some physical quantity: energy, momentum, etc.) transmitted in an actual transaction and the particles, which are operators acting on the initial and final states. The quanta are vehicled by the particles in an actual transaction, but the existence of the particles does not depend on them. The particles are the materialization of the evolution operator.

3. OBSERVATIONS

(1) The inversion of a path, $\Gamma \to \Gamma^*$, brings on the sign change of its phase, $\omega(\Gamma)$. If this phase is interpretable as a classical action, this means that the product (energy) $\times$ (time) changes the sign. Since the time axis is not inverted, it is the energy that changes sign. Therefore, if the energy is positive for the λ^+-paths, it shall be negative for the λ^--paths (in agreement with Cramer's interpretation). As we have seen in the introduction, this implies a time oriented dynamics of the observables.

(2) Let us consider two spin-1/2 particles in a single state, both coming from a decay event A and successively detected in two spatially separated point events $B_1, B_2 \in F_A$. The transaction probability depends on the coupling, occuring in A, between the advanced field φ^* emitted by the detection events and the offer. Therefore A "knows" the situation in B_1, B_2 (for instance the orientation of the analysers met in flight), and the conditions which permit the derivation of Bell's inequalities [8] no longer hold. Moreover, in B_1, B_2 another coupling occurs between the retarded field χ emitted by A and the confirmations; therefore the detection point events are connected with themselves via A (zigzagging [33]). This connection does not transport any quantum between B_1 and B_2.

(3) The final state ϕ and the final wave function χ are two completely different objects and no "collapse" $\chi \to \phi$ must be postulated. The cat paradox [6] does not take place, because ϕ is defined and therefore two mutually exclusive circumstances are possible: The transaction involving the transmission of a gamma ray from the radioactive nucleus to the Geiger counter actually occurs or not. If it occurs then a quantum is transmitted by the atom to the counter and the cat dies, otherwise the cat survives.

The system cat+counter is preparated in a state $\mid$ live, charged$>$; the evolution operator superposes it to the orthogonal state $\mid$ dead, uncharged $>$. When an observer interacts with the system a transaction occurs having final state $<$ live, charged $\mid$ or $<$ dead, uncharged $\mid$.

This transaction is not the atom-counter transaction which possbilty occured before; thus the observation does not determine the state of the cat.

(4) Let us consider a transaction converting two quanta A, B in two new quanta $C, D : A + B \to C + D$. The point events belonging to the paths on which the particles transporting the A, B, C, D quanta are defined can be connected through particles which do not transport quanta. The transition amplitude will contain these "virtual" particles too. We observe that no violation of the relativity principle necessarily occurs if these particles are defined on paths which are neither λ^+-paths or λ^--paths.

REFERENCES

1. J. A. Wheeler, and R. P. Feynman, *Rev. Mod. Phys.* **17**, 157 (1945); *Rev. Mod. Phys.* **21**, 425 (1949).
2. F. Hoyle and J. V. Narlikar, *Ann. Phys.* **54**, 207 (1969); *Ann. Phys.* **62**, 44 (1971).
3. J. G. Cramer, *Rev. Mod. Phys.* **58**(3), 647 (1986).
4. J. G. Cramer, *Int. J. Theor. Phys.* **27**, 227 (1988).
5. J. G. Cramer, *Phys. Rev. D* **22**, 362 (1980).
6. E. Schrödinger, *Naturwiss.* **23**, 807, 823, 844 (1935).
7. A. Einstein, B. Podolski and N. Rosen, *Phys. Rev.* **47**, 777 (1935).
8. Bell J. S., *Physics* **1**, 195 (1964).
9. L. Fantappie, *Rend. Accad. Italia* **IV**, 1 (1942).
10. C. W. Rietdijk, *Found. Phys.* **10**, 403 (1980).
11. C. W. Rietdijk, *Found. Phys.* **11**, 783 (1981).
12. C. W. Rietdijk, *Ann. Fond. L. de Broglie* **13** (2), 14 (1988).
13. H. P. Stapp *Nuovo Cimento* **29B**, 270 (1982).
14. W. C. Davidon, *Nuovo Cimento* **36B**, 34 (1976).
15. K. V. Roberts, *Proc. Roy. Soc. A* **360**, 135 (1978).
16. J. Rayski, *Found. Phys.* **9**, 217 (1979).
17. R. J. Sutherland, *Int. J. Theor. Phys.* **22**, 377 (1983).
18. N. Hokkio, *Found. Phys. Lett.* **1**, 293 (1988); **2**, 395 (1989).
19. D. J. Hoekzema, *Found. Phys.* **22**, (4), 467 (1992).
20. L. Chiatti, to appear in *Hadr. J.*
21. O. Costa de Beauregard, *C. R. Acad. Sci. (Paris)* **236**, 1632 (1953).
22. O. Costa de Beauregard, *Nuovo Cimento* **42**, 41 (1977).
23. O. Costa de Beauregard, *Nuovo Cimento* **51B**, 267 (1979).
24. O. Costa de Beauregard, *Phys. Rev. Lett.* **50**, 867 (1983).
25. O. Costa de Beauregard, *Found. Phys.* **15**, 871 (1985).
26. O. Costa de Beauregard, in "New Techniques and Ideas in Quantum Measurement Theory," D. Greenberger, ed., *Ann. N. Y. Acad. Sci.* **480**, 317 (1987).
27. O. Costa de Beauregard, in *Bell's Theorem, Quantum Theory and Conceptions of the Universe*, M. Kafatos, ed. (Kluwer Academic, Dordrecht, 1989), pp. 117–125.
28. O. Costa de Beauregard, in *Microphysical Reality and Quantum Formalism*, A. van der Merwe, F. Selleri, and A. Garuccio, eds. (Kluwer, Academic, Dordrecht, 1988), pp. 219–232.
29. O. Costa de Beauregard, *Found. Phys. Lett.* **5**, 291 (1992).
30. O. Costa de Beauregard, in this volume.
31. R. P. Feynman and A. R. Hibbs, *Quantum Mechanics and Path Integrals* (McGraw-Hill, New York, 1965).

32. R. P. Feynman, *Q.E.D.: The Strange Theory of Light and Matter* (Princeton University Press, Princeton, 1985).
38. O. Costa de Beauregard, *Found. Phys.* **17**, 775 (1987).

INTERSUBJECTIVITY, RELATIVISTIC INVARIANCE, AND CONDITIONALS (CLASSICAL AND QUANTAL)

O. Costa de Beauregard

76 rue Murger
77780 Bourron-Marlotte, France

Mutual dependence of probabilities of two physical occurences implies existence of an interaction or causal link between them. Axiomatized along these lines, the formalization of conditionals comes out as identical to that of transition probabilities.

Key words: joint, conditional, and transition probabilities.

1. LOGIC, SPACETIME INVARIANCE, AND PROBABILITY

The contention here is that probability is not the affair of logic and algebra alone. Insofar as *chance occurrences* are *events* observable in space and time, *geometry* is implied also, and so the *invariance* required in the description is exigible of the probability scheme itself.

The *joint probability* of non-independent chance occurrences expresses their *physical interaction.* In saying this are we (as Jaynes [1] would put it) "unduly confusing epistemology and ontology"? Nay: Logical inference is a thought causality following, so to speak, a thought telegraphic wire. For example, if at Berlin a ball is placed in one of two boxes sent the one to Atlanta and the other to Cape Town, the inference drawn by each recipient as to what the other one finds follows the spacetime ABC or CBA zigzag; therefore geometric invariance is required in logical inference no less than it is in physical causality.

So, as emphasized by Accardi [2], the concepts of a *joint* or of a *conditional* probability on the one hand, of a *transition* probability on the other hand, are tied to each other and should be amenable to a common formalization.

As *causality* and existence of an *interaction* are synonymous, the *propagation of causality* should be displayed in the very formalization of probabilities—and it will be so in a geometrically invariant one.

Thus emerges the concept of a *geometric theory of physical probabilities* which, very happily, comes out as identical to the existing one of *transition probabilities.*

Finally, as there is agreement between observers of an event, *intersubjectivity* rather than (questionable) objectivity or than (inappropriate) subjectivity qualifies the *physical* probabilities.

2. THE GEOMETRIC THEORY OF PHYSICAL PROBABILITIES

The well-known Bayes-Laplace expression of the (essentially symmetric) *joint probability* of two non-independent occurrences A and C, in terms of their *inverse conditional probabilities* and of their *prior probabilities* is (dropping for brevity the familiar P symbols)

$$|A) \cap (C| = |C) \cap (A| = |A|C)(C| = |A)(A|C| = |C|A)(A|. \qquad (1)$$

Unexpressed in this formula is a most important information: *geometric invariance* and *action-reaction reciprocity* of causality.

Therefore, rather than two inverse or "relative" conditional probabilities, let us use *one intrinsic reciprocal conditional probability* "of A if C or of C if A:

$$(A|C) = (C|A), \qquad (2)$$

and have *both* priors displayed together in the formula; and, indeed, is not the very epithet "joint" an invitation to do so?

Thus, we replace the expression (1) of the joint probability by that of the (un-normalized) *joint number of chances*

$$|A) \cdot (C| = |C) \cdot (A| = |A)(A|C)(C| = |C)(C|A)(A|. \qquad (3)$$

Identification of the concepts expressed in (1) and (3) is forbidden by normalization constraints. In both cases, (1) and (3), the prior probabilities $|A)$ and $(C|$ we normalize to unity. Then, if in (1) the inverse conditionals are normalized via

$$S_A\,|A|C) = 1, \quad S_C\,|C|A) = 1, \qquad (4)$$

the joint probability comes out as normalized via

$$SS\,|A)(C| = 1. \qquad (5)$$

In (3), instead, we normalize our "reciprocal conditional" via

$$S_A(A|C) = S_C(C|A) = 1; \tag{6}$$

identification of (1) and (3) is unacceptable as entailing $(A|C) = \delta_{AC}$, that is, independence of A and C.

This formalism, which contains the group-generating formula

$$(A|C) = S(A|B)(B|C) \tag{7}$$

(where the summation is over mutually exclusive states) is none other than the one of *physical transition probabilities.*

In a spacetime picture, (7) is the propagators composition formula; $(A|B)$ is the "mutual cross section", $|A)$ and $(B|$ are the "occupation probabilities" (initial one of an initial state, final one of a final state), and $|A)\cdot(B|$ the "dressed collision probability."

Expressed either in a spacetime or a momentum-energy picture, this formalism has "manifest Lorentz invariance." It also has PT invariance à la Loschmidt, *here derived directly from probability reversal.* It also has prediction-retrodiction symmetry; and, for chained events $ABCD\ldots$, it has topological, or "zigzagging causality invariance."

Concatenations with more than two propagators attached to a vertex can be used as guidelines for computing collision probabilities.

Let us illustrate by two examples the "manifestly covariant" and "intersubjective" handling of correlated chance occurrences, and how "conditionality" is built in the transition probability concept.

Two Spacelike Separated Occurrences: In 1987, 15 neutrinos of the swarm emitted by a galactic supernoval explosion were detected in laboratories, some of which are thousands of miles apart. What sense does it make to express the joint probability of two such detections by formula (1)? Quite the contrary, use of formulas (3) and (7) expresses all right, by $(B|A)$ and $(B|C)$, the propagation of neutrinos from the source B to the receptors A and C, and by $|A)$ and $(C|$ the detection efficiencies. Conditionality consists in that the formula holds *if* appropriately adjusted detectors are present at A and C—and it is reciprocal.

Two Timelike Separated Occurrences: What is the joint probability that, having seen a lightning, I will hear the corresponding thunder roar? Bayes' formula (1) gives for it two "inverse" answers, a predictive and a retrodictive one—not severing, however, propagation from detection of the signals. Lightning is not the cause of thunder: both proceed from a common earlier cause, an electric discharge emitting "jointly" photons and phonons.

Having sighted a flash of "summer lightning," I will not hear the thunder, if only because, when the phonons arrive, I will have lost interest in them. And if indoors, at night, I hear the thunder, I will not have seen the lightning. My eye and ear acutenesses are both expressed by the priors; and reciprocal conditionality is expressed as $(A|C) = (C|A)$. Using a pre-field-theoretic idea of "causality," Bayes' formula ignores these points.

In view of the following, strong emphasis is needed on one point of this logic of classical probabilities: The intermediate summations $|B)(B|$ in formula (7) are thought of as ranging over (possible) "real hidden states." *This is a trait inherent in the realistic natural philosophy.*

That a built-in "correspondence" exists between this renovated probability scheme, and the radically novel "wavelike" one invented in quantum mechanics by Born and by Jordan, I deem highly significant.

3. WAVELIKE PROBABILITIES, CPT INVARIANCE, QUANTAL NONSEPARABILITY

Einstein and de Broglie's wave-particle dualism, a Copernican breakthrough, together with its "statistical interpretation" by Born [3] and by Jordan [4], is a wonderful problem-solving toolkit. But as a paradigmatic revolution it causes headaches to the natural philospher.

The Born-Jordan wavelike calculation recipe is: Add partial, and multiply independent, *amplitudes* rather than probabilities. Interpretors quarrel here, as Bitsakis [5] summarizes. One school says this is no revolution at all, that only an appropriate handling of conditionals is needed. The other school, to which I adhere, holds that Born and Jordan's is a *non-Laplacian probability scheme.*

Born's seemingly innocent proposal that the wave's *intensity* expresses the particle's position *probability density* undermines the very concept of a reality. This it does via *wavelike interference*, associated with (strong) *spacetime reversibility.*

Corresponding to the "reversible transition probability" (2) there is the *self-adjoint transition amplitude*

$$< A|C >=< C|A > *. \tag{8}$$

Strong spacetime reversal is cryptically formalized in this Hermitian symmetry. In the spacetime picture, $< A|C >$ is a propagator, the PT reversal of which obviously is

$$PT: \;\; < A|C >\rightleftharpoons< C|A > . \tag{9}$$

Particle-antiparticle exchange can be expressed as

$$C: \quad <A|C> \rightleftharpoons <A|C>*. \tag{10}$$

Thus, the geometric rendering of the *Hermitian symmetry* (8) is the *strong spacetime reversal*: $CPT = 1$.

Corresponding to the "joint number of chances" (3), there is the joint amplitude

$$|A> \cdot <C| = |A><A|C><C|, \tag{11}$$

and corresponding to the generating formula (7) of Markov chains, there is that of Landé [6] chains

$$<A|C>= S <A|B><B|C>. \tag{12}$$

Concatenations with more than two propagators at one vertex are known as "Feynman graphs."

Algebraic *non-separability* and geometric *non-locality*—a paradigmatic revolution—result from the expression of the (reversible) transition probability

$$(A|C) =<A|C><C|A>= |<A|C>|^2; \tag{13}$$

insertion of (12) in it generates cross or "interference" terms. This *interference of probability amplitudes* precludes that the "intermediate sums" over $|B><B|$'s be conceived as over "real hidden states"—as were the classical $|B)(B|$'s.

For this very reason, there is no quantum analog to the one ball and two boxes problem: the logical inference drawn from Atlanta to Cape Town or vice-versa does indeed follow the ABC or CBA zigzag, but at Berlin not a ball, only an Alice-in-Wonderland smile-of-a-ball was hidden!

Wheeler's [7] metaphor for this is the "smoky dragon." For example, a photon prepared as polarized along A and measured as polarized along C has inbetween a "smoky" amplitude $\cos B = \cos A \cos C + \sin A \sin C$, with $B = C - A$. This is easily tested by inserting, and arbitrarily rotating, a birefringent crystal. Clearly, the measured "polarization state" $|C>$ *cannot* have preexisted to its measurement. Thus, "delayed choice" of the angle C does display "retrocausation."

Similarly, in an EPR [8] correlation, linear polarizations measured the one as $|A>$, the other as $|C>$, did not preexist as such in the source B; there was, coiled in B, a (twin mouthed) smoky dragon.

A time-reversed EPR correlation is analogous to a two-slit interference; as one cannot retrodict from which slit came any detected photon, coiled in the sink B waits a (twin tailed) dragon.

These are three topologically equivalent cases, with respective spacetime $<$ (or C), V, or $\wedge$ shapes of the ABC zigzag.

I feel quite sure [9] that the forthcoming "timed EPR experiments" will evidence the zigzagging, reversible, CPT invariant microcausality: Varying independently the AB and BC distances will display the spacetime geography of the ABC zigzag.

Retrocausation, as evidenced in the delayed-choice experiments, does of course not imply that "one can kill his grandfather in his cradle"! Lawlike reversible microcausality is one thing; factlike irreversible macrocausality is something else. Non-separability, nevertheless, is fatal to the concept of a macro-reality.

There have been significant advances in the so-called measurement problem, emphasizing the role of "destructive interference." Zurek [10], and others [11], have shown how deliberate ignorance of those apparatus variables that are uncoupled to the measured magnitudes blurs the off-diagonal terms.

Of course it is trivial [12,13] that assignment of a probability depends upon what one chooses to know or not, to control or not. And, or course, statistical averaging generates no more than the symbol of a *trompe-l'oeil* sort of reality. Here the implication is still more radical.

So, after all, even the grasped tail and the biting mouth of Wheeler's dragon may not "really" hang "down here"! It may be that physical inter-subjectivity, being equally distant from objectivity and subjectivity, has, in the generation of chance occurrence, a more active role than is presently recognized.

4. RECIPROCAL CONDITIONALITY VERSUS HIDDEN REALISM

Let us go back to Einstein's [8] reality criterion: "If, without in any way disturbing a system, one can [exactly] predict [teledict would be more apposite] the value of a physical quantity [by measuring a strictly correlated one], then there must exist a [corresponding] element of reality." The technical tool EPR used was the non-relativistic Schrödinger formalism; the example chosen was position x or momentum p measurement for particles a and b correlated via the commuting operators $x_a - x_b$ and $p_a + p_b$. Instead, we choose here linear polarizations of paired photons, because then the formalism can easily be made Lorentz and CPT invariant.

Experimentation disproves that finding at A the result $|A>$ reveals existence at C of the strictly correlated state $|C>$: Linear polarizations *can* be *independently* measured at A and C, and the amplitude $<A|C>$ of two YES answers is A and C *symmetric*. So, which of the two measurements collapses the other state?

Einstein's realistic assumption thus contradicts the very spirit of quantum physics. What is at stake is a doubly conditional probability

$(A|C)$: *If* the question $|A>$ is put at A, *and if* the question $|C>$ is put at C, *then* the amplitude $<A|C>$ is expressed by formula (7). *Reciprocal conditionality, not "hidden realism", is operational.*

5. REVERSIBLE CAUSALITY AS IDENTIFIED WITH RECIPROCAL CONDITIONALITY

The hint implicit in Laplace's 1774 "Memoir on the Probability of Causes" finds an answer in relativistic quantum mechanics. To such an identification Jaynes [1] objects that it "unduly confuses ontology and epistemology," adding that "in pure deductive logic, if A implies B, not $-B$ implies not $-A$; [but] if we tried to [identify implication with physical causation] "we could hardly accept that not $-B$ is the physical cause of not $-A$." What Jaynes objects to is retrocausation; defining not $-A$ as *observation of not A* (not the weaker "A non-observed") it is clear that, in a dichotomic context, if A causes B, not $-A$ causes not $-B$. So, to falsify Jaynes' statement, it suffices to produce an example where there is both dichotomy and reversibility. Here is one: A low intensity laser beam crosses in succession two birefrigent crystals a and b with parallel axes; by A and not $-A, B$ and not B, we denote the two possible answers.

If the two crystals are independently oriented, although the context is no more deterministic, reversibility still holds. If the beam is very long, there is plenty of time for fixing the orientation at b after a photon has passed a; that is, a delayed choice is possible among *incompatible* questions put at b. This is operational proof that, in the non-Laplacian scheme, retro-inference expresses retrocausation.

In the quoted paper, Jaynes comes very near to accepting (after work done by Gull) that in today's state of the art the only acceptable hidden-variable theories are those where causality is time-reversible, so that "exorcism of the superluminal spook" appears as canonization of the "teleological spook." Let me recall that an (informal) group of theoretists is playing, since quite a few years, with a yet unrefuted model combining these two features [14].

6. TO CONCLUDE

The contention here is that logic and spacetime kinematics are not separable from each other, so that *probabilistic inference* is a symbolic rendering of *causality.*

We have argued that the idea of *conditionality* is implied in the *transition probability* concept of physics, which we propose as an improvement over the Bayesian joint probability concept.

A Landé-style correspondence [3] exists between this scheme and the Born-Jordan wavelike probability scheme of quantum mechanics. Beyond obvious incidences concerning the idea of reality, this may have also significance concerning the handling of probability in general: "The greatest simplicity and elegance is attained when only square integrable functions are admitted." [15]

Chance, as it seems to me, is a *physical* concept. If the basic chance game is the quantum one, then (as Landé also argues) the most advanced form of the "calculus of probabilities" is Born and Jordan's "wavelike" one.

These ideas I have presented a few times [16,17], but two key elements, emphasized in this paper, I did not then make sufficiently clear.

The one has to do with Jayne's [1] radical distinction between "epistemology" and "ontology," which I do question insofar as chance *occurrences* are observable, physical *events*.

The other point having raised questions [18] is that substituing the *transition probability* to the Bayesian *joint probability* modifies not only the formalization, but also the conceptualization.

Comments made in Paris, at the MAXENT 1992 Conference after my presentation, by Skilling, Garrett, Gull, and also by Frohner have been extremely helpful to me regarding these two points.

The philosophy underlying this work is that physics is essentially probabilistic, so that there is equivalence and reciprocity of the *information* and the *negentropy* concepts. Then Wigner's [19] assumption of a two-way mind-matter interaction makes sense. But this I will not delve into here.

REFERENCES

1. E. T. Jaynes, "Clearing up mysteries," in *Maximum Entropy and Bayesian Methods*, J. Skilling, ed. (Kluwer Academic, Dordrecht, 1989), pp. 1–27.
2. L. Accardi, "The probabilistic roots of the quantum mechanical paradoxes," in *The Wave-Particle Dualism*, S. Diner *el al.*, eds. (Reidel, Dordrecht, 1984), pp. 297–330.
3. M. Born, *Z. Phys.* **38**, 803–827 (1926).
4. P. Jordan, *Z. Phys.* **40**, 808–838 (1926).
5. E. T. Bitsakis, "Classical and quantum probabilities," in *The Concept of Probability*, E. T. Bitsakis and C. A. Nicolaides, eds. (Kluwer Academic, Dordrecht), 1989), pp. 335–352; cf footnote 2.
6. A. Landé, *New Foundations of Quantum Mechanics* (Cambridge University Press, Cambridge, 1964), Chap. 6.
7. W. A. Miller and J. A. Wheeler, "Delayed choice experiments and Bohr's elementary quantum phenomenon," in *Foundations of Quantum Mechanics in the Light of New Technology*, S. Kamefuchi *et al.*, eds. (Physical Society of Japan, Tokyo, 1984), pp. 140–152.

8. A. Einstein, B. Podolsky, and N. Rosen, *Phys. Rev.* **47**, 777 (1935).
9. O. Costa de Beauregard, *Found. Phys. Lett.* **5**, 291 (1992).
10. W. Zurek, in *New Techniques and Ideas in Quantum Measurement Theory*, D. Greenberger, ed., *Ann. N.Y. Acad. Sc.* **480** (1986).
11. M. Namiki and S. Pascazio, *Phys. Rev. A* **44**, 39 (1991).
12. J. Bertrand, *Calcul des Probabilities* (Gauthier-Villars, Paris, 1988), pp. 4–5.
13. E. T. Jaynes, *Papers on Probability, Statistics and Statistical Physics*, Rosenkranz, ed. (Reidel, Dordrecht, 1983).
14. O. Costa de Beauregard, *Compt. Rend. Acad. Sci. Paris* **236**, 1632 (1953); *Phys. Rev. Lett.* **50**, 867 (1983). H. P. Stapp, *Nuovo Cim.* **29B**, 270–276 (1982). W. C. Davidon, *Nuovo Cim.* **36B**, 34–40 (1976). K. V. Roberts, *Proc. Roy. Soc.* J. C. Cramer, *Rev. Mod. Phys.* **58**, 845–887 (1986). C. W. Rietdijk, *Found. Phys.* **11**, 783–790 (1981). Sutherland, . N. Hokkyo, *Phys. Rev. Lett.* **1** 293 (1988); **2** 395 (1989).
15. W. Feller, *An Introduction to Probability Theory*, (1966).
16. O. Costa de Beauregard, *Time, The Physical Magnitude* (Kluwer Academic, Dordrecht, 1987), pp. 107–118.
17. O. Costa de Beauregard, "Causality as identified with conditional probability and the quantal nonseparability," in *Microphysical Reality and Conceptions of the Universe*, A. van der Merwe *et al.*, eds. (Kluwer Academic, Dordrecht, 1988), pp. 219–232; "Relativity and Probability, Classical and Quantal," in *Bell's Theorem and Conceptions of the Universe*, M. Kafatos, ed. (Kluwer Academic, Dordrecht, 1989), pp. 117–126.
18. C. I. J. M. Stuart, *Found. Phys. Lett.* **4** 37 (1991).
19. E. P. Wigner, *Symmetries and Reflections* (M.I.T. Press, Cambridge, Mass., 1967), pp. 181–184.

QUANTIZATION AS AN INHOMOGENEOUS WAVE EFFECT

Patrick Cornille

12 rue M. Ravel
74440 Santeny
France

This paper attempts to clarify the origin of quantization defined by the Planck's constant $\hbar$. By introducing the concept of inhomogeneous standing waves, one can derive the Planck-Einstein relation $E = \hbar\omega$ and the de Broglie relation $\boldsymbol{p} = \hbar\boldsymbol{k}$ and explain why quantization can be interpreted as a wave effect.

Key words: quantization, Planck's constant, inhomogeneous wave.

1. THE CONCEPT OF INHOMOGENEOUS WAVE

A homogeneous wave or Fourier mode $\Psi e^{j\varphi}$ with a phase function $\varphi(\boldsymbol{r}, t) = \omega t - \boldsymbol{k} \cdot \boldsymbol{r}$ and $\Psi = 1$ cannot convey information or transport finite energy. Therefore we have to deform the wave in order to encode the wave with information of finite energy. It is well known that there are two ways to deform a wave:

— Amplitude modulation, with $\Psi(\boldsymbol{r}, t) \neq 1$ and $\omega, \boldsymbol{k}$ constant.
— Frequency modulation, where the phase function is given by

$$\varphi(\boldsymbol{r}, t) = \int_{t_0}^{t} \omega(\boldsymbol{r}, t) d\tau - \int_{\boldsymbol{r}_0}^{\boldsymbol{r}} \boldsymbol{k}(\boldsymbol{s}, t_0) \cdot d\boldsymbol{s}, \tag{1}$$

which defines a pure inhomogeneous wave for $\Psi = 1$.

The concept of scalar inhomogeneous wave is a very useful concept, since we can explain special relativity theory [1] by assuming the existence of inhomogeneous standing waves in vacuum. From this concept, we can also derive Maxwell's equations, as shown in a series of papers [2–4]. Moreover, it has been demonstrated that all second-order hyperbolic equations [5–11] do admit soliton solutions, contrary to common belief. However these solutions [10,11] cannot be directly constructed from homogeneous waves and are composed from inhomogeneous waves. Finally, we will show in this paper the necessity of quantization for the deformation of inhomogeneous waves. In fact the concept of inhomogeneous wave is used thoroughly in many area in physics without it being stated plainly, as in the Aharonov-Bohm effect [12], in the Hamilton ray equations [13], in the classical Doppler effect [14], and in the Sagnac effect, as demonstrated by Post [15].

2. CONTINUITY VERSUS DISCONTINUITY

Nature provides us with two kinds of quantities: a first kind which occurs in integer units, and a second kind with a continuous variation. As an example, we can quote the digital/analog description of signals in electronics. We also know that matter and radiation present the same duality, since both can behave either as waves or as particles, depending on the experimental conditions. The wave-particle duality is still the central unresolved problem of modern quantum theory. Now if we want to describe the existence and propagation of matter and radiation in the context of a wave theory, we have to explain how quantization can play a role in such approach.

The method of separation of variables is the most commonly used method to get an explicit solution of some partial differential equations. For example, a general solution of the wave equation in spherical coordinates is of the form

$$e^{j\omega t} Y_n^m(\theta,\varphi) Z_{n+1/2}(kr)/(kr)^{1/2}, \tag{2}$$

where Y_n^m are the spherical harmonic functions and $Z_{n+1/2}$ are the spherical Bessel functions. This solution describes a set of homogeneous waves or modes, characterized by a set of numbers called separation constants, which are the wave number k and the integers m and n. A particular solution of a problem will be obtained through the Fourier method by forming a linear combination of the modes that satisfies the initial conditions and with a selection of the modes that are appropriate to the boundary conditions.

We also know that the energy and the angular momentum of an electron in an hydrogen atom are quantized in a manner which depends on some integer numbers related to the equation above and Planck's constant $\hbar$. Therefore the quantum numbers of the quantum theory are perfectly understandable from a classical wave point of view. However, there remains the question concerning the meaning of the quantization associated with the Planck's constant. Mathematically the wave-particle duality of matter and radiation [16,17] is described by the same Planck-Einstein and de Broglie relations:

$$E = \hbar\omega, \quad \boldsymbol{P} = \hbar\boldsymbol{k} \; . \tag{3}$$

These relations associate the energy and momentum of a particle with the angular frequency ω and wave vector $\boldsymbol{k}$ of a plane wave. The Planck's constant implies a quantization of the energy and momentum of the particle, which is often viewed as some kind of discontinuity in the physical behavior of the particle. This is unexpected, since a wave is a continuous phenomena by definition. A given mode, defined by the variables $\boldsymbol{k}, \omega$ represents what we call an homogeneous wave. By contrast, an inhomogeneous wave is a wave whose variables k, ω depend upon time and space.

If there are space or time symmetries in a problem, the solution will depend on a set of harmonic homogeneous waves characterized by the separation constants. Generally we are interested in a stationary problem, but the question arises: What happens during the time we jump from one stationary state to another? This problem has recently been addressed by Boudet [18,19] for the Lamb-shift calculation, where a time-periodic current is created during the passage from one state to another state. The calculation is done by using the Maxwell theory in an exact manner without renormalization and thus avoiding the occurrence of divergent integrals. Planck's constant is introduced in the source term as a conversion factor at the end of the calculation.

The occurrence of two quantum constants, Planck's constant $\hbar$, indicating the quantization of action, and the electronic charge q, defining the quantization of charge, is one of the puzzles of modern physics. These constants are numerically interrelated in the electromagnetic coupling constant, the so-called fine-structure constant $\alpha = q^2/\hbar c$, which is a pure number. They are at the basis of the three main theories governing the physical world, namely: quantum mechanics, special relativity, and electromagnetism. Therefore, the existence of the fine structure constant indicates that the three theories are deeply connected. We think that these constants define properties of a "waving" vacuum. Indeed, we will demonstrate hereafter that quantization may be interpreted as a wave effect, provided we work with inhomogeneous waves in vacuum.

3. THE WAVE PACKET CONCEPT

In classical physics, we can consider material particles as localized solutions of field equations; however it is believed that linear field equations cannot support continuous non-singular solutions. The concept of wave packet as representing a massive particle is not new, since it was first advocated by de Broglie, who proposed this approach in order to combine the wave and the corpuscular aspects of matter. The origin of matter waves was investigated by de Broglie by considering the uniform motion of a wave packet guided by a pilot wave in the vacuum. However, a difficulty arises since matter waves are dispersive in vacuum, and consequently a wave packet built with such waves will rapidly spread. Therefore, the research concerning a wave packet moving with a constant profile was rapidly abandoned. Note that the present interpretation of quantum mechanics faces the same problem, since a quantum mechanical wave packet for a massive particle disperses while it evolves in time. However, the possibility to use non-singular solutions for representing the physical reality of a localized particle has been recently demonstrated by several authors [5–11,20–22]. They have proposed non-dispersive wave packet solutions of the Klein-Gordon and Dirac equations. This may come as a surprise, since finite-energy soliton solutions are usually associated with nonlinear equations. Most of these wave packets keep a constant profile only for uniform velocity. This is the case for the Barut [20] and Schwinger [21] wave packets, which are solution of a Klein-Gordon equation

$$\Delta\Phi = \frac{1}{c^2}\frac{\partial^2\Phi}{\partial t^2} = k_0^2\Phi \; , \tag{4}$$

which gives a dispersion law of the form:

$$\omega(\boldsymbol{k}) = c\left(\boldsymbol{k}^2 + k_0^2\right)^{1/2} \; . \tag{5}$$

The solution obtained by Barut does not spread and has a finite field energy. However he considered a wave packet built with homogeneous waves consistent with a constant group velocity. For the Schwinger wave packet, the quantity k_0^2 is not constant but depends on spacetime coordinates. Therefore such wave packets cannot be built from homogeneous waves associated with the above dispersion law where the quantity k_0^2 is constant. The approach of Schwinger has been generalized by Beil [22] for an extended particle moving with a non-uniform velocity.

Our approach will however differ from previous work by considering inhomogeneous waves instead of homogeneous waves taken by most authors to build theirs wave packets as a model for a material particle. We postulate that a particle is indeed a wave packet built with classical scalar waves, and we examine what happens to one mode of this wave packet when the particle accelerates. It is interesting to note that Allis and Müller [23] were the first to conceive of an electron as a wave packet built with scalar inhomogeneous waves. More recently, Ginzburg [24, p.325)] proposed a similar approach with Fourier modes $e^{j\varphi}$, with a phase defined by the relation

$$\varphi(\boldsymbol{r},t) = \int_{t_0}^{t} \omega(\boldsymbol{r},\tau)d\tau - \int_{\boldsymbol{r}_0}^{\boldsymbol{r}} \boldsymbol{k}(\boldsymbol{s},t_0)\cdot d\boldsymbol{s} \; . \tag{6}$$

One can show [2–4] that a Fourier inhomogeneous mode yields a dispersion law given by Eq. 5, with the quantity k_0^2 constant, and a Lorentz force $q[\boldsymbol{E} + \boldsymbol{U} \wedge \boldsymbol{B})/c]$, which is zero. However, if we introduce an integrating factor in the phase definition

$$\varphi(\boldsymbol{r},t) = \int_{t_0}^{t} \alpha(\boldsymbol{r},\tau)\omega(\boldsymbol{r},\tau)d\tau - \int_{\boldsymbol{r}_0}^{\boldsymbol{r}} \alpha(\boldsymbol{s},t_0)\boldsymbol{k}(\boldsymbol{s},t_0)\cdot d\boldsymbol{s} , \tag{7}$$

then the dispersion law reads

$$\alpha^2 \left(\frac{\omega^2}{c^2} - \boldsymbol{k}^2\right) = k_0^2 = \alpha^2 k_1^2 , \tag{8}$$

and the Lorentz force $q[\boldsymbol{E} + (\boldsymbol{U} \wedge \boldsymbol{B})/c]$ becomes different from zero.

4. QUANTIZATION

It is a well-known fact that in the equation of motion of an electron in a solid, only external forces are considered as applied forces to the wave packet. The force which is originated by the lattice periodic field remains hidden in the so-called electron effective mass dyadic. The electron wave packet obeys a law of motion that is given by the equation

$$\overleftrightarrow{M} \cdot \frac{d\boldsymbol{U}}{dt} = \boldsymbol{F} , \tag{9}$$

where $\overleftrightarrow{M}$ is the effective mass and $\boldsymbol{F}$ the external force. The wave packet velocity $\boldsymbol{U}$ equals

$$\boldsymbol{U} = \frac{1}{\hbar}\frac{\partial E}{\partial \boldsymbol{k}} , \tag{10}$$

with $E(\boldsymbol{k})$ the electron energy in terms of the wave vector $\boldsymbol{k}$. The components of the mass dyadic are

$$M_{ij}^{-1} = \frac{1}{\hbar^2}\frac{\partial^2 E}{\partial k_i \partial k_j} . \tag{11}$$

In vacuum, the motion of a massive particle, with a rest mass m_0, subjected to a Lorentz force $\boldsymbol{F}$ is described by the relativistic dynamic equation

$$\frac{d}{dt}(m_0 \gamma \boldsymbol{U}) = \boldsymbol{F} , \tag{12}$$

with the definition $\gamma = (1 - U^2/c^2)^{-1/2}$, which gives the following relation:

$$\frac{d\gamma}{dt} = \frac{\gamma^3}{c^2}\frac{d}{dt}\left(\frac{U^2}{2}\right) . \tag{13}$$

By using the preceding relation, we can rewrite Eq. 10 in the following dyadic form:

$$\overleftrightarrow{M} \cdot \frac{dU}{dt} = F \quad \frac{dU}{dt} = \overleftrightarrow{M}^{-1} \cdot F \,, \tag{14}$$

with the direct and inverse mass dyadics given by the definitions

$$\begin{aligned} \overleftrightarrow{M} &= m_0\gamma \left(\overleftrightarrow{I} + \frac{\gamma^2}{c^2} UU \right), \\ \overleftrightarrow{M}^{-1} &= \frac{1}{m_0\gamma} \left(\overleftrightarrow{I} - \frac{1}{c^2} UU \right), \end{aligned} \tag{15}$$

where $\overleftrightarrow{I}$ denotes the unit dyadic. Therefore, the relativistic mass dyadic of a moving electron in vacuum indicates that the electron is acted upon by an applied force from the oscillating vacuum. This internal force can be understood as a resistive force arising from the vacuum.

In an interesting paper, Crawford [25] derives the de Broglie relation from the Doppler effect by considering the elastic collision of an electron with a massive mirror. In this approach, Crawford brought out the concept of inhomogeneous wave without saying so specifically. Our approach to deduce the relativistic de Broglie relation will differ from Crawford's calculation by our reasoning applying directly to the wave concept, independently of any conservation law of energy and momentum. By definition, the group velocity of a wave satisfying the dispersion law 8 is

$$U = \frac{\partial \omega}{\partial k} = c^2 \frac{k}{\omega} \,. \tag{16}$$

It follows immediately that $\gamma = \omega / ck_1$, which imply the identity $\gamma U = ck/k_1$, therefore the identification $\hbar = m_0 c/k_1$ does not justify that the quantity $\hbar$ is a constant since it depends on two independent parameters m_0 and k_1. The preceding calculation cannot be taken as a demonstration of the Planck and de Broglie relations 3, since such a calculation cannot justify the necessity of the Planck's constant $\hbar$. Therefore we proceed with our demonstration by calculating the quantity

$$\frac{\partial U_i}{\partial k_j} = \frac{c^2}{\omega} \left(\delta_{ij} - \frac{1}{c^2} U_i U_j \right) \,. \tag{17}$$

By definition, we have $dU_i = \sum_j \frac{\partial U_i}{\partial k_j} dk_j$ and hence

$$dU = \frac{c^2}{\omega} \left(\overleftrightarrow{I} - \frac{1}{c^2} UU \right) \cdot dk \,. \tag{18}$$

Now, using the relation 13, we get

$$d(\gamma \boldsymbol{U}) = \gamma \left(\overleftrightarrow{\boldsymbol{I}} + \frac{\gamma^2}{c^2} \boldsymbol{U}\boldsymbol{U} \right) \cdot d\boldsymbol{U} \ . \tag{19}$$

The substitution of Eq. (18) into Eq. 19 gives

$$d(\gamma \boldsymbol{U}) = \frac{\gamma c^2}{\omega} d\boldsymbol{k} \ . \tag{20}$$

With the definition given by Eq. 16, we can rewrite the preceding equation in the following form:

$$d(\gamma U_i) = \frac{\gamma c^2}{\omega} dk_i = \gamma U_i \frac{dk_i}{k_i} \tag{21}$$

Therefore we get the differential equation $d\,[\mathrm{Log}(\gamma U_i/k_i)] = 0$, whose integration gives the relation $\gamma \boldsymbol{U} =$ (constant) $\boldsymbol{k}$. We can use the relation $\gamma U = ck/k_1$ to prove that the constants Ct_i are the same, it follows the identity $\gamma U = Ctk$ where the integrating constant Ct does not depend on k_1 and can now be chosen equal to $\hbar/m_0$ in order to recover de Broglie's formula which shows the quantization relation $m_0 \gamma \boldsymbol{U} = \hbar k$. The above differential equation implies the quantization of velocity during the acceleration process, which is confirmed by Jennison's experiment [26].

To obtain the second-quantization relation, it suffices to calculate the variation of the kinetic energy with respect to ω, which is given by the equation

$$\frac{d}{d\omega} \frac{U^2}{2} = \frac{d}{d\omega} \frac{c^4 k^2}{2\omega^2} \ . \tag{22}$$

We therefore get a first equation:

$$\gamma^3 d \left(\frac{U^2}{2} \right) = \gamma c^2 \frac{d\omega}{\omega} \ . \tag{23}$$

Relation (13) gives a second equation:

$$\gamma^3 d \left(\frac{U^2}{2} \right) = d(\gamma c^2) \ . \tag{24}$$

We can subtract the two preceding equations to obtain the differential equation $d\,[\mathrm{Log}(\gamma c^2/\omega)] = 0$, which implies the second quantization relation $m_0 \gamma c^2 = \hbar\omega$.

We note that the Planck-Einstein and de Broglie relations obtained in the preceding calculation can now be used to recover the well-known formulas of solid state physics: For example, Eq. 11 leads to the inverse mass dyadic of a particle, which can be rewritten in the form

$$M_{ij}^{-1} = \frac{1}{\hbar}\frac{\partial^2\omega}{\partial k_i \partial k_j} = \frac{c^2}{\hbar\omega}\left(\delta_{ij} - \frac{1}{c^2}U_i U_j\right) . \tag{25}$$

The second quantization relation is consistent with the above formula, which allows one to obtain the definition (15). The analogy between the vacuum and solid-state physics is a fruitful concept, which has also been used to explain the Dirac sea of electrons by regarding the vacuum as a close analog of a semi-conductor [27], with two bands separated by a $2m_0c^2$ gap.

Now let us postulate that k and ω are implicit functions of time, then our wave packet is built with inhomogeneous waves, and the acceleration of the wave packet implies that the waving medium is dispersive, since we have

$$\frac{dU_i}{dt} = \sum_j \frac{\partial^2\omega}{\partial k_i \partial k_j}\frac{dk_j}{dt} \neq 0 . \tag{26}$$

The acceleration of the particle will be different from zero only if the matter waves are dispersive. The quantization relations we have obtained are precisely the Planck-Einstein and de Broglie relations, provided we use the well-known formula of ray theory:

$$\boldsymbol{U} = \frac{d\boldsymbol{r}}{dt} = \frac{\partial\omega}{\partial\boldsymbol{k}} . \tag{27}$$

This equation shows that the group velocity of matter waves is just equal to the velocity of the particle whose motion they govern. This indicates that de Broglie's postulate is internally consistent. However, it is not well known that the equality (27) is not verified in the general case, as shown by Molcho [13].

We therefore have found the Planck-Einstein and de Broglie relations by considering the acceleration of a material particle represented by inhomogeneous standing waves. The quantization results from the fact that the deformation of a wave cannot be of any form if we want to preserve the oscillating behavior of the inhomogeneous wave, implying the existence of operators of the form $d\,[\mathrm{Ln}(A * B)] = 0$ or $d\,[\mathrm{Ln}(A/B)] = 0$, which give the quantizations of the variable A with respect to B, namely: $A * B = C^t$ or $A = C^t B$, where the Planck constant $\hbar$ is a mere integration constant.

Since the quantum $\hbar$ is dimensionally an energy multiplied by time, it can be associated with the time deformation of the waves and therefore with motion, while the quantum q^2, being dimensionally an energy multiplied by length, can be related to the space deformation of the waves. It has been suggested recently by Wolter *et al.* [28] that the quantum Hall effect in solid state physics might have its origin in the spatial inhomogeneity of the electron density across the sample. The concept of inhomogeneous waves provides us with a continuous wave theory on a microscopic space-time scale which may appear discontinuous at a macroscopic space- time scale with the quantum constants $\hbar$ and q as the links between the two scales in order to allow a minimum space or time for the propagation and deformation of these waves.

5. CONCLUSION

We have been able to deduce the Planck-Einstein and de Broglie relations as the result of a wave effect if we assume that wave packets are constructed from inhomogeneous standing waves in vacuum. The similarity of the quantization between classical scalar waves and quantum mechanical waves may suggest a new possible interpretation of quantization.

The approach defined in this paper concerning quantization has brought the concept of inhomogeneous wave to light, which is a very important concept in physics, since we succeeded in deriving Maxwell's equations [2–4] from it and demonstrated the link with matter waves [1,11].

6. REFERENCES

1. P. Cornille, "On the meaning of special relativity in the earth frame," *Physics Essays* **5**, 262 (1992).
2. P. Cornille, "On the propagation of inhomogeneous waves," *J. Phys. D* **23**, 129 (1990).
3. P. Cornille, A. Lakhtakia, ed., *Essays on the Formal aspects of Electromagnetic Theory*, (World Scientific, Singapore, 1993).
4. P. Cornille, "Wave hydrodynamics of matter and radiation," submitted for publication.
5. I.M. Besieris, A.M. Shaarawi, and R.W. Ziolkowski, "A bidirectional traveling plane wave representation of exact solutions of the scalar wave equation," *J. Math. Phys.* **30**, 1254 (1989).
6. A.M. Shaarawi, I.M. Besieris, and R.W. Ziolkowski, "A novel approach to the synthesis of nondispersive wave packet solutions to the Klein-Gordon and Dirac equations," *J. Math. Phys.* **31**, 2511 (1990).
7. P. Hillion, "Nondispersive solutions of the Klein-Gordon equation," *J. Math. Phys.* **33**, 1817 (1992).
8. P. Hillion, "Nondispersive solutions of the Dirac equation," *J. Math. Phys.* **33**, 1822 (1992).

9. P. Hillion, "The Courant-Hilbert solutions of the wave equation," *J. Math. Phys.* **33**, 2749 (1992).
10. P. Hillion, "Approximation of the scalar focus wave modes," *J. Opt. Soc. Am.* **9**, 137 (1992).
11. P. Cornille, "Is the physical world built upon waves?" to appear in *Physics Essays* **6** (1993).
12. S. Olariu and I.I. Popescu, "The quantum effects of electromagnetic fluxes," *Rev. Mod. Phys.* **17**, 339 (1985).
13. J. Molcho and D. Censor, "A simple derivation and an example of Hamiltonian ray propagation," *Am. J. Phys.* **54**, 351 (1986).
14. P. Cornille, "La quantification est-elle un effet ondulatoire?" *Ann. Fond. L. de Broglie* **18**, 295 (1993).
15. E.J. Post, "Sagnac effect," *Rev. Mod. Phys.* **39**, 475 (1967).
16. L. De Broglie, *Tentative d'interprétation causale et non linéaire de la mécanique ondulatoire* (Gauthier-Villars, Paris, 1956).
17. L. De Broglie, *La réinterprétation de la mécanique ondulatoire. Collection discours de la méthode* (Gauthier-Villars, Paris, 1971).
18. R. Boudet, "The role of Planck's constant in Dirac and Maxwell theories," *Ann. Phys.*, supplement to **14**(6), 27 (1989).
19. R. Boudet, "La théorie classique du champ et le décalage de Lamb," *Ann. Fond. L. de Broglie*, **14**, 119 (1989).
20. A.O. Barut, "$E = \hbar\omega$," *Phys. Lett. A* **143**, 349 (1990).
21. J. Schwinger, "Electromagnetic mass revisited," *Found. Phys.* **13**, 373 (1983).
22. R.G. Beil, "The extended classical charged particle," *Found. Phys.* **19**, 319 (1989).
23. W.P. Allis and H. Muller, "A wave theory of the electron," *J. Math. Phys.* **VI**, 119 (1927).
24. V.L. Ginzburg, *The propagation of Electromagnetic Waves in Plasmas*, (Pergamon, New York, 1970).
25. F.S. Crawford, "Derivation of the de Broglie relation from the Doppler effect," *Am. J. Phys.* **50**, 269 (1982).
26. R.C. Jennison, "What is an electron?," *Wireless World* **6**, 42 (1979).
27. R.E. Prange and P. Strance, "The semiconducting vacuum," *Am. J. Phys.* **52**, 19 (1984).
28. R. Wolter *et al.*, J. P. Andre, *Europhys. Lett.* **2**, 149 (1986).

IN QUEST OF DE BROGLIE WAVES

J. R. Croca

Departamento de Física
Faculdade de Ciências, Universidade de Lisboa
Campo Grande, Ed. C1, 1700 Lisboa, Portugal

An overall view of the problematic and work done recently on the waves of de Broglie is presented. In addition an improved experiment is discussed. This new experiment may, if the experimental requirements are met, clear the longstanding problem of the existence or non-existence of de Broglie waves.

Key words: quantum waves, wave-particle dualism, de Broglie's waves, two-photon interferometry, foundations of quantum mechanics.

1. INTRODUCTION

Since the birth of quantum mechanics de Broglie and many others supposed that some kind of physical reality must somehow be related with the quantum waves. Still only after the theorem of von Neumann was shown inconsistent with the physical reality it was possible to build a true causal model for the quantum particle.

The model proposed by de Broglie[1] includes into a beautiful causal synthesis both the localized corpuscle effect and the undulatory wave extended properties of the quantum particle. In this picture what is usually called a quantum particle must be thought as a kind of an extended, yet finite, wave with almost no energy plus a well localized singularity practically with all the energy of the particle. It is also assumed that the fundamental property of these waves is the aptitude of guiding the singularity through a nonlinear process to the points of higher wave intensity. Within the framework of de Broglie theory almost all quantum experimental phenomena can be explained in a causal way.

Still the problem of deciding which is the good theory for explaining and predicting new phenomena is not an easy one. As is well known, the non causal explanation proposed by Bohr and followers stands as the one accepted by the great majority of the scientific community. Yet some people, even if they adopt it as a kind of working tool, find it hard to accept the implicit epistemological consequences for two principal reasons:

One results from a certain discomfort in accepting the statement that the usual quantum theory is a definite complete theory. If that is the case, then we have reached, at the fundamental level, the maximum allowed to the human capacity. What is left, using the words Lord Kelvin expressed about a century ago, relative to classical physics, is *better determination of the value of the constants and the development of the practical applications.* Nevertheless, as is also well known, classical physics was not a dead end for the human potentialities of understanding nature as was, at those times, supposed.

The other reason is related with the problem of causality. Is it true or not that nature, at the basic level of understanding, can not be comprehended in terms of the concepts of space and time? The fact that we have been used to those concepts does not necessary imply that they are no more than mere auxiliary tools we have been employing for too long time. So, a more complete theory should rule them out of the picture. Even if that statement is accepted in line of principle, there still remains the problem of knowing if, at the quantum level, the usual concepts of space and time ceased already to be useful.

With the aim of clarifying the problem of whether at the quantum level causality must be given up, many attempts have been made to propose and make experiments that could test the two interpretations. The attack followed two main lines: One connected with the EPR paradox and Bell's theorem that shall not be discussed here. The other related with the existence or non-existence of the real quantum waves proposed by de Broglie.

Some experiments to test the existence of de Broglie waves have been made mainly by Mandel[2] and his collaborators at Rochester. Still due to the fact that the apparatuses used in the experiments are real, that is imperfectly described by the theory, one never, till now, reached the vicinity of the ideal situation required in the different experimental proposals so that they could be conclusive. Therefore the quest for the waves of de Broglie continues.

Situations of this kind are not new in science. One has just to remember the heliocentric model for the solar system. This model, as it is known, proposed by Aristarchus of Samos as early as about 310 B.C. was later retaken by Copernicus. One big problem with that model was the prediction of the existence of the stellar parallax, a very small effect, that after Copernicus took about three hundred hears to be put into experimental evidence. The search for the stellar parallax was a real saga in the history of experimental physics. Only in the second quarter of the nineteen century both theoretical and experimental techniques were good enough for revealing it.

Mostly all experiments proposed with the aim of detecting the waves of de Broglie are based in two complementary approaches. These processes are based on one of the basic postulates of usual quantum mechanics, the reduction of the wave function by a measure. Assuming that the wave function is a probability wave, then if the initial wave ψ is split into many waves by a physical apparatus so that one has $\psi = \psi_1 + \psi_2 + \cdots + \psi_n$, when the actual measure is done the original wave collapses to one of these function $\psi \longrightarrow \psi_k$. After the measurement, the other waves corresponding to the different probabilities vanish into nothing. If, following de Broglie, it is assumed that the quantum waves are real, then after the measurement they still keep going on being real, supposing that no physical action is exerted upon them.

2. AUTO-COLLAPSE OF THE WAVE FUNCTION

This method was proposed by Koh and Sasaki[3] for material particles, but the principle can be generalized to any quantum particle. Consider the situation in which the by means of a physical device, say a Mach-Zehnder interferometer, the function ψ is split into two waves ψ_1 and ψ_2 at the first beamsplitter, following each independent well separated paths, that overlap only at the second beamsplitter. The idea of those authors is that the fringe visibility may decrease as the length of the interferometer arms increases. This assumption can perfectly well be interpreted in terms of the Broglie pilot wave theory[4] as shown in Fig.1.

It is assumed that the empty wave θ going along one of the two possible independent paths as it interacts with the subquantum medium pro-

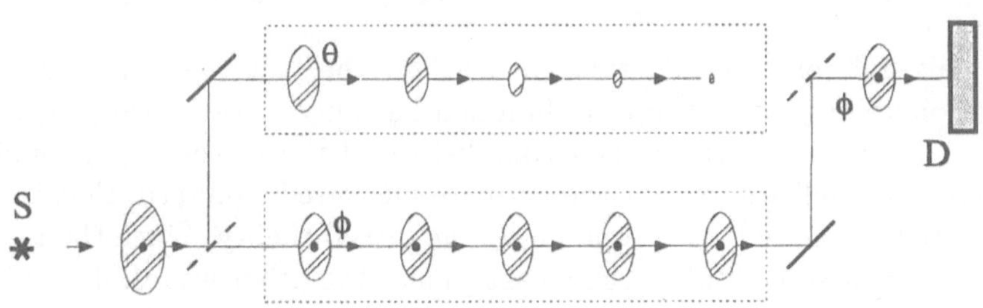

Fig.1. Koh-Sasaki experiment interpreted in the light of the Broglie waves. If the interferometer arms are large enough the intensity of the empty wave going along one of the arms decreases so that at the second beamsplitter only the full wave arrives, therefore no interference is seen.

gressively looses its amplitude till it vanishes completely. The full wave ϕ on the contrary does not disappears because the singularity it carries keeps regenerating its wave all the time. In such conditions at the overlapping region of the interferometer only the full wave ϕ arrives, so no interference pattern is to be seen.

For the usual interpretation no change in the fringe visibility is to be expected, for any increase or decrease of the interferometer's arms, as long as it is kept stable.

Naturally the results of that experiment, implying a kind of autoreduction of the wave packet, depend on the validity of the assumed assumptions. In what concerns photons, Fransom's[5] experiment done with an interferometer with the arms with a length of about 45 m, seems to indicate that, at least the photonic empty waves can travel that distance without change in its intensity. De Broglie assumed that a free photonic empty wave could travel kilometers without significant loose of intensity. Jeffers and Sloan[6], have recently done a very interesting experiment, still due to the fact that the source they used was not monophotonic the results are not conclusive. Nevertheless in an interference experiment with electrons, done by Mölenstedt and Bayth[7], with a biprism, fringes were reported to a maximum length of 52 cm, which may be an indication that after that distance the free electronic empty wave looses practically its intensity.

3. COLLAPSE OF THE WAVE FUNCTION BY THE MEASUREMENT

In the previous proposed experiment there is a kind of autoreduction of the wave function without any measurement which is not allowed by the current theory. Now I shall discuss other possible experiments based on the opposite situation.

This different method takes in account the fact that the usual interpretation assumes that the wave function has only a statistical probabilistic nature. So when a measure is accomplished all the waves corresponding to different possibilities, except the actually measured, collapse, that is vanish completely. Differently is proposed by the causal theory. Since the reality of the waves is assumed, after the measurement the other waves, if no physical action is practiced upon them, still keep existing.

The overall method can be summarized in two steps as follows:

- Production of the θ empty waves.
- Detection of the empty waves.

3.1. Production of Empty Waves

For the construction of a device able to produce empty waves, also called an empty wave generator EWG, it is necessary to dispose of a source emitting quantum particles one at a time. Without sources of that kind it is not possible to make an empty wave generator. For most quantum particles the problem can be considered experimentally solved. In the case of monophotonic sources the devices built by Mandel[8], Aspect[9] and others meet the criteria. For the case of massive particles the problem seems easier. For instance the Rauch team's experiments[10] confirm the possibility of making single neutron sources.

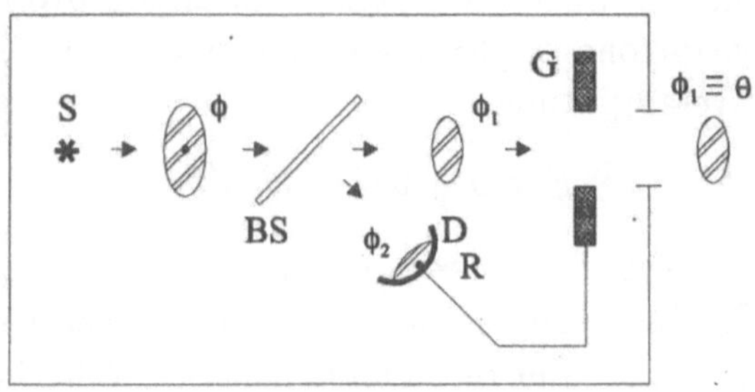

Fig.2. Empty wave generator. When the singularity is detected at D_R the gate G is opened for a short time to allow the wave θ to leave.

Consider Fig.2 where a monoparticle source S emits quantum particles represented by the function $\psi \equiv \phi$ which upon impinging on the beamsplitter Bs is divided into two wave packets ϕ_1 and ϕ_2, each following well separated trajectories. When the detector D_R placed in the path of the wave ϕ_2 is triggered, this means, according to the current interpretation, that the particle chose that particular path. In such conditions, if the wave function is considered only a probability wave, the probability of being at the other path turns to zero.

Quite different is the answer given by de Broglie school. As stated before, in this theory, the quantum particle is composed of a real wave plus the singularity. So that when the wave strikes the beamsplitter gives origin to two real waves one of them carrying the singularity. Whenever the detector D_R is triggered that means that a wave plus a singularity followed that particular path. At the other pathway travels a wave devoid of singularity, unaffected by the far-away interaction corpuscle detector. The detector D_R is electronically connected to the fast gate G which opens for a very short time, just sufficient, to allow the empty wave $\phi_1 = \theta$ to pass. Had the singularity followed the other way, at the detector only the a wave missing the singularity without sufficient energy to activate the detector would arrive. In that case no signal for opening the gate would be sent.

In such conditions the apparatus sketched in Fig.2 behaves as a generator of empty waves in the de Broglie model, while in the current interpretation it produces nothing. Now the problem is to find a way to make sure if something really leaves it.

3.2. Detection Based on the Guiding Principle

The only property attributed by de Broglie to the empty wave is the ability of guiding the singularity, through a non linear process, preferentially to the regions of higher wave intensity. That is the probability of localizing the singularity, for one single singularity, is proportional to the square modulus of the wave amplitude

$$P(x, x + dx) \propto |\theta|^2 dx, \tag{1}$$

where θ represents the total wave resulting from the overlapping of the waves in which the single singularity is immersed. It is also assumed that those waves may came from the same or different sources. Whether the waves, that overlap, are coherent or not has nothing to do with the guiding principle. If the overlapping waves are coherent, the position of the interference fringes for each total wave, arriving one at a time at the detection region, is constant in time. So in a long run of similar experiments the singularities shall be distributed according to the same interference pattern. The distribution of those singularities, the only that can be seen by the common detectors, shows therefore a stable interference pattern. In the opposite case, when the overlapping waves are not coherent, the fringe position for each arriving total wave packet may change randomly from case to case. So at the end no observable interference pattern for the singularities is to be seen.

In the guiding principle was inspired the first proposal of experiment by J. and M. Andrade e Silva[11] that was later developed into a feasible experiment by the author[12]. The idea can be understood with the aid of a modified Young type interferometer where under the usual conditions, at the detection region, an interference pattern with a visibility nearly one is seen. When an empty wave, produced by the empty wave generator, is mixed with the other two waves in the interferometer, see Fig.3, if the above assumptions about the reality of the empty waves θ and of the guiding principle are correct, the interference patter must change.

In those proposals the two independent sources, S_1, common source, and S_2 source of empty waves θ, are incoherent. In such conditions, following de Broglie, at the detection zone the total wave is the sum of the three waves, two coherent common waves ϕ plus the incoherent empty wave θ:

$$\phi_t = \phi_1 + \phi_2 + \theta. \tag{2}$$

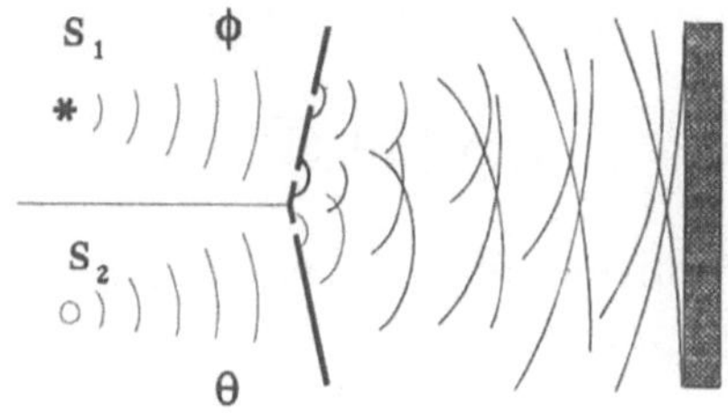

Fig.3. Schematic representation of an experiment for detection of the empty waves θ. When the empty wave is mixed with the other waves the interference pattern changes.

Therefore, following the guiding principle, the expected causal intensity I_c shall be proportional to the total wave intensity:

$$I_c \propto |\phi_1 + \phi_2 + \theta|^2 \tag{3}$$

or

$$I_c \propto |\phi_1 + \Phi_2|^2 + |\theta|^2, \tag{3'}$$

because the two sources are incoherent. For the case of equal wave amplitude one gets

$$I_c \propto (1 + \frac{2}{3}\cos\delta), \tag{4}$$

δ being the relative phase shift between waves ϕ_1 and ϕ_2.

If the empty waves do not exist, as stated by the usual interpretation, then the expected intensity I_u shall be given by

$$I_u \propto |\phi_1 + \phi_2|^2, \tag{5}$$

that is,

$$I_u \propto (1 + \cos\delta). \tag{6}$$

The expected visibility is unity, whereas the causal prediction states that it should be 2/3. This implies that the presence of the empty wave is manifested by the addition of a kind of noise smoothing down the interference pattern.

If the two sources are coherent[13] or quasi-coherent[14], then the principle of the experimental setup would be simpler, see Fig.4. In this case, the usual theory predicts no fringe pattern since only one wave is present:

$$I_u \propto |\phi|^2, \tag{7}$$

while the causal intensity depends also on the empty wave θ:

$$I_c \propto |\phi + \theta|^2. \tag{8}$$

Taking, as usual, the amplitude of the waves equal, this formula leads to

$$I_c \propto (1 + \cos\delta). \tag{9}$$

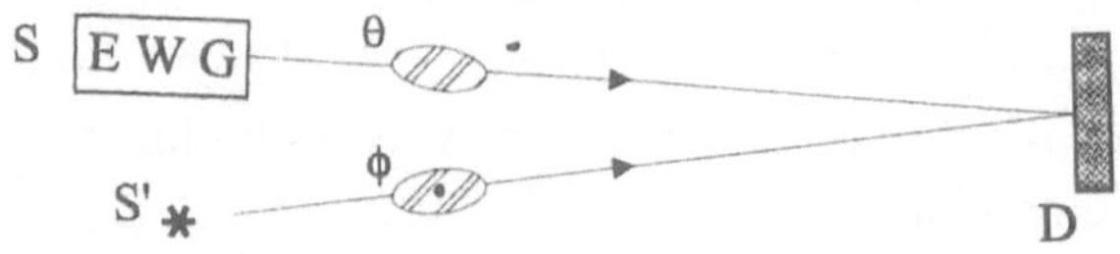

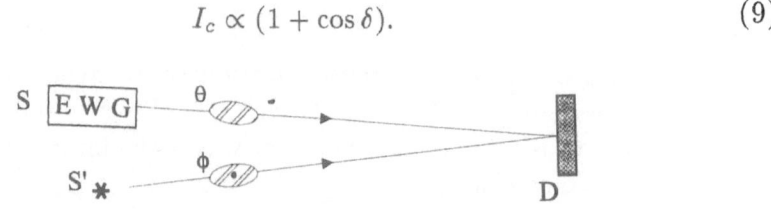

Fig.4. Setup for coherent detection of empty waves. S, source of empty waves, S' source of common waves, the two sources are coherent or quasi coherent,

Many possible experimental setups, based in those assumptions, have been presented[4,15], being all them variants of same principles. Recently Mandel and his coworkers, have carried auto a delicate experiment based on a concrete proposal of experiment[16], presented by J.R. Croca, A. Garuccio, V.L. Lepore, and R.N. Moreira, to test the eventual existence of the empty waves. Their experimental apparatus[2] being slightly different from the one proposed in Ref.16, but the principle being essentially the same. The theoretical expected visibility for the causal theory was 50%, whereas the usual theory predicted no interference. According to the authors of the experiment, the results are not favorable to the existence of the empty waves. Still, when one looks at an interference experiment[17] made by Mandel and his students with similar equipment, temporal resolution interference filters, and so, one sees that, while the theoretical visibility was 100%, they got in the best case only 30%. This significant reduction of the fringe visibility was justified on the basis that in a real experiment the light is not monochromatic, the apparatuses are not perfectly aligned, there are always random instabilities, imperfect overlapping of the different wave packets and many other unknown causes. If the same kind of argumentation can be applied to the experiment they performed for detection of the empty waves the observed visibility would not for sure be the ideal theoretical 50%, but only about 15% . A close look at the results[2] indicates that, due to the poor statistics of the experiment (6 counts per second), a visibility of that order of magnitude is not completely out of question. This means that, if the above remarks have some validity, the experiment cannot be considered conclusive and therefore do not rule out the possibility of the existence of de Broglie waves.

The main problem with that experiment lies in the fact that the resolution time of the system, electronics and detectors, was of about 10^{-8}s, whereas the coherence time of the wave packets about 10^{-13}s. That is, in the minimum resolution time of the system it is possible to got approximately 10^5 photonic wave packets side by side. In an experiment of this kind which depends so critically on the overlapping of incoherent waves, if one wants to draw secure conclusions, it is necessary to guarantee a reasonable degree of average overlapping of the respective wave packets. With the aim of settling those experimental objections a proposal of new experiment is presented.

In this experiment, instead of interference filters, that do not allow the narrow bandwidth required, it is convenient to use Fabri-Perot etalons, so that the coherence time of the wave packets can be made of the same order of magnitude of the resolution time of the detection system of about 10^{-9}s. Since this requirement implies a decreasing in the number of counts, more efficient detectors and more intense sources must be used. Meeting those experimental requirements is extremely important if one really wants to clear this problem.

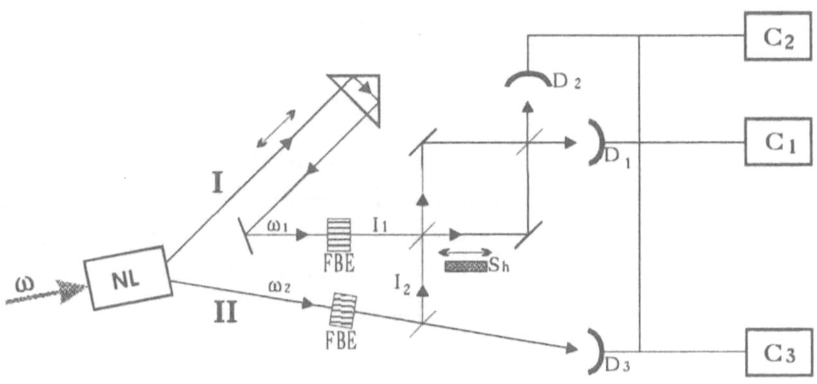

Fig.5. Outline of the experimental setup for detection of the empty waves. FPE, Fabri-Perot Etalon with a bandwidth of about 10^{-9} s.

The proposed experiment, shown in Fig.5, is a modification of one[18] presented some time ago. NL, represents a non-linear parametric-down converter crystal, where an U.V. pump beam of frequency ω is injected generating pairs of photons of frequency ω_1 and ω_2, so that $\omega = \omega_1 + \omega_2$, the Fabri-Perot etalon FPE, of a bandwidth of about 10^{-9}, is set for $\omega_1 = \omega_2$. The Mach-Zehnder interferometer is set so that when the beam I_2 is blocked all light goes to detector D_1, if I_1 obstructed the light from II goes to D_2. Also the intensities from the two beams, I_1 and I_2, entering the interferometer are fixed to be equal ($I_1 = I_2 = I_0$). After having set the interferometer according to the above specifications, (C_1, C_3) and (C_2, C_3) must be put in coincidence, so that one looks for the output of the interferometer at C_1 and C_2 only when the photon from II is revealed at D_3. In such circumstances no light from II penetrates the interferometer; all photons go to D_3, only the empty waves, if they exist, may have a chance to enter the interferometer.

The output of the interferometer, at C_1 and C_2, predicted by the usual theory is independent of the shutter S_h position, whether it blocks I_2 or not and is given by

$$I_u^1 = I_0, \quad I_u^2 = 0. \tag{10}$$

In the causal theory, the empty wave goes through the interferometer, mixing with the common waves from path I. The net effect of this mixing is a change in the distribution so that the expected intensity is[18,19]

$$I_c^1 = \frac{1}{2} I_0, \quad I_c^2 = \frac{1}{2} I_0. \tag{11}$$

The presence of the empty wave in the interferometer changes the probability of the output of the photons from path I. The photons instead of going only to C_1 go also to C_2. Naturally those predictions are valid only if there is a constant total overlapping of the wave packets. To take into account the possibility of imperfect overlapping Eqs. (11) must be written

$$I_c^1 = (1 - \frac{1}{2}\varepsilon) I_0, \quad I_c^2 = \frac{1}{2}\varepsilon I_0, \tag{12}$$

where ε, with $0 \leq \varepsilon \leq 1$, is the average overlapping factor of the wave packets from paths I and II. If there is no overlapping ($\varepsilon = 0$), the causal relations (12) transform into the usual ones (10). For a perfect overlapping ($\varepsilon = 1$), they change into (11).

It is convenient to present the experimental results in a way that best show the possible effect of the empty waves. So let us define Δ as the difference between the two simultaneous intensities registered at C_1 and C_2:

$$\Delta = I^1 - I^2. \tag{13}$$

In such case one has:

$$\text{Usual theory:} \qquad \Delta_u = I_u^1 - I_u^2 = I_0. \tag{14}$$

$$\text{Causal theory:} \qquad \Delta_c = I_c^1 - I_c^2 = (1-\varepsilon)I_0. \tag{15}$$

These predictions are shown in Fig.6 for the particular case of $\varepsilon = 1/2$.

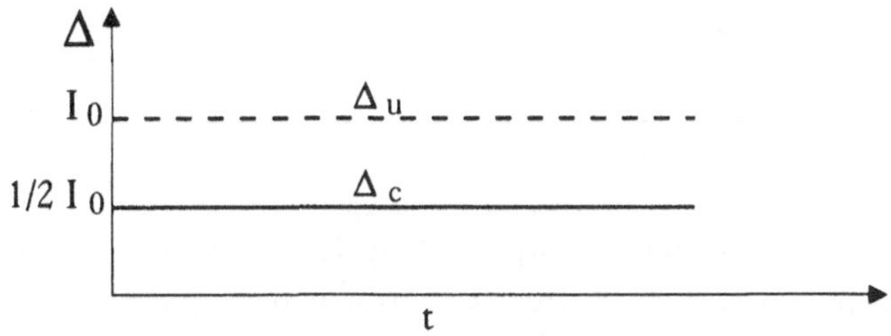

Fig.6. Predictions of the two theories for the particular case of $\varepsilon = 1/2$. Broken line usual theory. Solid line causal theory.

3.3. Detection Assuming Other Properties for the Empty Waves

Other proposals of experiments not based in the properties de Broglie explicitly attributed to the empty waves have been presented and some were even carried out experimentally.

One very interesting one was presented by Selleri[20], where the process of detection of the empty waves is based on the idea that they may *modify the decay probability of unstable systems.* The empty waves, see Fig.7, emitted by the empty wave generator, are injected in a laser gain tube where they may reveal their existence by generating a zero energy-transfer stimulated emission. A positive answer from this experiment would prove the existence of some real physical entity, unseen with a common detector, coming from the empty wave generator.

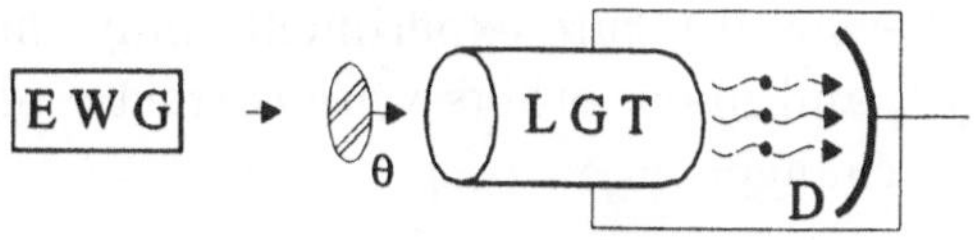

Fig.7. Selleri's experiment. LGT - laser gain tube, D - photomultiplier, EWG - empty wave generator.

Another possible effect is that the empty waves may have the capacity of inducing coherence in a non-linear crystal in a parametric-down conversion. In fact an experiment[21] to test that hypotheses, whose outline is presented in Fig.8, was carried out by Mandel and his group.

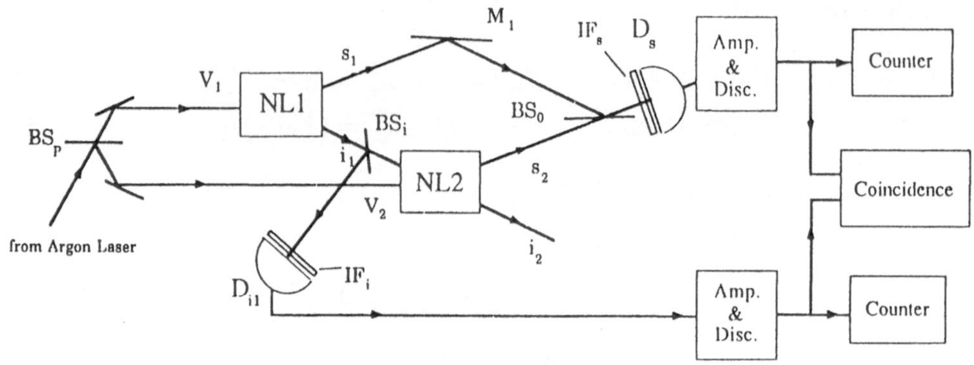

Fig.8. Mandel's setup to test the capacity of the empty waves to induce coherence in a parametric-down converters.

When D_{i1} is not set in coincidence with D_s interference effects are seen. If (D_{i1}, D_s) are put in coincidence, the empty wave θ_{i1} enters NL2 where it could have induced coherence. The results of the experiment indicate that the empty wave has not, by itself, the power of inducing coherence in the parametric-down converter.

3. CONCLUSION

We have seen that experiments have been proposed and done to see if the empty waves of de Broglie exist or not, the final aim of those experiments is to test the validity of the quantum theory. It may be argued that the usual theory has proven to be correct so many times that it shall be a loss of time keeping proposing and doing new experiments. Nevertheless it is known that for each new experiment the deep understanding of the theory expands. Without the debate about the EPR and Bell's theorem many implications of the theory would be completely overlooked. These facts show that the research on the foundations of the quantum phenomena pays off. One the other hand future experiments may show that de Broglie, Einstein, Schrödinger, and many others were correct and so that there is in fact a real need for a change.

ACKNOWLEDGEMENTS

The author wants to thank Professor L.Mandel for the kindness he showed when of my stay at Rochester, where part of this work was done; Professor A.Barbosa, for many discussions and helpful hints concerning some experimental subjects; the Fundação Luso Americana, whose financial support made the visit to Rochester possible. I can not forget my dear friends F. Selleri, A. Garuccio and V.L. Lepore who fight for the truth.

REFERENCES

1. L. de Broglie, *The Current Interpretation of the Wave Mechanics* (Elsevier, Amsterdam, 1964).
2. L.J. Wang, X.Y. Zou, and L. Mandel, *Phys. Rev. Lett.* **66**,111 (1991).
3. Y. Koh and Sasaki, in *Microphysical Reality and Quantum Formalism*, A. van der Merwe, F. Selleri and G. Tarozzi, eds. (Kluwer Academic, Dordrecht, 1988).
4. J.R. Croca, in *The Concept of Probability*, E.I. Bitsakis and C. A. Nicolaides, eds. (Kluwer Academic, Dordrecht, 1989).
5. J.D. Franson and K.A. Potocki, *Phys. Rev. A* **37**, 2511 (1988).
6. S. Jeffers and J. Sloan, private communication.
7. G. Mölenstedt and W. Bayh, *Naturwiss* **48**, 400 (1961).
8. C. Hong and L. Mandel, *Phys. Rev. Lett.* **56**, 58 (1986).
9. P. Grangier, G. Roger, and A. Aspect, *Europhys. Lett.* **1**, 173 (1986).
10. H. Rauch, in *Open Questions in Quantum Physics*, G. Tarozzi and A. van der Merwe, eds. (Reidel, Dordrecht, 1985).
11. J. and M. Andrade e Silva, *Compt. Rend. Acad. Sci.* **290**, 501 (1980).
12. J.R. Croca, in *Microphysical Reality and Quantum Formalism*, A. van der Merwe, F. Selleri. and G. Tarozzi, eds. (Kluwer Academic, Dordrecht. 1988).
13. A. Garuccio, V. Rapisarda, and J.P. Vigier, *Phys. Lett. A* **90**, 17 (1982).
14. J.R. Croca, in *Quantum Measurements in Optics*, P. Tombesi and D.F. Walls, eds. (Plenum, New York, 1992).
15. J.R. Croca, in *Problems in Quantum Physics*, L. Kostro, A. Posiewnik, J. Pykacz, and M. Zukowski, eds. World Scientific, Singapore, 1988).
16. J.R. Croca, A. Garuccio, V.L. Lepore, and R.N. Moreira. *Found. Phys. Lett.* **6**, 557 (1990).
17. X.Y. Zou, L.J. Wang, and L. Mandel, *Phys. Rev. Lett.* **67**, 318 (1991).
18. J.R. Croca, *Ann. Fond. Louis de Broglie* **14**, 323 (1989).
19. J.R. Croca, in *Philosophical Foundations of Quantum Theory*, R. Nair, ed.(World Scientific, Singapore, in print).
20. F. Selleri, *Found. Phys.* **12**, 1087 (1982); **17**, 739 (1987).
21. X.Y. Zou, T. Grayson, L.J. Wang, and L. Mandel, *Phys. Rev. Lett.* **25**, 3667 (1992).

WHY LOCAL REALISM?*

James T. Cushing

Departments of Physics and of Philosophy
University of Notre Dame
Notre Dame, 46556 Indiana

I consider why it has so long been the case historically, and remains largely so today, that local realism, as a constraint or condition on any world view to be deemed satisfying, exercises such a hold on our attempts at explaining and understanding physical phenomena. While I here take as relatively unproblematic the demand of realism (that is, the objective existence of an observer-independent reality), the requirement of locality (here something akin to a first-signal principle for any physical interaction or influence) is much more problematic. The basic question I address is whether the great appeal of locality is based on our inherent patterns of understanding or simply on acclimation through a series of successful explanatory discourses in the history of science. I also discuss the relative importance of (event-by-event) causality and of locality in producing a sense of understanding.

Key words: quantum theory, locality, realism.

1. INTRODUCTION

What we take to be an adequate or satisfying explanation of a class of physical phenomena depends in large measure on our view of the goals of science. Since this conference commemorates the 100th anniversary of Louis de Broglie's birth, it is perhaps appropriate for me to begin with a brief summary of de Broglie's personal perspective on the purpose of science. We know from de Broglie's own preface to the *Essais* of Emile Meyerson that he greatly admired Meyerson both personally and professionally and that he was significantly influenced by Meyerson's philosophy.[1] Meyerson's major works argued that the mind always and in all areas of knowledge works in the same fashion, and this commonality established, so Meyerson claimed, an element of the *a priori* in its functioning. For him, the goal of science is explanation and understanding of the disparate phenomena of nature, as opposed to an instrumentalist or positivist mere aggregation of these phenomena under a set of formal laws.

Waves and Particles in Light and Matter, Edited by
A. van der Merwe and A. Garuccio, Plenum Press, New York, 1994

Scientific theories seek realistic and causal explanations and, for this, the goal of science is an *ontological* one.[2] Meyerson's causality is a form of logical identity, an essentially atemporal state of affairs.[3] However, nature will not yield totally to such a representation and, hence, the real remains only partly intelligible. Meyerson himself felt that the Copenhagen interpretation of quantum mechanics was an aberration that would eventually be replaced by more traditional causal, realistic views.[4]

We can see elements of Meyerson in de Broglie's own goals. For de Broglie the issue at stake was not (classical) determinism, but rather the possibility of a precise space-time representation for a clear picture of micro-processes.[5] In this, his expectations were similar to Einstein's. Although de Broglie had been much impressed with Henri Poincaré's writing on the philosophy of science, he parted company on the thesis of radical underdetermination. De Broglie did believe that *one* theory should best conform to nature.[6] He felt that the classical Hamilton-Jacobi formalism provided an embryonic theory of the union of waves and particles, all in a manner consistent with a realist conception of matter.[7] With the concept of the quantum potential, he came to believe, one could provide a *model* for fundamental processes.[8] For de Broglie, as for Meyerson, the Copenhagen interpretation had effectively sidestepped the difficult issue of delineating a coherent ontology. Even earlier, de Broglie had, under the influence of Henri Bergson, turned from the positivistic trend in science, although he never embraced Bergson's radical disdain for positive science and his criticism of mathematical method in science.[9] For Bergson, continuity at the most fundamental level was an essential feature of nature. We see the imprint of this on de Broglie's views as well.

Of course, de Broglie was not alone among the founders of quantum theory in valuing realism over formalism. Uppermost for Erwin Schrödinger, too, was a concern for conceptual matters and for physical models.[10] Similarly, "allegorical" pictures of the physical situation and clear physical interpretations were important for Schrödinger.[11] Even more long-standing and central was Schrödinger's concern with physical pictures that undergird the formal mathematics of physical theories.[12] In the fall of 1926 Schrödinger wrote to Wilhelm Wien:

> Quite certainly, the point of view [using] visualizable pictures, which de Broglie and I assume, has not been carried through nearly far enough in order to render an account of the most important facts [of atomic theory]. · · · [F]or me the comprehensibility of the external processes in nature is an axiom · · · .[13]

On the other hand, Bohr's "Copenhagen" school tenaciously held to the position that accounts of atomic phenomena must be formal rather than realistic.[14]

While instrumentalism is a logically consistent *possibility* as a minimal demand on scientific theories, people in general, and even most scientists, require more – namely a story about the furniture of the world, an ontology. The basic issue here is a belief in the (at least effective) uniqueness of a correct scientific theory, with the selection process being "objective" and not involving in any ineliminable fashion "subjective" criteria such as coherence ("beauty"), simplicity or the like.[15] Scientists typically take for granted the practical uniqueness of successful scientific theories (in any given era). For instance, Albert Einstein allowed for the *theoretical* (*i.e.*, logical) possibility of more than one empirically adequate theory, but then went on to make the (rather startling) declaration of faith that at any given time the "world of phenomena" (which sounds pretty objective) uniquely determines one theory as superior to all others.[16]

A less sanguine view on the actual impact of the underdetermination of a theory by its empirical base (the so-called Duhem-Quine thesis) has been given by Quine himself when he says that the world intrudes first as a surface irritation and remains thereafter as a constraint on our imaginations in constructing scientific theories.[17] That is, our theories of the world are creations of our imaginations and sense input is *negative* only in this regard as checks and balances on these constructions of ours. This is not, of course, to claim that just *any* theory can be made to work, but rather to emphasize that logic and physical phenomena *alone* are not enough to rule out or reject some theories as viable candidates.

2. WHAT CONSTITUTES AN EXPLANATION?[18]

It is in attempting to answer this question that one's prior philosophical commitments and/or predilections come into play. For the purpose of discussion, let me parse a (modern) scientific theory into a formalism and an interpretation. The claim is that these are conceptually separable, even if they are often entangled in practice. What I mean by a formalism is a set of equations and a set of calculational rules for making predictions that can be compared with experiment (*i.e.*, "getting the numbers right"). The (physical) interpretation refers to what the theory tells us about the underlying structure of these phenomena. In terms of such a matrix, I consider three levels on which scientific theories function: (i) empirical adequacy, (ii) (formal) explanation and (iii) understanding.

The basic distinction I am making in (ii) and (iii) is that between a logical/mathematical explanation versus a physical explanation (in terms of causes and a coherent ontology). Let me consider specifically the situation for quantum phenomena as illustrated by the well-known Bohm version of the Einstein-Podolsky-Rosen (EPR) thought experiment. John Bell's original work showed that no determinate, local theory can account for the outcome of the EPR-Bohm (EPRB) experiment.[19] Much subsequent analysis has sharpened our understanding of precisely what must be assumed to obtain Bell-type inequalities.[20] In particular, Arthur Fine has established that, for the EPRB situation, the Bell inequalities are the necessary and sufficient conditions on the joint distributions for a common-cause explanation of the actually observed outcome of the experiments.[21] Since these inequalities are violated, no common-cause explanation is possible. A reasonably exhaustive list of explanatory resources for the observed results is (i) mere chance or coincidence; (ii) a direct causal link between the two spatially separated stations; (iii) a common-cause explanation located in the light cone of the common past events of the system and apparatus. The last (iii) is blocked by the violation of the Bell inequalities and (ii) (apparently) by the first signal principle of special relativity interpreted as prohibiting any type of instantaneous action at a distance.[22] We put aside (i) for the moment as no explanation at all.

In the EPRB case, the distinctions I have in mind among empirical adequacy, explanation and understanding can be seen by starting with the joint probabilities for the experiment. Given the observed data for a series of such experiments, a modern-day Ptolemy – or, perhaps better, Kepler – (without use of quantum mechanics or any other theory) might find, within experimental errors, the following purely phenomenological representation of the data

$$p^{AB}(++|\,\theta,\phi) = p^{AB}(--|\theta,\phi) = \frac{1}{2}\sin^2\left(\frac{\theta-\phi}{2}\right),$$

$$p^{AB}(+-|\,\theta,\phi) = p^{AB}(-+|\theta,\phi) = \frac{1}{2}\cos^2\left(\frac{\theta-\phi}{2}\right).$$

This is surely empirically adequate. In fact, it is little more than equivalent to the data itself. At this stage, though, we have neither an explanation of these results nor any understanding of how they come about. The situation is similar to Kepler's laws for planetary motion. Next, if someone were to give us the formalism of quantum mechanics, then we could formally explain these results by performing a standard and familiar quantum-mechanical calculation. We would then have a formal explanation of these joint distributions, but we would have no understanding of how these events have been produced in nature. In other words, an intelligible *interpretation* is missing. We have no understanding what the formalism means, in terms of underlying physical processes.

I have indicated some moves – specifically, traditional direct causal links and common-cause explanations – that are blocked in any attempt to produce an understanding of quantum phenomena. That is a negative message. There, of course, does exist a positive explanatory framework (quantum mechanics) for quantum events. To provide some understanding of this explanatory scheme is a very desirable goal, one pursued mainly by the scientific realist. We certainly have no guarantee that such a goal is attainable. Producing some sense of understanding is the purpose of engaging in the explanatory discourse of an interpretation. It seems to be a part of our human nature (at the very least in more recent Western tradition) to want this something extra that goes beyond mere formal explanation. There have been many, often realist, attempts at interpreting standard quantum mechanics in a broadly "Copenhagen" spirit. However, simply introducing new terminology, such as "ontic blurring," "quantum particle" or "relational holism" does not in itself produce in us any sense of understanding of the physical phenomena. If we say that quantum realism is that realism required by quantum mechanics, we have not thereby helped anyone to comprehend just what quantum realism *is* as a picture of the world. Many of these interpretative moves are little more than attempts to paper over our ignorance with words.

In a *very* different type of move, a philosopher of science whom I admire greatly, Arthur Fine, has taken the point of view that, since traditional direct causal links and common causes are impossible for quantum correlations, these quantum correlations themselves ought to be taken as the irreducible brute facts or the primitive givens upon which we build any further explanations.[23] If, as Fine suggests, we must buy wholeheartedly into an indeterministic world view, it may still be that we cannot understand this world view.

The point is not that any explanation must stop somewhere so that there must always exist irreducible primitives in an explanation. Rather, not all sets of primitives are equally acceptable for producing understanding. It is a matter of fashioning a comprehensible or intelligible explanatory discourse versus giving mere redefinitions of terms. It is not always easy to distinguish between these two exercises. I do not see how an understanding of physical processes is possible if the move to a causal story is blocked. Of course, this may be a difficulty peculiar to just me. If so, it is of no general interest and will be taken care of by the "Planck effect" (*i.e.*, the present generation will die off and a new one will come along that has no such conceptual difficulty). One must evaluate whether it is worth arguing about understanding in terms of causal explanations as a necessary or even a legitimate goal of science.

My argument, of course, turns on the claim that understanding for us is possible for physical processes only when we can tell a causal picture story of those processes. If this is so, then we are, by definition, doomed, according to the standard ("Copenhagen") interpretation of quantum mechanics, to be unable to understand quantum phenomena (and quantum mechanics). Now the world *may* just be that way. In our human intellectual history, we have been forced to acknowledge that we are unable to fulfill certain perceived intellectual (psychological?) needs. My basic position here is that, not an *a priori*, but rather a historicist argument shows that picturable models or explanations are necessary for producing in us a sense of understanding.[24] This necessity is rooted in the way we think about and understand physical phenomena.

A line of argument that modes of thought are permanent, or at least endure for centuries, was used by Meyerson.[25] His *principle of lawfulness*, related to our need to know *how* phenomena are interconnected, can be rooted in the advantage (or survival value) of purposeful action or prevision. His *principle of causality*, related to our desire to know *why* certain observed regularities obtain, is a manifestation of a fundamental and universal characteristic of human thought and modes of understanding. For Meyerson, they have the status of contingent but universal principles of the human mind. His was a search for such principles that the human mind uses in framing theories. He anchored these, especially causality, in our demand for something permanent and self-identical (*e.g.*, conservation laws). Although Meyerson accounted only for lawfulness, but not for causality, on the basis that it simplifies our process of comprehending the phenomena of the world (*i.e.*, survival value), it would seem that causality can also confer the advantage of our being able simply and efficiently to reduce (and hence, to deduce thereafter) our laws to the understanding of basic causal mechanisms. In his *Identity and Reality*, he tells us that "Science also wishes to make us *understand* nature" as a reflection of man's causal instinct.[26] His belief in these innate urges or needs of the intellect is again reflected in "[A]tomism has its roots in the depth of our spirit", as *prior* to experience.[27] An important question for us here is whether we can ever hope to purge mankind of this need for a causal (I would claim, picturable) explanation of physical phenomena. Are explanations based on mechanisms and picturable processes essential to produce understanding in us or are they just a temporary resting place?

In a similar vein, Gaston Bachelard in his *The New Scientific Spirit* emphasized the dialectic of science, stressing that psychology and epistemology are equally important components in a representation of the functioning scientific enterprise.[28] While citing Meyerson on the stability of modes of thought that endure for aeons, Bachelard disagreed that there are any permanent forms of rationality and did consider the possibility of changes in these patterns or modes, but only on a time scale characteristic of or limited by the rate of evolution of the human brain.[29]

If the nature of reality is such that we cannot satisfy our need for understanding or intelligibility, then there would be an essential mismatch between the metaphysics and the epistemology of our world view. We may need theories that are *able* to be interpreted realistically in order to find them intelligible, even though we may not find such an interpretation *justifiable* (nor always even possible).[30] The scientific realist is an optimist in believing that such a meshing is possible, even though he has at times in the past been forced into a grudging retreat by the nature of physical reality (*e.g.*, the hope that true theories could be known to be true). If the nature of the world is such that we cannot understand it in a fundamental sense, then we must accept that fact no matter how unpleasant. However, we should not confuse such resignation toward a fact with any sense of progress toward understanding.

3. LOCAL REALISM

Like de Broglie and Einstein, I shall here take the demand of an objective, observer-independent reality to be an evident requirement and focus my attention on the causality and locality issues. Previously I have argued for a local causal story as the ideal in producing in us a sense of understanding for physical processes. Furthermore, I have taken as the only really coherent notion of causality event-by-event causality, as opposed, say, to "statistical causality" (whatever that may mean). Such causality is, for my present purposes, sufficiently akin to determinism that I do not care to discuss any distinction between the two concepts. However, Bell's theorem surely blocks such an ideal – deterministic, local theories do not offer us much hope. And we do have an empirically adequate deterministic quantum theory – David Bohm's. So, I ask that we think a bit about locality and its hold on us. I do this in spite of the strong feelings in support of locality held by our energetic conference organizer and gracious host, Franco Selleri. Just as harsh experience over the centuries has disabused us of what John Watkins so felicitously termed the Bacon-Descartes ideal (*i.e.*, the belief that the proper goal of science is to find truth and to *know* that truth has been found), so we may be forced to yield on locality as a reasonable demand on any successful quantum theory.[31]

We can locate the roots of what we today term the locality/nonlocality debate as far back as the origins of our Western philosophical tradition.

This debate may be seen as a continuation of attempts to grapple with a very old philosophical problem – that of the one and the many.[32] Later, in the long debate about action at a distance versus contact action we also find a concern with the effective individuation of particles or subsystems. Ernan McMullin catalogues various attempts throughout history to explain distant action.[33] The basic motivating factor in demanding contact action was that of *intelligibility*. Frans van Lunteren, in his extensive recent study of conceptions of gravity in the eighteenth and nineteenth centuries, demonstrates that a quest for intelligibility was uppermost for several major physicists who attempted aether-type explanations for action at a distance.[34] Maxwell put the matter quite succinctly when he argued for "a force of the old school – a case of *vis a tergo* – a shove from behind."[35]

I shall now turn to current attempts to cope with the nonlocality dilemma of quantum phenomena. No doubt the boldest move is a frontal attack in which one simply denies that there is *any* empirical evidence that blocks local realism. This is just what Emilio Santos has claimed in his analysis of the experimental results that are usually interpreted as violating the Bell inequalities.[36] That is, one directs the refutation not against local realism, but against the auxiliary assumptions – such as "fair sample" or "no enhancement" – that are needed to bring prediction and observation into direct contact. However, unfortunately for such a move, the fact is that just any *one* of several different and apparently independent auxiliary assumptions will do to produce the conflict with experiment. Franco Selleri, in a recent analysis of the situation for local realism, adopts a similar stance when he says that "our results imply that local realism is *a priori* incompatible with the 'additional assumptions' made in the analysis of Bell-type experiments."[37] This usually leads to desperate moves, such as Fine's prism models which verge on a cosmic conspiracy.[38] One recent version has the efficiencies of detectors (*any* type of detector) be a *universal* function of the quantum state of the system being detected.[39] This has become a "research programme" in the Lakatosian sense and locality serves as the methodologically agreed-upon hard core protected by a belt of auxiliary assumptions.[40] I do not feel this is a very promising avenue of escape and so I shall not pursue it further here.

It may appear more fruitful to turn to other types of attempts to block *inferences* to nonlocality. There have been two major categories of attacks on any *necessary* inference to nonlocality: one has been to modify the formalism of quantum mechanics (*e.g.*, drop the projection postulate), the other to question the existence of joint probability distributions (jpd) (this following largely on Arthur Fine's decade-old result that jpd are equivalent to factorizability).[41] However, an obvious question is *what* keeps these joint probabilities from being defined, if not some type of nonlocality?[42] And, of course, one concrete example of an empirically adequate nonlocal, hidden-variable theory that we have – namely, David Bohm's – does *not* have such joint distributions defined.

In spite of several such attempts to block the inference from long-range quantum correlations to physical nonlocality, *some* type of nonlocality does seem to be a feature of our world. Perhaps its is more promising to accept nonlocality, but to remove its sting. Let us see whether one version of "benign" nonlocality – that associated with David Bohm's theory – is really so painful.

4. BOHMIAN MECHANICS

As is well known, all of the usual statistical predictions of the standard ("Copenhagen") theory are recovered from Bohm's quantum-potential theory provided *all* three of the following mutually consistent assumptions are made: (i) that the ψ-field satisfies Schrödinger's equation; (ii) that the particle momentum is restricted by the *guidance condition* ($\boldsymbol{p} = \nabla S(\boldsymbol{x})$, where S is the phase of ψ); (iii) that we do not predict or control the precise location of the particle, but have, in practice, a statistical ensemble with probability density $P(\boldsymbol{x}) = |\psi(\boldsymbol{x})|^2$. We shall term (iii) *quantum equilibrium*. The use of statistics here is not inherent in the conceptual structure, but merely a consequence of our ignorance of the precise initial conditions of the particle.[43] Bohm was then able to rewrite the equations of quantum mechanics in a Newtonian form ($\boldsymbol{F} = m\, \boldsymbol{a}$) for the dynamical equation determining the trajectories of the particles.

Recently Anthony Valentini has given an insightful discussion of the relation among the quantum equilibrium distribution for P, the uncertainty relations and the possibility of superluminal signaling, all within the framework of a stochastic mechanics based on Bohm's pilot-wave formulation.[44] This is conceptually very similar to how we typically envision the molecules in a large sample of gas being driven toward the (Maxwell-Boltzmann) equilibrium distribution through random interactions among the gas molecules themselves. Valentini shows that the impossibility of instantaneous signaling ("signal locality") and the (Heisenberg) uncertainty principle are related and turn on the quantum equilibrium condition being satisfied.[45] On this view, the world would be fundamentally nonlocal in its structure, yet possess signal locality as a *contingent* fact once quantum equilibrium has been reached.[46] Initially, P can be an arbitrary probability distribution on this configuration space. Valentini establishes a quantum analogue of Boltzmann's H-theorem to argue that random interactions will drive P to the quantum equilibrium distribution.[47]

By considering in detail how one extracts, from a larger system, a specific *measured* result for a given particle, Valentini demonstrates explicitly that once the probability distribution for the "universe" has been driven to its equilibrium distribution, *any* subsystem will *necessarily* satisfy the quantum equilibrium distribution. Finally, he shows with specific calculations that, if we have two interacting systems, whose overall state is an *entangled* one, then a change on one subsystem (accomplished, say, by modifying the Hamiltonian for that subsystem) can produce an instantaneous change in the probability distribution P for the other system *if and only if* quantum equilibrium is *not* satisfied for the interacting system.[48] Essentially, he establishes (again) a no-signaling theorem when $P = |\psi|^2$, but illustrates how to signal *instantaneously* for an entangled state when $P \neq |\psi|^2$. He is also able to relate quantum equilibrium to the Heisenberg uncertainty principle. Here we have a realistic ontology of actually-existing particles and events in which several remarkable features of our world – the uncertainty relations, no instantaneous signalling, even Lorentz covariance – all are given a unified and essentially understandable explanation. It would seem unreasonable to hold any of this *against* a causal interpretation.

5. CONCLUSIONS

As in the fall from the Bacon-Descartes ideal, occasioned in no small part by the underdetermination of theory by any empirical data base, so we may again be disabused of a cherished goal and be forced to readjust our demands on what are deemed to be acceptable scientific theories. This is a practical matter since we do the best we can, but that best sometimes turns out to be far less than we had hoped for.

I have argued that, due to our innate modes of understanding and to historical acclimation, we have a felt need for local, causal explanations to produce in us a sense of truly understanding physical processes. Lacking any coherent notion of statistical causality, we are left with event-by-event causality, something essentially akin to determinism. Since Bell's theorem effectively blocks *local* deterministic theories, we are forced to relent on the *desiderata* of locality. But, upon further reflection, we find that the actual nonlocality demanded by nature is of a fairly benign variety: we cannot signal with it and it does not so entangle the world as to prevent us from doing science as we have traditionally known it. And, I have claimed, we are able to construct a less incomprehensible, more nearly picturable, representation of the physical universe with "Bohm" than with "Copenhagen." The choice is ours to make purely on such practical or pragmatic grounds – no realism/antirealism debates here.

De Broglie has reminded us of Voltaire's observation that only a fool never changes his mind.[49] So, while we may still pursue the ideal goal of local realism, I have asked what readjustments would be demanded if locality has to go. It is fitting that I should give David Bohm the last word here:

> For several centuries, there has been a strong feeling that non-local theories are not acceptable in physics. It is well known, for example, that Newton felt uneasy about action-at-a-distance and that Einstein regarded this action as "*spooky*". One can understand this feeling, but if one reflects deeply and seriously on this subject one can see nothing basically irrational about such an idea. Rather it seems to be most reasonable to keep an open mind on the subject and therefore to allow oneself to explore this possibility. *If the price of avoiding non-locality is to make an intuitive explanation impossible, one has to ask whether the cost is not too great.*[50]

REFERENCES

Aerts, D., and Reignier, J. (1991), "On the problem of non-locality in quantum mechanics," *Helv. Phys. Acta* **64**, 527-547.

Bachelard, G. (1934), *Le nouvel esprit scientifique* (Presses Universitaires de France, Paris); A. Goldhammer (1984) translator, *The New Scientific Spirit* (Beacon Press, Boston).

Bell, J. S. (1964), "On the Einstein-Podolsky-Rosen paradox," *Physics* **1**, 195-200.

Bell, J. S. (1966), "On the problem of hidden variables in quantum mechanics," *Rev. Mod. Phys.* **38**, 447-452.

Ben-Menahem, Y. (1990), "Equivalent descriptions," *Brit. J. Phil. Sci.* **41**, 261-279.

Bohm, D. (1952), "A Suggested interpretation of the quantum theory in terms of 'hidden' variables, I and II," *Phys. Rev.* **85**, 166-193.

Bohm, D., Hiley, B. J., and Kaloyerou, P. N. (1987), "An ontological basis for the quantum theory," *Phys. Rep.* **144**, 321-375.

Brown, H. (1991), "Nonlocality in quantum mechanics," *The Aristotelian Society, Suppl. Vol.* **LXV**, 141-159.

Cushing, J. T. (1991), "Quantum theory and explanatory discourse: endgame for understanding?", *Phil. Sci.* **58**, 337-358.

Cushing, J. T. and McMullin, E. eds. (1989), *Philosophical Consequences of Quantum Theory: Reflections on Bell's Theorem* (University of Notre Dame Press, Notre Dame).

de Broglie, L. (1953), *The Revolution in Physics* (Noonday Press, New York).

de Broglie, L. (1962), *New Perspectives in Physics*, A. J. Pomerans, translator, (Oliver & Boyd, Edinburgh).

de Broglie, L. (1970), "The reinterpretation of wave mechanics," *Found. Phys.* **1**, 5-15.

de Broglie, L. (1973), "The beginnings of wave mechanics," in Price *et al.*, pp. 12-18.

Edwards, P. (1967), *The Encyclopedia of Philosophy*, 8 Vols. (Macmillan, New York).

Einstein, A. (1954), *Ideas and Opinions* (Dell, New York)

Feuer, L. S. (1974), *Einstein and the Generations of Science* (Basic Books, New York).

Fine, A. (1982a), "Hidden variables, joint probability, and the Bell inequalities," *Phys. Rev. Lett.* **48**, 291-295.

Fine, A. (1982b), "Some local models for correlation experiments," *Synthese* **50**, 279-294.

Fine, A. (1982c), "Antinomies of entanglement: The puzzling case of the tangled statistics," *J. Phil.* **79**, 733-747.

Fine, A. (1989), "Do Correlations need to be explained?", in Cushing and McMullin, pp. 175-194.

Jarrett, J. P. (1984), "On the physical significance of the locality conditions in the Bell arguments," *Noûs* **18**, 569-589.

Kamefuchi, S., Ezawa, H., Murayama, Y., Namiki, M., Nomura, S., Ohnuki, Y., and Yojima, T., eds. (1984), *Proceedings of the International Symposium on the Foundations of Quantum Mechanics* (Physical Society of Japan, Tokyo).

Lahti, P. and Mittelstaedt, P., eds. (1985), *Symposium on the Foundations of Modern Physics* (World Scientific, Singapore).
Lahti, P. and Mittelstaedt, P., eds. (1987), *Symposium on the Foundations of Modern Physics 1987* (World Scientific, Singapore).
Lakatos, I. (1970), "Falsification and the methodology of scientific research programs," in Lakatos and Musgrave, pp. 91-196.
Lakatos, I., and Musgrave, A., eds. (1970), *Criticism and the Growth of Knowledge* (Cambridge University Press,London).
MacKinnon, E. (1980), "The rise and fall of Schrödinger's interpretation," in Suppes, pp. 1-57.
MacKinnon, E. (1982), *Scientific Explanation and Atomic Physics* (University of Chicago Press, Chicago).
Maxwell. J. C. (1890), *The Scientific Papers of James Clerk Maxwell*, 2 Vols., W. D. Niven, *ed.* (Cambridge University Press, Cambridge).
McMullin, E. (1989), "The explanation of distant action: historical notes," in Cushing and McMullin, pp. 272-302.
Mehra, J. (1987), "Niels Bohr's discussions with Albert Einstein, Werner Heisenberg, and Erwin Schrödinger: the origins of the principles of uncertainty and complementarity," in Lahti and Mittelstaedt, pp.19-64.
Meyerson, E. (1908), *Identité et réalité* (Libraries Félix Alcan et Guillaumin Reúnies, Paris).
Meyerson, E. (1936), *Essais* (Librairie Philosophique J. Vrin., Paris).
Price, W. C., Chissick, S. S. and Ravensdale, T., eds. (1973), *Wave Mechanics: The First Fifty Years* (Butterworths, London).
Quine, W. V. (1960), *Word and Object* (The MIT Press, Cambridge).
Santos, E. (1991), "Does quantum mechanics violate the Bell inequalities?", *Phys. Rev. Lett.* **66**, 1388-1390.
Selleri, F., ed. (1988a), *Quantum Mechanics Versus Local Realism* (Plenum Press, New York).
Selleri, F. (1988b), "Even local probabilities lead to the paradox," in Selleri (1988a), pp.149-174.
Selleri, F. (1992), "Einstein-de Broglie Waves and Two-Photon Detection," in van der Merwe *et al.*, pp. 422-427.
Shimony, A. (1984), "Controllable and uncontrollable non-locality," in S. Kamefuchi *et al.*, pp.225-230.
Suppes, P., ed. (1980), *Studies in the Foundations of Quantum Mechanics* (Philosophy of Science Association, East Lansing, Michigan).
Valentini, A. (1991a), "Signal-locality, uncertainty, and the subquantum H-theorem. I," *Phys. Lett. A* **156**, 5-11.
Valentini, A. (1991b), "Signal-locality, uncertainty, and the subquantum H-theorem. II," *Phys. Lett. A* **158**, 1-8.
van der Merwe, A., Selleri, F., and Tarozzi, G., eds. (1992), *Bell's Theorem and the Foundations of Modern Physics* (World Scientific, Singapore).
van Fraassen, B. C. (1985), "EPR: When is a correlation not a mystery?", in Lahti and Mittelstaedt, pp.113-128.
van Lunteren, F. H. (1991), "Framing hypotheses: Conceptions of gravity in the 18th and 19th Centuries," unpublished Ph. D. dissertation, University of Utrecht.
Watkins, J. (1984), *Science and Skepticism* (Princeton University Press, Princeton).

Wessels, L. (1979), "Schrödinger's route to wave mechanics," *Stud. Hist. Phil. Sci.* **10**, 311-340.

Wessels, L. (1980), "The intellectual sources of Schrödinger's interpretations," in Suppes, pp.59-76.

NOTES

* The research on which this paper is based has been supported in part by the National Science Foundation under grants DIR 89 08497 and SBE 91 214779.

1. Meyerson, 1936, vii-xiv.
2. Edwards, 1967, Vol. 5, 307.
3. Meyerson, 1908.
4. Edwards, 1967, Vol. 5, 308.
5. de Broglie, 1962, vii-viii.
6. Feuer, 1974, 206.
7. de Broglie, 1973, 12.
8. de Broglie, 1970, 12.
9. Feuer, 1974, 208-212.
10. Wessels, 1979, 313.
11. MacKinnon,1980, 16, 19.
12. Wessels, 1980, 62-63.
13. Mehra, 1987, 44.
14. MacKinnon, 1982, 246.
15. Ben-Menahem, 1990, 267.
16. Einstein, 1954, 221-222.
17. Quine, 1960.
18. The arguments in this section consist largely of a summary of the more extensive discussion given in Cushing (1991).
19. Bell, 1964, 1966.
20. See, for example, Jarrett, 1984.
21. Fine, 1982a. As Fine (1989, 182, note 3) himself points out: "A common-cause explanation · · · is what the foundational literature calls a factorizable stochastic hidden-variables model." That is, this factorizability is the condition for a common cause .
22. Jarrett, 1984; Shimony, 1984.
23. Fine, 1989. A similar position that the EPRB correlations do not need an explanation has been developed by van Fraassen (1985). There he considers various types of explanation and effectively rules out all but causal ones, which are excluded as a possibility by Bell's theorem (and by subsequent work on its implications).
24. For more details and specific historical examples, see Cushing, 1991.
25. Meyerson, 1908.
26. Meyerson, 1908, 384.
27. Meyerson, 1908, 395.
28. Bachelard, 1934.
29. Bachelard, 1934, 54, 175.
30. That is, the great ("psychological") appeal of scientific realism may be, in large part, accounted for as satisfying our need for understandable explanations of physical phenomena.

31. On the Bacon-Descartes ideal, see Watkins, 1984, Sec. 4.2.
32. I thank J. B. Kennedy for having suggested this to me.
33. McMullin, 1989.
34. van Lunteren, 1991.
35. Maxwell, 1890, 480-490.
36. Santos, 1991.
37. Selleri, 1988b, 150.
38. Fine, 1982b, 1982c.
39. Selleri, 1992.
40. Lakatos, 1970.
41. Fine, 1982a.
42. Brown, 1991.
43. Bohm, 1952, 171.
44. Valentini, 1991a, 1991b.
45. Valentini, 1991a, 1991b.
46. Aerts and Reignier (1991, 545) also suggest that nonlocality may pose a difficulty for us largely because of the way we have come to view physical processes in an arena of Euclidean space and that the actual problem may not be nonlocality but rather that "one has to explain why the strongly correlated quantum entities get local when organized into more complex entities."
47. Valentini, 1991a.
48. Valentini, 1991b.
49. de Broglie, 1953, 237.
50. Bohm, Hiley, and Kaloyerou, 1987, 331. (My emphasis.)

BOHM'S INTERPRETATION OF QUANTUM FIELD THEORY

C. Dewdney

Department of Applied Physics
University of Portsmouth
Portsmouth PO1 2DZ, United Kingdom

In this contribution we review some calculations in Bohm's approach to quantum theory. In particular, we discuss the interaction of a quantized cavity scalar field with firstly one and then two particles confined in infinite potential wells.

Key words: nonlocality, quantum measurement, Bohm, quantum field theory, photon.

1. INTRODUCTION

Bohm's interpretation of non-relativistic-particle quantum theory [1] ought, by now, to be well known, but perhaps it is not so well known that his approach can be extended to quantum field theory [2]. Conceptually, Bohm's approach to quantised fields is very different from his approach to particle systems; although the formal mathematics is rather similar when the field theory is formulated in the Schrödinger picture.

The aim of this contribution is to review the essential features of some detailed calculations that seek to clarify the manner in which the Bohm approach to quantum fields is able to recover the usual results predicted by quantum field theory, and yet provide a model in which well-defined fields evolve continuously in space and time [3,4]. In this paper we do not treat the electromagnetic field since, as argued by Bohm Hiley and Kaloyerou [2], the essential features of Bohm's approach can be brought out in the context of scalar field theory. We first discuss field/matter interactions using the Jaynes-Cummings model [5] and, having shown how a single photon is absorbed from the transition inducing field during an individual transition, we then proceed to outline how the anti-correlation

Waves and Particles in Light and Matter, Edited by
A. van der Merwe and A. Garuccio, Plenum Press, New York, 1994

of two detectors in interaction with a single photon field can be explained in Bohm's approach. Before embarking on this task it may be useful if what we may call the Bohm programme for the interpretation of quantum theory is briefly summarised.

The work described here extends that already carried out in which detailed modelling has been used to show exactly what happens, according to Bohm's approach to the quantum mechanics of particle systems, in two-slit intereference [6], quantum tunelling [7], neutron interferometry [8], Stern-Gerlach measurements of orbital angular momentum and Einstein-Podolsky-Rosen (EPR) angular momentum correlations [9], quantum statistics [10], spin measurements [11] and spin superposition [12], EPR spin correlations [13,14] and quantum transitions (in the semi-classical case) [15].

2. THE BOHM PROGRAMME

We shall consider a simple model system consisting of scalar particles of mass m, moving in one-dimension, with coordinates x_i and a quantized cavity scalar field $\phi(x)$, also in one dimension with normal mode coordinates q_i. The particle wavefunctions are solutions of the Schrödinger equation, whilst the classical scalar field obeys the massless Klein-Gordon equation. In Bohm's approach the motion of the particles in the system is described by a set of trajectories. The situation for the quantum field is rather different: the quantum field is not described in terms of "photon" trajectories, indeed the "photon" is not at all to be thought of as a particle with well-defined coordinates. Instead, boson fields, in Bohm's approach, have definite amplitude and phase at all points in space and the photons of the field are simply mode excitations. The ontology of Bohm's theory is founded on the classical notions of continuous fields and discrete particles as distinct fundamental entities, but that is not all there is at the ontological level. In addition one has the configuration-space wavefunction, $\psi(x_i, q_i)$ the evolution of which governs the motion of both the particles and the field. In fact, it is this latter feature of the ontology that accounts for the wave-like behaviour of matter and the particle-like behaviour of fields, that is, the quantum mechanical behaviour of the particle-field system.

To develop a grasp of Bohm's approach to quantum fields let us, for the moment, just consider the field alone. As is well known, a classical cavity field can be described in terms of the cavity modes. Any field configuration can be written as a Fourier sum of the normal modes and the coefficients in the expansion of the field in normal modes (the normal mode coordinates q_i), evolve according to the equation for an harmonic oscillator of unit mass. Hence, classically, the motion of the field can be described in terms of the motion of an infinite set of ficticious, independent harmonic oscillators. The independence of the oscillators reflects the fact that the field evolves locally. The evolution of the field can be described in terms of a trajectory within the configuration space spanned by the normal mode coordinates.

One can move to the quantum description of the field by quantizing the ficticious normal mode oscillators. (This is the standard, but not often used, Shrödinger-picture approach to quanutm fields). Once having quantized the field, according to the orthodox approach the most complete description of the behaviour of each normal mode coordinate is given in terms of its wavefunction. The normal mode coordinates no longer have a definite value and consequently the quantum field no longer has a definite form. The quantum state of the field is described by a configuration space wavefunction and, in the usual approach the sole interpretation of the wavefunction is that $|\psi|^2$ gives the probability of a particular field configuration.

A system comprised of both the scalar field and a set of particles can then be described in terms of a configuration space spanned by both the particles' coordinates x_i and by the normal-mode coordinates of the field q_i. The state of the system is described by a configuration-space wave function over both particle and field coordinates. For convenience we shall refer to the collection of field and particle coordinates through the use of a single symbol η_i. The wavefunction obeys a"super" Schrödinger equation (in the following we take $\hbar = 1$.

$$i\frac{\partial\Psi}{\partial t}(\ldots,\eta_n,\ldots,t) = H(\ldots,\eta_n,\ldots,t)\Psi(\ldots,q_n,\ldots,t), \tag{1}$$

which governs the behaviour of both the scalar field, $\phi(x,t)$ and the particles and H is the hamiltonian for the field/matter system, including any interactions. To develop Bohm's approach for this system of field and particles one writes the wavefunction in the polar form

$$\Psi(\ldots,\eta_n,\ldots,t) = R(\ldots,\eta_n,\ldots,t)e^{iS(\ldots,\eta_n,\ldots,t)}, \tag{2}$$

where R and S are real fields. Two equations are obtained when this form of Ψ is substituted into the Schrödinger Eq. (1). Firstly the equation

$$\frac{\partial S}{\partial t} + \frac{1}{2}\sum_n^\infty\left[\left(\frac{\partial S}{\partial q_n}\right)^2 + k_n^2 q_n^2\right] + \frac{1}{2}\sum_n \frac{1}{m_n}\left(\frac{\partial S}{\partial x_n}\right)^2 + V(\ldots,\eta_n,\ldots,t)$$
$$+ Q(\ldots,\eta_n,\ldots,t) = 0 \tag{3}$$

and secondly the equation

$$\frac{\partial P}{\partial t} + \sum_n^\infty \frac{\partial}{\partial q_n}\left[P\frac{\partial S}{\partial q_n}\right] + \sum_n \frac{\partial}{\partial x_n}\left[P\frac{1}{m_n}\frac{\partial S}{\partial x_n}\right] = 0. \tag{4}$$

The quantity $P = |\Psi|^2$ is the probability density for a particular field and matter configuration and the quantity Q, given by

$$Q(\ldots,\eta_n,\ldots,t) = -\frac{1}{2}\sum_n^{\infty}\frac{1}{R}\frac{\partial^2 R}{\partial q_n^2} - \frac{1}{2}\sum_n \frac{1}{m_n R}\frac{\partial^2 R}{\partial x_n^2} \tag{5}$$

is the sytem's quantum potential. V represents any classical potentials acting on the particles alone and also any field matter interaction terms. In the absence of the quantum potential, Q, Eq. (3) is simply the Hamilton-Jacobi equation for the classical field and particle system. In Bohm's theory the rate of change of each of the coordinates is given by

$$\frac{dq_n}{dt} = \frac{\partial S(\ldots,\eta_n,\ldots,t)}{\partial q_n}, \tag{6}$$

for the field, and by

$$\frac{dx_n}{dt} = \frac{1}{m_n}\frac{\partial S(\ldots,\eta_n,\ldots,t)}{\partial x_n} \tag{7}$$

for the particles. If we assume that at some point in time the probability density, $P = |\Psi|^2$, gives the actual distribution of the coordinates, q_n and x_n in an ensemble then $|\Psi|^2$ will continue to give the distribution for all subsequent times, and Eq. (4) just expresses the conservation of probability. By making these two assumptions we can obtain a self-consistent, causal interpretation of the scalar field/matter system. The equations of motion in Newtonian form are

$$m_n\frac{d^2x_n}{dt^2} = -\left[\nabla_n V(\ldots,\eta_n,\ldots,t) + \nabla_n Q(\ldots,\eta_n,\ldots,t)\right], \tag{8}$$

for the particle coordinates, and

$$\frac{d^2q_n}{dt^2} = k_n^2 q_n + \frac{\partial Q(\ldots,\eta_n,\ldots,t)}{\partial q_n} \tag{9}$$

for the field normal-mode coordinates.

The quantum potential terms in both (8) and in (9) account for the quantum behaviour of both field and matter and both can only be derived from the from of the configuration-space wavefunction: they are not pre-assigned functions of the system's coordinates. It is these terms that determine the nonlinear and nonlocal behaviour of the system. Evidently the motion of a given particle can depend on the coordinates of the other particles in the system and also on the entire field configuration (i.e., not just the field in its vicinity). Similarily the evolution of a particular field-mode coordinate can depend on all the other field-mode coordinates and also all the particle positions.

Notice that in general none of the system coordinates evolves independently, but if the wave function factorises into a product of single-coordinate functions then each coordinate is independent of the others. Thus the ground state and the energy eigenstates (or number states) of the scalar field are all simple products of wavefunctions for each of the mode coordinates which consequently behave independently; further, in these cases the scalar field is stationary. For superposition states (entangled states) the wavefunction of the field is not a simple product: the field is no longer stationary, the normal-mode coordinates are not independent and the field evolves nonlocally.

Given the initial configuration $(\ldots, \eta_n, \ldots)$, the initial wavefunction $\Psi(\ldots, \eta_n, \ldots, 0)$ and the system hamiltonian, the evolution of the composite field/matter syatem is completely determined and can be represented by a unique trajectory in configuration space. It is this motion in configuration space which determines the evolution of both the scalar field and the particles in real space. In Bohm's approach to quantum theory a given initial configuration has a unique evolution, so, for example, if one has an excited atom surrounded by detectors some distance from the atom, then, given the (experimentally unavailable) initial configuration of the particles and the field, one could calculate which detector will absorb the"photon". Whereas Bohr insisted on the indivisibilty of quantum phenomena, Bohm's approach demonstrates that the analysis of a quantum system into parts reveals a nonlocal and nonlinear behaviour which expresses the essential unity of the quantum system.

It is our purpose in the rest of this paper to demonstrate exactly how Bohm's approach works in specific simple models of cavity-field/matter interactions.

3. THE JAYNES-CUMMINGS MODEL

The simplest model of a quantum field interacting with matter is furnished by the Jaynes-Cummings model, in which a tuned cavity field interacts with a two-level material system. In our version both field and matter are confined to one dimension. The cavity field is tuned so that one of its modes has exactly the right frequency to cause transitions between the two levels of the matter system. In fact, for the purposes of illustration, we take the material system to be a particle confined to a box, with the parameters chosen so that there is a resonant interaction between one particular field mode q and one particular transition of the particle in the box (we chose the ground state to the third excited state). Because of the exact resonance condition the particle in the box will behave essentially as a two-level system.

We assume that initially the field mode is in its first excited state, that is, the resonant mode contains one photon, whilst the particle in the box is intially in its ground state. All non resonant field modes are taken to be in the ground state (no photons) and they are then ignored in the calculation, however together they would contribute a spatially random but static noise to the form of the scalar field. With this initial state it is easy to show that the system evolves in such a way that the excitation is transfered periodically between the field mode and the two-level system. (Just as happens in a classical system of coupled resonant oscillators). The reason for chosing the motion to take place in one dimension is that the configuration space of the system can be taken to be, for the purposes of illustration, two dimensional, spanned by the resonant field mode-coordinate q and the two level system coordinate x. The configuration space wavefunction can be written

$$\chi(q,x,t) = U_1(q,x)\cos(\Omega t) - i\sin(\Omega t)U_2(q,x); \qquad (10)$$

here Ω is determined by the coupling between field and matter and we have used the notations

$$U_1(q,x) = \theta_1(q)\psi_0(x) \qquad (11)$$

and

$$U_2(q,x) = \theta_0(q)\psi_1(x), \qquad (12)$$

where $\theta_0(q)$ is the ground state of the resonant field mode and $\theta_1(q)$ is the first excited, or single photon, state of the field mode; similarily $\psi_0(x)$ and $\psi_1(x)$ are the ground state and first excited state of the two-level system.

The variation in probability density, $P(q,x,t)$, from time $t = 0$ to $\frac{\pi}{2\Omega}$, during the transition from state $\theta_1(q)\psi_0(x)$ to state $\theta_0(q)\psi_1(x)$ is shown in Figs. 1(a)-(f). Figure 1(a) shows the initial probability density when the system wavefunction is in the state $\theta_1(q)\psi_0(x)$; the most likely initial system-point values, at the two maxima in P, are at $(q,x) = (\pm 8.22, 0)$ and the coordinate q cannot be initially equal to zero. Figure 1(e) shows the probability density when the system is in an equal superposition of the two states $\theta_0(q)\psi_1(x)$ and $\theta_1(q)\psi_0(x)$, and Fig. 1(f) shows the probability density when the system is entirely in the state $\theta_0(q)\psi_1(x)$. During the time interval $t = \frac{\pi}{2\Omega}$ to $\frac{\pi}{\Omega}$, the system wavefunction and the probability density return to the form which they possessed initially.

A set of possible configuration-space/time trajectories was calculated by numerical integration for the time period $t = 0$ to $t = \frac{\pi}{2\Omega}$, and is

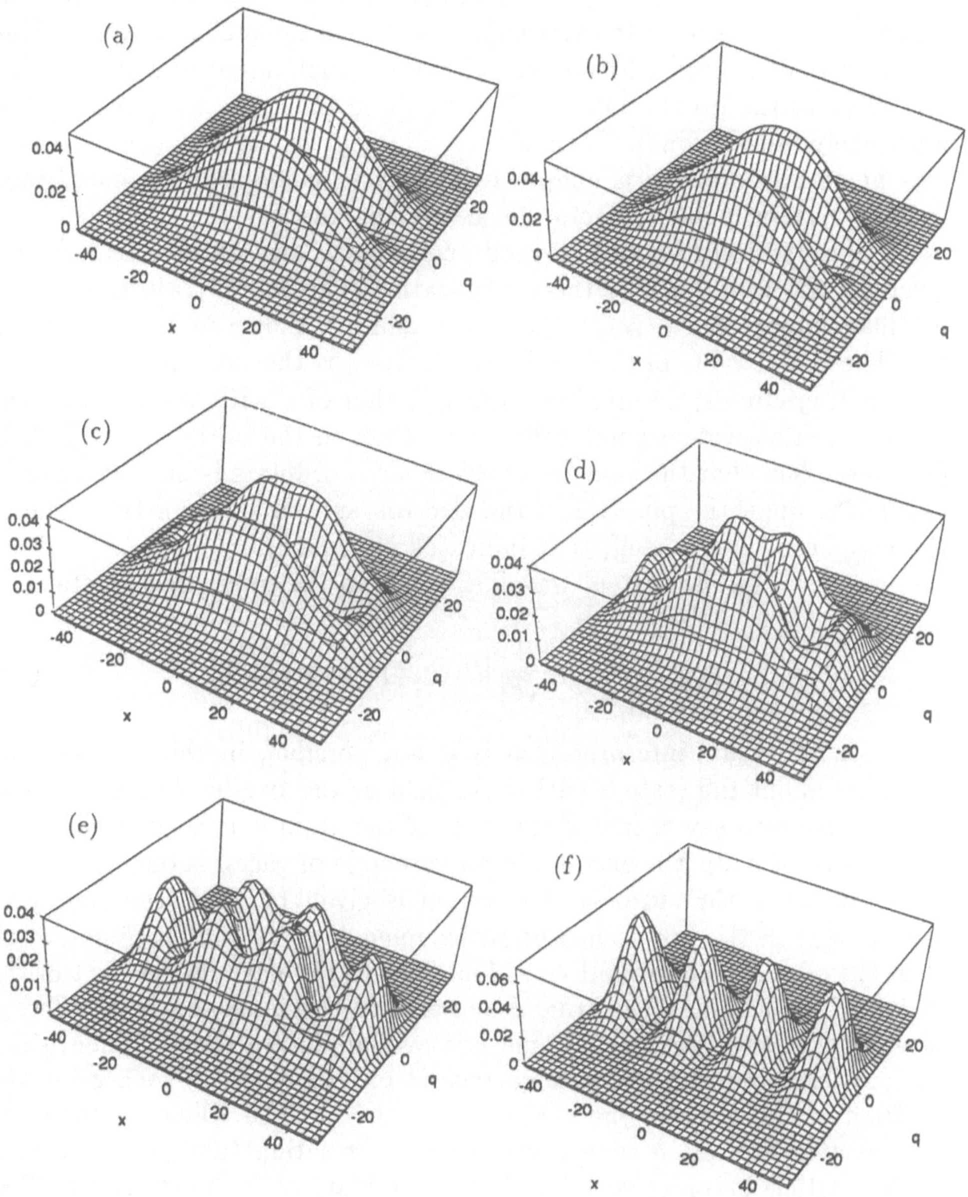

Fig. 1. The probability density at time t equal to (a) 0, (b) $\frac{\pi}{16\Omega}$, (c) $\frac{\pi}{8\Omega}$, (d) $\frac{3\pi}{16\Omega}$, (e) $\frac{\pi}{4\Omega}$ and (f) $\frac{\pi}{2\Omega}$, when the scalar field $\phi(x,t)$ is initially in a single-mode single-"photon" state, $\theta_1(q,t)$, and the "atom" is initially in its ground state, $\psi_0(x)e^{-iE_0t}$. During this period of time the "atom" absorbs a quantum of energy, ω, from the scalar field and the system undergoes a transition to the state $\theta_0(q)\psi_1(x)e^{-iE_1t}$.

shown in Fig. 2(a). Each one traces out a unique evolution of the configuration of the system as the system-point (the point representing the instantaneous configuration of the system) moves in the configuration space. The set consists of sixteen trajectories, with the initial values of x distributed according to the probability density and an initial value for q of 8.22 (its most likely initial value). Each of these different initial values for the system point gives rise to a unique trajectory in configuration-space/time, with no system- point trajectory ever crossing another.

The trajectories of the mode coordinate and of the electron are given by the projection of the configuration-space/time trajectories onto the planes (q,t) and (x,t), respectively, and are shown in Figs. 2(b) and (c). The initial value of the mode coordinate q is the same for each of the sixteen trajectories; despite this, the variation of q with time is different for each initial system-point value, even close to the initial time. This is due to the fact that the velocity of the mode coordinate is instantaneously dependent upon the position of the electron as well as upon the value of the mode-coordinate itself, as is indicated by Eq. (6).

At $t = \frac{\pi}{4\Omega}$ the system wavefunction is in the superposition state

$$\chi(q,x,\frac{\pi}{4\Omega}) = \frac{1}{\sqrt{2}}\left[U_1(q,x) - iU_2(q,x)\right]. \tag{13}$$

In the usual interpretation it is not possible, in this case, to say anything about the state of either the field or the two-level system alone, all that one may say is that if the state of the atom is measured, then the probability of finding it excited, in an ensemble of cases, is one- half.

If, when the state of the system is given by (13), the energy of the particle in the box is measured by removing the confining potential, then the excited and ground wave functions of the particle flow out of the region of the box, and the ground and excited states do this at a different rate, so that they tend to become non-overlapping in space. Then, since the particle's position becomes correlated to its energy, by measuring the position of the particle one can infer its energy. The whole particle and field system evolves in such a way that the position (and hence energy) of the particle becomes correlated with the state of the field. Notice that until this measurement is carried out on the particle, neither the particle nor the field can be considered to have separately a definite state. In the usual approach this is because the wave function is still a superposition of both possibilities: the excitation energy can not be associated with either the field or the particle. In Bohm's approach it is because the coordinate of the particle is not associated with a unique wave function (the ground and excited wavefunctions overlap in space); the energy of both particle and mode-coordinate are well defined but fluctuating. On removing the potential confining the particle the particle wave functions separate in space

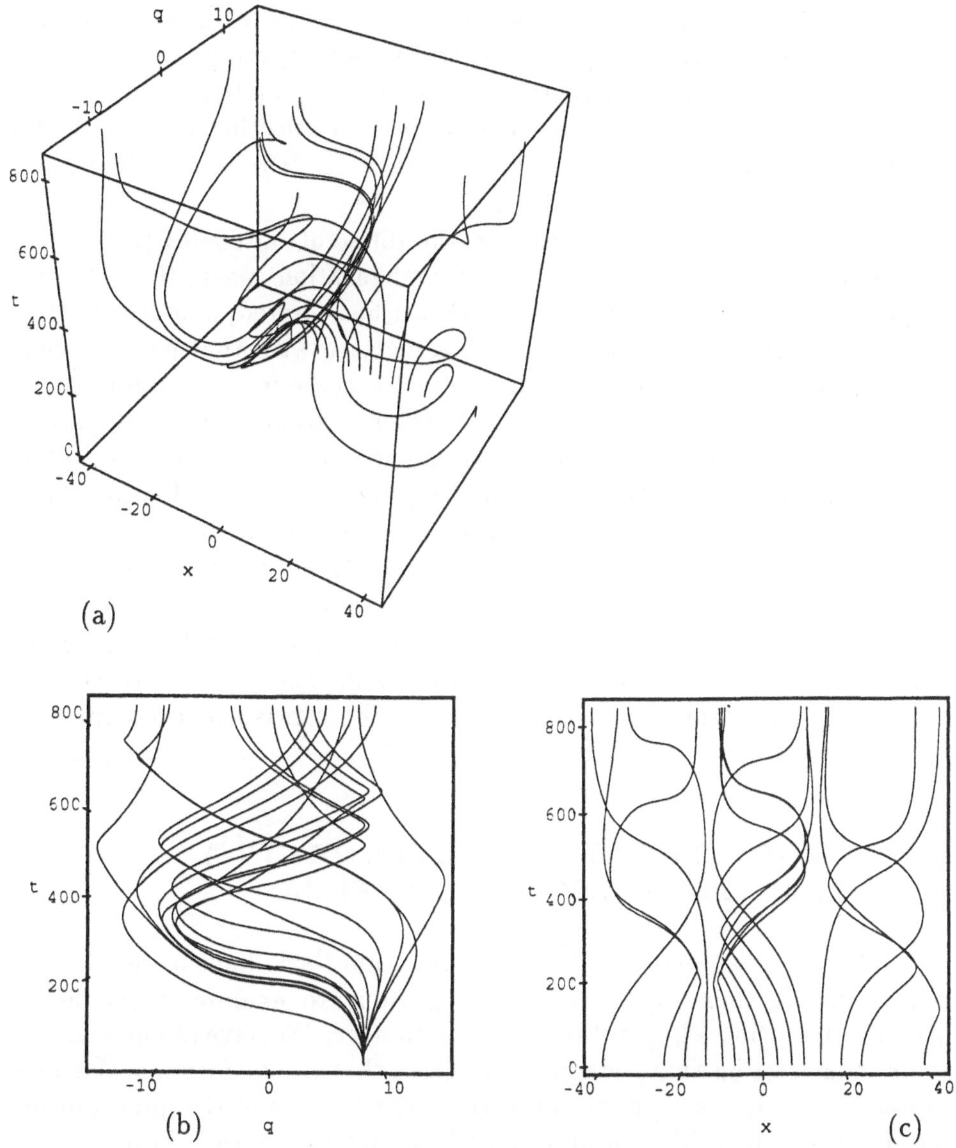

Fig. 2. (a) shows a set of possible configuration-space/time trajectories during the time period $t = 0$ to $\frac{\pi}{2\Omega}$, for the interaction of the "atom" with a single-mode single-"photon" state, $\theta_1(q,t)$. The initial positions of the "atomic" electron are distributed according to the probability density and the initial value of the mode coordinate q is 8.22, its maximum value. The projection of the configuration-space/time trajectories onto the (q,t) plane and the (x,t) plane are shown in Figs. (b) and (c). These show the motion of the electron and the mode coordinate, q, respectively. Figures (d)-(f) show the effect on the configuration-space/time trajectories, and their projections, of reducing the initial value of the mode coordinate to $q = 2.74$.

and, in Bohm's approach, the actual particle position becomes associated uniquely with one component of the superposition state 13, and the field energy and the particle energy both become definite; the particle and the field adopt the eigenvalues associated with either U_1 or U_2.

In the usual approach one has no way of selecting one particular branch of the superposition 13 and so there is no definite outcome in an individual case.

If the packets corresponding to the different outcomes (associated with U_1 or U_2) remain separated in the configuration space, then the system point remains within whichever packet it has entered, and the effective state of the system will remain the one associated with that packet alone. Before the electron reaches a particle detector there is a possibility of recombining the wavepackets separated in configuration space, and causing the wavefunction acting upon the system point to be a superposition again, by reflecting the wavepackets $\theta_1\psi_0$ and $\theta_0\psi_1$ back towards each other; however, the interaction of the electron with its detector causes entanglement of the wavefunction $\chi(q,x,t)$ with the wavefunction of the particle detector, and the original system configuration space must therefore expand to include the coordinates of the electron detector as well. Reflection of the wavepackets in the subspace spanned by q and x,subsequent to the detection of the electron, can not alone cause the overlap in the expanded configuration space required for further interference between these packets.

4. ANTI-CORRELATION OF DETECTORS IN A SINGLE-PHOTON SINGLE-MODE FIELD

If one has an onotlogy in which quantum fields have a well-defined form in space, then the interesting behaviour to explore is the field's particle-like motion when it interacts with matter. We have already seen in the previous section that, in spite of the fact that the quantum scalar field is spread out over space, it can nevertheless, by virtue of the quantum potential, cause a transition in an atom which interacts with it in one region only: one does not need to think in terms of the usual informal description of the process in which a"photon" is emitted or absorbed by the particle in a discontinuous jump. Rather, the whole process proceeds in continuous and deterministic fashion according to Bohm's account.

Perhaps even more surprising is the fact that Bohm's approach can account for the anti-correlation of two detectors placed in a single photon field in a similar way (such as one might have when a single-photon field is split at a semi-transparent mirror, near which are two detectors, one in each of the reflected and transmitted paths). The calculation we present is a simple extension of the one described above. We take exactly the same situation as in the previous section but now we have two identical particles in their respective identical boxes interacting with the tuned cavity field mode. If the system is initiated with a single"photon" in the field mode, whilst the particles are initially in their ground state, how does it come about that only one atom may later be found to be excited? After all the system is set up so that the well-defined cavity field acts on each equally.

If the ground and excited state wavefunctions of the two atoms are denoted by $\psi_0(x_{1,2})$ and $\psi_1(x_{1,2})$, respectively, and the ground and first-excited states of the resonant field-mode are represented by $\theta_0(q)$ and $\theta_1(q)$, then (in the absence of a field/matter coupling) there are three degenerate wavefunctions

$$\begin{aligned} U_1 &= \theta_1(q)\psi_0(x_1)\psi_0(x_2), \\ U_2 &= \theta_0(q)\psi_1(x_1)\psi_0(x_2), \\ U_3 &= \theta_0(q)\psi_0(x_1)\psi_1(x_2). \end{aligned}$$

When field and matter are coupled, if the system is in the state $U_1(x_1, x_2, q)$ at t=0, then the subsequent development of the system wavefunction during the period of interaction is given, from degenerate perturbation theory, by

$$\Psi(x_1, x_2, q, t) = U_1(x_1, x_2, q)\cos\Omega t - \frac{i}{\sqrt{2}}\sin\Omega t\,[U_2(x_1, x_2, q) + U_3(x_1, x_2, q)], \tag{14}$$

where Ω depends on the overlap integrals between the unperturbed states, for the interaction. When Ωt is an integer multiple of π (case A), the system is entirely in the state U_1. This is normally interpreted as corresponding to an excited field and two de-excited atoms since, at this time, a measurement of the energy of each atom will yield the ground level eigenvalues. When Ωt is a half-integer multiple of π (case B), the system is in a superposition of the states U_2 and U_3. This is usually interpreted as corresponding to a de-excited field and either atom 1 or atom 2 being excited, since if a measurement of momentum is performed on both atoms at this time, one of the atoms (the probability of it being atom 1 or atom 2 is equal) will be observed as having the ground level energy, while the other will have the momentum eigenvalue corresponding to the excited state. The wavefunction of the two atoms, for these values of time, is

$$\Psi(x_1, x_2, q, t) = \frac{\pm i}{\sqrt{2}}[U_2(x_1, x_2, q) + U_3(x_1, x_2, q)], \tag{15}$$

which is an entangled state, involving the two separated atomic systems ie: the type of wavefunction required to observe EPR-type correlations.

For all other values of t, the system wavefunction is a superposition of the two states corresponding to cases A and B, and observations associated with either case are made with probabilities that vary as a function of time, as given by Eq. (14). For these values of time it is not possible, in the usual approach, to give any description of the individual behaviour of the field or the atoms.

In Bohm's approach the wave function 14 determines a unique set of trajectories for the evolution of the system's configuration. As the wave function evolves from case A to case B, the coordinates of the field mode and of the two atoms trace out correlated paths, their energy fluctuating in a correlated way as they do so. When case B pertains we know that the field has gone into its ground state, whilst the atoms are in an entangled state in which their energies are correlated. At this stage the energy of both particles is fluctuating in a correlated manner (regardless of their separation) and neither can be considered to be in an energy eigenstate.

Consider now the same process of measuring one of the particle's energy (say particle one) as was used in the previous section (removing its confining potential). Since the overall state of the field/matter system is a product of a function for the field and a function for the particles, we can ignore the field coordinate, which remains static and independent of the motion of the particle coordinates. On removing the potential confining particle one, $\psi_1(x_1)$ and $\psi_0(x_1)$ evolve and tend to separate in space, this causes the functions U_2 and U_3 to separate in the system's configuration space, and the actual configuration of the system becomes associated with either U_2 or U_3 alone. Then the energy of both particles becomes defnite. If, for example, particle one runs out of the box quickly, and thus becomes associated with $\psi_1(x_1)$, then particle two, which is separated and"classically" isolated from particle one, attains its ground state, $\psi_0(x_2)$,with the appropriate energy. (The detailed trajectories are given in [4].)

Note that in this description there is no flow of energy through the (arbitrary amount of) space between the two atoms as their energies both tend towards the correlated and static eigenvalues predicted by 15; rather, they are each acted on by a nonlocal and time-dependent quantum potential which brings about the appropriate correlation of the energies. That is, the quantum force that acts on particle two, as particle one is measured, depends not only on its own coordinate but on the coordinate of particle one as well. The distribution of the values of dynamical quantities amongst the component parts of a composite quantum system is determined, not by pre-assigned local (or nonlocal) interactions in space and time, as we used to in classical physics, but by the evolution of the configuration space wavefunction.

The dependence of the motion of the individual components of a composite quantum system on the evolution of the configuration-space wavefunction is the ground of the essential unity of all quantum phenomena. Although Bohm's interpretation of quantum mechanics allows for the analysis of a composite quantum system in terms of well-defined and deterministic individual motions of its constituent parts in the usual three-dimensional space, the true arena in which the system's motion is determined (and hence the arena for understanding the system's motion and the interrelations of its parts) is configuration space. This is the essential feature of quantum mechanics which transcends the paradigm of classical physical description and expresses the essential unity of quantum phenomena [16].

REFERENCES

1. D. Bohm, *Phys. Rev.* **85**, 166, 180 (1952).
2. D. Bohm, B. J. Hiley and P. Kaloyerou, *Phys. Rep.* **144**, 6, 349 (1987).
3. M. M. Lam and C. Dewdney, "Cavity quantum scalar field dynamics: the Bohm approach: Part I - the free field, Part II - the interaction of the field with matter," *Found. Phys.* **24**(1) (1994).
4. C. Dewdney, G. Horton, M. M. Lam, Z. Malik, M. Schmidt, *Found. Phys.* **22**, 10 (1992).
5. E. T. Jaynes and F. W. Cummings, *Proc. IEEE* **51**, 89 (1963).
6. C. Philippidis, C. Dewdney, and B. J. Hiley, *Nuovo Cimento* **52B**, 15 (1979).
7. C. Dewdney and B. J. Hiley, *Found. Phys.* **12**, 27 (1982).
8. C. Dewdney, *Phys. Lett.* **109A**, 377 (1985).
9. C Dewdney and Z Malik, "Angular-momentum measurement and non-locality in Bohm's interpretation of quantum theory," *Phys. Rev. A*, to appear.
10. C. Dewdney, A. Kyprianidis, and J. P. Vigier, *J. Phys. A* **17**, L741 (1984).
11. C. Dewdney, P. R. Holland, and A. Kyprianidis, *Phys. Lett.* **119A**, 259, (1986).
12. C. Dewdney, P. R. Holland, and A. Kyprianidis, *Phys. Lett.*, **121A**, 105, (1987).
13. C. Dewdney, P. R. Holland, and A. Kyprianidis, *J. Phys. A* **20**, 4717 (1987).
14. C. Dewdney, *J. Phys A:Math. Gen.* **25**, 3615-3626 (1992).
15. C. Dewdney and M. M. Lam, in *Information Dynamics*, (NATO ASI Series B), H. Atmanspacher and H. Scheingraber, eds. (Plenum, New York, 1990).
16. D. Bohm and B. J. Hiley, *The Undivided Universe* (Routledge, London 1993).

PROPOSAL FOR AN EXPERIMENT TO DETECT MACROSCROPIC QUANTUM COHERENCE WITH A SYSTEM OF SQUID'S

G. Diambrini-Palazzi

Dipartimento de Fisica
Università di Roma "La Sapienza"
and INFN sezione di Roma

The proposed experiment seeks to obtain the following results: (1) Discover the existence of a pure macroscopic quantum coherent state (MQCS) by measuring the quantum states of an rf SQUID with two symmetrical potential wells. (2) Once this first goal is accomplished, a set of measurements could distinguish the part of the MQCS damping due to the interaction with the environment from that possibly due to the presence of wormholes, as, for example, predicted by some theoretical approach to the quantum gravity.

Key words: macroscopic, quantum coherence, SQUID, wormholes.

1. INTRODUCTION

The quantum mechanics (QM) behaviour of *microscopic* quantum states has been successfully established after about 70 years of experimental results obtained mainly in the field of elementary particle physics.

But problems concerning consistency of QM predictions with causality, relativity, and macroscopic behaviour have been raised and discussed just since the beginning of the QM. In particular it was suggested that the locality and causality of interactions could be found to be in conflict with the basic interpretation of the QM, in particular with the coherent quantum behaviour of the wave functions and their collapse as a consequence of the observational process (according to the "Copenhagen school"). The most famous reference to this problem is of course the paper wherein Einstein, Podolski, and Rosen proposed their *gedanken* experiment (1935).

"Realistic" theories, such as those of "hidden variables" proposed to solve the conflict, became "falsifiable" only after the Bell inequalities have been established (1964). In fact, soon after the Bell theorem was published, a number of experiments were carried out (e.g., Clauser and Shimony, 1978; Aspect, 1982) which have shown a violation of the Bell inequalities, as predicted by the QM, essentially only for the single particle behaviour.

Recently in several theoretical papers it was proposed that the *macroscopic quantum coherence* could be violated (Hawking, 1988; Coleman, 1988; Ellis et al., 1990; Ghirardi, 1988, Ellis et all, 1991).

Recently, the possibility to realize experiments for detecting the quantum behaviour of macroscopic coherent states was investigated in a number of papers (Leggett, 1980, 1985; Ellis et al., 1990).

For this aim the quantum behaviour of an rf SQUID (a superconducting ring with one Josephson junction) has been analysed, and it has been shown that, in principle, it should be possible to perform an experiment with this superconducting device for detecting the coherent quantum behaviour of a "macroscopic" number of Cooper pairs (up to 10^{22}), performing "tunneling oscillations" through a Josephson junction with two symmetric potential wells.

In fact, we want to underline that the existence of one superconducting quantum state is not at all, in itself, a proof of the existence of a MQC state.

The experiments performed up to now were able to test the quantum decay (Clarke et al., 1988), or the *incoherent* tunnelling oscillation (J. Lapoint et al., 1990), in rf SQUID *but not* the macroscopic coherent behaviour due to the superposition principle of macroscopic quantum states.

We believe that the technology for the preparation of SQUID's and the basic physics knowledge of their behaviour is now so advanced that such an experiment may be proposed and finally performed, as far as we have been able to reach all the intermediate needed steps, as will be explained in the following.

2. AN MQC EXPERIMENT BY A SYSTEM OF SQUIDS

First let us give a description of the basic experimental method. Consider a system of 3 SQUIDs, as in Fig. 1. The rf SQUID is the source of two quantum states of the magnetic flux ϕ. The SQUID has to be prepared for having symmetric tunnelling oscillation of a circulating superconducting current, associated with the two quantum states of the magnetic flux. We want to detect the effect of the superposition of these two-state wave functions as follows.

In order to make a measurement of the fluxoid direction at a certain

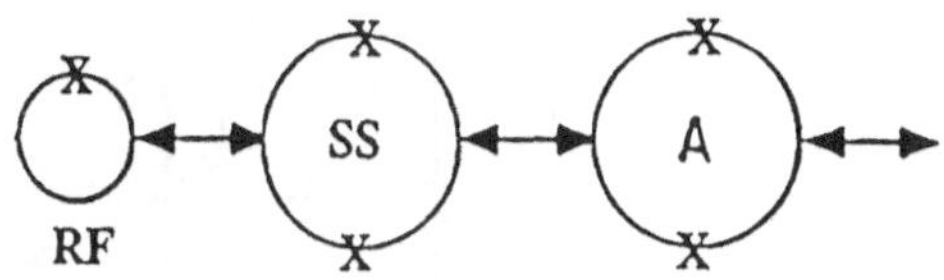

Fig. 1.

time t_1, we can use a SQUID switch [SS] arranged in order to have the following behaviour. First of all, it can be activated for the time Δt by an impulsive current given to it at a prefixed time t_i. When we do so, then the following cases can happen:

Either the the SS has a transition to the normal state or nothing happens and the SS remain essentially unperturbed; in both cases we can infer the sign of the flux of the rf SQUID.

3. THE RF SQUID

The rf SQUID is a superconducting ring interrupted by a Josephson junction (JJ in the following). This system can be described with good approximation by four macroscopic parameters (Barone and Paternò, 1982), which can be classically measured (Martinis, Devoret and Clarke, 1989, and related references): L, i_c, R, C. L is the value of the geometric inductance of the superconducting ring; i_c is the critical current of the junction, i.e., the maximum current that can pass through the junction with no voltage appearing across the junction itself. R and C are, respectively, the equivalent resistance and the capacity associated with the junction. In the following we will see that particular care must be taken in defining the proper value of R to be used in the various part of the experiment.

The other two quantities that must be defined to describe the time evolution of this system are the thermodynamic temperature T of the heath bath which is in thermal equilibrium with the junction through the resistance R and Φ_e, the flux of the external magnetic field $\boldsymbol{B}$ through the superconducting ring of inductance L. The macroscopic variable to be studied is the total flux Φ of the magnetic field through the ring, defined as

$$\Phi = \Phi_e + Li,$$

where i is the current circulating in the ring.

Now, the Hamiltonian associated with this system is

$$H[\phi, \dot{\phi}] = \frac{1}{2} C \dot{\phi}^2 + \frac{(\phi - \phi_e)^2}{2L} - \frac{i_c \phi_e}{2\pi} \cos \frac{2\pi\phi}{\phi_e},$$

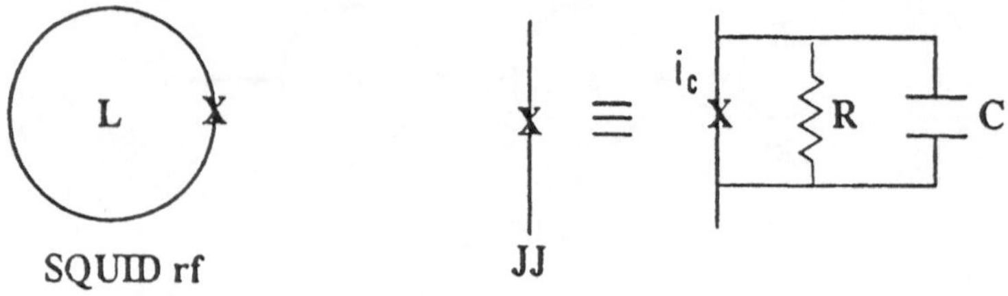

Fig. 2. Scheme of an rf SQUID and of a Josephson junction.

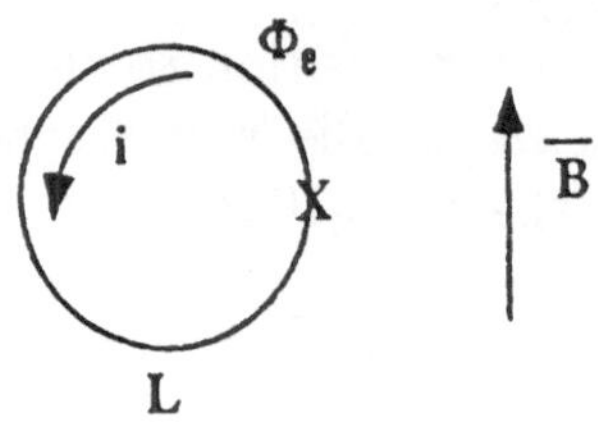

Fig. 3.

where the first term on the right-hand side represents the energy of the electric field across the gap, the second term of the magnetic field through the ring, and the third is the coupling energy of the junction (Josephson, 1962, Leggett, 1980). Suppose we apply an external field B such that $\Phi_e \cong \Phi_0/2$, and if the adimensional parameter $\Lambda = 2Li_e/\Phi_0$ is such that

$$1.5 > \Lambda > \pi^{-1},$$

then the effective potential takes the shape of a double well with two minima in Φ_-, Φ_+ separated by a barrier of amplitude

$$\Delta U \leq \frac{i_c \phi_0}{2\pi}$$

The two solutions Φ_-, Φ_+ represent two currents equal in modulus but circulating in opposite directions:

$$I_\pm = \pm \left| \frac{\phi - \phi_e}{L} \right| .$$

Each one of these two states $|- >; |+ >$ has a frequency of oscillation $\omega_p = 1/\sqrt{LC}$ in the bottom of the well and a probability of transition

$$| < \phi_+ | \phi_- > |^2 = P_\pm = \left(\frac{\Delta U}{\hbar \omega_p} \right)^{1/2} \exp \left(\frac{-\Delta U}{\hbar \omega_p} \right)$$

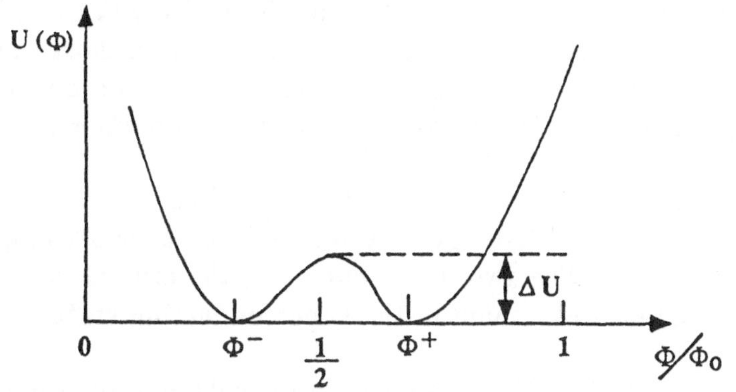

Fig. 4. Potential of the rf SQUID as function of the total flux.

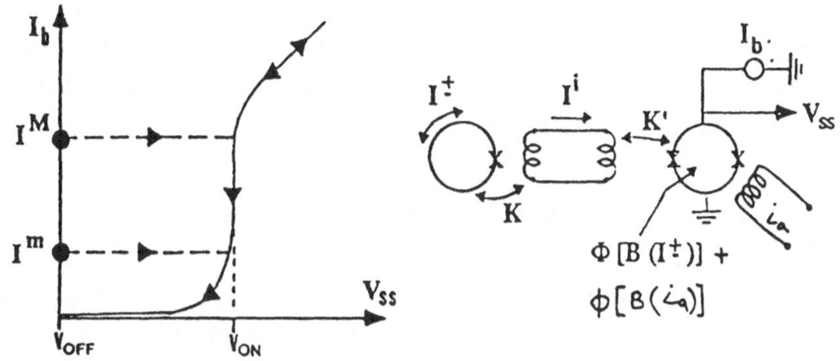

Fig. 5. Current/voltage characteristic of the switch and scheme of the inductive coupling with the rf SQUID.

4. THE SUPERCONDUCTING SWITCH

The superconducting switch is an hysteric dc SQUID (a superconducting ring interrupted by two JJ, where the hysteresis parameter $\beta_c = 2\pi R^2 I_c C/\Phi_e >> 1$) inductively coupled to the rf SQUID. Its peculiarity is to switch between two different voltage states (ON/OFF)=(V_{on}/V_{off}). depending on the bias current I_b and on the total applied magnetic field.

The corresponding current/voltage characteristic is shown in Fig. 5. The maximum (superconducting) current that can pass through the SQUID, with no voltage across the junctions is modulated with the continuity by the flux of the magnetic field between the two values I^M and I^m, corresponding, respectively, to $\phi(B) = 0$, and $\phi(B) = \phi_e/2$. Now, suppose we apply a constant flux $\phi_e \approx \phi_0/4$ such as to obtain $I_e(\phi_0/4) \approx (I^M + I^m)/2$ and to "turn on" the system with a current $I_b < I_c$ for a short time Δt; and suppose that during this time a flux Φ_x of amplitude $\phi_0/4$, but of unknown sign, is applied to the SQUID. We can have then two different situations:

I:$\hat{\phi}_x = \phi_e$. The two fluxes add, the critical current will go below I_b, the Switch becomes resistive and at the output we have a voltage pulse.

II:$\hat{\phi}_x = -\hat{\phi}_e$. The two fluxes subtract, the critical current will go up, and the system will remain in the superconducting state.

According to the QM, as a result of the coherent quantum tunneling between the two wells, the following doublet of states is obtained:

$$\psi_\pm = \frac{1}{\sqrt{2}}[\psi_L \pm \psi_R].$$

Suppose now we start the rf SQUID at time zero in the left well. Then the system will oscillate between the two wells, and the difference $P(t)$ between the probabilities P_L and P_R to find the system in the left or in the right well, respectively, will be given by

$$P(t) = P_L - P_R = \cos^2\frac{\omega_0 t}{2} - \sin^2\frac{\omega_0 t}{2} = \cos\omega_0 t.$$

Therefore, by performing the measurement at $t = \pi/2\omega_0$, one has to find $P(t) = 0$, so $P_L = P_R$ and $P_L = 50\%$, $P_R = 50\%$. By repeating the measure many times, exactly with the same procedure, these values can be checked. By performing, always with the same procedure, a set of measurements at $t = \pi/\omega_0$, one gets:

$$P(t) = P_L - P_R = -1 \quad \text{and thus} \quad P_L = 0\% \text{ and } P_R = 100\%,$$

i.e. the flux will be found always in the right state.

Of course, with no quantum interference, but with a mixture of states, we always should find $P_L = 50\%, P_R = 50\%$.

5. MEASURE OF THE MQCS DAMPING

Following theoretical approach by S. W. Hawking, S. Coleman (1988), and J. Ellis (1990), the influence of wormholes would be to multiply the previous expression by an exponential factor as follows:

$$P(t) = \cos\omega_0 t^e e^{-2\beta^2 N^2 t^2},$$

where

$$\beta = \lambda \left(\frac{e^2}{2m_e} \right) \varepsilon, \quad \varepsilon = \omega_0^2, \quad \omega_0 = \frac{2\pi}{\sqrt{LC}} = 10^{11} Hz.$$

Here λ is the coupling constant of the wormholes, which may be estimated to be of the order of $\frac{m_e}{m_p}$, where m_p is the Planck mass.

From these values, one finds $\beta = 1.6\ 10^{-3}\ \sec^{-1}$; therefore the collapse of the coherent quantum state, due to wormholes, occurs in a time $t = (1/\beta N) = 5$ sec.

Here we assume a total number of Cooper pairs $N = 1.3 \cdot 10^{22}$, the one due to a massive SQUID, whose volume is nearly 1 cm^3 and whose relative density of pairs is 23% of the atomic density (Wood and Van Vechten, 1992).

Therefore, at least in principle, if we could repeat the measurements with the method explained above after such a delay, we should find $P(t) = 0$.

Moreover one must keep into account the thermal damping with the surroundings, calculated by Caldeira and Leggett (1983) and expressed by a characteristic damping constant α. In order to estimate in a first approximation this effect, it is enough to multiply by the exponential factor $e^{-\alpha t}$ all the four density matrix elements. Therefore, at time t, one obtains

$$\rho(t) = \begin{vmatrix} \cos^2 \frac{\omega_0}{2} t \cdot e^{-\alpha t} & \cos \omega_0 t \cdot e^{-\alpha t} \cdot e^{-\beta^2 N^2 t^2} \\ \cos \omega_0 t \cdot e^{-\alpha t} \cdot e^{-\beta^2 N^2 t^2} & \sin^2 \frac{\omega_0}{2} t \cdot e^{-\alpha t} \end{vmatrix}.$$

It is evident that the measure of the diagonal term damping allows us to measure the constant α. Therefore we may divide the nondiagonal terms by $e^{-\alpha t}$, and at the end we may calculate the possible collapse time due to wormholes.

REFERENCES

A. Aspect et al., *Phys. Rev. Lett.* **49** (1982) 1804.

A. Barone and G. Paternò, *Physics and Applications of the Josephson Effect* (Wiley, New York, 1982).

J. Clarke et al., *Science* **239** (1988) 992.

J. F. Clauser and A. Shimony, *Rep. Prog. Phys.* **41** (1978) 1882.

S. Coleman, *Nucl. Phys. B* **307** (1988) 867.

J. Ellis, S. Mohanty, and D. V. Nanopulos, *Phys. Lett. B* **235** (1990) 305.

S. W. Hawking, *Phys. Rev. D* **37** (1988) 904.

A. J. Leggett, *Suppl. Prog. Theor. Phys.* **69** (1980) 80.

A. J. Leggett and A. Garg, *Phys. Rev. Lett.* **54** (1985) 857.

J. M. Martinis, M. Devoret, and J. Clarke, *Phys. Rev. D* **35** (1987) 4682.

ON LONGITUDINAL FREE SPACETIME ELECTRIC AND MAGNETIC FIELDS IN THE EINSTEIN-DE BROGLIE THEORY OF LIGHT

M. W. Evans

Department of Physics
University of North Carolina
Charlotte, North Carolina 28223

In the Einstein-de Broglie theory of light, the photon has finite rest mass. It is shown that longitudinal solutions of the Einstein-de Broglie-Proca (EBP) equation give longitudinal magnetic and electric fields, $\boldsymbol{B}^{(3)}$ and $i\boldsymbol{E}^{(3)}$, respectively, which can be used to devise a number of novel experimental tests of the theory. It is shown that special relativity does not prohibit finite photon rest mass (m_0), because m_0^2 is the first Casimir invariant of the Poincaré group (inhomogeneous Lorentz group). It is the invariant m_0^2 that appears in the EBP equation. It is shown that gauge transformations of the first and second kind remain valid for finite m_0, provided $A_\mu A_\mu = 0$, where A_μ is the vector potential of the de Broglie- Proca field. This condition implies that the scalar potential is equal to the magnitude of the vector potential. For finite photon mass, the Coulomb gauge is therefore inconsistent, and the Lorentz condition always holds, i.e., is a mathematical consequence of the EBP equation. Experimental evidence [17] for finite photon mass is briefly reviewed, evidence which removes the numerous physical obscurities inherent in the notion of a zero photon rest mass. Within the structure of the Einstein-de Broglie theory, the photon becomes a massive boson, electric and magnetic fields become manifestly covariant four vectors, the quantization of the de Broglie-Proca field becomes straightforward, leading to a particle interpretation (the photon) without the difficulties and obscurities involved in the quantization of the massless electromagnetic field, and longitudinal fields $\boldsymbol{B}^{(3)}$ and $i\boldsymbol{E}^{(3)}$ emerge consistently. The acceptance of finite photon mass removes the physical obscurity introduced by the need to use the Euclidean E(2)

Waves and Particles in Light and Matter, Edited by
A. van der Merwe and A. Garuccio, Plenum Press, New York, 1994

for the little group of the Poincaré group. The Lie algebra becomes that of a massive boson, described naturally by a rotation group.

1. INTRODUCTION

It is usually concluded in the electrodynamical literature [1 – 16] that the photon is massless and that the range of the electromagnetic field is infinite. This conclusion is not, however, supported by experimental data. To the contrary, Vigier [17] has recently reviewed a substantial amount of evidence that leads to the conclusion of finite photon rest mass. These data include, to take two of many examples, the direction dependent anisotropy of the frequency of light in cosmology and frequently observed anomalous red shifts.

In papers and correspondence circa 1916 to 1919 [17], Einstein [18] proposed a photon rest mass [19], which can be estimated from the Hubble constant to be about 10^{-68} kg. An immediate consequence is that the d'Alembert equation is replaced by the Einstein-de Broglie-Proca (EBP) equation, which can be expressed [20, 21] in the form

$$\Box A_\mu = -\xi^2 A_\mu , \tag{1}$$

where $\xi = m_0 c/\hbar$; ; here m_0 is the photon rest mass, c the speed of light, the universal constant of special relativity, and $\hbar$ the reduced Planck constant. The potential four-vector A_μ of the de Broglie-Proca field is manifestly covariant, and has four, physically meaningful, components, one timelike ((0)) and three spacelike, of which two are transverse ((1) and (2)) and one is longitudinal ((3)). From Eq. (2), the range ξ^{-1} of the field becomes 10^{26} m; cosmic in dimensions, but finite. Equation (1) is an expression of the Einstein-de Broglie theory of light [17] and implies that gauge transformations of the first and second kind [20, 21] can no longer be interpreted as implying zero photon rest mass. It is well known that the EBP equation implies mathematically [20, 21] the Lorentz condition

$$\frac{\partial A_\mu}{\partial x_\mu} = 0 \tag{2}$$

for the massive boson. If the photon has rest mass, it is always described therefore by the Lorentz condition. Experimental evidence [17] for finite photon rest mass implies that gauge invariance must be re-interpreted fundamentally, and this is part of the purpose of this paper, in which it is shown that finite m_0 is consistent with gauge invariance of the first and second kind if and only if

$$A_\mu A_\mu = 0, \quad m_0 \neq 0, \tag{3a}$$

a condition which implies

$$\phi = c|\mathbf{A}|, \tag{3b}$$

where, in S.I. units,

$$A_\mu = \epsilon_0 (\mathbf{A}, \frac{i}{c}\phi); \tag{3c}$$

herein ϕ is the scalar potential, ϵ_0 the permittivity in vacuo, and $\mathbf{A}$ the vector potential of the de Broglie-Proca field. The condition (3a) is consistent with the Lorentz condition (2) but inconsistent with a massless gauge such as the traditional Coulomb gauge [1 – 16].

It is well known, furthermore [20, 21], that the notion of zero photon rest mass leads to considerable physical obscurity, for example, in the quantization of the Maxwellian electromagnetic field. The traditional theory abandons the longitudinal and timelike field polarizations as being "unphysical," and in so doing inevitably loses manifest covariance. Another traditional difficulty [20] is that the little group of the Poincaré group [20] for the massless photon becomes the Euclidean E(2), which is physically obscure. The Lie algebra for the Maxwellian electromagnetic field on the other hand is that of the Lorentz group. These difficulties are accepted because it is traditionally thought that special relativity implies zero photon mass, and that gauge invariance of the first and second kind can be interpreted only in terms of zero photon rest mass. In this paper it is shown that both of these traditional viewpoints are flawed, and that in consequence, the Einstein-de Broglie theory of light is consistent with: (a) special relativity and (b) gauge transformation. We recall for reference that the massless electromagnetic field is summarised in the d'Alembert equation

$$\Box A_\mu = 0. \tag{4}$$

Quantization [20] of Eq. (1) is straightforward, but that of Eq. (4) is beset with considerable difficulty. From quantization of Eq. (1), for the massive boson, the conclusion is reached that the massive boson is a particle (the photon) with finite mass and three physically meaningful spacelike polarizations, ((1), (2), and (3)). Quantization [20] of Eq. (4) traditionally proceeds in the Coulomb or Lorentz gauge. To quote from Ryder [20]:

> Quantisation of the electromagnetic field suffers from difficulties posed by gauge invariance. The quantisation procedure is outlined in both the radiation (Coulomb) gauge, in which there appear only the two physical (transverse) polarisation states, and in the Lorentz gauge, in which all four polarisation states appear, the formalism being Lorentz covariant. The resulting difficulties are resolved by the method of Gupta and Bleuler.

The reader is referred to Ryder [20] for an excellent account of these difficulties. The Coulomb gauge is inconsistent, furthermore, with a non-zero photon rest mass, so that conversely, finite m_0 implies immediately that the notion of there being only two physically meaningful photon polarization states must be abandoned. One is led ineluctably to the conclusion that there are four physically meaningful photon polarizations ((0) to (3)).

Lorentz gauge quantization [20] in the limit $m_0 \to 0$ is possible only with the Gupta Bleuler condition [22], which leads to the conclusion that admixtures of timelike and longitudinal spacelike photon polarizations are physical states [20]. In a diametrically self-contradictory procedure, the traditional theory abandons these physical states as "unphysical."

This procedure is logically untenable, and recently [23 – 28] this has become clear through the discovery of a simple relation between longitudinal and transverse solutions of Maxwell's equations in vacuo:

$$\boldsymbol{B}^{(3)} = \frac{\boldsymbol{E}^{(1)} \times \boldsymbol{E}^{(2)}}{E_0 c i} = \frac{\boldsymbol{B}^{(1)} \times \boldsymbol{B}^{(2)}}{B_0 i} = B_0 \boldsymbol{k}, \tag{5}$$

where $E_0 = cB_0$ are the scalar free-space amplitudes. Equation (5) comes directly from the original Maxwell equations, without the introduction of scalar and vector potentials; and is an entirely novel relation between physically meaningful electric and magnetic components of the electromagnetic field in vacuo. It can be derived without reference to gauge theory, but is consistent with gauge invariance. Here $\boldsymbol{E}^{(1)}$ and $\boldsymbol{E}^{(2)}$ are the oscillating transverse components of the electric field, taken to be a plane wave in vacuo. The vector product in Eq. (5) is defined by the Stokes parameter S_3:

$$\boldsymbol{E}^{(1)} \times \boldsymbol{E}^{(2)} = -iS_3\boldsymbol{k}. \tag{6}$$

In a light beam in which there is some degree of circular polarization, therefore, S_3 is always non-zero, implying that the longitudinal magnetic field $\boldsymbol{B}^{(3)}$ is non-zero in vacuo. The transverse components in Eq. (5) are the usual vacuum plane wave solutions of Maxwell's equations:

$$\begin{aligned} \boldsymbol{E}^{(1)} &= \frac{E_0}{\sqrt{2}}(\boldsymbol{i} - i\boldsymbol{j})e^{i\phi}, & \boldsymbol{E}^{(2)} &= \frac{E_0}{\sqrt{2}}(\boldsymbol{i} + i\boldsymbol{j})e^{-i\phi}, \\ \boldsymbol{B}^{(1)} &= \frac{B_0}{\sqrt{2}}(i\boldsymbol{i} + \boldsymbol{j})e^{i\phi}, & \boldsymbol{B}^{(2)} &= \frac{B_0}{\sqrt{2}}(-i\boldsymbol{i} + \boldsymbol{j})e^{-i\phi}, \end{aligned} \tag{7}$$

where the phase is $\phi = \omega t - \boldsymbol{\kappa} \bullet \mathbf{r}$. Here ω is the angular frequency at an instant t and $\boldsymbol{\kappa}$ the wave vector at a point $\mathbf{r}$. It can also be shown [23] that the concommitant longitudinal electric field $\boldsymbol{E}^{(3)}$ exists in vacuo, and is related to $\boldsymbol{B}^{(3)}$ by

$$\boldsymbol{E}^{(3)} \times \boldsymbol{B}^{(2)} = \boldsymbol{B}^{(3)} \times \boldsymbol{E}^{(2)}, \tag{8}$$

so that $\boldsymbol{E}^{(3)}$ is imaginary if $\boldsymbol{B}^{(3)}$ is real. The real part of the field $\boldsymbol{E}^{(3)}$ can be expressed as

$$\boldsymbol{E}^{(3)} \propto E_0\boldsymbol{k}. \tag{9}$$

It is worth demonstrating explicitly that $\boldsymbol{B}^{(3)}$ and $\boldsymbol{E}^{(3)}$ are solutions in vacuo of the Maxwell equations, because:

$$\begin{aligned} \nabla \times \boldsymbol{E}^{(3)} &= 0, & -\frac{\partial \boldsymbol{B}^{(3)}}{\partial t} &= 0, \\ \nabla \times \boldsymbol{B}^{(3)} &= 0, & \frac{1}{c^2}\frac{\partial \boldsymbol{E}^{(3)}}{\partial t} &= 0, \\ \nabla \cdot \boldsymbol{E}^{(3)} &= 0, & \nabla \cdot \boldsymbol{B}^{(3)} &= 0. \end{aligned} \tag{10}$$

These relations follow from Eqs. (5) and (9), i.e., $\boldsymbol{B}^{(3)}$ and $\boldsymbol{E}^{(3)}$ are solenoidal and phase-independent.

In this paper we show that $\boldsymbol{B}^{(3)}$ and $\boldsymbol{E}^{(3)}$ are natural consequences of the Einstein-de Broglie theory of light, and are physically meaningful magnetic and electric fields. Experiments to detect them would support the theory of Einstein and de Broglie. Equations (5) and (9) are therefore relations between longitudinal and transverse field components in the massless limit of the Einstein-de Broglie theory. This conclusion is consistent with the recent development [24] by the present author of manifestly covariant electrodynamics, using electric and magnetic four-vectors. This development is equivalent to the Einstein-de Broglie theory in the massless limit ($m_0 \rightarrow 0$), and is a direct consequence of the existence of $\boldsymbol{B}^{(3)}$ and $\boldsymbol{E}^{(3)}$ defined by Eqs. (5) and (9), respectively. It is impossible to reconcile the existence of Eqs. (5) and (9) with traditional thinking, in which $\boldsymbol{B}^{(3)}$ and $\boldsymbol{E}^{(3)}$ are abandoned as unphysical. Clearly, $\boldsymbol{B}^{(3)}$ and $\boldsymbol{E}^{(3)}$ are formed from physical quantities such as the Stokes parameter S_3. In the Einstein-de Broglie theory, on the other hand, $\boldsymbol{B}^{(3)}$ and $\boldsymbol{E}^{(3)}$ are physical fields, components of the four-vectors E_μ and B_μ in vacuo. A longitudinal solution of Eq. (1) for $\boldsymbol{B}^{(3)}$ is given in Sec. 2; where it is shown that $\boldsymbol{B}^{(3)}$ is an exponentially decaying function of ξ in the propagation axis z of the light beam. The divergence of $\boldsymbol{B}^{(3)}$ is non-zero for finite m_0, and is given by $-\xi B^{(0)}$, a magnetic monopole in vacuo. The numerical value of ξ ($10^{-26}\ m^{-1}$) is so small that for all practical purposes, and for laboratory dimensions, $\boldsymbol{B}^{(3)}$ is a constant magnetic field, independent of distance and time. Section 3 derives general solutions of Eq. (1) for the transverse and longitudinal fields of the electromagnetic plane wave in vacuo. A discussion follows of the role of $\boldsymbol{B}^{(3)}$ and $\boldsymbol{E}^{(3)}$ in various experimental tests of the Einstein-de Broglie theory of light, taking into account experimental evidence [17] for finite photon mass.

2. LONGITUDINAL SOLUTIONS OF THE EBP EQUATION IN VACUO

In quantum optics interpreted by Einstein and de Broglie [17], light is constituted by real Maxwellian waves which co-exist in spacetime with moving particles – photons. In the Copenhagen interpretation of Bohr, Schrödinger, Pauli, Glauber, and others, on the other hand, light is made up of waves of probability, which cannot co-exist in spacetime with photons. In the interpretation of the Einstein-de Broglie school, the photon is massive, in that of the Copenhagen school, it is not necessarily so. The basic electrodynamical equations are therefore (1) and (4) respectively. Although it is frequently asserted [1 – 16, 20, 21] that the photon is massless in its rest frame, there is no supporting experimental evidence. Indeed, it appears to be impossible to test the hypothesis of zero m_0, because it is impossible to test the implication that the range of electromagnetic radiation is infinite. On the other hand, finite m_0 leads [17] to such observable implications as anomalous red shifts, reported on numerous occasions, and tired-light phenomena. Einstein [18], some years after his theory of special relativity (1905), and during his development of general relativity, proposed that the photon's rest mass is finite, i.e., that the mass of the photon is finite in a frame of reference moving at the speed of light. This leads [17 – 19] to Eq. (1). It is clear therefore that Einstein saw no contradiction with special relativity in his proposal, i.e., Eq. (1) is Lorentz covariant, *even though the photon rest mass m_0* is non-zero. Several conclusions flow immediately from this proposal.

Firstly, the notion that the photon is massless in the frame of the observer (laboratory frame) because it travels at the speed of light is incorrect if the photon rest mass m_0 is non-zero. In the contemporary description [20] of special relativity, the reason for this is that the quantity

$$C = P_\mu P_\mu \tag{11}$$

is the first (or "mass") Casimir invariant of the Poincaré (inhomogeneous Lorentz) group; here P_μ is the generator of spacetime translations, first introduced by Wigner in 1939 [29]. A spacetime translation is defined by the operation

$$x'_\mu = x_\mu + q_\mu , \tag{12}$$

where x_μ is the distance-time four-vector of Minkowski spacetime. P_μ does not appear in the homogeneous Lorentz group [20], i.e., in a group made up only of boost transformations and Lorentz rotations. The quantity m_0^2 (the square of the rest mass) is therefore invariant under Lorentz transformations, i.e., is the same in the rest frame of the photon (which travels at the speed of light) and in the observer frame. The invariant m_0^2 appears in Eq. (1), which is Lorentz covariant, i.e., fully consistent with special relativity. The latter theory does not imply, therefore, that the photon rest mass is zero. It is clear that Einstein himself [17, 18] saw no inconsistency with special relativity in his proposal of finite m_0, and contemporary theory also shows that m_0^2 is an invariant of the Poincaré group. The Einstein-de Broglie theory of light is therefore consistent with special relativity. This means that the rest-frame momentum of the photon (a massive boson) is timelike, not lightlike, and that the photon has rest energy m_0c^2, i.e., that the energy of the photon in its own frame of reference, which moves at the speed of light, is m_0c^2, about 10^{-57} J. The spacelike momentum of the photon in its own rest frame is zero, because it does not move relative to this rest frame. In its rest frame, the photon is thus described by a four-vector:

$$q_\mu = (0, 0, 0, im_0c) = (0, 0, 0, i\frac{En_0}{c}) \tag{13}$$

in Minkowski spacetime. In the laboratory frame of the observer, however, the photon's momentum is finite, and the vector (13) is transformed into

$$p_\mu = L_{\mu\nu}q_\nu , \tag{14}$$

where $L_{\mu\nu}$ denotes a Lorentz transformation [20] which transforms q_ν into p_μ. Clearly, the latter is observed in the laboratory. Wigner [29] showed that this transformation can be described from a knowledge of the rotation group and that the little group for q_μ is a rotation group.

As discussed by Vigier [17], the consequences are that photons *slow down* in the laboratory frame of an observer, although the rest frame must move at the speed of light, which is a universal constant of special relativity. Photons in the frame of the observer behave like relativistic non-zero mass particles, with rest mass $m_0 \sim 10^{-68}$ *kg*. The energy-momentum four-vector in the observer frame is p_μ, with components [17]:

$$\begin{gathered} p_\mu := (\boldsymbol{p}, i\frac{En}{c}) , \\ En = h\nu = m_0c^2(1 - \frac{v^2}{c^2})^{-\frac{1}{2}} , \\ |\boldsymbol{p}| \sim \frac{h\nu}{v} \sim \frac{h\nu}{c} . \end{gathered} \tag{15}$$

The velocity of the photon in the observer frame is therefore not c, but v, defined from the guidance theorem of de Broglie, the basis of wave mechanics:

$$En_0 = h\nu_0 = m_0c^2 \tag{16}$$

In other words, the energy of the photon in the rest frame is

$$En_0 = m_0c^2 = h\nu_0 \tag{17}$$

and its energy in the observer frame is

$$En = m_0c^2\left(1 - \frac{v^2}{c^2}\right)^{-\frac{1}{2}} = h\nu \tag{18}$$

so that there is a change in the *frequency* of light from one frame to the other. This is the origin of *observed* distance proportional shifts [17], the "tired-light" of Hubble and Tolman. There are photons, therefore, which move at low velocities and contribute to the mass of the universe. Clearly, this is a consequence of the fact that the field has a finite range, of about 10^{26} m, as discussed in the introduction. This conclusion does not contradict the principle of conservation of energy, because in special relativity the quantity $P_\mu P_\mu$ is invariant to Lorentz transformation. Appendix 1 discusses the implications of tired-light for the equation of state in electromagnetism.

Therefore special relativity does not imply that the rest mass of the photon is zero, as in the traditional interpretations [1 – 16, 20, 21].

Secondly, if m_0 is not zero, the traditional interpretation [20, 21] of gauge transformations must be revised fundamentally, because it leads to the conclusion that the photon rest mass m_0 is zero and therefore contradicts the Einstein-de Broglie theory and experimental evidence [17] for finite photon mass. Traditional considerations of gauge transformations also lead to the principle of gauge invariance (Eicheninvarianz Prinzip) which holds if and only if the photon mass is identically zero. For these reasons, we consider carefully the basic, Lagrangian, formalism of gauge theory, and modify its interpretation in order to make it consistent with finite m_0. The result of our considerations is Eq. (3a) of the introduction.

Geometrically, a gauge transformation of the first kind [20, 21] is a rotation in the (1,2) plane of the "vector" field

$$\boldsymbol{\phi} = \phi_1 \boldsymbol{i} + \phi_2 \boldsymbol{j} \tag{19}$$

through an angle Λ. Under such a rotation, Noether's theorem leads to conserved charge Q in a volume V,

$$Q = i\int (\phi^* \frac{\partial\phi}{\partial t} - \phi \frac{\partial\phi^*}{\partial t})\, dV, \tag{20}$$

and a conserved current

$$J_\mu = i(\phi^* \frac{\partial\phi}{\partial x_\mu} - \phi \frac{\partial\phi^*}{\partial x_\mu}) \,. \tag{21}$$

The existence of Q and J_μ is based on the invariance of action. When the action is real, the Lagrangian is [20]

$$\mathcal{L} := (\frac{\partial\phi}{\partial x_\mu})(\frac{\partial\phi^*}{\partial x_\mu}) - m^2\phi^*\phi, \tag{22}$$

where m is a mass associated with the complex field ϕ, defined by

$$\phi = \frac{(\phi_1 + i\phi_2)}{\sqrt{2}}, \qquad \phi^* = \frac{(\phi_1 - i\phi_2)}{\sqrt{2}} \tag{23}$$

Since Λ is a constant (an angle in (1,2)), the gauge transformation of the first kind, which can be expressed [20] as

$$\phi \rightarrow e^{-i\Lambda}\phi, \qquad \phi^* \rightarrow e^{i\Lambda}\phi, \tag{24}$$

is the same at all points in spacetime, so that at an instant t the same rotation occurs for all points in space. This contradicts special relativity [20], whose universal constant is the speed of light and which implies that action at a distance is impossible. Electrodynamics cannot therefore be invariant to a gauge transformation of the first kind. In order to comply with special relativity, Λ is made an arbitrary function of spacetime:

$$\Lambda := \Lambda(x_\mu)\,, \tag{25}$$

thus defining a gauge transformation of the second kind. For $\Lambda \ll 1$, electrodynamics is invariant to the gauge transformation of the second kind:

$$\phi \rightarrow \phi - i\Lambda(x_\mu)\phi\,. \tag{26}$$

Condition (25) implies however [20] that $\partial\phi/\partial x_\mu$ does not transform in the same way as ϕ, i.e., does not transform covariantly; so that the action is no longer invariant [20,

21]:

$$\delta\mathcal{L} = J_\mu \frac{\partial\Lambda}{\partial x_\mu} \neq 0 , \tag{27}$$

with $\mathcal{L}$ defined by Eq. (22). In order to preserve the invariance of action under (26), the potential four-vector A_μ is introduced through

$$\mathcal{L}_1 := -e J_\mu A_\mu . \tag{28}$$

This implies the need for two more conditions [20]:

$$A_\mu \rightarrow A_\mu + \frac{1}{e}\frac{\partial\Lambda}{\partial x_\mu} \tag{29}$$

and

$$\mathcal{L}_2 := e^2 A_\mu A_\mu \phi^* \phi . \tag{30}$$

Equation (28) to (30) imply [20]

$$\delta\mathcal{L} + \delta\mathcal{L}_1 + \delta\mathcal{L}_2 = 0 . \tag{31}$$

In fundamental gauge theory, therefore, A_μ of the conventional d'Alembert equation is introduced to produce Eq. (31) in association with the extra term (30). *At this stage in the development, nothing has been said about the need for zero mass. We note that, if,* $A_\mu A_\mu = 0$, $\mathcal{L}_2$ is automatically zero.

As is well known [20], the field A_μ itself makes a contribution to the Lagrangian, implying the need for an additional $\mathcal{L}_3$ [20] to maintain a zero overall action:

$$\mathcal{L}_3 := -\frac{1}{4} F_{\mu\nu} F_{\mu\nu} , \tag{32}$$

where

$$F_{\mu\nu} := \frac{\partial A_\nu}{\partial x_\mu} - \frac{\partial A_\mu}{\partial x_\nu} , \tag{33}$$

the four-curl of A_μ, is the well known electromagnetic field four-tensor [20], an invariant under (29). The complete Lagrangian is therefore

$$\mathcal{L}_{tot} = \mathcal{L} + \mathcal{L}_1 + \mathcal{L}_2 + \mathcal{L}_3 . \tag{34}$$

If the mass, m_0, associated with the electromagnetic field is not zero, then the form of the Lagrangian is changed from (32) to

$$\mathcal{L}_4 = -\frac{1}{4} F_{\mu\nu} F_{\mu\nu} + \frac{1}{2} m_0^2 A_\mu A_\mu , \tag{35}$$

and this is invariant under the transformation (29) if and only if

$$m_0^2 A_\mu A_\mu = 0 . \tag{36}$$

If $m_0 \neq 0$,

$$A_\mu A_\mu = 0 , \quad A_\mu \neq 0 \tag{37}$$

is the only alternative possible, as described in the introduction. Conventionally, it is *asserted* [20, 21] *that the invariance* of $\mathcal{L}_4$ under (29) means that $m_0 = 0$. However, in the Einstein-de Broglie theory, Eq. (37) is consistent with Eq. (29), and m_0 is quantized as the photon rest mass. Equation (37) is also consistent with Eq. (31) of fundamental gauge theory, because [20] $\mathcal{L}_2 = 0$ if $A_\mu A_\mu = 0$. This implies that

$$\delta\mathcal{L}_2 = 2eA_\mu (\frac{\partial\Lambda}{\partial x_\mu}) \phi^*\phi = 0 , \tag{38}$$

so that

$$\delta\mathcal{L} + \delta\mathcal{L}_1 = -\delta\mathcal{L}_2 = -2eA_\mu (\frac{\partial\Lambda}{\partial x_\mu}) \phi^*\phi = 0 , \tag{39}$$

i.e., the action is conserved as in Eq. (31). The condition (37) for finite m_0 is one in which the EBP equation is invariant to the gauge transformation (29), which is implied by the need to conserve action under the gauge transformation of the second kind, Eq. (26).

We therefore conclude that gauge theory does not imply that photon rest mass is zero.

If $m_0 \neq 0$, the quantity $A_\mu A_\mu$ vanishes, implying that $\phi = c|\mathbf{A}|$, where $A_\mu := \epsilon_0 (\mathbf{A}, i\phi/c)$. This condition is furthermore consistent with Eqs. (1) and (2), which reads

$$\nabla \cdot \mathbf{A} + \frac{1}{c^2} \frac{\partial\phi}{\partial t} = 0 , \tag{40}$$

thus giving

$$\nabla\cdot\mathbf{A} = -\frac{1}{c}\frac{\partial|\mathbf{A}|}{\partial t}. \tag{41}$$

Additionally, using the Lorentz condition (40), one gets

$$(\nabla^2 - \frac{1}{c^2}\frac{\partial^2}{\partial t^2})\,|\mathbf{A}| = 0, \tag{42a}$$

$$(\nabla^2 - \frac{1}{c^2}\frac{\partial^2}{\partial t^2})\phi = 0, \tag{42b}$$

whose solutions are the well known [1 – 16, 20] Liénard-Wiechert potentials. Clearly, Eqs. (42a) and (42b) become the same if $\phi = c|\mathbf{A}|$ in S.I. units.

It is clear that fundamental gauge theory does not imply that the photon rest mass must be zero, contrary to much of the current literature [1 – 16, 20, 21]. Secondly, special relativity, as pointed out by Einstein [17, 18], also does *not* imply zero photon rest mass. Thirdly, there is experimental evidence [17], for non-zero m_0, and none for zero photon rest mass. Fourthly, the transverse, radiation, or Coulomb gauge [1 – 16, 20] is inconsistent with $\phi = c|\mathbf{A}|$ because in that gauge $\phi = 0$, $\mathbf{A} \neq 0$. The Lorentz gauge and Dirac gauge [17] are on the other hand consistent with $m_0 \neq 0$.

Having argued in some detail in this way, it becomes natural to see that much of the obscurity in the current thought on electromagnetism is due to the notion that the photon is massless and travels at the speed of light. Both statements contradict experimental evidence [17]. These notions result in "too much gauge freedom," in that Eqs. (26) and (29) can be satisfied with $m_0 = 0$ by the Coulomb, Lorentz, and other gauges. For $m_0 \neq 0$, as in the Einstein-de Broglie theory, the Coulomb gauge is invalidated, but the Lorentz gauge is a direct mathematical consequence of the EBP equation (1). The excess gauge freedom for $m_0 = 0$ results in well known and severe [20, 21] difficulties of quantization of the electromagnetic field, whereas quantization of the EBP equation, (a wave equation) is straightforward [20], leading to *longitudinal*, physically meaningful, spacelike photon polarization, as well as the two transverse spacelike polarizations. It is natural to expect that a particle, the photon, should have three spacelike polarizations in three physical dimensions, x, y, and z.

The $m_0 = 0$ assertion is conventionally associated with the notion that the electromagnetic field is a massless gauge field with two independent components, customarily identified with left and right circular polarization. However, even in the limit $m_0 = 0$, the same Maxwellian field is covariantly described by the *four* components of A_μ. The Bohm-Aharonov effect [20] shows that A_μ is physically meaningful. Recent work [23 – 28], leading to Eq. (5), shows conclusively that there is a well defined relation between the transverse ((1) and (2)) and longitudinal ((3)) components of solutions of Maxwell's field equations in vacuo. It is straightforward to show that the three magnetic field components form a classical cyclic permutation in the circular basis, (1), (2), and (3) with $B^{(0)} = B_0$:

$$\boldsymbol{B}^{(1)} \times \boldsymbol{B}^{(2)} = iB^{(0)}\boldsymbol{B}^{(3)*}, \qquad \boldsymbol{B}^{(3)*} = \boldsymbol{B}^{(3)}, \tag{43a}$$

$$\boldsymbol{B}^{(2)} \times \boldsymbol{B}^{(3)} = iB^{(0)}\boldsymbol{B}^{(1)*}, \qquad \boldsymbol{B}^{(1)*} = \boldsymbol{B}^{(2)}, \tag{43b}$$

$$\boldsymbol{B}^{(3)} \times \boldsymbol{B}^{(1)} = iB^{(0)}\boldsymbol{B}^{(2)*}, \qquad \boldsymbol{B}^{(2)*} = \boldsymbol{B}^{(1)}. \tag{43c}$$

Furthermore, there exist classical permutations involving $\boldsymbol{E}^{(3)}$. If we assert $\boldsymbol{E}^{(3)} := E^{(0)}\boldsymbol{k}$, these are, algebraically,

$$\boldsymbol{E}^{(1)} \times \boldsymbol{E}^{(2)} = iE^{(0)}c\boldsymbol{B}^{(3)*}, \tag{44}$$

$$\begin{aligned}
\boldsymbol{E}^{(2)} \times \boldsymbol{E}^{(3)} &= -E^{(0)}c\boldsymbol{B}^{(1)*}, \\
\boldsymbol{E}^{(3)} \times \boldsymbol{E}^{(1)} &= E^{(0)}c\boldsymbol{B}^{(2)*}, \\
\boldsymbol{E}^{(1)} \times \boldsymbol{B}^{(2)} &= B^{(0)}\boldsymbol{E}^{(3)*}, \\
\boldsymbol{E}^{(2)} \times \boldsymbol{B}^{(3)} &= iB^{(0)}\boldsymbol{E}^{(1)*}, \\
\boldsymbol{E}^{(3)} \times \boldsymbol{B}^{(1)} &= -B^{(0)}\boldsymbol{E}^{(2)*},
\end{aligned} \tag{45}$$

$$\begin{aligned}
\boldsymbol{E}^{(1)} \times \boldsymbol{B}^{(1)} &= 0, \\
\boldsymbol{E}^{(2)} \times \boldsymbol{B}^{(2)} &= 0, \\
\boldsymbol{E}^{(3)} \times \boldsymbol{B}^{(3)} &= 0,
\end{aligned} \tag{46}$$

and are reminiscent of the well-known Lie algebra of the Lorentz group [20], a classical commutator algebra which is built up with boost and rotation generators defined in Minkowski spacetime. However, all the Eqs. (45) violate $\hat{T}$ symmetry, which is a consequence of the fact that $\boldsymbol{E}^{(3)}$ has no real part (i.e., is pure imaginary) and cannot be derived from $\boldsymbol{E}^{(1)}$, $\boldsymbol{E}^{(2)}$, $\boldsymbol{B}^{(1)}$ or $\boldsymbol{B}^{(2)}$.

Before proceeding, therefore, to the derivation of $\boldsymbol{B}^{(3)}$ for non-zero m_0, the purpose of this section, it is demonstrated that this Lie algebra also applies to the electric and magnetic components of electromagnetic radiation in vacuo (the Maxwellian field), provided that these components are defined as classical field operators directly proportional, respectively, to the boost and rotation generators of the Lorentz transformation. This is a mathematical demonstration of the fact that if the longitudinal spacelike components of these fields be unphysical (i.e., zero), then the Lie algebraic structure of the Lorentz group is contradicted. This means that the Lorentz transformation itself is incorrectly defined, in that the longitudinal (z) boost and rotation generator components are incorrectly asserted to be zero. This is equivalent to destroying the geometrical structure of Minkowski spacetime. Even in the Maxwellian limit $m_0 \to 0$, therefore, the assertion that $\boldsymbol{B}^{(3)}$ and $\boldsymbol{E}^{(3)}$ are zero results in a mathematical *reductio ad absurdum*. We have shown that $\boldsymbol{B}^{(3)}$ is real and $\boldsymbol{E}^{(3)}$ is imaginary, both being *non-zero*.

It is well known [20, 21] that the Maxwell equations in vacuo are the Lorentz-covariant equations

$$\frac{\partial F_{\mu\nu}}{\partial x_\mu} = 0, \qquad \frac{\partial \tilde{F}_{\mu\nu}}{\partial x_\mu} = 0, \tag{47}$$

where $\tilde{F}_{\mu\nu}$ is the dual of $F_{\mu\nu}$, the electromagnetic field four-tensor. The latter is antisymmetric under Lorentz transformation and its structure can be displayed in Ryder's notation [20, 21] as:

$$F_{\mu\nu} := \begin{bmatrix} 0 & -E_1 & -E_2 & -E_3 \\ E_1 & 0 & -cB_3 & cB_2 \\ E_2 & cB_3 & 0 & -cB_1 \\ E_3 & -cB_2 & cB_1 & 0 \end{bmatrix}. \tag{48}$$

We note that this structure is identical with that of the Lie algebra of the Lorentz group, defined [20] by the well-known dimensionless boost generator $\hat{K}_i$, an operator, and the rotation generator $\hat{J}_i$, also a dimensionless operator. The Lie algebra of the Lorentz group can be displayed as [20, 21]:

$$\hat{J}_{\mu\nu} = \begin{bmatrix} 0 & \hat{K}_1 & \hat{K}_2 & \hat{K}_3 \\ -\hat{K}_1 & 0 & \hat{J}_3 & -\hat{J}_2 \\ -\hat{K}_2 & -\hat{J}_3 & 0 & \hat{J}_1 \\ -\hat{K}_3 & \hat{J}_2 & -\hat{J}_1 & 0 \end{bmatrix}, \tag{49}$$

i.e., as

$$\hat{J}_{ij} = -\hat{J}_{ji} = \epsilon_{ijk}\hat{J}_k, \qquad i,\ j,\ k = 1,\ 2,\ 3, \tag{50}$$
$$\hat{J}_{i0} = -\hat{J}_{0i} = -\hat{K}_i\ .$$

Equations (49) and (50) are condensed representations of the well-known [20] classical commutator (Lie) algebra of the Lorentz group:

$$[\hat{J}_x,\ \hat{J}_y] = i\hat{J}_z, \qquad \text{and cyclic permutations}, \tag{51a}$$

$$[\hat{K}_x,\ \hat{K}_y] = -i\hat{J}_z, \qquad \text{" " "}, \tag{51b}$$

$$[\hat{K}_x,\ \hat{J}_y] = i\hat{K}_z, \qquad \text{" " "}, \tag{51c}$$

$$[\hat{K}_x,\ \hat{J}_x] = 0, \qquad \text{etc.} \tag{51d}$$

The geometrical equivalence of (48) and (49) means that, in a circular basis:

$$\hat{B}^{(1)} = -B^{(0)}\frac{\hat{J}^{(1)}}{\hbar}e^{i\phi},$$
$$\hat{B}^{(2)} = -B^{(0)}\frac{\hat{J}^{(2)}}{\hbar}e^{-i\phi}, \tag{52}$$
$$\hat{B}^{(3)} = iB^{(0)}\frac{\hat{J}^{(3)}}{\hbar},$$

where $\hat{B}_i$ are magnetic field operators; In particular, the longitudinal operator $\hat{B}^{(3)}$ is the elementary quantum of magnetic flux density in the propagation axis. In the Cartesian basis (x, y, z), we obtain the formal relations:

$$[\hat{B}_x,\ \hat{B}_y] = iB^{(0)}\hat{B}_z, \qquad \text{and cyclic permutations}, \tag{53a}$$

$$[\hat{E}_x,\ \hat{E}_y] = -icB^{(0)}\hat{B}_z, \qquad \text{" " "}, \tag{53b}$$

$$[\hat{E}_x,\ \hat{B}_y] = iB^{(0)}\hat{E}_z, \qquad \text{" " "}, \tag{53c}$$

$$[\hat{E}_x,\ \hat{B}_y] = 0, \qquad \text{etc.} \tag{53d}$$

Equations (53) represent a classical operator equivalent of the vector products in Eqs. (36), where the Maxwellian fields are vectors in space, and not operators defined in spacetime.

The result (52) is based on the fundamental Lie algebraic structure of Minkowski spacetime, and implies the following:

(1) The classical electric field operator $\hat{E}_i$ is a boost generator, and the classical magnetic field operator $\hat{B}_i$ is a rotation generator in Minkowski spacetime.

(2) If the longitudinal component operators $\hat{B}_z$ and $\hat{E}_z$ are asserted to be zero, or "unphysical," the structure of the Lie algebra is destroyed in Eqs. (53). For example, if $\hat{B}_z = \hat{E}_z = \hat{0}$, $[\hat{B}_x, \hat{B}_y] =? \hat{0}$ and from the structure of $\hat{J}_i$ in Eq. (52), *this is mathematically incorrect*. Explicitly,

$$[\hat{B}_x, \hat{B}_y] = B^{(0)2} [\hat{J}_x, \hat{J}_y] = B^{(0)2} \hat{J}_z \neq 0, \tag{54}$$

because [20], in Ryder's notation,

$$\hat{J}_x := -i \begin{bmatrix} 0 & 0 & 0 & 0 \\ 0 & 0 & 0 & 0 \\ 0 & 0 & 0 & 1 \\ 0 & 0 & -1 & 0 \end{bmatrix}, \tag{55a}$$

$$\hat{J}_y := -i \begin{bmatrix} 0 & 0 & 0 & 0 \\ 0 & 0 & 0 & -1 \\ 0 & 0 & 0 & 0 \\ 0 & 1 & 0 & 0 \end{bmatrix}, \tag{55b}$$

$$\hat{J}_z := -i \begin{bmatrix} 0 & 0 & 0 & 0 \\ 0 & 0 & 1 & 0 \\ 0 & -1 & 0 & 0 \\ 0 & 0 & 0 & 0 \end{bmatrix}. \tag{55c}$$

(3) The Maxwell equations (47) are seen to be relations between boost and rotation generators defined in spacetime:

$$\frac{\partial J_{\mu\nu}}{\partial x_\mu} = 0, \qquad \frac{\partial \tilde{J}_{\mu\nu}}{\partial x_\mu} = 0, \tag{56}$$

and are thus given a precise geometrical interpretation. In

this light, it is seen that the d'Alembert equation (4) is also geometrical in nature:

$$\Box \hat{\mathcal{L}}_\mu = 0 , \tag{57}$$

where $\hat{\mathcal{J}}_{\mu\nu}$ is the four-curl of $\hat{\mathcal{L}}_\mu$:

$$\hat{\mathcal{J}}_{\mu\nu} = \frac{\partial \hat{\mathcal{L}}_\nu}{\partial x_\mu} - \frac{\partial \hat{\mathcal{L}}_\mu}{\partial x_\nu} . \tag{58}$$

(4) It may be seen precisely that the conventional notion that the Maxwellian $\hat{B}_z$ (and $\hat{E}_z$) be unphysical is equivalent to the *geometrically incorrect assertion*

$$\hat{\mathcal{J}}_z =? \begin{bmatrix} 0 & 0 & 0 & 0 \\ 0 & 0 & 0 & 0 \\ 0 & 0 & 0 & 0 \\ 0 & 0 & 0 & 0 \end{bmatrix}, \tag{59}$$

which by implication habitually [1 – 16, 20, 21] replaces the correct rotation generator (55c).

There is of course no experimental evidence for Eq. (59), *even in the massless limit $m_0 \to 0$ conventionally associated with the Maxwellian field.*

Our geometrical interpretation of the Maxwell field equations is a direct logical consequence of the geometry of Minkowski spacetime itself and of the theory of special relativity. This is consistent with the fact that Einstein's considerations of the Maxwell equations led to his formulation of special relativity. If it is asserted that longitudinal solutions of Maxwell's equations be unphysical, special relativity is contradicted; and the structure of the Lorentz group and its associated Lie algebra destroyed. *There is no experimental evidence whatsoever that the longitudinal solutions of Maxwell's equations in vacuo are unphysical, and there is no evidence for $m_0 = 0$.*

The commutator relations (51a) and (53a) lead to a method of quantization of the Maxwellian field simply by noting the ordinary angular momentum commutator relations of quantum mechanics. In the Cartesian basis:

$$[\hat{\mathcal{J}}_x, \hat{\mathcal{J}}_y] = i\hbar \hat{\mathcal{J}}_z \tag{60}$$

are structurally identical with Eq. (51a) except for $\hbar$ (which has the units of angular momentum). In quantum mechanics, the $\hat{J}$ operators in Eq. (60) are angular momentum operators. Quantized angular momentum is therefore a consequence of the classical rotation generator [20], as is well known. The quantized equivalent of Eq. (53a) must therefore be

$$[\hat{B}_x, \hat{B}_y] = i\hbar \left(\frac{B^{(0)}}{\hbar}\right) \hat{B}_z \tag{61}$$

in order to balance units, symmetries, and dimensions on the left- and right-hand sides. This implies

$$\hat{B} = B^{(0)} \frac{\hat{J}}{\hbar}, \tag{62}$$

which is identical with the result obtained recently by the present author [25] using an independent method of derivation. Therefore

$$\hat{B}_z = B^{(0)} \frac{\hat{J}_z}{\hbar} \tag{63}$$

is the elementary longitudinal component of the quantized Maxwellian magnetic field in vacuo. In the same way that $\hbar$ is the archetypical elementary quantum of angular momentum, $B^{(0)}$ is the elementary quantum of magnetic flux density of the Maxwellian field in vacuo.

The eigenvalues of $\hat{J}_z$ in Eq. (63) may be identified with those of a massless boson (the "conventional" photon), i.e., $\hbar M_J$, where $M_J = \pm 1$; so that the classical limit of Eq. (63) is

$$\boldsymbol{B}_z = B^{(0)} \boldsymbol{k}, \tag{64}$$

which is Eq. (5) in a Cartesian basis instead of a circular basis. *Equation (5) is therefore geometrically consistent with the Lie algebra of the Lorentz group.*

The generalization of our development to $m_0 \neq 0$ is now straightforward.

Having considered in some detail the geometrical structure of the Lorentz group, we revert to a simpler development of the EBP equation (1), solving it as a classical eigenvalue equation with the differential operator

$$\Box := -\nabla^2 + \frac{1}{c^2}\frac{\partial^2}{\partial t^2} . \tag{65}$$

The order of magnitude of ξ is such that

$$\Box A_\mu \sim 10^{-52} A_\mu , \tag{66}$$

which closely approximates the d'Alembert equation (4). It is clear therefore that the classical interpretation of the EBP field closely approximates the Maxwellian field. However, in the EBP field, the Coulomb gauge is inconsistent with Eq. (66), which must be written in terms of the space-like **A** as

$$(\nabla^2 - \frac{1}{c^2}\frac{\partial^2}{\partial t^2})\mathbf{A} = \xi^2 \mathbf{A} , \tag{67}$$

with $\phi = c|\mathbf{A}|$. In the non-relativistic limit, this equation becomes

$$\nabla^2 \mathbf{A} = \xi^2 \mathbf{A}. \tag{68}$$

Using the relation

$$\mathbf{B} = \nabla \times \mathbf{A}, \tag{69}$$

it can be seen that the equation

$$\nabla^2 \mathbf{B} = \xi^2 \mathbf{B} \tag{70}$$

is the same as Eq. (68), because:

$$\nabla^2 (\nabla \times \mathbf{A}) = \xi^2 \nabla \times \mathbf{A}, \tag{71a}$$

$$\nabla \times \nabla^2 \mathbf{A} = \nabla \times \xi^2 \mathbf{A}. \tag{71b}$$

In considering the limit $c \rightarrow \infty$, we have removed the time dependence in the solution for **B** of Eq. (70). Furthermore, since

$$\nabla^2 \mathbf{B} \sim 10^{-52} \mathbf{B} \sim 0 \tag{72}$$

describes the magnetic component in vacuo of an electromagnetic field closely resembling the Maxwellian field, we know that the time-independent solution to Eq. (70) must be the longitudinal component, defined in the propagation axis z. The solution to Eq. (70) in the Cartesian basis is therefore:

$$\boldsymbol{B} = B^{(0)} \exp(-\xi z)\, \boldsymbol{k}, \qquad |\boldsymbol{B}| = B_z, \tag{73}$$

and, since $\xi \sim 10^{-26}$ m^{-1}, this is for all practical purposes identical with Eq. (64) of the Maxwellian field. Several physical consequences flow from Eqs. (64) and (73):

(1) The longitudinal solution for B of the EBP field, Eq. (73), is for all practical purposes identical with the corresponding Maxwellian solution, Eq. (64). By the caveat "for all practical purposes" we imply laboratory dimensions and time scales. On a *cosmic* scale, in which z ~ $1/\xi$, Eq. (73) is different from Eq. (64) in general. In the "tired-light" terminology of Hubble [17], **B** becomes a "tired-field" if z is big enough (ca. 10^{26} m).

(2) Physically meaningful, practically identical, and longitudinal solutions exist for **B** from the EBP and Maxwell equations, the former being considered as a classical wave equation. To assert **B** = **0** in Eq. (64) is mathematically incorrect in the Maxwellian field, because it corresponds to the assertion (59) in spacetime. For all practical purposes, therefore, this assertion is incorrect in the EBP field. Quantization of the EBP field [20] confirms this conclusion, leading to a physically meaningful longitudinal photon polarization.

(3) Since the EBP and Maxwellian fields are practically (i.e., in the laboratory) identical, the EBP field obeys the various commutator relations of this paper for all practical purposes; and the transverse EBP solutions are practically those of Eqs. (7). In the cosmology of light from distant sources, however, this simple classical interpretation is no longer tenable.

It is well known [20] that quantization of the EBP field is straightforward, whereas that of the Maxwellian field is obscure. Although the rest mass m_0 of the photon is very small, it is therefore *essential* that it be rigorously non-zero to maintain a logical and self-consistent, physically meaningful, structure for the quantized electromagnetic field in vacuo. If this is done, quantization results in a consistent particle interpretation [20] in terms of a massive boson, with eigenvalues $M_J \hbar$, $M_J = -1, 0, +1$. The three polarization vectors of the quantized EBP field are orthonormal and spacelike, i.e., there are physically meaningful longitudinal and transverse components. The little group of Wigner [29] is a physically meaningful rotation group, utilizing the three dimensions of space. If $m_0 = 0$ on the other hand, the constraint $A_\mu A_\mu = 0$ is conventionally lost, resulting in "too much gauge freedom." The two Casimir invariants [20] of the Poincaré group vanish for $m_0 = 0$, meaning that physical quantities that are invariant under the most general type of Lorentz transformation must vanish identically for the massless gauge field. This implies $A_\mu A_\mu = 0$, if $A_\mu A_\mu$ is to be an invariant of the Poincaré group, diametrically contradicting the conventional use of gauge freedom for a massless particle, i.e., contradicting the conventional assertion that $A_\mu A_\mu \neq 0$ for $m_0 = 0$. Thus, the conventional assertion $A_\mu A_\mu \neq 0$ for $m_0 = 0$ is geometrically unsound, i.e., contradicts the geometry of Minkowski space-time, a geometry that requires $A_\mu A_\mu$ to be an invariant of the Poincaré (inhomogeneous Lorentz) group. We are forced to conclude that the widespread use of the Coulomb gauge, in which $A_\mu A_\mu = 0$, is relativistically incorrect.

The conventional assertion that m_0 must be zero because $A_\mu A_\mu$ is non-zero is also basically incorrect, because $A_\mu A_\mu$ is always zero in vacuo.

It is the customary use of the Coulomb (or "transverse") gauge that leads more than any other factor to the conventional assertion that the electromagnetic field can have no longitudinal solution which is physically meaningful. The Coulomb gauge is relativistically incorrect, and is inconsistent with finite photon rest mass, for which there is experimental evidence [17]. The widespread use of the Coulomb gauge [1 – 16, 20, 21] should therefore be viewed with caution. It is obvious that quantization in the Coulomb gauge cannot be consistent with special relativity, because its use is equivalent to the incorrect assertion (59). These difficulties are frequently compounded in the literature by a series of mis-statements, traceable to the relativistically incorrect assertion $A_\mu A_\mu \neq 0$. For example, it is frequently asserted that the Lorentz gauge does not define A_μ uniquely. This is true if and only if $A_\mu A_\mu \neq 0$. If $A_\mu A_\mu = 0$, the Lorentz condition defines A_μ uniquely. Quantization in the Coulomb gauge is therefore a mathematically incorrect procedure, and we discard its results as meaningless. In other words it is meaningless to assert that the Maxwellian field has only two transverse polarizations.

Quantization of the Maxwellian field in the Lorentz gauge [20] retains manifest covariance, but is physically obscure. It also relies on the notion that the gauge field is massless, so that quantization of the field must lead to a massless photon. In consequence, the internally inconsistent notion $A_\mu A_\mu \neq 0$ is habitually retained in the Lorentz gauge. This immediately leads to the difficulty that the

Lagrangian has to be modified with a gauge fixing term, a procedure which leads to a non-Maxwellian equation of motion [20]. Even with this artifice, the conjugate momentum field Π^0 [20] vanishes; and the traditional method is forced to assert that the Lorentz condition, within whose framework the method is developed, cannot hold as an operator identity. This difficulty is habitually resolved by the method of Gupta and Bleuler [20], a method which results in the conclusion that admixtures of timelike and longitudinal spacelike photon polarizations are physical states [20]. Despite this conclusion, these states are abandoned as unphysical in order to comply with the results of Coulomb gauge quantization, which, as we have just seen, are incorrect. Quantization of the Maxwellian field, regarded as a massless gauge field, is therefore inconsistent and physically obscure.

In considerations of the Poincaré group, the notion of a massless gauge field, customarily associated with the Maxwellian field, leads to the little group [20, 29] E(2), the Euclidean group of rotations and translations in a plane. *It is well known [20] that the physical significance of this little group is obscure*. Its Lie algebra does not correspond to that of a rotation group, but it is the group that is needed to maintain a lightlike vector invariant under the most general Lorentz transformation. This suggests that the notion of a massless field is physically meaningless. The traditional line of reasoning, however, considers a massless particle travelling in the propagation axis (z) described by a lightlike four-vector k_μ. Invariance of k_μ under the most general type of Lorentz transformation leads to the Lie algebra:

$$[\hat{L}_1, \hat{L}_2] = 0, \qquad [\hat{J}_3, \hat{J}_1] = i\hat{L}_3, \qquad [\hat{L}_2, \hat{J}_3] = i\hat{L}_1, \tag{74}$$

where

$$\hat{L}_1 := \hat{K}_1 - \hat{J}_2, \qquad \hat{L}_2 := \hat{K}_2 + \hat{J}_1.$$

Thus

$$[\hat{L}_1, \hat{L}_2] = [\hat{K}_1, \hat{K}_2] + [\hat{K}_1, \hat{J}_1] - [\hat{J}_2, \hat{K}_2] - [\hat{J}_2, \hat{J}_1] = 0. \tag{75}$$

In the Cartesian basis, x = 1, y = 2, z = 3; and if we attempt to apply to Eq. (75) the Lorentz group algebra of Eqs. (51):

$$[\hat{K}_1, \hat{K}_2] = [\hat{K}_x, \hat{K}_y] = -iJ_z, \tag{76a}$$

$$[\hat{J}_2, \hat{J}_1] = -[\hat{J}_x, \hat{J}_y] = -i\hat{J}_z, \tag{76b}$$

$$[\hat{K}_1, \hat{J}_1] = [\hat{K}_x, \hat{J}_x] = 0, \tag{76c}$$

$$[\hat{J}_2, \hat{K}_2] = [\hat{J}_y, \hat{K}_y] = 0, \tag{76d}$$

we obtain

$$[\hat{L}_1, \hat{L}_2] = 2i\hat{J}_z \neq \hat{0}. \tag{77}$$

Since, in the Lorentz group,

$$\hat{J}_z \neq \hat{0} \tag{78}$$

in general, Eq. (77) contradicts Eq. (75).

Therefore the most general Lorentz transformation that leaves the lightlike momentum vector k_μ invariant cannot be described by the Lie algebra of the Lorentz group. This implies that the notion of lightlike momentum (a massless particle travelling at the speed of light) is not relativistically self-consistent. This is another way of demonstrating that the quantization of a massless field into a massless particle is beset with obscurity, i.e., we are led to the conclusion that the Maxwellian field has no meaning in quantum theory. Attempts to impose a meaning lead into physical obscurity as we have described. In the Einstein-de Broglie theory of light, the quantization of the EBP field leads directly and without difficulty [20] to a particle interpretation of light in terms of a massive boson. The quantum-classical equivalence in the EBP field is therefore clear. The only physically meaningful and consistent interpretation is to accept the photon as a massive boson whose classical field is described by the classical limit of the EBP equation. The mathematical limit of this field for zero mass is the Maxwellian field. Direct quantization of the Maxwellian field, regarded as a classical massless gauge field, is physically obscure. The quantized Maxwellian field must therefore be defined as being for all practical purposes the quantized EBP field, with which it is practically identical because photon mass is numerically very small.

In Appendix 2, we consider the consequences for unified and grand unified field theory of finite photon mass and longitudinal solutions of Proca and Maxwell equations in vacuo.

3. TRANSVERSE SOLUTIONS IN VACUO FOR FINITE PHOTON MASS

The EBP equation can be written in terms of the tensor $F_{\mu\nu}$, defined in Eq. (33), as

$$\frac{\partial F_{\mu\nu}}{\partial x_\mu} = -\xi^2 A_\nu \sim 0\,; \tag{79}$$

and, as we have seen, the Lie algebra associated with $F_{\mu\nu}$ is given by Eqs. (43) – (46). Therefore, transverse solutions of the EBP equation in its classical limit obey the classical cross products in Eqs. (43) – (46). Using Eq. (43a) with the longitudinal solution of the EBP equation (73), we obtain

$$\boldsymbol{B}^{(1)} = \frac{B^{(0)}}{\sqrt{2}}\,(i\boldsymbol{i} + \boldsymbol{j})\,e^{i\phi}e^{-\frac{\xi z}{2}}, \qquad \boldsymbol{B}^{(2)} = \boldsymbol{B}^{(1)*}, \tag{80}$$

and its complex conjugate. The difference between this solution and the equivalent, Eq. (7c), for the Maxwell equations can be expressed by replacing the wave vector of the Maxwell equations by $\kappa_z \rightarrow \kappa_z - (i\xi)/2$, for polarization (1), and $\kappa_z \rightarrow \kappa_z + (i\xi)/2$, for polarization (2). At visible frequencies, the order of magnitude of the Maxwellian $\boldsymbol{\kappa}$ in vacuo is given by

$$\kappa_z = \frac{\omega}{c} \sim \frac{10^{15}}{10^{8}} \sim 10^{7}\ m^{-1}, \tag{81}$$

so that at these frequencies κ_z is about thirty three orders of magnitude greater than ξ. For all practical purposes, therefore, the transverse solutions of the classical limit of the EBP equation are identical with those of the Maxwell equations.

This is a simple demonstration in the classical limit that the fields associated with the EBP and Maxwell equations contain physically meaningful longitudinal as well as transverse components in vacuo. In the next section we discuss several experimental consequencies of physically meaningful longitudinal fields when electromagnetic radiation interacts with matter. Firstly, however, we review the available experimental evidence for finite photon mass, following a recent account by Vigier [17].

4. DISCUSSION

There is available an increasing amount of evidence for

finite photon rest mass, upon which is based the theory of Einstein and de Broglie. A recent experiment by Mizobuchi and Ohtake [17] has demonstrated for single photons the simultaneity of classical wave and particle behaviour in light. This has demonstrated for the first time that the Copenhagen interpetation cannot be valid, but supports the Einstein-de Broglie interpretation as reviewed recently by Vigier [17], an interpetation which implies for example that photons are emitted from a source in quanta of energy with well defined directionality. The wave associated with a single photon has a physical reality. Light consists of massive bosons (photons) controlled or piloted [17] by real surrounding spin one fields. The motion of the photon is thus controlled by a quantum potential. The photons are the only directly observable elements of light and behave in Minkowski spacetime as relativistic particles with finite mass. Light is also constituted in the Einstein-de Broglie theory by physically meaningful fields (waves) which, as we have seen, obey Maxwell's equations for all practical purposes, essentially because the photon rest mass is finite (10^{-68} kg) but small. These fields are described by complex vector waves which also describe photon motion. Thus, if there is a longitudinal photon polarization, there must be a longitudinal field polarization, as already described. Longitudinal field solutions of the EBP equation were first derived by Schrödinger and de Broglie and, in general, the EBP equation has longitudinal and transverse *wave* solutions [17]. Since these are also wave solutions of Maxwell's equations for all practical purposes, it becomes clear that Maxwell's equations must have physically meaningful longitudinal solutions. The relation of these to the transverse solutions has only recently become clear [23 – 28], as described in Secs. 2 and 3 of this paper.

Following Vigier's recent description [17], there are several consequences of finite photon mass. The r dependence of the Coulomb potential is replaced by that of the Yukawa potential

$$V_y \propto \frac{exp(-\xi z)}{z}. \tag{82}$$

There exist low velocity photons (i.e., photons travelling at considerably less than the speed of light c), whose small but finite mass contributes to that of the universe. There is, thirdly, a red shift proportional to $exp(-\xi z)$, which can be applied to explain recent astronomical observations of anomalous red shifts from several distant sources, such as quasars. These "tired-light" phenomena originate in the EBP equation and may account for observed anomalies in double star motions, galaxy clusters, observed variations of the Hubble "constant," and other evidence reviewed in the literature [17]. A photon with finite rest mass behaves relativistically in the frame of observation, leading to the expectation [17] of a direction-dependent anisotropy in the frequency of light in the observer frame. Such an anisotropy has been observed experimentally by Hall *et. al.* [30] in the direction of the apex of the 2.7 K background of microwave radiation. These Boulder experiments are currently being repeated in Copenhagen by Poulsen and co-workers [17]. Experimental evidence for the Einstein-de Broglie theory of light has also been reviewed by Vigier [17] in the following areas:

(1) super-luminal action at a distance, a facet of Einstein's interpretation of light;
(2) the question of locality or non-locality of the quantum potential;
(3) the direct experimental testing of Heisenberg's uncertainty principle using single photons;
(4) experimental testing for the existence of particle trajectories in light (Einweg-Welcherweg);
(5) testing the existence of physically meaningful waves without the presence of particles, for example the recent experimental observation by Bartlett and Corle [31] of the Maxwell displacement current in vacuo;
(6) testing directly the existence of the quantum potential with intersecting laser beams and laser induced fringe patterns.

There is therefore a considerable amount of experimentation in progress concerning the existence of finite photon mass, and it is no longer tenable to assert [1 - 16, 20, 21] that the photon mass is zero.

Similarly, it is not reasonable to assert that $\boldsymbol{B}^{(3)}$ and $\boldsymbol{E}^{(3)}$ must be zero, "irrelevant," "unphysical" or similar, as happens in much of the contemporary literature. It is in fact implied, but not specifically so, in the work of de Broglie and Schrodinger [17] that $\boldsymbol{B}^{(3)}$ and $\boldsymbol{E}^{(3)}$ must exist. They exist, as we have seen, both for finite photon mass and in the Maxwellian limit, but a finite photon rest mass is essential for a natural quantization of the electromagnetic field. For all intents and purposes, therefore, evidence for $\boldsymbol{B}^{(3)}$ and $\boldsymbol{E}^{(3)}$ is evidence for finite photon mass, and corroboration for other sources of evidence quoted already. The present author has proposed a number of different magneto-optic experiments [23 – 28] which would test for $\boldsymbol{B}^{(3)}$ through its interaction with matter, using its characteristic square-root dependence on light intensity I_0 (watt m^{-2}). In free space, fundamental electrodynamics leads to [23 – 28]

$$|\boldsymbol{B}^{(3)}| \sim 10^{-7} I_0^{\frac{1}{2}} ; \tag{83}$$

and, assuming that $\boldsymbol{B}^{(3)}$ acts as a magnetic field whose time average is non-zero, it is to be expected [23 – 28] that there exist the following effects (collected details in Ref. [26] proportional to the square root of laser intensity, provided that the laser is circularly polarised: (1) inverse Faraday effect (magnetization due to $\boldsymbol{B}^{(3)}$); (2) optical Faraday effect (azimuth rotation due to $\boldsymbol{B}^{(3)}$); (3) effects of $\boldsymbol{B}^{(3)}$ in NMR (preliminary observations reported in Ref. [32], and ESR spectroscopy; (4) Cotton-Mouton effect due to $\boldsymbol{B}^{(3)}$; (5) forward-backward birefringence due to $\boldsymbol{B}^{(3)}$; (6) reinterpretation of antisymmetric light scattering and similar phenomena in terms of $\boldsymbol{B}^{(3)}$.

Finally, in this paper we propose the Bohm-Aharonov effect due to $\boldsymbol{B}^{(3)}$ of a circularly polarised laser, which replaces the solenoid, or iron whisker [20] of the conventional Bohm-Aharonov effect.

It is well known that the Bohm-Aharonov effect [20] indicates that the vector potential in quantum mechanics is physically meaningful, and that the vacuum has a non-trivial topology. It is therefore one of the most incisive effects in contemporary electrodynamics. The experiment has been repeated independently several times and consists of placing a small solenoid between two slits, which are used to generate interference fringes due to electron beams. The magnetic flux density **B** is confined within the solenoid and is inaccessible to the interfering electrons passing through the two slits. Despite this, the solenoid is observed experimentally [20] to produce a shift in the interference pattern (or fringes) set up by the electrons. This shift is due to the curl of the vector potential **A** set up outside the solenoid. Essentially, **A** changes the electron wave function:

$$\psi = |\psi| \, exp\left(i \frac{\boldsymbol{p} \cdot \boldsymbol{r}}{\hbar} \right) , \tag{84}$$

because **p**, the electron momentum, is changed to **p** - e**A**, where e is the electronic charge. This does not occur in classical mechanics, but in quantum theory the electronic wave function, and thus the electron, is influenced by **A** even though it travels in regions where the magnetic flux density **B** is zero. This means that there is non-locality in the integral $\oint \boldsymbol{A} \cdot d\boldsymbol{r}$ [20]. The Bohm-Aharonov effect is therefore evidence for this type of non-locality.

The shift is given in metres by

$$\Delta x = \frac{L\lambda}{d}\frac{e}{h}\Phi \tag{85}$$

where λ is the wavelength of the electron beam entering the two slits, L is the distance between the screen containing the two slits and the detector plane, d is the distance between the two slits, and

$$\Phi := \int \boldsymbol{B} \cdot d\boldsymbol{S} = \oint \boldsymbol{A} \cdot d\boldsymbol{r} \tag{86}$$

is the magnetic flux in webers.

It is clear that if the solenoid is replaced by a thin, circularly polarised, laser beam, there should be a Bohm-Aharonov effect due to $\boldsymbol{B}^{(3)}$, in which this field shifts the interference pattern of the electrons, with **B** of Eq. (85) replaced by $\boldsymbol{B}^{(3)}$. This shift should: (a) be proportional to the square root of the laser intensity; (b) reverse with the sense of circular polarization of the laser (because $\boldsymbol{B}^{(3)}$ changes sign); (c) disappear if the laser is linearly polarized or incoherently polarized. This laser induced fringe displacement would be a particularly interesting investigation of the nature of $\boldsymbol{B}^{(3)}$ and of its concommitant $\boldsymbol{A}^{(3)}$. Presumably $\boldsymbol{B}^{(3)}$ is confined to the radius of the laser beam, and $\boldsymbol{A}^{(3)}$ exists outside this beam, as in a solenoid generating a conventional, longitudinal, magnetostatic field. The experiment would prove: (a) the existence and (b) the non-locality of $\boldsymbol{A}^{(3)}$.

For all practical purposes, the Maxwellian

$$\boldsymbol{B} = \nabla \times \boldsymbol{A}, \tag{87a}$$

$$\boldsymbol{E} = -\nabla\phi - \frac{\partial \boldsymbol{A}}{\partial t} \tag{87b}$$

can be used to find time independent $\boldsymbol{B}^{(3)}$ and $\boldsymbol{E}^{(3)}$ from the

Proca equation. Equation (87a) shows that if $\boldsymbol{B}^{(3)}$ is time independent, then so is $\boldsymbol{A}^{(3)}$. Therefore,

$$\frac{\partial^2 \boldsymbol{A}^{(3)}}{\partial t^2} = 0, \tag{88}$$

in the Proca equation, and

$$\boldsymbol{E}^{(3)} = -\nabla\phi^{(3)} \tag{89}$$

in Eq. (87b). Using the gauge condition (3b), necessitated by finite photon mass, $\phi^{(3)}$ is the magnitude of $c\boldsymbol{A}^{(3)}$ and must be time-independent if $\boldsymbol{A}^{(3)}$ is so. Therefore

$$\frac{\partial^2 \phi^{(3)}}{\partial t^2} = 0. \tag{90}$$

In order to obtain from Eq. (87b) a $\boldsymbol{B}^{(3)}$ along the z axis, the required solution to the Proca equation

$$\nabla^2 \boldsymbol{A}^{(3)} - \frac{1}{c^2}\frac{\partial^2 \boldsymbol{A}^{(3)}}{\partial t^2} = \xi^2 \boldsymbol{A}^{(3)} \tag{91}$$

and Eq. (85) is

$$\boldsymbol{A}^{(3)} = \frac{B_0}{2}(-x\boldsymbol{i} + y\boldsymbol{j})\, e^{-\xi z}, \tag{92}$$

where $R^2 = x^2 + y^2$ is a constant. Therefore,

$$\phi^{(3)} = c|\boldsymbol{A}^{(3)}| = \frac{1}{2}E_0 R e^{-\xi z} \tag{93}$$

and

$$\boldsymbol{E}^{(3)} = E^{(0)}\boldsymbol{k} = -\frac{\partial \phi^{(3)}}{\partial z}\boldsymbol{k} = \frac{1}{2}\xi E_0 R e^{-\xi z}\boldsymbol{k}, \tag{94}$$

$$\boldsymbol{B}^{(3)} = B^{(0)}\boldsymbol{k} = B_0 e^{-\xi z}\boldsymbol{k}. \tag{95}$$

Since $\xi \sim 10^{-26}\ \mathrm{m}^{-1}$, Eqs. (94) and (95) show that, in the Proca description of electrodynamics,

$$|\boldsymbol{B}^{(3)}| \gg |Re(\boldsymbol{E}^{(3)})| \tag{96}$$

in free space. In Maxwell's description, $\xi \rightarrow 0$ and $\boldsymbol{E}^{(3)}$ vanishes but $\boldsymbol{B}^{(3)}$ remains finite. This is consistent with our earlier finding that $\boldsymbol{B}^{(3)}$ is real and $\boldsymbol{E}^{(3)}$ is imaginary, with no real part in Maxwellian theory. Significantly, $\boldsymbol{B}^{(3)}$ can be obtained from cross products of $\boldsymbol{B}^{(1)}$ and $\boldsymbol{B}^{(2)}$, but the real part of $\boldsymbol{E}^{(3)}$ cannot without $\hat{T}$ violation.

ACKNOWLEDGEMENTS

Professor J.-P. Vigier is thanked for a preprint of reference [17] and for his suggestion in a letter of January 4, 1993, that $\boldsymbol{B}^{(3)}$ and $\boldsymbol{E}^{(3)}$ can be accommodated naturally within the Einstein-de Broglie theory of light.

APPENDIX 1. TIRED-LIGHT AND THE CONSERVATION OF ELECTROMAGNETIC ENERGY

The concept of tired-light rests on the assertion that the mass of the photon is not zero in its rest frame. This leads to the Proca equation, to which there belong longitudinal oscillatory solutions. The effect of these on the Poynting theorem was reviewed by Goldhaber and Nieto [33], who showed that tired-light can be accommodated with the laws of conservation of electromagnetic energy. De Broglie [34] and Bass and Schrödinger [35] showed that there exists a conserved energy-momentum density which depends on photon mass squared. For all practical purposes therefore the Poynting theorem is unaffected by finite photon mass in the laboratory. However, on a cosmic scale, tired-light affects the fundamental law of electromagnetic energy conservation. Vigier has shown [17] that photon mass is also expected to increase the mass of the universe, providing an answer to the "missing mass" paradox.

Goldhaber and Nieto [33] have provided a thorough discussion of the problems posed by photon mass for conservation of electromagnetic energy and have suggested several experimental methods for the resolution of this problem. They conclude [33] that, in addition to the basic symmetry principles of special relativity and the assumption that fields are linear functions of currents, a new postulate of energy conservation is needed to deduce that classical electromagnetic theory can be modified in only one way, by introducing finite photon mass and replacing the Maxwell equations by the Proca equation. This new postulate maintains that there exists a locally conserved energy-momentum density such that the total energy and momentum of a system of charges and fields is conserved [33]. On the basis of this, Goldhaber and Nieto arrive at the interesting conclusion that the only way to maintain energy conservation is to insist that there are two different types of photon, though they act on charge in the same way. The accumulating experimental evidence for tired-light appears to suggest [33] that basic electrodynamics must be modified in this kind of way.

It is already clear, however, that a finite photon mass leads to a third degree of freedom for the photon [33]: longitudinal polarization. The existence of tired-light (and finite photon mass) would therefore appear to suggest [33] a fifty percent increase in the stored energy of a system of photons at thermodynamic equilibrium with a reservoir at a temperature T. Since Planck's law is precise in the laboratory, such an effect does not seem to be present [33]. Bass and Schrödinger [35] have shown, however, that the approach to equilibrium of longitudinal photons (for example of tired-light from a far distant cosmic source) is very slow, comparable with the age of the universe. Wigner has argued [33] that, even with the highest densities and longest time scales known at present, longitudinal photons have a negligible effect on thermodynamic systems and therefore do not affect the Planck radiation law. More accurately, the law is not affected on time scales comparable with the age of the universe. Stellar matter is expected [33] to be transparent to longitudinal photons, and the chances of observing thermal longitudinal photons are remote. Reaction and emission rates for longitudinal photons are suppressed in comparison to those for transverse counterparts by a factor proportional to photon mass over longitudinal photon frequency all squared. For this reason, longitudinal photons have apparently not been observed to date in atomic and molecular spectroscopy.

However, as shown in this paper, spectral effects due to $\boldsymbol{B}^{(3)}$ are expected to be consistent with finite photon mass. Several of these effects, such as the inverse and optical Faraday effects, have been observed experimentally [36 – 42] but interpreted in terms of the conjugate product rather than the fundamental $\boldsymbol{B}^{(3)}$ field. If the latter were zero, however, the conjugate product $iB^{(0)}\boldsymbol{B}^{(3)}$ would vanish, and these magneto-optic effects would not exist. Finite photon mass means that $\boldsymbol{B}^{(3)}$ is replaced by the closely similar $\boldsymbol{B}^{(3)}e^{-\xi z}$, as explained in the text. It is expected that there are effects due to $\boldsymbol{B}^{(3)}$ at first order as well as those at second order due to $iB^{(0)}\boldsymbol{B}^{(3)}$. Anomalous red shifts and other cosmological phenomena due to photon mass have been reviewed recently by Vigier [17].

It is easily shown that the existence of $\boldsymbol{B}^{(3)}$ does not upset the conventional Poynting theorem, because the intensity of light is given in circular or linear polarization by

$$I_0 = \frac{1}{2}\epsilon_0 c\left(|S_0| + |S_3|\right). \tag{A1}$$

Here S_0 and S_3 are the zero'th and third Stokes parameters, defined by

$$\begin{aligned} S_0 &= c^2 B^{(0)2} = E_0^2, \\ S_3 &= \left|-ic^2\boldsymbol{B}^{(1)}\times\boldsymbol{B}^{(2)}\right| = c^2\left|B^{(0)}\boldsymbol{B}^{(3)}\right| = c^2\boldsymbol{B}^{(3)}\cdot\boldsymbol{B}^{(3)} = E_0^2. \end{aligned} \tag{A2}$$

It is seen that $\boldsymbol{B}^{(3)}$ enters into the Poynting theorem through S_3. The definition (A1) in terms of the absolute, positive, magnitudes of S_0 and S_3 ensures that the intensity of light is the same in circular and linear polarization. (For elliptical polarization we must include the first and second Stokes parameters.) Note that S_3 itself changes sign between right and left circular polarizations, meaning that the sign of $\boldsymbol{B}^{(3)}$ is reversed. Therefore $\boldsymbol{B}^{(3)}$ contributes to the electromagnetic energy density in free space through S_3, and is accounted for in Poynting's theorem, the law of conservation of electromagnetic energy in classical, Maxwellian, electrodynamics.

APPENDIX 2. THE EFFECT OF FINITE PHOTON MASS ON GWS AND SU(5)

We have seen that the key to conventional gauge theory is Eq. (36) of the text, from which it is conventionally asserted that the photon mass is zero. We have replaced this assertion by the alternative, Eq. (37), which also satisfies the conventional gauge condition (36) for invariance of the second kind. Huang [43] has discussed the consequences of finite photon mass in unified (GWS) and grand unified (SU(5)) field theories. In the present paper the structure of the conventional U(1) (electromagnetic) sector is maintained intact, because the conventional gauge invariance of the second kind is maintained through Eq. (37) for finite m_0. This means that the U(1) sector enters GWS and SU(5) in the same way as in conventional theory, where m_0 is zero and $A_\mu A_\mu$ is non-zero in general. However, as in all theories of finite photon mass, we lose gauge freedom, a consequence which is summarised in our Eq. (3b). This loss of gauge freedom is compensated by the fact that A_μ becomes manifestly covariant, with four physically meaningful components. Equation (3b) is exactly what is expected from a physical vector A_μ in the lightlike condition. The fact that A_μ is taken to be manifestly covariant means that it is physically meaningful, as in the Bohm-Aharonov effect discussed in the text. The loss of gauge freedom which accompanies the Proca equation is a well known consequence [33] of finite photon mass.

Huang [43] has provided an interesting discussion of finite photon mass in GWS and SU(5), showing that it leads to a finite electron lifetime which is expressible in terms of the finite photon mass. Huang [43] mentions that the Proca equation can be consistent with charge non-conservation, another consequence of finite photon mass. However, the presence or absence of photon mass does not in itself guarantee non-conservation or conservation of charge [43]. Huang maintains that what is important ultimately in complicated theories such as GWS or SU(5) is whether the fundamental interactions between the various fields commute with the total diagonal charge operator, which is proportional to the number operators of the relevant fields. Huang allows for the vacuum expectation value of both the charged and neutral Higgs field to be non-zero and avoids in this way problems of re-normalization due to longitudinal massive spin-1 states. After the symmetry is spontaneously broken, the Huang theory [43] allows no residual local U(1) symmetry through the use of two Higgs fields in the GWS model. Mass terms then arise from Lagrangian densities that are bilinear in the gauge fields. The interaction terms that allow the electron to decay into a photon and an electron-neutrino are produced by interaction terms from minimal coupling of the leptons to the gauge fields. In this way, finite photon mass is worked into the theory self consistently.

Huang's theory reduces to standard GWS and SU(5) when photon mass is zero, and in this limit U(1) sector properties are also regained. In the U(1) sector in this limit, the field $\boldsymbol{B}^{(3)}$ emerges naturally from Huang's theory. This is therefore an example of how finite photon mass can be worked into GWS and SU(5). In the text of this paper we have restricted consideration to the U(1) sector, and have maintained the traditional gauge invariance of the second kind through the condition (3b).

REFERENCES

[1] J. D. Jackson, *Classical Electrodynamics* (Wiley, New York, 1962).

[2] R. M. Whitner, *Electromagnetics* (Prentice Hall, Englewood Cliffs, 1962).

[3] A. F. Kip, *Fundamentals of Electricity and Magnetism* (McGraw-Hill, New York, 1962).

[4] L. D. Landau and E. M. Lifshitz, *The Classical Theory of Fields*, 4th edn. (Pergamon, Oxford, 1975).

[5] M. Born and E. Wolf, *Principles of Optics*, 6th edn. (Pergamon, Oxford, 1975).

[6] W. M. Schwartz, *Intermediate Electromagnetic Theory* (Wiley, New York, 1964).

[7] P. W. Atkins, *Molecular Quantum Mechanics*, 2nd edn. (Oxford University Press, Oxford, 1983).

[8] C. Cohen-Tannoudji, J. Dupont-Roc, and G. Grynberg *Photons and Atoms: Introduction to Quantum Electrodynamics* (Wiley, New York, 1989).

[9] L. D. Barron, *Molecular Light Scattering and Optical Activity* (Cambridge University Press, Cambridge, 1982).

[10] B. W. Shore, *The Theory of Coherent Atomic Excitation*, Vols. 1 and 2 (Wiley, New York, 1990).

[11] D. E. Soper, *Classical Field Theory* (Wiley, New York, 1976).

[12] E. L. Hill, *Rev. Mod. Phys.* **23**, 253 (1951).

[13] S. S. Schweber, *An Introduction to Relativistic Quantum Field Theory* (Harper & Row, New York, 1962).

[14] N. N. Bogoliubov and D. V. Shirkov, *Introduction to the Theory of Quantised Fields*, 3rd edn. (Wiley Interscience, New York, 1980).

[15] D. Lurie, *Particles and Fields* (Wiley Interscience, New York, 1968).

[16] R. Jost, in M. Fierz and V. F. Weisskopf, eds., *Theoretical Physics in the Twentieth Century* (Wiley Interscience, New York, 1960).

[17] J. P. Vigier, "Present experimental status of the Einstein-de Broglie theory of light." conference reprint, communication to the author of January 4, 1993 from Université Pierre et Marie Curie, Paris.

[18] A. Einstein, e.g., *Verh. Deutsch. Phys. Ges.* **18**, 318 (1916); *Mitt. Phys. Ges. Zürich* **16**, 47 (1916); *Phys. Z.* **18**, 121 (1917); letters to Besso, August 8, 1916 and September 6, 1916.

[19] L. de Broglie, *La Mécanique Ondulatoire du Photon* (Gauthier-Villars, Paris, 1936).

[20] L. S. Ryder, *Quantum Field Theory*, 2nd edn. (Cambridge University Press, Cambridge, 1987).
[21] L. S. Ryder, *Elementary Particles and Symmetries* (Gordon & Breach, London, 1986).
[22] See e.g., W. Heitler, *The Quantum Theory of Radiation*, 3rd edn. (Clarendon, Oxford, 1954).
[23] M. W. Evans, *Mod. Phys. Lett.* **6**, 1237 (1992).
[24] M. W. Evans, *Physica B* **182**, 237 (1992).
[25] M. W. Evans, *Physica B* **182**, 227 (1992).
[26] M. W. Evans, *The Photon's Magnetic Field* (World Scientific, Singapore, 1993).
[27] F. Farahi and M. W. Evans, *Phys. Rev. E*, in press, March 1993.
[28] M. W. Evans, *Physica B* **183**, 103 (1993).
[29] E. P. Wigner, *Ann. Math.* **40**, 149 (1939).
[30] R. D. Hall *et al.*, *Phys. Rev. Lett.* **60**, 81 (1988).
[31] D. F. Bartlett and T. R. Corle, *Phys. Rev. Lett.* **55**, 59 (1985).
[32] W. S. Warren, D. Goswami, S. Mayr, and A. P. West, Jr., *Science* **255**, 1681 (1992).
[33] A. S. Goldhaber and M. M. Nieto, *Rev. Mod. Phys.* **43**, 277 (1971).
[34] L. de Broglie, *Mécanique Ondulatoire du Photon et Théorie Quantique des Champs*, 2nd edn. (Gauthier-Villars, Paris, 1957), 62.
[35] L. Bass and E. Schrödinger, *Proc. R. Soc.* A**232**, 1 (1955).
[36] J. P. van der Ziel, P. S. Pershan, and L. D. Malmstrom, *Phys. Rev. Lett.* **15**, 190 (1965).
[37] J. Deschamps, M. Fitaire, and M. Lagoutte, *Phys. Rev. Lett.* **25**, 1330 (1970).
[38] N. Sanford, R. W. Davies, A. Lempicki, W. J. Miniscalco, and S. J. Nettel, *Phys. Rev. Lett.* **50**, 1803 (1983).
[39] V. P. Kaftandjian and L. Klein, *Phys. Lett. A* **62A**, 317 (1977).
[40] P. F. Liao and G. C. Bjorklund, *Phys. Rev. Lett.* **36**, 584 (1976).
[41] B. A. Zon, V. Ya. Kuperschmidt, G. V. Pakhomov, and T. T. Urazbaev, *JETP Lett.* **45**, 273 (1987).
[42] T. W. Barrett, H. Wohltjen, and A. Snow, *Nature* **301**, 694 (1983).
[43] J. C. Huang, *J. Phys. G: Nucl. Phys.* **13**, 273 (1987).

TIME-DELAYED INTERFEROMETRY WITH NUCLEAR RESONANCE

Yuji Hasegawa[1] **and Seishi Kikuta**

Department of Applied Physics
University of Tokyo
7-3-1 Hongo, Bunkyo-ku, Tokyo 113, Japan

Experiments are reported where interference of beams with some time delay, due to the nuclear resonance, was observed. The interference arises from the photons emitted by different nuclei with some time interval after the absorption of a single incident photon. High visibility interference oscillations verified the coherent superpositions as well as the complete coincidence in the time domain of the beams after nuclear resonance. The collective nuclear resonances are discussed with regard to the no-reduction of the detection probability.

Key words: x-ray interferometry, nuclear resonance, time-delay, coherence.

Since the early stage of quantum physics, a double-slit experiment has served as the example of the strange features of quantum phenomena. Feynman stated on this situation that the double-slit experiment is a phenomenon "which has in it the heart of quantum mechanics: in reality it contains the *only* mystery" [1]. There the complementarity of the path (particle) and the interference (wave) is seen very clearly. This complementarity lies at the core of quantum mechanics. In practice, double-slit experiments have been investigated with many types of interferometers, (IFMs) in the visible light region. To make this kind of experiment in an x-ray region, the Bonse-Hart type IFM will be the most appropriate candidate, in which beams are split, reflected and recombined by use of Laue reflections [2–5].

Coherent phenomena have been observed in many aspects of nuclear resonant scattering: A nuclear Bragg reflection stems from the spatial coherence in crystal lattice of resonant scatterers [6–9], whereas the

Waves and Particles in Light and Matter, Edited by
A. van der Merwe and A. Garuccio, Plenum Press, New York, 1994

quantum beats appearing in subsequent decays reflect the coherently excited hyperfine split energy levels of nuclei [10–14]. What effects are expected on the beams, when both split beams suffer nuclear resonsances? Are they still coherent? Do absorption and emission in nuclear resonances from a single photon really happen in both beams? We have performed the experiments to make such questions clear.

The experiments on the direct observation of interference of the resonantly scattered beams were carried out by use of an undulator source at the 6.5-GeV accumulation ring for TRISTAN at the National Laboratory for High Energy Physics (KEK) [15]. The fast detector with TAC enabled us to count the delayed yields of the nuclear resonances. The incident beam was monochromatized through successive asymmetrically and symmetrically channel-cut monochromators in dispersive arrangement placed downstream of the Si double-crystal premonochromator.

An LLL-IFM made of Si single crystal was adjusted to give 220 reflections (Fig. 1). The width, the height, and the thickness of each plate are 70mm, 40mm, and 1.8mm, respectively. The cross section of the incident beam was reduced to $5.0\times0.2\text{mm}^2$. Fe foils enriched in ^{57}Fe of about 4μm in thickness were inserted parallel between the first and the second plates of the IFM, so that pulsed beams in both split paths do not go through the foils simultaneously. This made the time difference over the coherence time of the incident beam. A parallel-sided Si wafer of 290μm in thickness was inserted between the second and the third plate of the IFM as a phase shifter. A relative phase was shifted by rotation of this Si wafer. The fast detector was set for the interfering O-beam.

In determining the delayed gate time, the phase shifter was set so that the intensity of the interfering beam was maximum. The time spectrum after the IFM on the resonant energy is shown in Fig. 2. Although the peak around 28 nsec is seen due to the partially filled (10^{-4}) electron bunch, the delayed yield with quantum beats of the peak around 43 nsec arises from resonant forward scattering. The delayed gate time in measuring the interference oscillations was set between 37 and 89 nsec, so that only the resonantly scattered yields were counted with removing the prompt peak and the background. Under these conditions, the interference oscillations were measured by shifting the relative phase. The data with the least square fits are shown in Fig. 3. Though only the fluctuation of the background is seen for counts off resonance, almost complete interference oscillation above the background is seen for counts on resonance.

The observed interference oscillations arise from the interference between two split beams with an ^{57}Fe foil. Only photons emitted through the nuclear resonances, even at a distance of macroscopic scale and with some time intervals, are responsible for these oscillations. In our x-ray

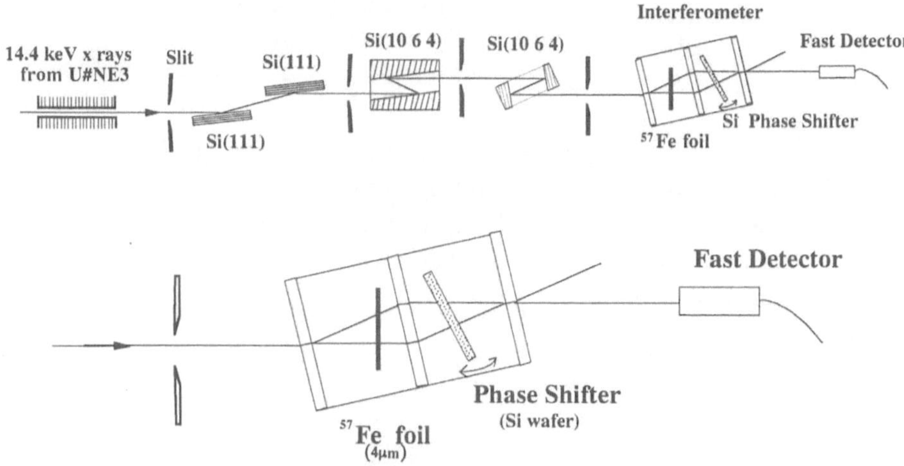

Fig. 1. Schematic view of the interferometry experiments. The incident beam is monochromatized to ΔE=8meV and falls on the interferometer. In the LLL-interferometer, the stainless-steel foil enriched in ^{57}Fe are inserted to cause the nuclear resonances in both beam paths in addition to the auxiliary phase shifter, made of a Si wafer.

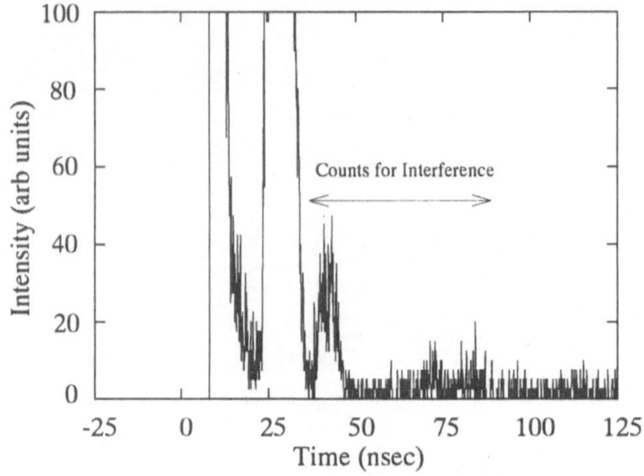

Fig. 2. Time spectrum of forward nuclear resonant scattering after the interferometer. The time-delayed yields within the time range indicated by the arrow are counted for the interference oscillation measurements.

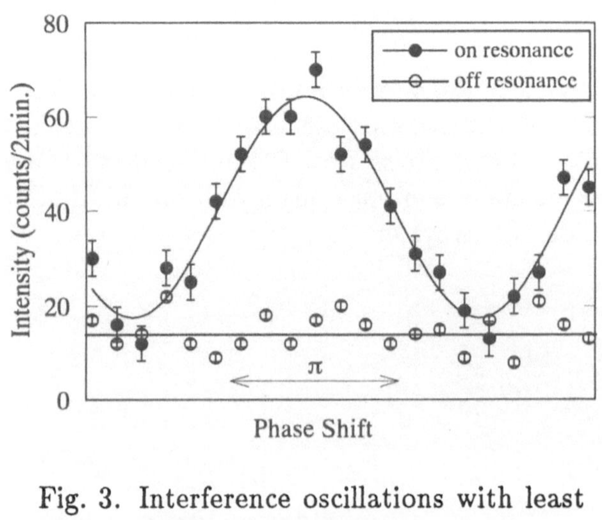

Fig. 3. Interference oscillations with least squares fits for the energy on and off resonance. A high visibility interference oscillation is observed on resonance, while only the background is observed off resonance.

experiments, photons can be considered to come one by one and the self-interference was observed. The present results indicate that (i) two split beams after nuclear resonances are coherent and that (ii) even an incident beam with a single photon is split into two paths and causes nuclear resonances in both paths. Moreover, the visibility of the interference oscillations with delyaed gate is as high as that without gate, which demands that delayed scattered wave fields in both beam paths are completely coincident in the time domain. In other words, emitted photons get exactly the same delay through the nuclear resonances.

In an x-ray experiment, photons can be considered to come one by one, even if SR is utilized. On the other hand, it is concluded from the study of the emission and absorption of the radiation field, i.e., the photoelectric effect, that "electromagnetic waves can change their energy only in units hν" [16]. In addition, the anti-coincident detection is observed in a single-photon beam splitting experiment, which shows that the detection in one path will reduce the probability of detecting the photons in the other paths [17,18]. Our experimental results, however, showed that even an incident beam with a single photon can bring forth the nuclear resonances in both paths, and those two beams, which are emitted through the nuclear resonances, are superposed. The nuclear resonances in one of the paths cannot reduce the probability of detecting the photons in the other paths. This suggests that the absorptions of photons need not be accompanied by the reduction. These nuclear resonances should be regarded as the scatterings, after which all the beams—scattered and transmitted, including the forward scattered part—are superposed. The noteworthy difference from the conventional scattering by electrons is that the absorbed photons can be regarded as emitted not instantaneously but continuously.

We may know Bohr's statement of "no elementary phenomenon is a phenomenon until it is a registered phenomenon" or Wheeler's concept of the smoky dragon [19], which symbolizes the fact that we can only get a counter reading without the right to know how it comes. It can or cannot be true. But we can conclude from the experimental results above that, although absorption is necessary for the registration, the absorption in itself does not mean the registration. Additional factors to and after the absorption of photons bring the registration and the mystery of quantum physics can be accompanied by them.

Let us write down the experimental situations in the IFM in a mathematical form. The state of the beam as well as the nuclei after the nuclear resonances, $|\Psi(t)>_{sc}$, is given by

$$|\Psi(t)>_{sc}= \alpha\{\beta(t)|\, g;1> +\gamma(t)|\, e;0>\}, \tag{1}$$

where $|g,1>$ and $|e,0>$ represent the state in which the nuclei are in the ground state and one photon exists, and the state when nuclei are excited and no photon exists, respectively; $\beta(t)$ and $\gamma(t)$ satisfy $\beta(0)^2+\gamma(0)^2 = 1$.

The state after going through the IFM, $|\Psi(t)>_{\rm int}$, can be written as

$$\begin{aligned} |\Psi(t)>_{\rm int} &= \frac{\alpha}{2}\{\beta_{\rm I}(t)\,|g;1>_{\rm I} +\gamma_{\rm I}(t)|\,e;0>_{\rm I}\} \\ &+ \frac{\alpha}{2}e^{i\phi}\{\beta_{\rm II}(t)\,|g;1>_{\rm II} +\gamma_{\rm II}(t)|\,e;0>_{\rm II}\}, \end{aligned} \tag{2}$$

where ϕ represents the phase shift by an auxiliary phase shifter and the subscripts denote the split beam paths I and path II in the IFM, respectively. The complete coincidence in the time domain demands $\beta_{\rm I}(t) = \beta_{\rm II}(t)$, so $\beta(t) \equiv \beta_{\rm I}(t) = \beta_{\rm II}(t)$ and $\gamma_{\rm I}(t) = \gamma_{\rm II}(t)$. The measured intensity $I_{\rm int}$ is given by

$$\begin{aligned} \mathrm{I}_{\rm int} &= \int_{t_1}^{t_2} {}_{\rm int}< \Psi(t)\left|\left(a^{\dagger}(t)a(t)\right)\right|\Psi(t)>_{\rm int}\, dt \\ &= (1-\cos\phi)\times\int_{t_1}^{t_2}\frac{|\alpha|^2}{2}|\beta(t)|^2dt, \end{aligned} \tag{3}$$

where $a^{\dagger}(t)$ and $a(t)$ are the creation and annihilation operators as a function of time t. In our experiments, t_1 and t_2 are set at 37 and 89 nsec, respectively, and a clear intereference oscillation due to ϕ was observed.

It must be emphasized that the interference demands the coherent superposition of the beams emitted in each path by nuclear resonances. Thus, the observed interference arises from beams emitted by different atoms. It is reasonable to assume that not a single but lots of atoms are concerned with the nuclear resonances in each of two paths. In other words, the absorption in scatterings in itself is a non-localized phenomenon and the collective excitations of nuclei are possible. It is seen in Eq. (2) that one is dealing with the superposition of excited and ground states of nuclei as well as the superposition of the beams of path I and path II. The former superposition is related to a continuous emission. One of the interpretations of the intermediate state of the continuous emission is the coexistence of emitted photon and emitting nuclei.

In summary, we have demonstrated the direct observation of interference of the resonantly scattered beams in an x-ray IFM, which are emitted from different atoms through the nuclear resonances. A clear interference oscillation was obtained with visibility as high as the instrumental property of the IFM. This clarifies that two beams emitted at different times by different atoms through nuclear resonances are coherently superposed and completely coincide in the time domain. In addition, it was shown that even a beam with a single photon can bring forth the nuclear resonances of collective atoms without the reduction of the detection probability.

ACKNOWLEDGEMENTS

This work was supported in part by a Grant-in-Aid for Specially Promoted Research from the Ministry of Education, Science, and Culture.

NOTE

1. JSPS Fellowship for Japanese Junior Scientists.

REFERENCES

1. R. P. Feynman, R. B. Leighton, and M. Sands, *The Feynman Lectures on Physics* (Wesley, Reading, Massachusetts, 1965), Vol. 3, p. 1.
2. U. Bonse and M. Hart, *Appl. Phys. Lett.* **6** (1965) 155.
3. U. Bonse and M. Hart, *Z. Phys.* **188** (1965) 154.
4. M. Hart, *Rep. Prog. Phys.* **34** (1971) 435.
5. Y. Hasegawa and S. Kikuta, *Jpn. J. Appl. Phys.* **30** (1991) L316.
6. P. J. Black and P. B. Moon, *Nature* **188** (1960) 481.
7. A. M. Afanasév Artemév and Yu. Kagan, *Sov. Phys. JETP* **21** (1965) 215.
8. A. I. Chechin, N. V. Andronova, M. V. Zelepukhin, A. N. Artemév, and E. P. Stepenov, *JETP Lett.* **37** (1983) 663.
9. E. Gerdau, R. Rüffer, H. Winker, W. Tolksdorf, and C. P. Kagan, *Phys. Rev. Lett.* **54** (1987)
10. G. T. Trammell and J. P. Hannon, *Phys. Rev.* **B18** (1987) 165.
11. E. Gerdau, R. Rüffer, R. Hollatz, and J. P. Hannon, *Phys. Rev. Lett.* **57** (1986) 1141.
12. G. Faigel, D. P. Siddons, J. B. Hastings, P. E. Haustein, J. R. Grover, J. P. Remeika, and A. S. Cooper, *Phys. Rev. Lett.* **58** (1987) 2699.
13. J. Arthur, G. S. Brown, D. E. Brown, and S. Ruby, *Phys. Rev. Lett.* **63** (1989) 1629.
14. J. P. Hastings, D. P. Siddons, U. van Bürck, R. Hollatz, and U. Bergmann, *Phys. Rev. Lett.* **66** (1991) 770.
15. S. Yamamoto, X. W. Zhang, H. Kitamura, T. Shioya, T. Mochizuki, H. Sugiyama, M. Ando, Y. Yoda, S. Kituka, and H. Takei, *J. Appl. Phys.* **74** (1993) 500.
16. D. Bohm, *Quantum Mechanics* (Prentice-Hall, Englewood Cliffs, 1951).
17. J. F. Clauser, *Phys. Rev. D* **9** (1974) 835.
18. P. Grangier, G. Roger, and A. Aspect, *Europhys. Lett.* **1** (1986) 173.
19. W. A. Miller and J. A. Wheeler, *Proceedings International Symposium on the Foundations of Quantum Mechanics* (Physical Society of Japan, Tokyo, 1983), p. 140.

A COUNTEREXAMPLE OF BOHR'S WAVE-PARTICLE COMPLEMENTARITY

Dipankar Home

Bose Institute
Calcutta 700 009, India

We discuss the recently performed two-prism experiment showing classical wave-like tunnelling of single photon states. In particular, it is explained in what sense this experiment provides a *counterexample* to Bohr's idea of *"mutual exclusiveness"* of *classical* wave and particle pictures.

Key words : Bohr's complementarity, wave-particle duality, optical tunnelling, two-prism experiment.

1. INTRODUCTORY REMARKS

One of the central tenets of Bohr's wave-particle complementarity, viz., mutually incompatible classical pictures such as wave and particle are called for *only* in mutually exclusive (ME) physical situations, is *not* justifiable through a completely general argument based on the mathematical formalism of quantum mechanics. In fact, Bohr's strategy was to argue for the ME hypothesis by illustrative analyses of specific positive instances. As recently emphasized by Scully et al. [1], it is *essentially* in an *interference type* experiment that the quantum mechanical formalism guarantees the validity of ME. Disappearance of an interference pattern is ensured whenever one has particle-like "which-path" information through a measurement interaction resulting in an *entanglement* between orthogonal and macroscopically distinguishable states of the measuring apparatus and corresponding interfering wave functions of the observed particle.

However, Bohr's wave-particle complementarity was meant to be a *general epistemological principle* of fundamental significance. If so, it should *not* be valid *only* in interference-type experiments. From this point of view, it is important to investigate the validity of the ME hypothesis in the context of *other* types of experiment as well, *atleast those* which are comprehensible in terms of *classical* pictures (leaving apart those experimental results which are "irreducibly" *non-classical*).

Like interference, *tunnelling* is also a characteristic signature of classical wave-like behaviour, for which, however, ME is *not* automatically enforced by the quantum mechanical formalism. It is this feature which motivated Ghose, Home, and Agarwal [2] to propose the two-prism experiment with single photon states, discussed as follows.

2. The Two-Prism Experiment

The basic idea of this experiment (Fig.1) is as follows. Light pulses are incident on a combination of two prisms placed opposite each other, with a *variable gap* between them. When the gap between the prisms is sufficiently *large* compared with the wavelength, the incident light pulse would suffer *total internal reflection* inside the first prism (registered by detector 2). When the gap is *less* than the wavelength, there is a possibility that the light pulse will *tunnel* across the gap and emerge from the second prism (registered by detector 1). The formalism of quantum optics predicts [2] that the two detectors 1 and 2 should click in *perfect anticoincidence* if the incident light pulses correspond to "ideal" single photon states. Here the key conceptual point is that, while the single photon states tunnel like a *classical wave, at the same time* perfect anticoincidence between the two detectors implies a *classical particle-like propagation* because of *which-path* information (reflected *or* transmitted). We therefore contend that if one wishes to comprehend (following the Bohrian outlook) the results of this experiment in terms of *classical pictures* this is possible *only* by using *both* classical wave and particle pictures.

The two-prism arrangement of Fig.1 is a *special tunable* beam splitter, where one has a control over the transmission mechanism by varying the gap between the prisms relative to the wavelength associated with the incident single photon states. The fact that the transmission probability decreases (exponentially) with the increase of the gap compared to the wavelength can *only* be understood in terms of a wave-like description. The two-prism device, therefore, serves to demonstrate

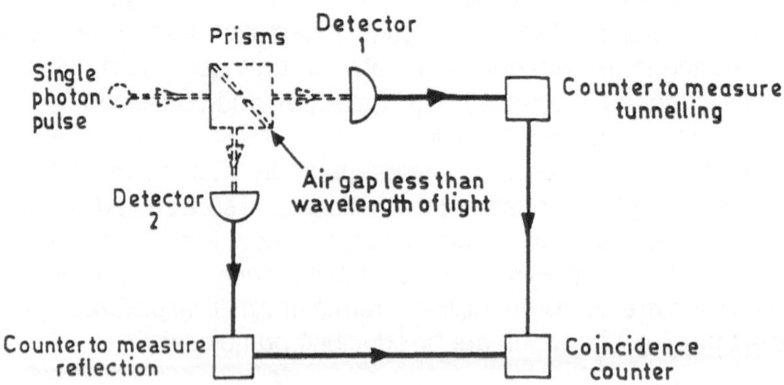

Fig.1: The two-prism device to show classical wave-like tunnelling of single photon states, *along with* "anticoincidence" between the detectors 1 and 2.

explicitly the *essentially wave-like* character of the transmission through this device. On the other hand, for an *arbitrarily chosen* beam splitter, transmission may *not necessarily* imply a wave-like behaviour; there remains a possibility of modelling it in such a way that certain specified fractions of incident *particles* are reflected and transmitted.

As in the case of a typical beam splitter, the state vector of the emergent single photon state from the two-prim arrangement, entangled with the vacuum states, can be written in the form

$$|\psi\rangle = \alpha|1,0\rangle + \beta|0,1\rangle, \quad (1)$$

where $|\alpha|^2$ gives the transmission probability and $|\beta|^2$ the reflection probability.

Perfect anticoincidence between the detectors 1 and 2 (Fig.1) is a *kinematic* feature, since it follows from the structure of the state vector (1). On the other hand, tunnelling in the quantum optical treatment follows from the *dynamics* of field propagation, because the Maxwell equations can be looked upon as Heisenberg equations with the classical fields replaced by quantum mechanical operators, and the boundary conditions now become those for the electric and magnetic field operators. Tunnelling identical to that obtained from classical electromagnetic theory is predicted, just as classical wavelike interference is predicted for single photon states.

The proposed experiment has recently been performed by Mizoubchi and Ohtake[3], who have verified the quantum optical prediction. It is found that the experimental data recorded in the counters of Fig.1 contain *both* classical wave-like *and* particle-like information about the propagation of *a given ensemble* of light pulses. While registrations by the counter 1 to measure the tunnelling probability pertain to a classical wave-like propagation of light pulses, the recordings by the coincidence counter (connected to detectors 1 and 2) showing anticoincidence for incident single photon states is comprehensible in terms of a classical particle-like propagation. The experiment of Fig.1, therefore, shows that there is a situation allowed by the formalism of quantum mechanics where the Bohrian notion of "mutual exclusiveness" is contradicted [4].

3. Some Critical Aspects

(A) In a "*single particle at a time*" interference experiment with, say, electrons/ neutrons, an interference pattern is built up by a gradual accumulation of *discrete* detection events registered as "spots" on a visual screen. The discrete nature of the detection events *per se* does *not* necessarily imply particle-like *propagation* of the detected entitles. In fact, they can entirely be regarded as originating from the discrete energy levels of the atomic constitutents of the detector. Since each individual "spot" does *not* correspond to any information about which of the possible parths an electron/neutron has followed, the building up of the final interference pattern can be interpreted as implying *classical wave-like propagation*, followed by discrete detection events. It may be worthwhile to stress that a key difference with the experiment of Fig.1 is that an "anticoincidence" between the detectors on an "interference screen" does *not* provide "which path" information.

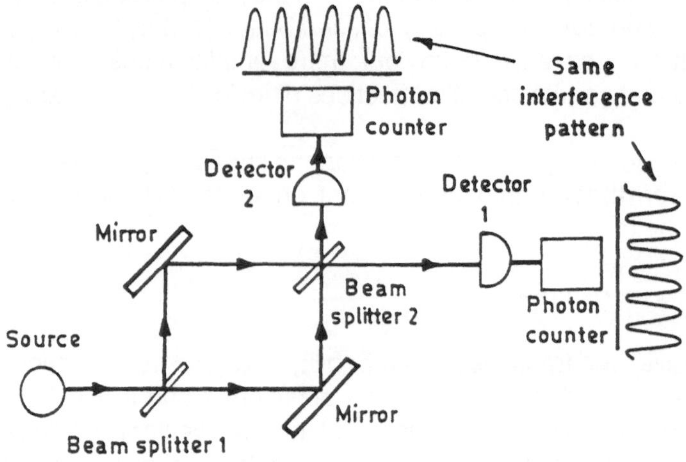

Fig.2: Anticoincidence between the detectors 1 and 2 and an interference pattern in each of them will be registered in this experimental arrangement, using single photon states.

(B) Let us consider an experimental arrangement as given by Fig.2. If a source of single – photon states is used and detectors 1 and 2 are connected with a counter, one would observe both "anticoincidence" between the detectors and an interference pattern in each of them. The observed interference will correspond to a classical wave-like propagation between the source and the beam splitter 2 (one does not know "which path" a photon has followed after the beam spilitter 1 *till* arriving at the beam splitter 2), while the "anticoincidence" will provide "which–path" information *essentially after* the beam splitter 2 (implying particle-like propagation from the beam splitter 2 to each of the two detectors 1 and 2). In contrast, if one considers the experiment of Fig.1, *both* classical wave *and* particle-like propagation are implied all through *in between* the source and the detectors 1 and 2.

(C) An important point concerning the experiment of Fig.1 is that here the coincidence rates and singles rates pertain to the *same* ensemble of light pulses incident on the two-prism arrangement, *unlike* the so-called "intermediate type interference experiments" in which, by using "partial which path" determination, the input ensemble is split into two [5]. One sub-ensemble gives rise to an interference pattern and the other sub-ensemble, corresponding to "which–path" information, does *not* contribute to the interference pattern. Here there is, therefore, no contradiction with Bohr's wave-particle complementarity.

(D) There could be a variant of the beam-splitter experiment which would be analogous to the two-prism experiment. By varying the orientation of the beam splitter (i.e., by changing the angle of incidence of the incident light pulse), transmission and reflection probabilities can be varied. This is a classicial wave-like property [6]. However, this effect is not as pronounced as the one due to the variation of the gap between the two prisms.

4. CONCLUDING REMARKS

Conceptual inadequacies of Bohr's wave-particle complementarity have been much debated. It is however equally important, I think, to investigate whether these *limitations* of the complementarity principle can be exposed through experiments. The two-prism experiment is an effort in this direction. While this experiment poses a problem for Bohrs wave-particle complementarity, its understanding is straightforward on the basis of the "realist" interpretations, such as *Bohm's causal interpretation* as applied to electromagnetic fields or the de Broglie picture of photons propagating as localized particles associated with objectively real waves.

Louis de Brogile wrote in 1964: "I have, for a long time, adopted the idea of complementarity in the realm of quantum physics, whilst at the same time realizing that it was inadequate. In recent years I have been led to regard the concept of complementarity with increasing suspicion" [7]. The two-prism experiment proposed by Ghose, Home and Agarwal and performed by Mizobuchi and Ohtake constitutes a *vindication* of de Broglie's "suspicion." It is therefore appropriate to dedicate this work to his memory on the occasion of his birth centenary.

ACKNOWLEDGMENTS

This research was supported by the Department of Science and Technology, Government of India.

REFERENCES

1. M. O. Scully, B. G. Englert, and H. Walther, *Nature* **351**, 111 (1991).
2. P. Ghose, D. Home, and G. S. Agarwal, *Phys. Lett. A* **153,** 403 (1991).
3. Y. Mizobuchi and Y. Ohtake, *Phys. Lett. A*, **168,** 1 (1992).
4. P. Ghose, D. Home, and G. S. Agarwal, *Phys. Lett. A* **168**, 95 (1992).
5. W. K. Wooters and W. H. Zurek, *Phys. Rev. D* **19**, 473 (1979); D. M. Greenberger and A. Yasin, *Phys. Lett. A* **128**. 391 (1988).
6. M. Born and E. Wolf, *Principles of Optics* (Pergamon, London, 1970), pp. 41-45.
7. L. de Broglie, *The Current Interpretation of Wave Mechanics* (Elsevier, Amsterdam, 1964), p.7.

CLASSICAL ELECTROMAGNETIC THEORY OF DIFFRACTION AND INTERFERENCE: EDGE, SINGLE-SLIT AND DOUBLE-SLIT SOLUTIONS

S. Jeffers,[a] R. D. Prosser,[b]
G. Hunter,[c] and J. Sloan[d]

[a] *Department of Physics and Astronomy and*
Centre for Research in Earth and Space Science
York University, North York, Ontario, Canada
[b] *16 Elaine Court, 123 Haverstock Hill*
London NW3 4RT, United Kingdom
[c] *Department of Chemistry and*
Center for Research in Earth and Space Science
York University, North York, Ontario, Canada
[d] *Department of Physics and Astronomy*
York University, North York, Ontario, Canada

Solutions to Maxwell's equations have been obtained for a variety of diffracting geometries (i.e., edge, single slit, and double slit), where the diffracting screen serves to impose boundary conditions on the solutions. Consequently, the full vector solution (amplitude, phase, and Poynting vector) is obtained for any distance from the diffracting screen. The spatial distribution of intensity has been computed and agrees exactly with the results of classical, scalar diffraction theory for distances from the diffracting aperture of more than several wavelengths. The full vector solution gives the field distribution very close to the diffracting screen.

For the double-slit solution, the Poynting vectors show undulations in the energy flow past the diffracting screen which reproduce interference effects. We have also studied solutions obtained by closing one of the slits. Thus we can study the flow of electromagnetic energy through the double-slit system for various ratios of slit width as well as the variation in the contrast of the fringes as the ratio of slit widths is varied.

Key words: classical electromagnetic theory, Maxwell's equations, double-slit interference.

1. INTRODUCTION

Classical scalar diffraction theory is essentially a geometrical construction based on physically unrealistic assumptions which gives predictions of the spatial distribution of the diffracted intensity in excellent agreement with measurements. By its very nature, it is incapable of dealing with polarized radiation or of giving accurate information on diffraction phenomena within a few wavelengths of the diffraction aperture. It is, as characterised by Marion and Heald [1], "a highly artificial theory" but, nevertheless, a remarkably successful one.

Physically realistic solutions to diffraction phenomena must involve solutions to Maxwell's equations subject to boundary conditions imposed by the diffracting aperture. Such solutions have been found for relatively simple cases, e.g., the diffraction of a plane electromagnetic wave by the edge of a semi-infinite plane of zero thickness and infinite conductivity (Sommerfeld [2]). Notwithstanding the idealised assumptions, the solution can be given in closed form and is valid for all regions of space. Prosser [3] has given similar solutions for the single- and double-slit cases for distances from the diffracting screens of up to eight wavelengths.

We have extended the work of Prosser [3] to allow for solutions for the cases of edge diffraction, single- and double-slit diffraction for an arbitrarily large range of distances. Consequently, these full vector solutions allow for detailed comparisons with classical scalar theory and for detailed predictions to be compared with experimental data. We have also extended the calculations in order to study the effect of closing one of the slits. In this situation, we have plotted the flow lines of electromagnetic energy through the double-slit system and also computed the variation of the contrast in the fringes as a function of the ratio of the slit widths.

2. SCALAR DIFFRACTION THEORY

Huygen's principle asserts that every point on a wave front can be regarded as a source of secondary wavelets. Wavefront propagation is treated as the linear superposition of the secondary wavelets. This construction is physically unrealistic, since the assumed, spherical secondary wavelets would also propagate in the backward direction. In Kirchoff's formulation, these backward propagating waves are mathematically suppressed by the introduction of the Stoke's inclination factor $1 + \cos\theta$. Kirchoff showed that the light wave (scalar) at any point in space can be expressed as an integral over a closed arbitrary surface surrounding the point. In actual calculations, the surface is chosen to include the diffracting aperture. To evaluate the integral, it is necessary to know the wave amplitude and its partial derivative with respect to the normal to the surface. However, these are not known and approximations have to be introduced in order to evaluate the integral. It is assumed that over the aperture, the wave amplitude and its partial derivative have the values appropriate to the absence of the aperture, which is equivalent to stating that the presence of the diffracting screen does not perturb the incoming wave. This is clearly unrealistic near the edge of the aperture but will be a reasonable approximation for apertures whose dimensions are large compared to the wavelength of the radiation. In the plane of the diffracting aperture, it is assumed that both the amplitude and its partial derivative vanish. For the remainder of the surface of integration, it is assumed that the surface is at such a large distance from the aperture that the wave amplitude and its partial derivative can be made arbitrarily small. These boundary conditions yield solutions which do not satisfy the imposed boundary conditions—the justification for the boundary conditions is that they yield solutions that agree very well with experimental data (but not close to the diffracting aperture). In the words of Sommerfeld [2], "It is amazing that the classical diffraction theory nevertheless yields for all practical purposes satisfactory results."

3. EXACT SOLUTIONS TO MAXWELL'S EQUATIONS

Maxwell's equations were solved exactly by Sommerfeld [2] on an infinitesimally thin, infinitely conducting sheet of semi-infinite extent bounded by a straight edge. Braunbek and Laukien [4] expressed Sommerfeld's solution in the form of intensity and phase distributions, which were computed for a region extending to one wavelength from the diffracting edge. Prosser [3] extended these calculations out to eight wavelengths, plotting amplitude, phase, and Poynting vector. The undulations seen in the Poynting vectors were interpreted as diffraction and interference effects but not in the conventional sense; i.e., in the case of the double slit the lines of energy flow (Poynting vectors) do not cross the axis of symmetry. It was thus concluded that no radiation which passes through one slit actually interferes with radiation passing through the other slit in the conventional interpretation. This construction is similar to the quantum potential account of the double slit for non-relativistic particles given by Bohm et al [5] and Vigier [6]. Prosser speculated that, if the calculations were extended to finite distances, the calculations would reproduce the results of scalar diffraction theory. We show below that this speculation is correct.

We follow the treatment given by Prosser [3] based on the work of Braunbek and Laukien [4]. The incident plane electromagnetic wave propagates in the positive y direction with the semi-infinite plane in the xz plane with the edge along the z axis. The magnetic vector is parallel to the diffracting edge with the electric vector in the xy plane.

For normal incidence, Sommerfeld's solution gives

$$H_z = A\left[F(\sigma)e^{-i\gamma} + F(\sigma')e^{i\gamma}\right], \tag{1.1}$$

wherein

$$\begin{aligned} \gamma &= \frac{2\pi r}{\lambda}\sin\theta, \quad \sigma = 2\left(\frac{r}{\lambda}\right)^{\frac{1}{2}}\left(\sin\frac{\phi}{2} - \cos\frac{\phi}{2}\right), \\ \sigma' &= 2\left(\frac{r}{\lambda}\right)^{\frac{1}{2}}\left(\sin\frac{\phi}{2} + \cos\frac{\phi}{2}\right), \quad F(\sigma) = \int_{-\infty}^{\alpha} e^{\frac{-i\pi\tau^2}{2}}\,d\tau, \end{aligned} \tag{1.2}$$

and the electric field

$$E_x = \frac{\lambda}{2\pi i}\left(\frac{\mu_0}{\epsilon_0}\right)^{\frac{1}{2}}\frac{\delta H}{\delta y}, \quad E_y = -\frac{\lambda}{2\pi i}\left(\frac{\mu_0}{\epsilon_0}\right)^{\frac{1}{2}}\frac{\delta H}{\delta x} \tag{1.3}$$

In these results, the time dependence $e^{i\omega t}$ has been dropped, λ denotes wavelength, and r and ϕ are polar coordinates in the xy plane.

Equation (1.1) may be written in amplitude and phase form as

$$H_z = H(r,\phi)\, e^{i\Psi(r,\phi)}. \tag{1.4}$$

The components of the Poynting vector are

$$S_x = \frac{1}{2}Re\left(H_z \cdot E_y^*\right), \quad S_y = -\frac{1}{2}Re\left(H_z \cdot E_x^*\right), \tag{1.5}$$

and the direction of the Poynting vectors are obtained from

$$\frac{\delta r}{\delta\phi} = \frac{r\,(S_x\cos\phi + S_y\sin\phi)}{(S_y\cos\phi + S_x\sin\phi)}\,. \tag{1.6}$$

This equation can be integrated by making substitutions from Eqs. (1.1), (1.3), and (1.5).

4. RESULTS

Figures 1(a), (b), and (c) show contours of constant intensity for distances from the diffracting edge of $40\lambda, 1550\lambda$, and $120{,}000\lambda$, respectively. Figures 2(a), (b), and (c) show the computed Poynting vector flow lines for the same range of distances. Figures 3(a), (b), and (c) show a perspective plot of the amplitude over these distances. Finally, Figs. 4(a)

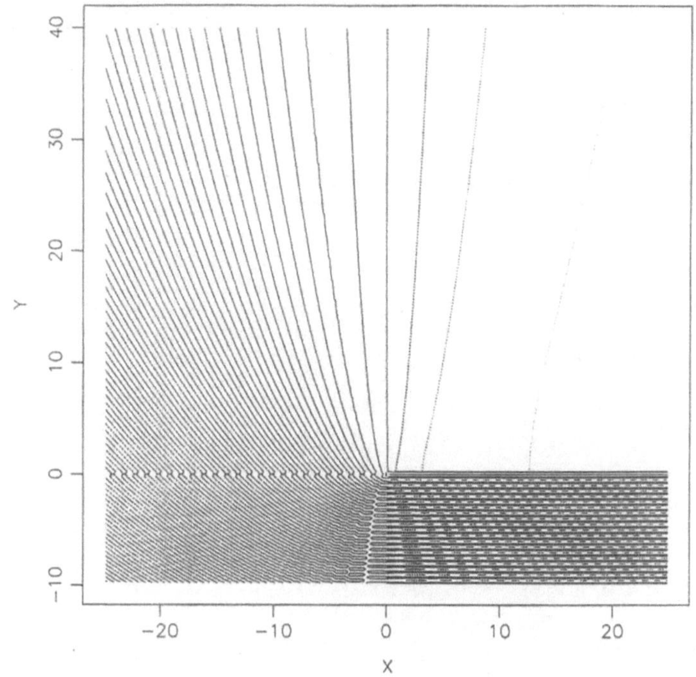

Fig. 1(a). Lines of constant intensity, edge.

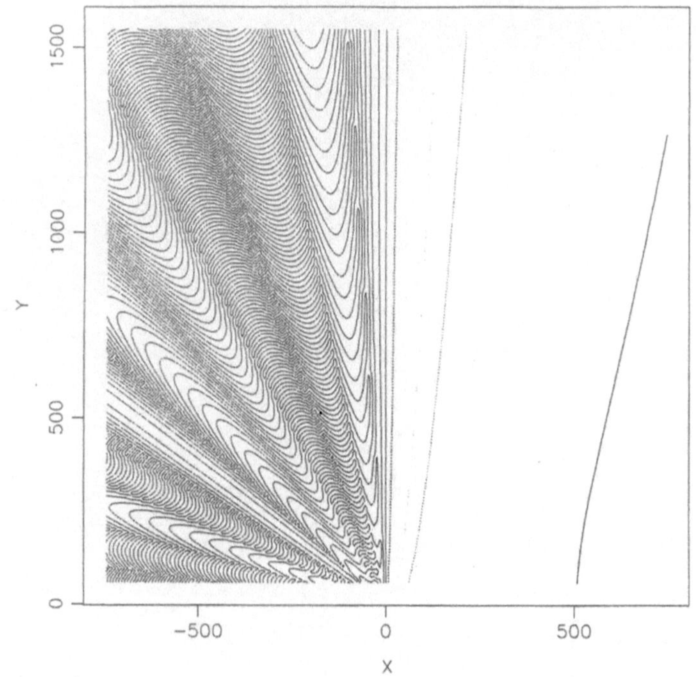

Fig. 1(b). Lines of constant intensity, edge.

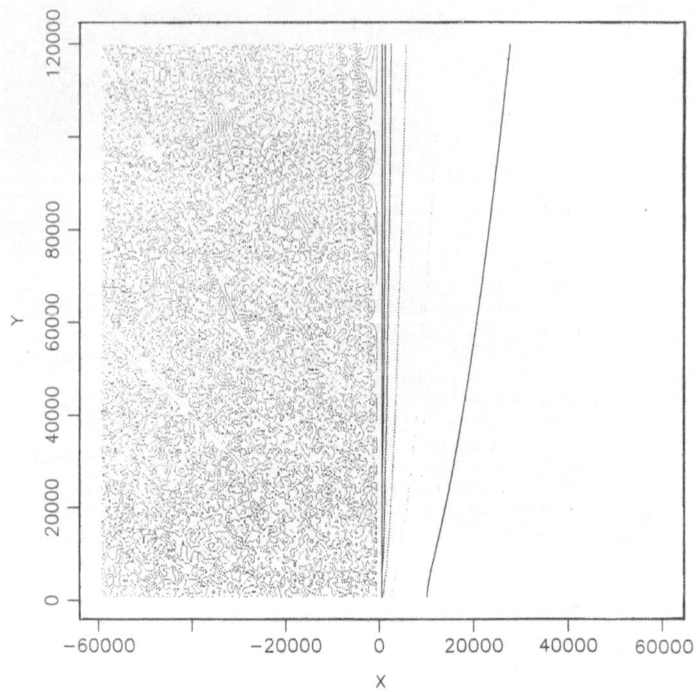

Fig. 1(c). Lines of constant intensity, edge.

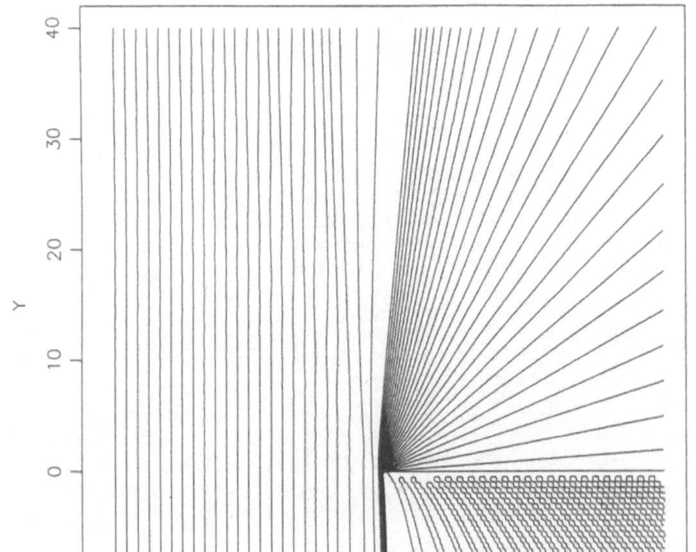

Fig. 2(a). Flow lines of Poynting vector, edge.

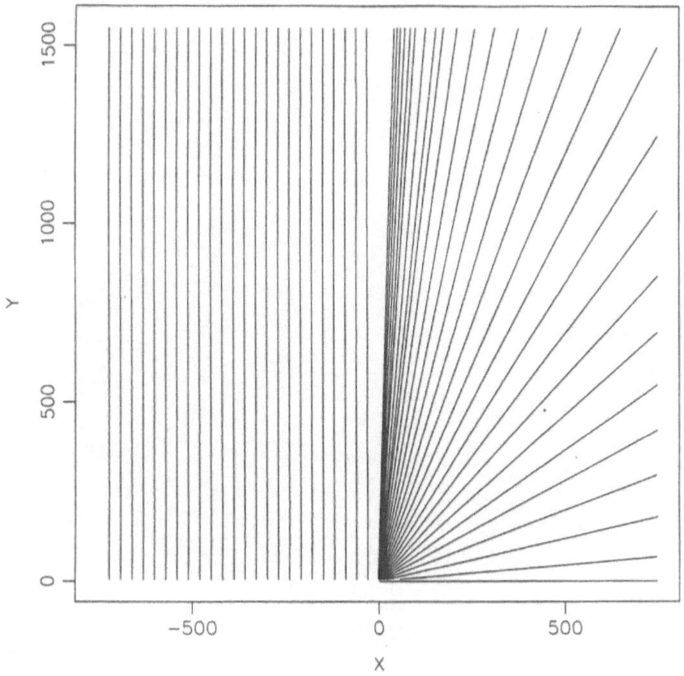

Fig. 2(b). Flow lines of Poynting vector, edge.

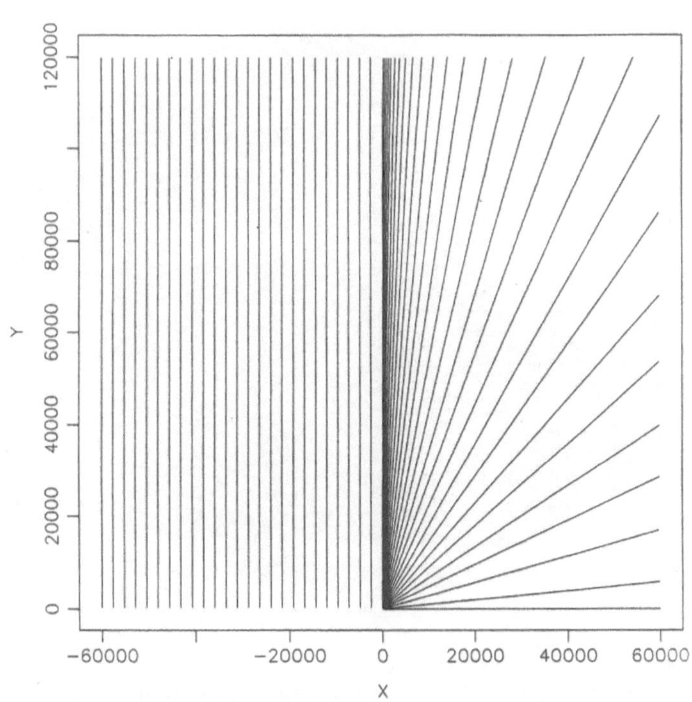

Fig. 2(c). Flow lines of Poynting vector, edge.

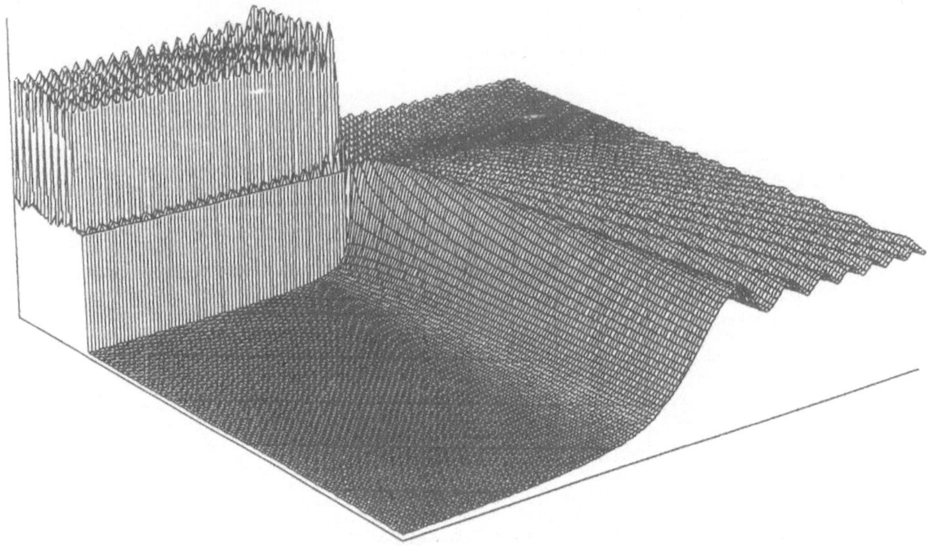

Fig. 3(a). Perspective plot of amplitude, edge.

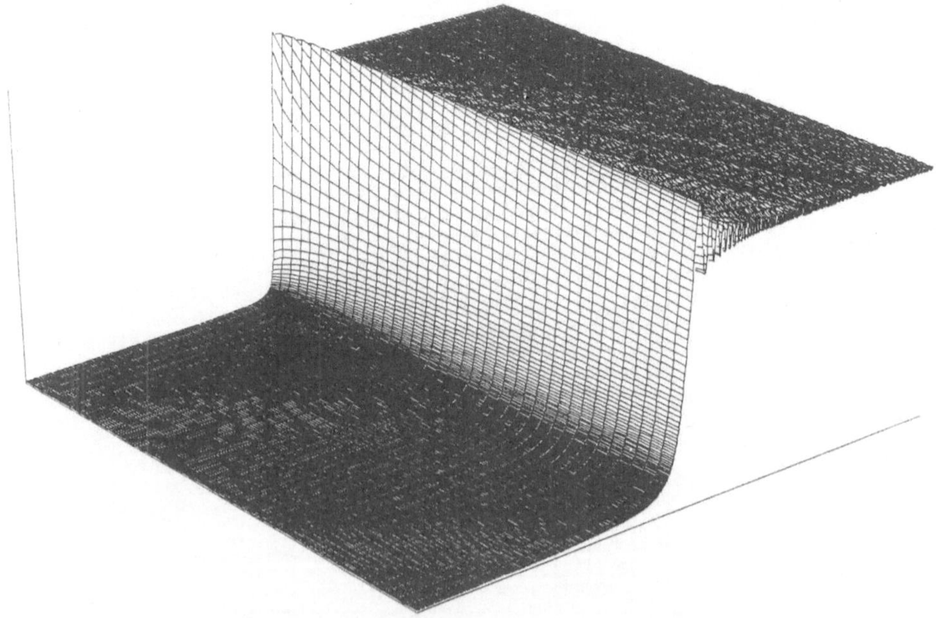

Fig. 3(b). Perspective plot of amplitude, edge.

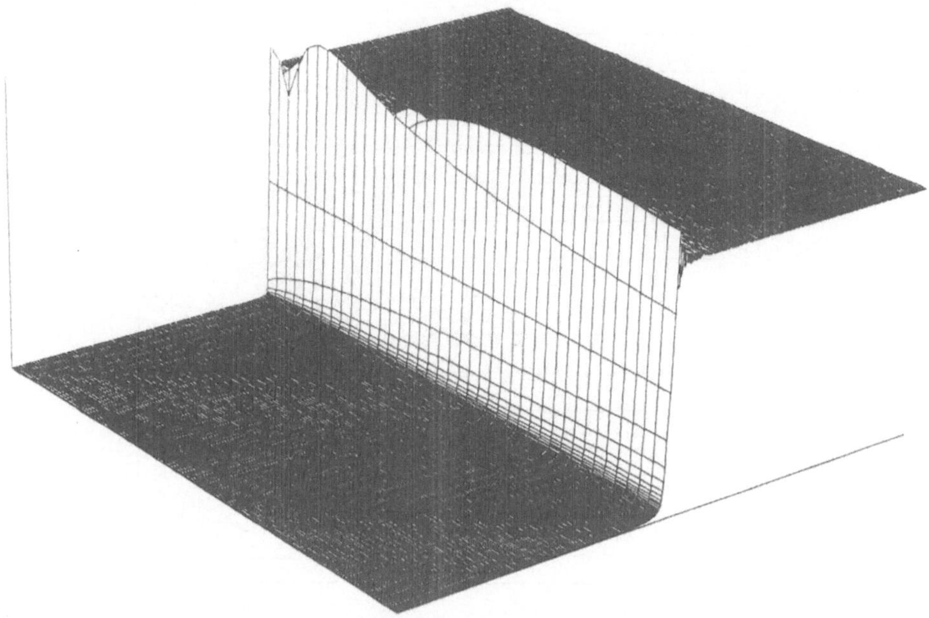

Fig. 3(c). Perspective plot of amplitude, edge.

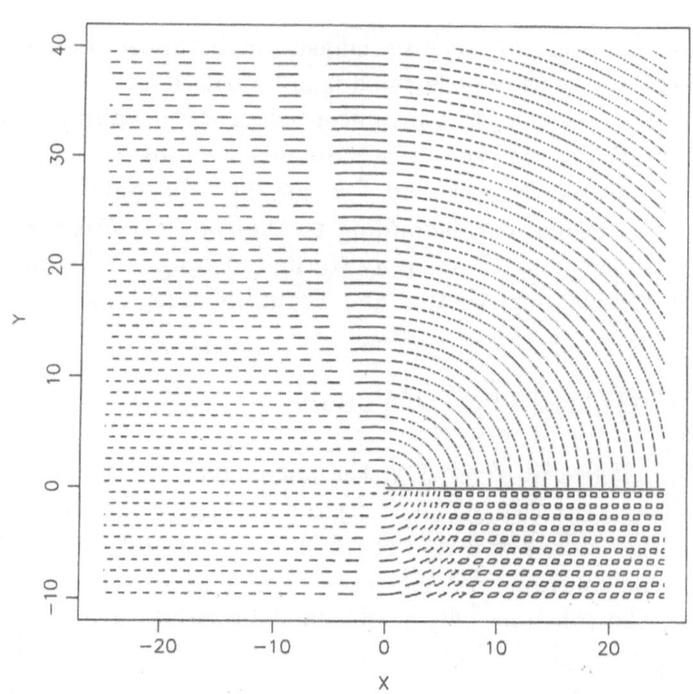

Fig. 4(a). Lines of constant phase, edge.

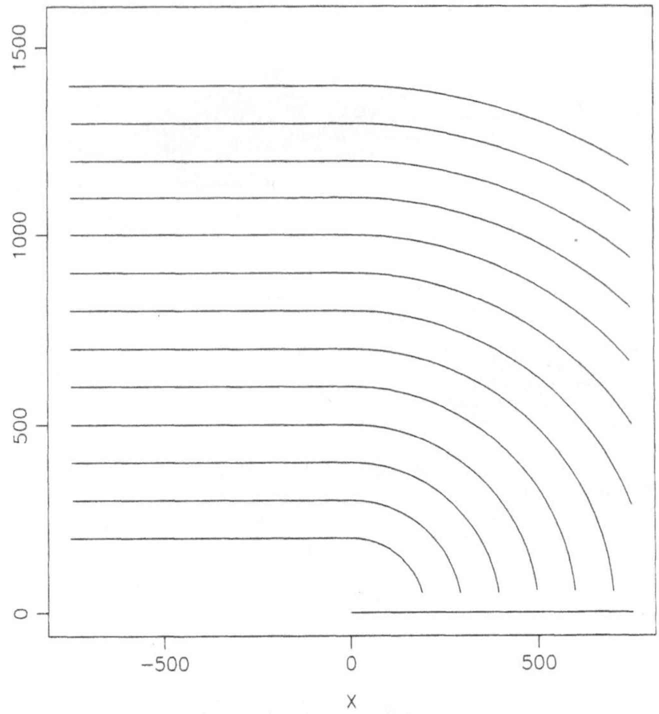

Fig. 4(b). Lines of constant phase, edge.

and (b) show contours of constant phase for distances of 40λ and 1550λ. All these calculations were performed using FORTRAN-77 on a 80486 IBM PC-compatible, with the plots produced using the graphing capabilities of the S-PLUS statistics package. The plots produced for the closest distances reproduce those already given by Prosser [3], who only calculated these quantities for distances corresponding to the classical Fresnel region. Our extension of these calculations apply to the full range of distances and thus include both the classical Fresnel and Fraunhofer regions.

5. THE SINGLE-SLIT SOLUTION

In the single-slit geometry, the slit edges are at $x = a, b$, running parallel to the z axis. Equations (1.3) and (1.4) give the solution for a half plane extending from $x = 0$ to $+\infty$. Following the notation of Prosser [3], we refer to this solution as ${}_{+}\Phi_0$. For a half plane extending from $x = a$ to $x = +\infty$, the solution ${}_{+}\Phi_a$ is obtained by substituting $x - a$ for x in ${}_{+}\Phi_0$. Similarly, for the half plane from $x = b$ to $x = -\infty$, the solution ${}_{-}\Phi_b$ is obtained by substituting $b - x$ for x. The free-space solution Φ_f is the limiting form of ${}_{+}\Phi_a$ as $a \to \infty$. The solution for the case of a perfectly conducting and infinite plane is represented by Φ_r and is obtained from

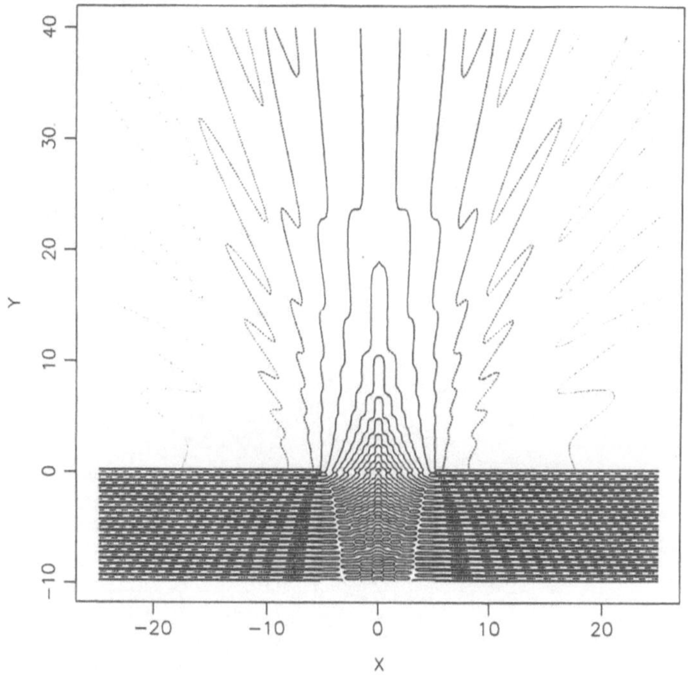

Fig. 5(a). Lines of constant intensity, single slit.

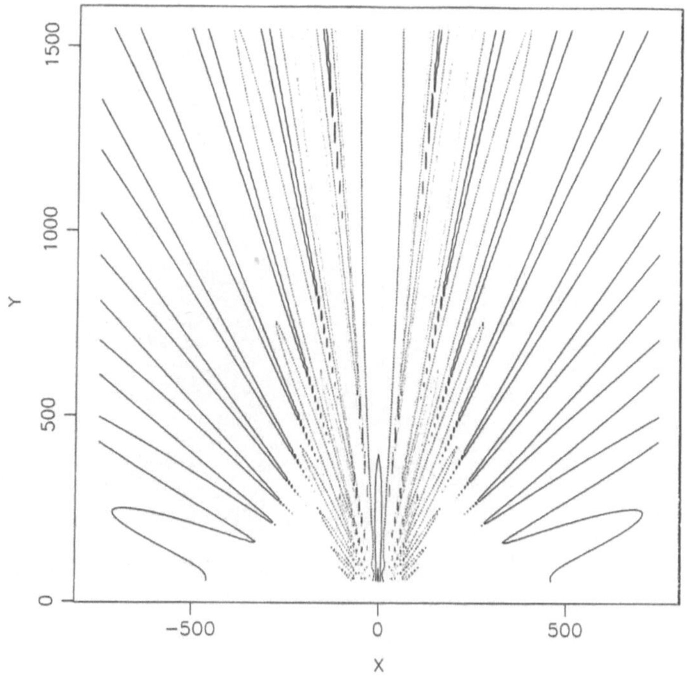

Fig. 5(b). Lines of constant intensity, single slit.

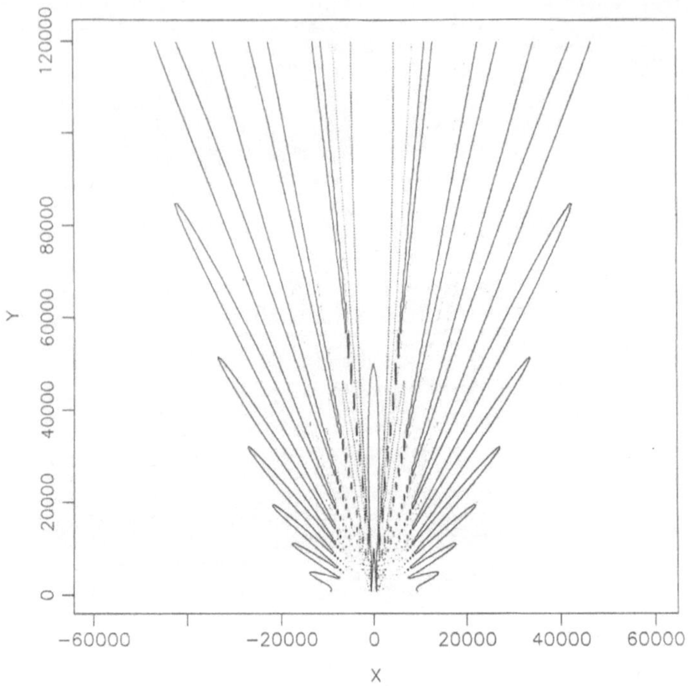

Fig. 5(c). Lines of constant intensity, single slit.

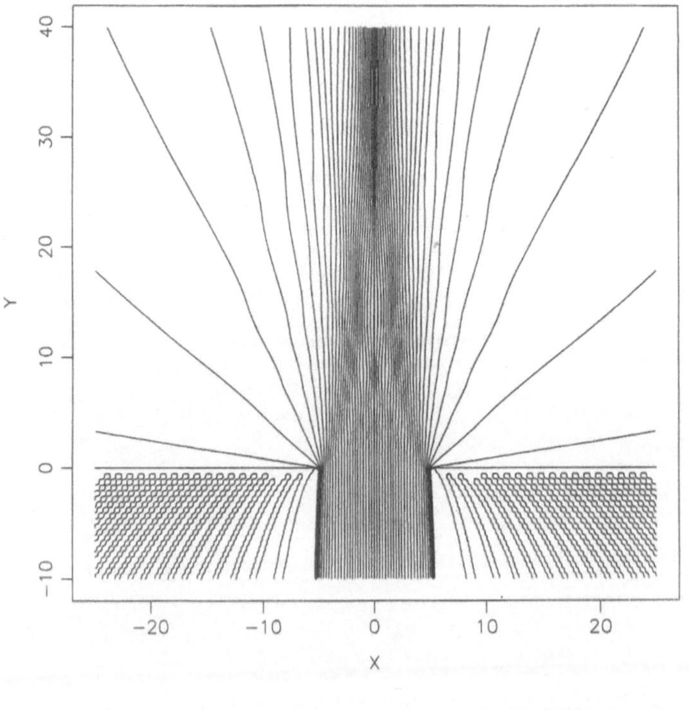

Fig. 6(a). Flow lines of Poynting vector, single slit.

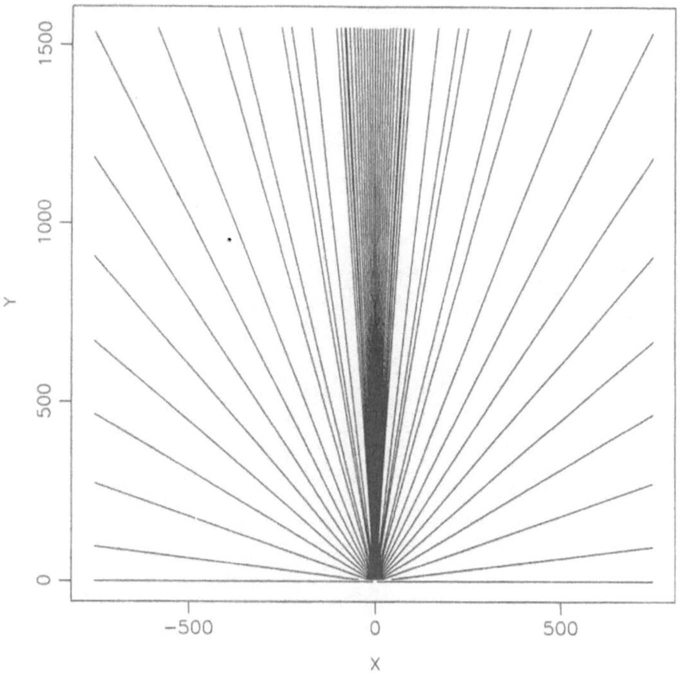

Fig. 6(b). Flow lines of Poynting vector, single slit.

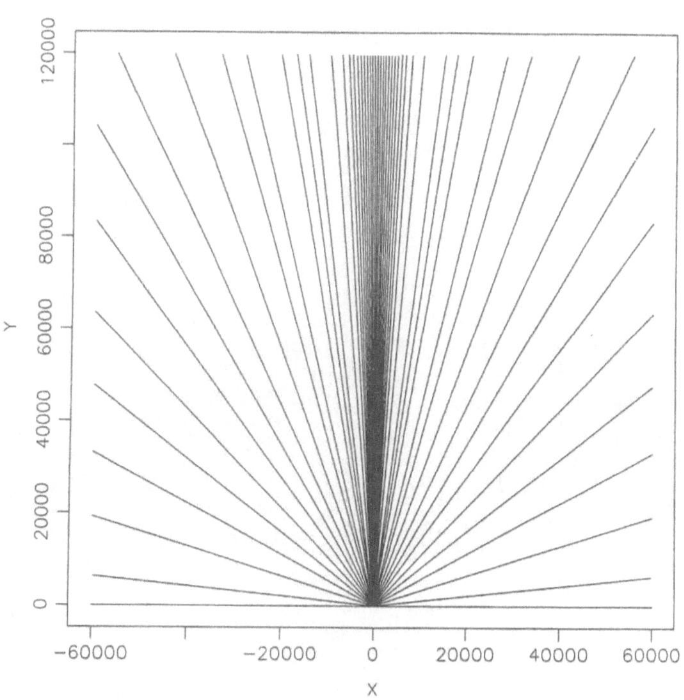

Fig. 6(c). Flow lines of Poynting vector, single slit.

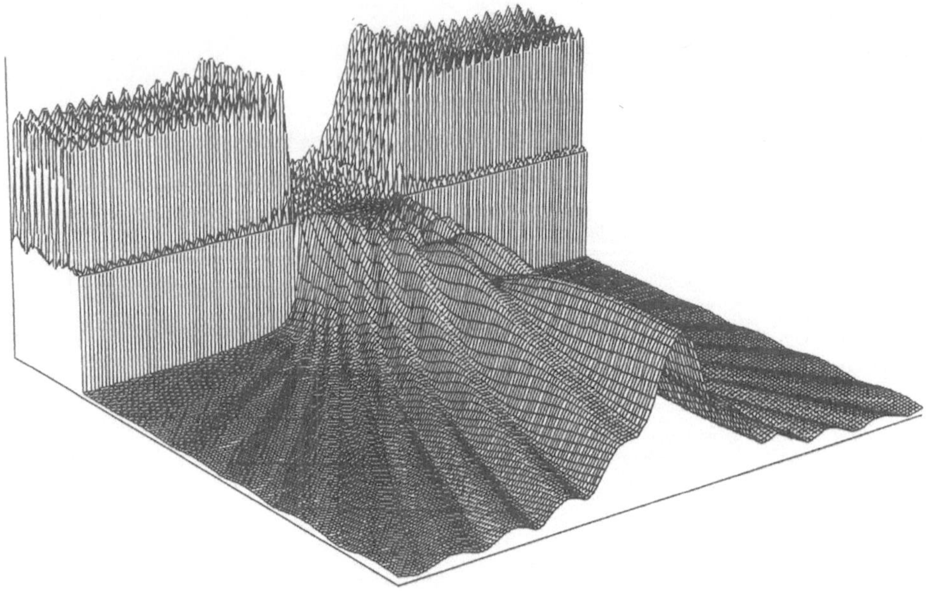

Fig. 7(a). Perspective plot of amplitude, single slit.

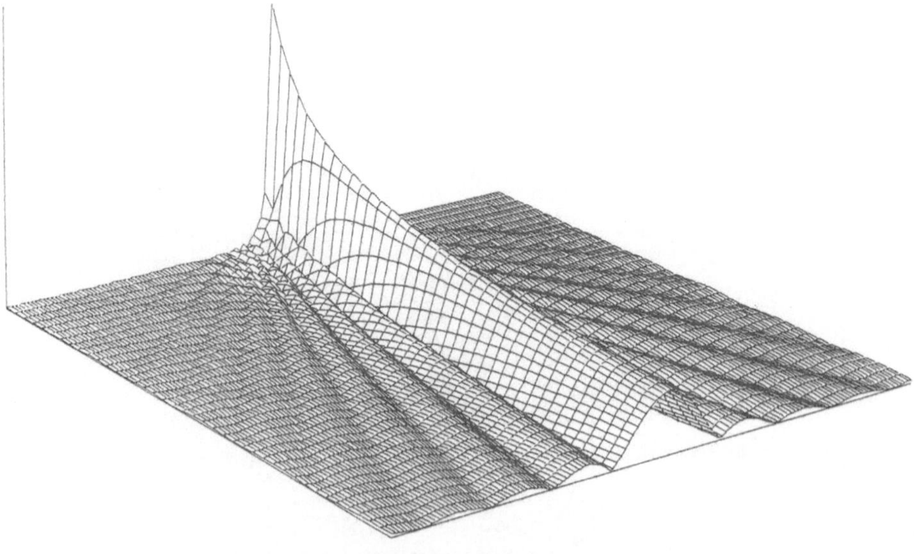

Fig. 7(b). Perspective plot of amplitude, single slit.

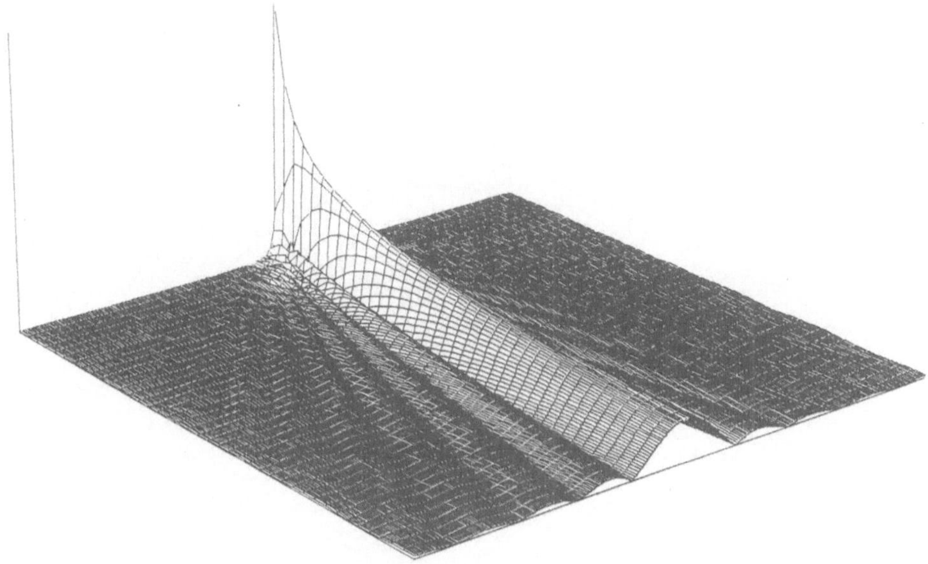

Fig. 7(c). Perspective plot of amplitude, single slit.

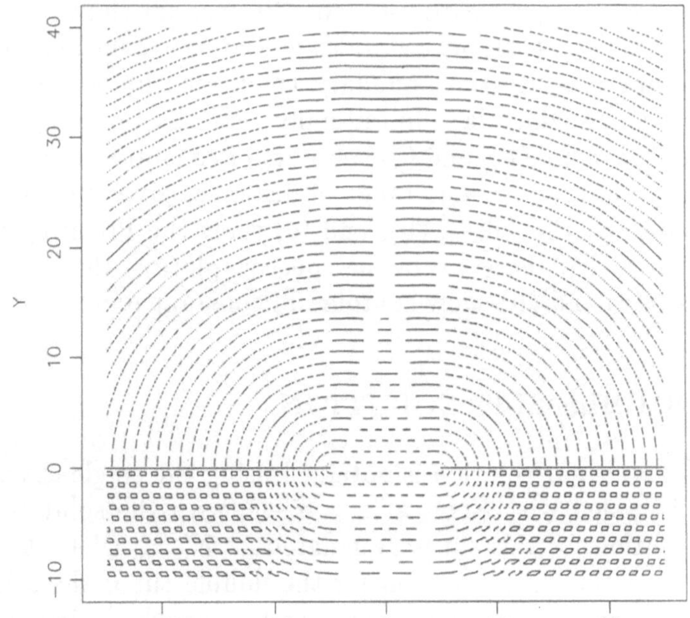

Fig. 8(a). Lines of constant phase, single slit.

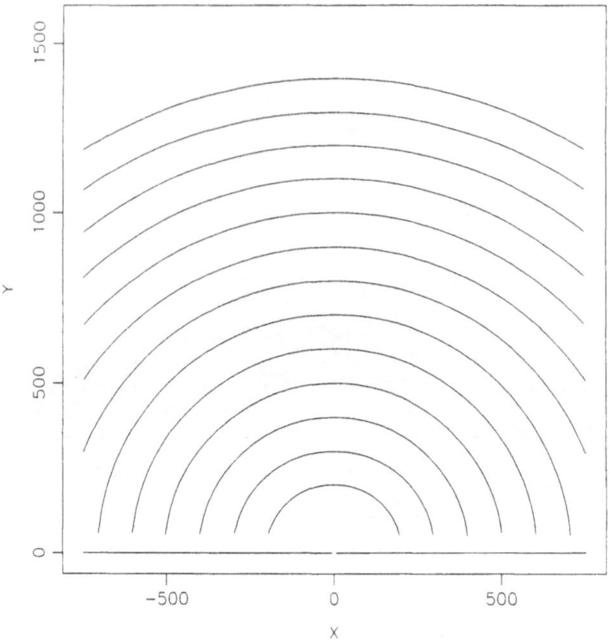

Fig. 8(b). Lines of constant phase, single slit.

the limiting form of ${}_{+}\Phi_a$ as $a \to -\infty$. Prosser [3] shows that the sum ${}_{+}\Phi' ={}_{+} \Phi_a +_{-} \Phi_b - \Phi_f$ represents an approximate solution to Maxwell's equations which satisfies the boundary conditions for a slit from a to b in an infinite plane which is infinitesimally thin and perfectly conducting. We show in Figs. 5(a), (b), and (c) contours of constant intensity obtained from this solution for the same range of distances given in Figs. 1(a), (b), and (c). Similarly, Figs. 6(a), (b), and (c) show the Poynting vector flow lines, Figs. 7(a), (b), and (c) show perspective plots of the amplitude and, finally, Figs. 8(a) and (b) show contours of constant phase.

6. THE DOUBLE-SLIT SOLUTION

The double-slit solution is obtained using the single-slit solution. A second slit is added with edges at $x = a, -b$, with the solution ${}_{-}\Phi'$. The single-slit solution for this slit is designated as $\Phi'' ={}_{+} \Phi' +_{-} \Phi' - \Phi_r$. The solution is then the exact solution for the double slit if the solutions ${}_{+}\Phi'$ and ${}_{-}\Phi'$ are exact single-slit solutions and where Φ_r is included to give continuity of the field components across the apertures. We include in Figs. 9(a–c), 10(a–c), 11(a–c), and 12(a–b) contours of constant intensity, Poynting vector lines, perspective plots of the amplitude, and contours of constant phase that correspond to the same distances for the edge and single-slit cases.

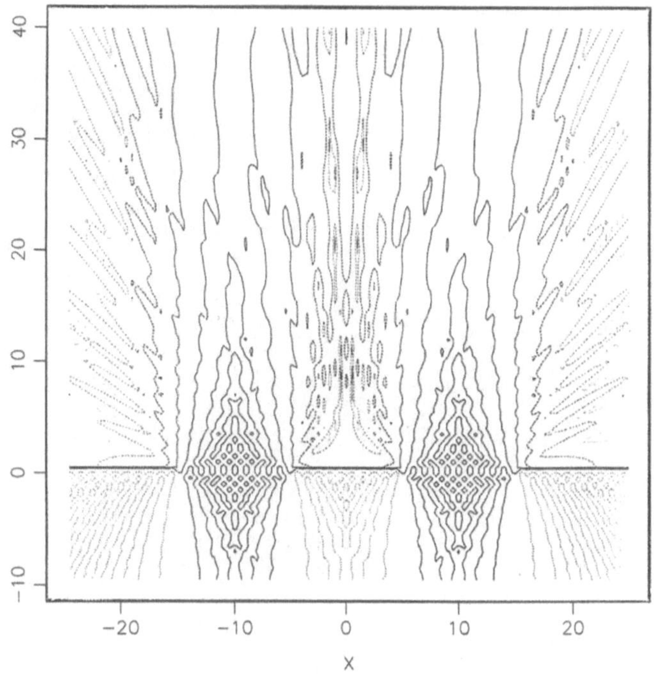

Fig. 9(a). Lines of constant intensity, double slit.

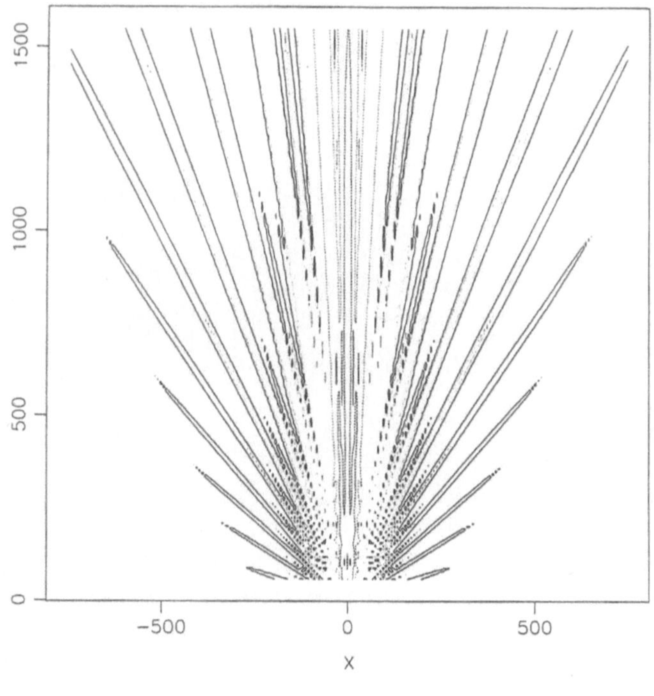

Fig. 9(b). Lines of constant intensity, double slit.

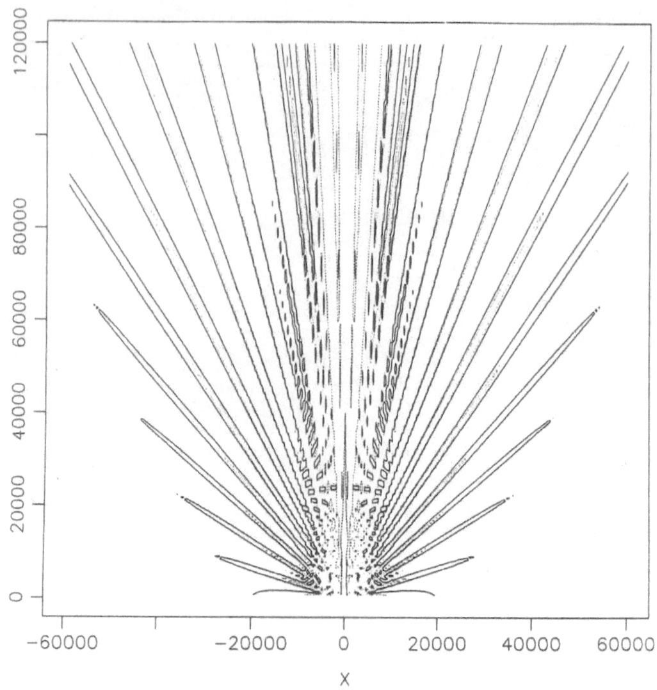

Fig. 9(c). Lines of constant intensity, double slit.

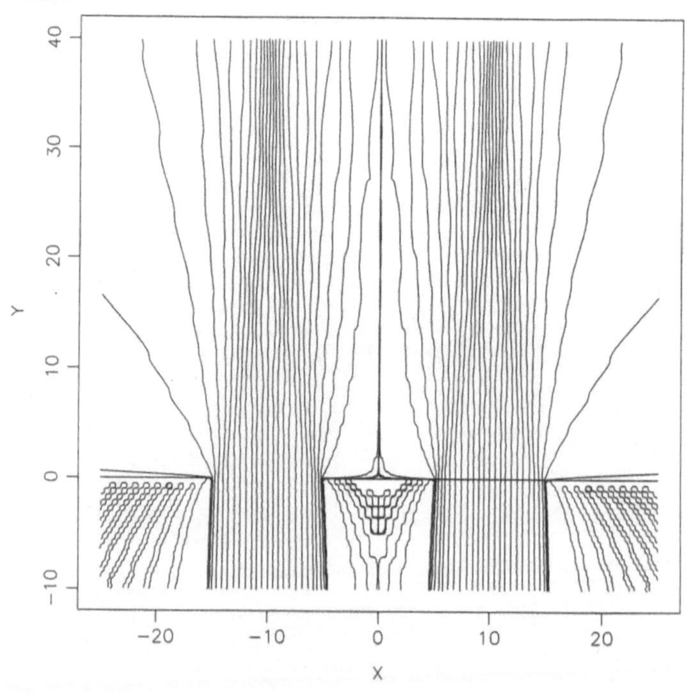

Fig. 10(a). Flow lines of Poynting vector, double slit.

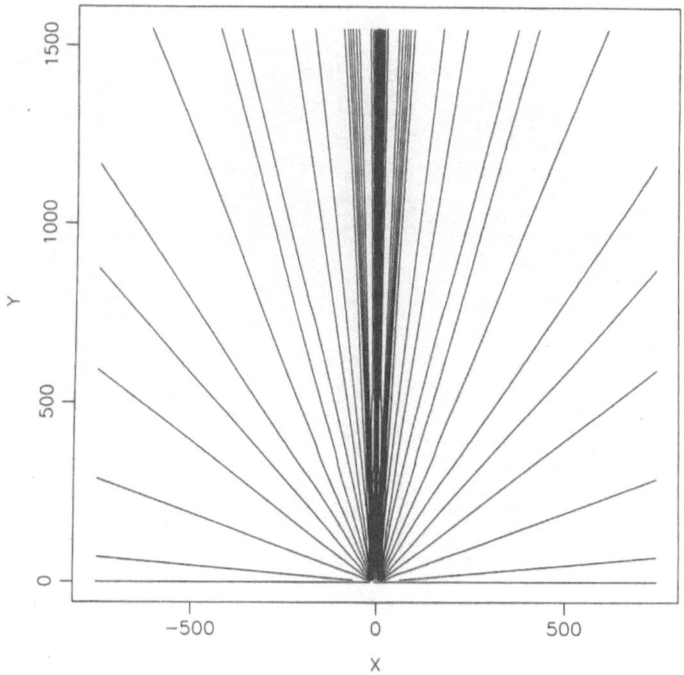

Fig. 10(b). Flow lines of Poynting vector, double slit.

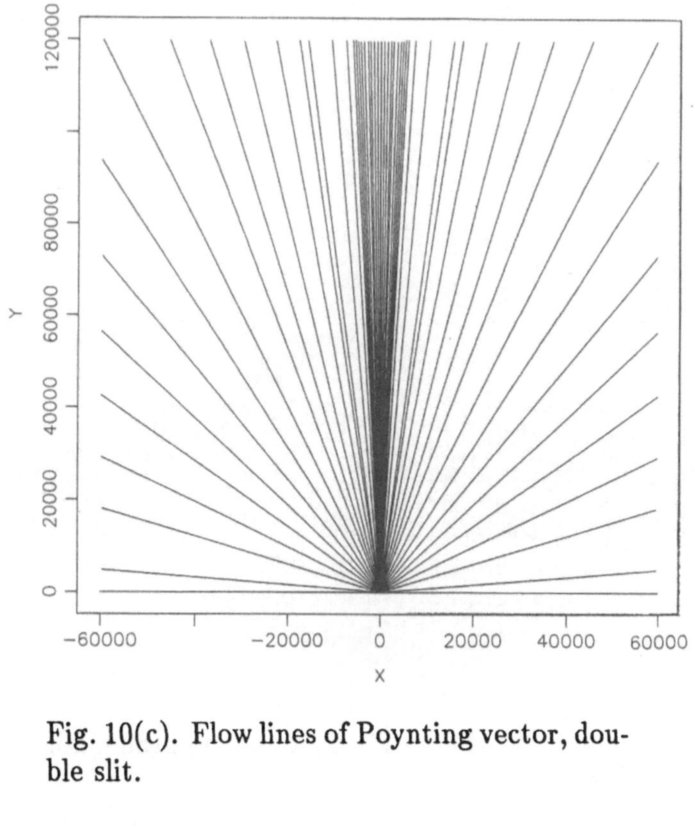

Fig. 10(c). Flow lines of Poynting vector, double slit.

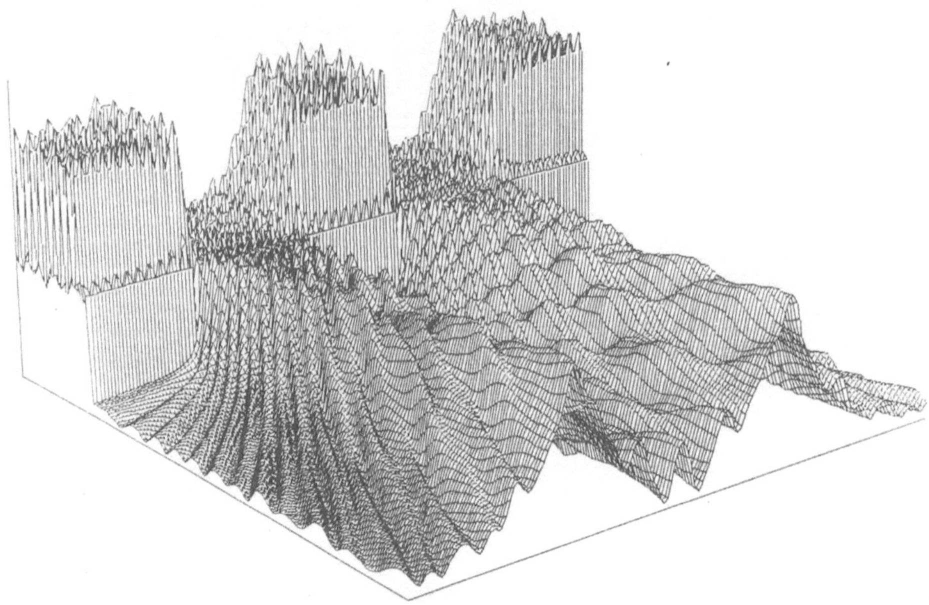

Fig. 11(a). Perspective plot of amplitude, double slit.

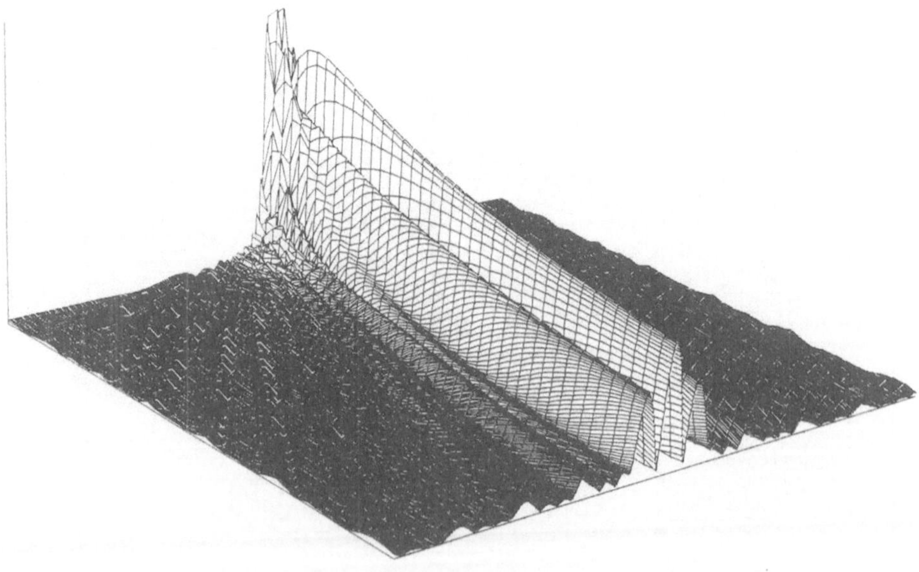

Fig. 11(b). Perspective plot of amplitude, double slit.

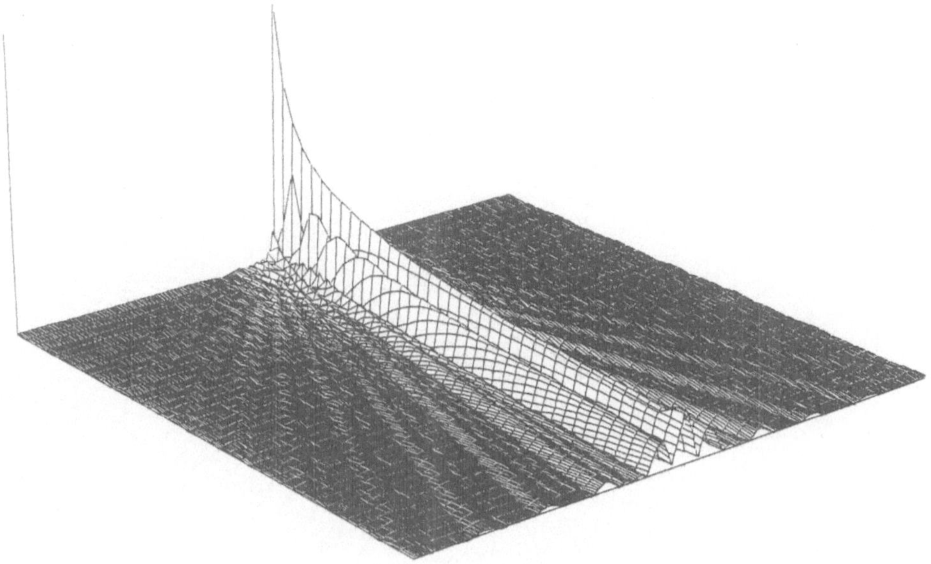

Fig. 11(c). Perspective plot of amplitude, double slit.

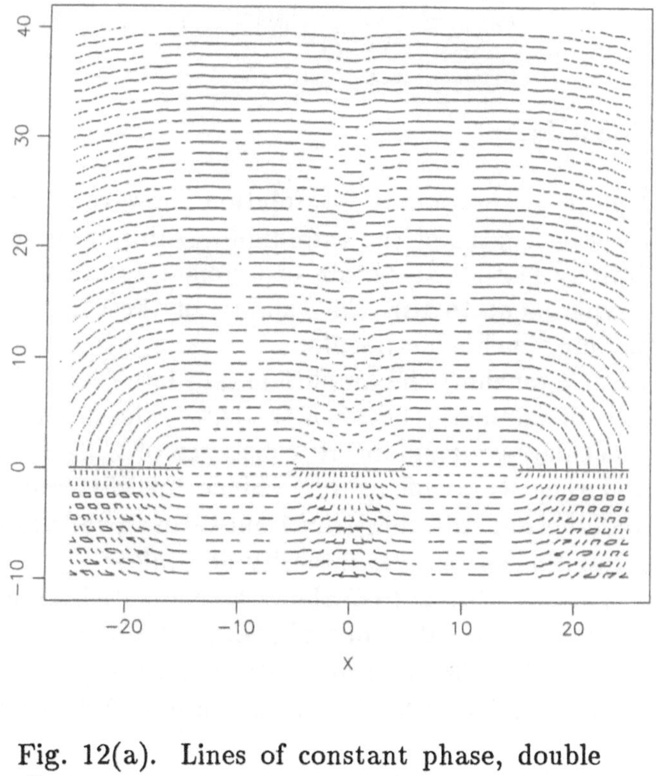

Fig. 12(a). Lines of constant phase, double slit.

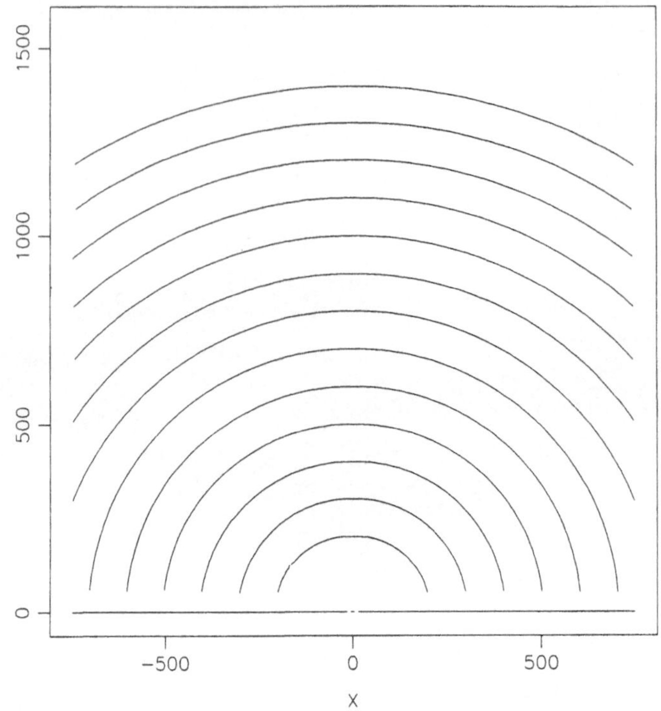

Fig. 12(b). Lines of constant phase, double slit.

7. DISCUSSION

The plots of the double-slit solution show considerable structure within a few wavelengths of the slits unobtainable from scalar theory. It would be entirely feasible to obtain experimental data from a microwave double-slit experiment to test the validity of these results, and we are currently planning such experiments.

In Fig. 13 we show the amplitude and intensity [i.e., (amplitude)^2 along the axis of symmetry between the slits. The intensity builds up as shown up, reaching a maximum at a distance of 270λ from the plane of the slits, and then drops off as $1/r$. Since the slits are infinitely long, the wavefronts beyond the slits travel outwards cylindrically rather than spherically, resulting in this $1/r$ intensity dependence.

Figures 14(a), (b), and (c) show the computed intensity across the interference pattern for a distance of 120λ, 360λ and $120,000\lambda$ from the double slits and also the intensity variation expected from scalar theory. These are indistinguishable for the case at $120,000\lambda$; however, at the shorter distances, it is clear that classical scalar theory breaks down.

Figures 2, 6 and 10 displaying the Poynting vectors reveal the paths of energy flow in much clearer detail than previously obtained by Prosser [3]. For the double-slit case, the energy flow is concentrated along paths normally interpreted as the interference pattern. However in Prosser's interpretation, there is no such thing as classical interference as no radiation actually crosses the axis of symmetry; so, according to this view, no radiation passing the top slit arrives at a point below the axis of symmetry and vice-versa.

Figure 15 shows the flow lines of electromagnetic energy when the left-hand slit is closed to 1.25 wavelengths relative to the right-hand slit of width 5 wavelengths. The plot illustrates the energy flow close to the slits. Figures 16 and 17 show the variation in intensity across the observation plane at a distance of 2×10^8 wavelengths for the standard Fraunhofer solution and for the full Maxwell's equation solutions. Note that, whereas the standard solution is perfectly symmetric, the Maxwell's equations solution shows that clearly more energy flows through the right-hand slit compared to the left-hand slit. We have these fringe distributions for a wide range of slit width ratios and for each distribution computed the contrast in the fringes. Since the Maxwell's equation's solutions give asymmetric fringes, we have computed the contrast of the right and left fringes and used the average value. Figure 18 shows the respective contrast as a function of slit width ratio for a distance of 120,000 wavelengths behind the slits. For this distance, there is no discernible difference between the standard and Maxwell's solutions. Finally, Fig. 19 shows contrast as a function of slit width ratio for a distance of only 1550 wavelengths behind the slits. At this distance discernible differences in the solutions start to appear as the standard solution starts to break down this close to the slit system.

8. CONCLUSIONS

Maxwell's equations have been solved for a variety of diffracting geometries for an incident plane polarized electromagnetic wave. Solutions have been obtained for the full range of distances behind the diffracting aperture for amplitude, phase, intensity and Poynting vector. These have been compared to the standard scalar diffraction theoretical results. The effect of changing the ratio of slit widths in the double-slit case has been studied to reveal the pattern of flow of electromagnetic energy near the slits and how the contrast of the resulting interference fringes changes as a function of slit width ratio.

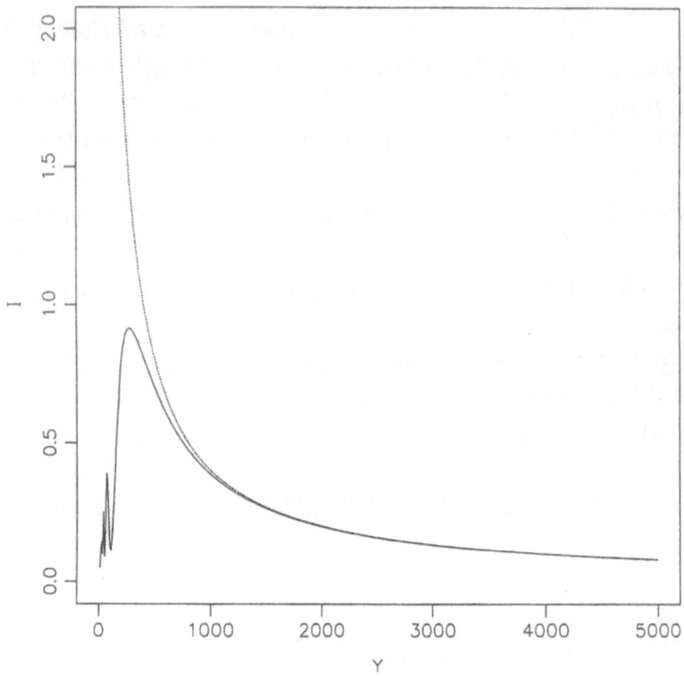

Fig. 13. Plot of intensity vs. Y at $X = 0.0$ (distances in wavelengths).

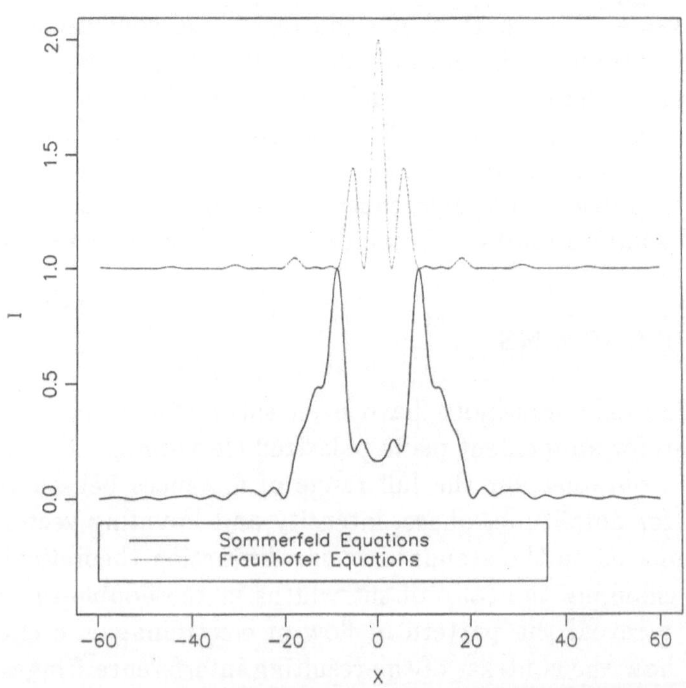

Fig. 14(a). Plot in intensity vs. X at 120 wavelengths from the double slit (distances in wavelengths; note that the upper plot has been displaced for clarity).

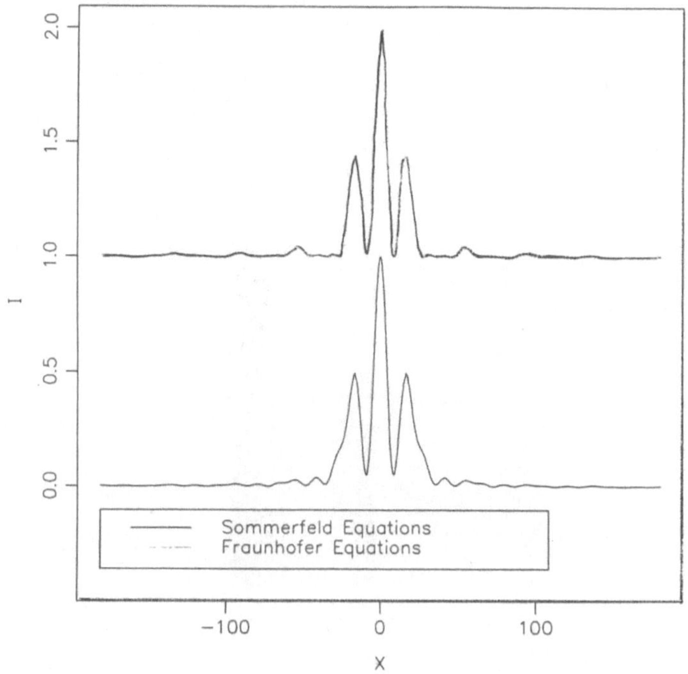

Fig. 14(b). Plot in intensity vs. X at 360 wavelengths from the double slit (distances in wavelengths; note that the upper plot has been displaced for clarity).

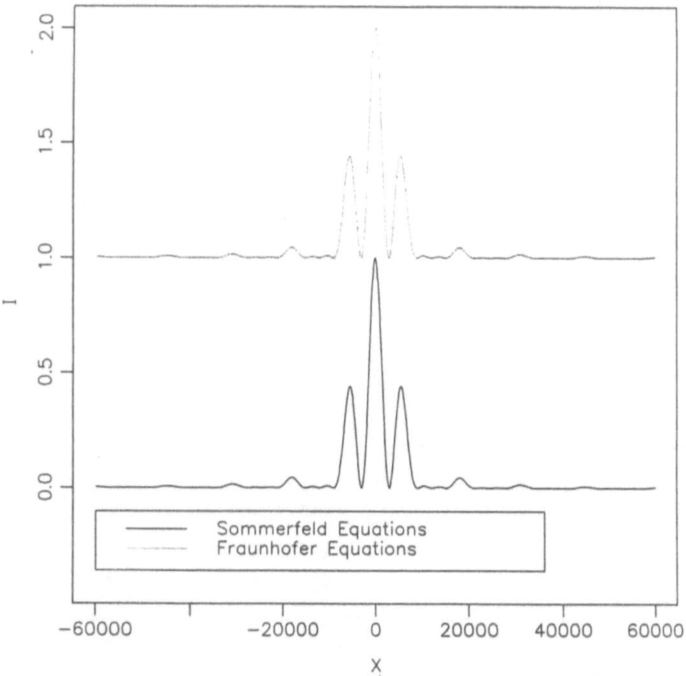

Fig. 14(c). Plot in intensity vs. X at 120,000 wavelengths from the double slit (distances in wavelengths; note that the upper plot has been displaced for clarity).

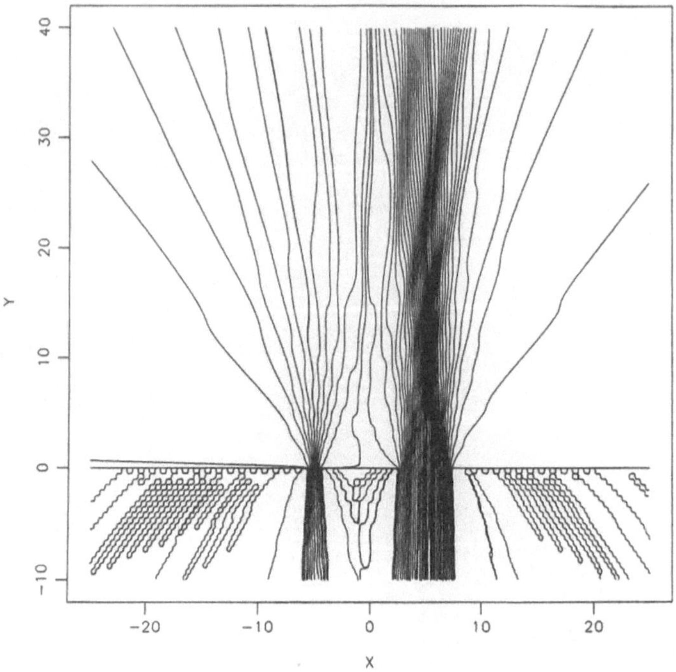

Fig. 15. Flow lines of Poynting vector. Left slit width=1.25 wavelengths, right slit width = 5 wavelengths, slit separation = 10 wavelengths (distances in wavelengths).

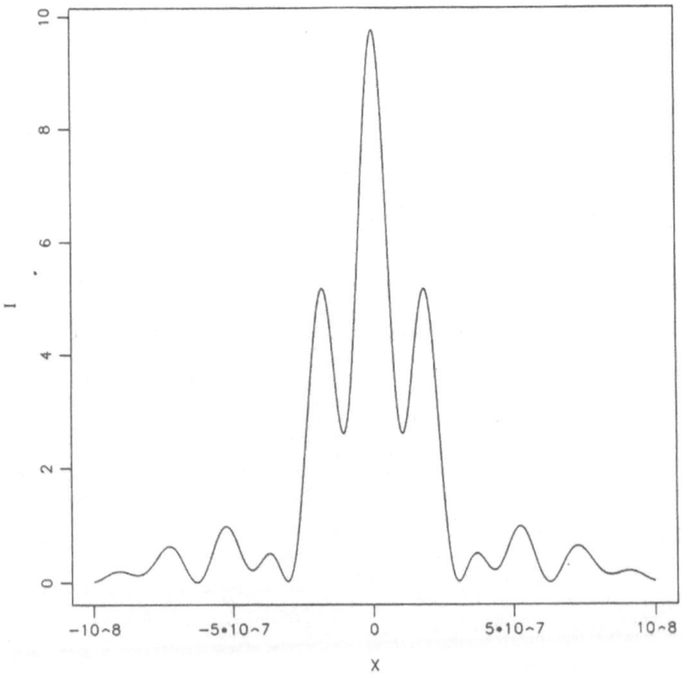

Fig. 16. Plot of intensity for Fraunhofer solution for distance $=2.0 \times 10^8$ wavelengths. Left slit width=1.25 wavelengths, right slit width = 5 wavelengths, slit separation = 10 wavelengths (distances in wavelengths).

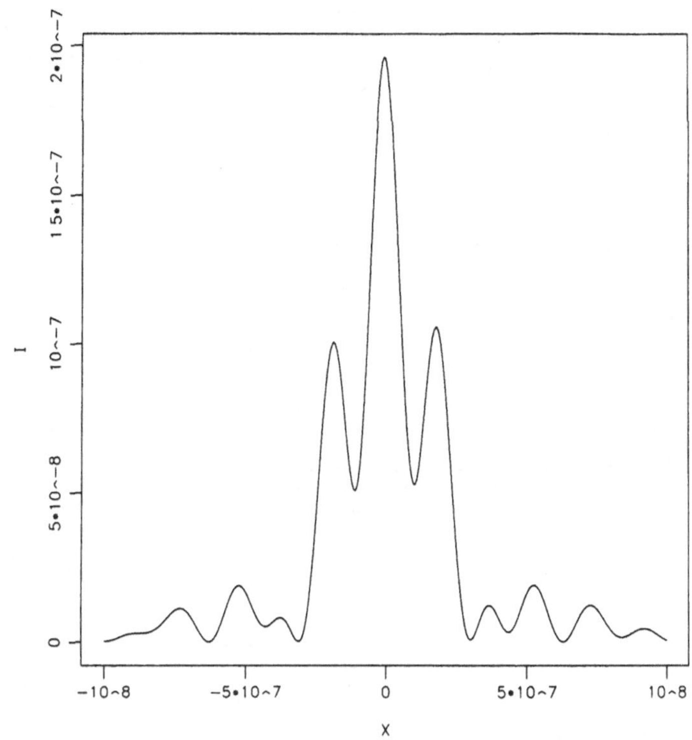

Fig. 17. Plot of intensity for Prosser solution for distance $=2.0 \times 10^8$ wavelengths. Left slit width=1.25 wavelengths, right slit width = 5 wavelengths, slit separation = 10 wavelengths (distances in wavelengths).

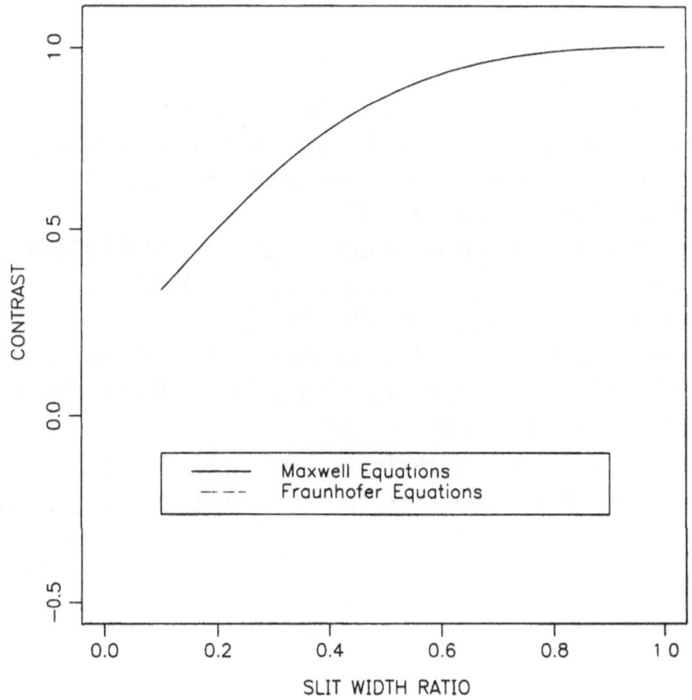

Fig. 18. Plot of contrast vs. slit width ratio at distance = 1.25×10^5 wavelengths.

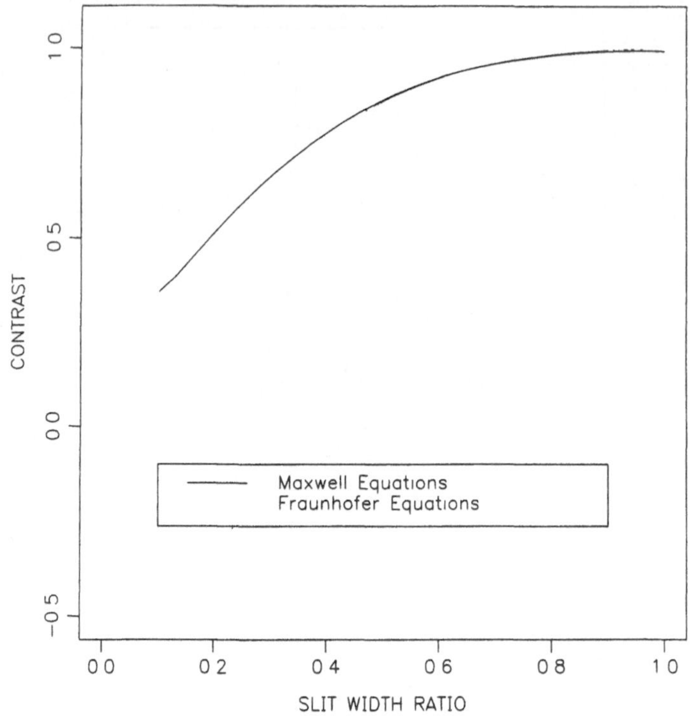

Fig. 19. Plot of contrast vs. slit width ratio at distance = 1.55×10^3 wavelengths.

REFERENCES

1. J. B. Marion and M. A. Heald, *Classical Electromagnetic Radiation*, 2nd edn. (Academic, Harcourt Brace Jovanovich, 1980), p. 369.
2. A. Sommerfeld, *Lectures in Theoretical Physics, Vol. 4* (Academic, New York, 1954), Chap. 5, pp. 247–72.
3. R. D. Prosser, *Int. J. Theor. Phys.* **15**(3), 169–180 (1976).
4. W. Braunbek and G. Laukien, *Optik* **9**, 174 (1952).
5. D. Bohm, *Phys. Rev.* **85**, 166, 180 (1952).
6. J. P. Vigier, in *Quantum Uncertainties—Recent and Future Experiment and Interpretations*, W. M. Honig, D. W. Kraft, and E. Panarella (Plenum, New York, 1987).
7. A. Zeilinger, in "New Techniques and Ideas in Quantum Measurement Theory," D. Greenberger, ed., *Ann. N.Y. Acad. Sc.* **480**, 164–174 (1986).

A QUASI-ERGODIC INTERPRETATION OF QUANTUM MECHANICS

A. Julg

Université de Provence
Place Victor Hugo
13331 Marseille-3, France

A system consisting of a particle subjected, on the one hand, to a deterministic force, on the other, to the electromagnetic damping force and to very brief electromagnetic bursts arising from the radiation of the other systems of the universe, exhibits a quasi-ergodic behavior, which can be transcribed into an operator formalism, very analogous to the quantum one. But the values of the various observables fluctuate, which corresponds to a quantum-like formalism where the operators are completely symmetrized with respect to q and p, the obtained values being the average values over sufficiently long time intervals. Spin appears at the relativistic level as an eigen-kinetic momentum attached to the particle. The Bell inequality is replaced by a weaker one, which is never violated.

Key words: Bell's inequality, ergodicity, interpretation, quantum mechanics, spin.

The goal of this paper is to show that quantum mechanics can be interpreted within a classical framework. The basic idea is that any system is never strictly isolated, but is unceasingly exchanging electromagnetic energy with the other systems of the universe [1].

1. GENERAL RELATIONSHIPS

Let us consider a particle of charge q and mass m, subjected, on the one hand, to a deterministic force $\boldsymbol{F}(\boldsymbol{r})$, on the other, to the electromagnetic damping force $\boldsymbol{f}$, and to the electromagnetic field created by all the systems that constitute the universe (we will call it the universe field). According

Waves and Particles in Light and Matter, Edited by
A. van der Merwe and A. Garuccio, Plenum Press, New York, 1994

to classical electrodynamics, the equation which governs its motion is the following:

$$-\boldsymbol{f} + \dot{\boldsymbol{p}} = \boldsymbol{F}(\boldsymbol{r}) + q\boldsymbol{E} + q\frac{\dot{\boldsymbol{r}}}{c} \wedge \boldsymbol{H}, \tag{1}$$

where $\boldsymbol{p}$ is the momentum of the particle, $\boldsymbol{E}$ and $\boldsymbol{H}$ the electric and magnetic fields, respectively. We will assume that the universe field consists of non-correlated very sudden and very brief *bursts*, travelling at the speed of light c, and randomly oriented in an isotropic manner, so that over a sufficiently long time, the average values of the components of this field are equal to zero.

Within the non-relativistic approximation ($v \ll c$),

$$\boldsymbol{f} = \left(2q^2/3c^3\right) \dddot{\boldsymbol{r}}, \tag{2}$$

so that Eq. (1) reduces to

$$-\tau\dddot{\boldsymbol{r}} + \ddot{\boldsymbol{r}} = \frac{\boldsymbol{F}(\boldsymbol{r})}{m} + \frac{q}{m}\boldsymbol{E}, \tag{3}$$

with $\tau = 2q^2/3mc^3$.

Starting from the point $M_0(\boldsymbol{r}_0)$ at the time $t = 0$, if the electric field was equal to zero, the particle would describe the arc $\boldsymbol{C}$ defined by

$$-\tau\dddot{\boldsymbol{C}} + \ddot{\boldsymbol{C}} = \frac{\boldsymbol{F}(\boldsymbol{r}_0 + \boldsymbol{C})}{m}, \tag{4}$$

with $\ddot{\boldsymbol{C}}_0 = \ddot{\boldsymbol{r}}_0$ and $\dot{\boldsymbol{C}}_0 = \dot{\boldsymbol{r}}_0$.

If the time origin is chosen such that at this time a burst begins, the actual motion is given by

$$-\tau(\dddot{\boldsymbol{C}} + \dddot{\boldsymbol{T}}) + (\ddot{\boldsymbol{C}} + \ddot{\boldsymbol{T}}) = \frac{\boldsymbol{F}(\boldsymbol{r}_0 + \boldsymbol{C} + \boldsymbol{T})}{m} + \frac{q}{m}\boldsymbol{E}; \tag{5}$$

that is, to second order in $\boldsymbol{C}$ and $\boldsymbol{T}$, and taking Eq. (4) into account,

$$-\tau\dddot{\boldsymbol{T}} + \ddot{\boldsymbol{T}} - \frac{\dot{\boldsymbol{F}}(\boldsymbol{r}_0)}{m}\boldsymbol{T} = \frac{q}{m}\boldsymbol{E} \tag{6}$$

($\boldsymbol{T}$ = jump between two arcs $\boldsymbol{C}$).

If we assume that the burst duration τ_0 is small with respect to τ, and that, after a very sudden variation, the electric field is very quickly damped, neglecting the ratio $\boldsymbol{F}'(0)\tau^2/m$, for the *entire* burst, we obtain the following variations

$$\begin{aligned} \delta\boldsymbol{T} &= -\frac{q\tau_0}{m}\int_0^{\tau_0} \boldsymbol{E}dt, \\ \delta\dot{\boldsymbol{T}} &= -\frac{q\tau_0}{m\tau}\int_0^{\tau_0} \left(1 - \frac{t}{\tau_0}\right) \boldsymbol{E}dt, \\ &\cong -\frac{q\tau_0}{m\tau}\int_0^{\tau_0} \boldsymbol{E}dt = \frac{1}{\tau}\delta\boldsymbol{T} \end{aligned} \tag{7}$$

On average, we obtain the remarkable relationship

$$\delta T_x \delta(m\dot{T}_x) = K \tag{8}$$

K being a constant, independent of q and m, therefore of the nature of the particle, and of that of the system under consideration.

1.1. Consequences

Starting from an arbitrary point M_0, under the effect of the successive bursts, the particle draws a very complicate trajectory Γ, which *passes and passes again* in the vicinity of any point of space, *whatever the initial conditions.*

This property entails that all the possible trajectories Γ are equivalent. Moreover, no property is constant on a given trajectory Γ. Indeed, if it were constant, it would be constant at all the points of space, which is physically absurd. Consequently, the system does not possess *first integrals.* All the dynamical properties fluctuate versus time around an average value, which becomes stable after a sufficiently long time. This is typically the character of a *quasi-ergodic* behavior.

One will notice that such a process implies a probability density $\rho(\boldsymbol{r})$, which will be reached after the time τ_e, called the *ergodicity time.*

Let us note some other properties following from our model:

(i) The average energy E is stationary with respect to a small arbitrary virtual modification of $\Gamma(\delta E = 0)$.
(ii) The virial theorem remains valid.
(iii) $m\tau\overline{\ddot{r}^2} = q\overline{\dot{r}E}$, which shows that the radiated energy is equal to the absorbed one, which justifies our model,
(iv) The fluctuation in energy allows the particle to jump over a potential barrier whose height is superior to its energy without it being necessary to invoke the tunnelling effect.

1.2. Remark

The theory also applies to a complex particle. Indeed, Eq. (1) is formally unchanged if we put

$$q^2 = \frac{\sum q_i^2/m_i}{\sum 1/m_i} \quad \text{and} \quad \frac{1}{m} = \overline{(m_i^{-1})},$$

q_i being the charge and m_i the mass of the fundamental constituents of the particle under consideration. Consequently, the universe field can be acting on a neutral particle such as neutron. The pion π^0 and the various quarkonia are exceptions.

2. THE QUANTUM FORMALISM

This dynamics can be transcribed into an operator formalism, as follows: By definition, we will assume that the time average value of a dynamical property G is expressed by

$$\bar{G} = \int \Psi^* \hat{G} \Psi dv = \langle \Psi \hat{G} \Psi \rangle, \tag{9}$$

$\hat{G}$ being a linear operator associated with G and $\Psi\Psi^* = \rho(\boldsymbol{r})$. $\bar{G}$ being real, it is necessary and sufficient that $\hat{G}$ be Hermitian. For $G = G(\boldsymbol{r}), \hat{G}$ is, obviously, identical to G, so that, more generally, we can put

$$\hat{G}(x, p_x \ldots) = G(x, \hat{p}_x, \ldots), \tag{10}$$

$\hat{p}_x$ being the operator associated with p_x.

In order to preserve the invariance of the speed of the particle under a translation of the reference frame, we must necessarily have

$$\hat{p}_x = \sum_{k>0} c_k \frac{\partial^k}{\partial x^k}; \tag{11}$$

thus, to the second order in k (p_x being odd with respect to x),

$$\hat{p}_x = \frac{C}{i} \frac{\partial}{\partial x} \tag{12}$$

Consequently, the average energy is given by

$$\bar{E} = \langle \Psi \left| -\frac{C^2}{2m} \nabla^2 + U \right| \Psi \rangle \equiv \langle \Psi \hat{H} \Psi \rangle \tag{13}$$

This energy being stationary with respect to any virtual variation of Γ, therefore to that of Ψ, we obtain

$$\hat{H} \Psi = \bar{E} \Psi. \tag{14}$$

This equation is formally analogous to that of Schrödinger, but it corresponds to the *average energy* of the system.

For the harmonic oscillator, Eq. (14) leads to $\bar{E} = \frac{C}{2}\omega$, while, from a direct examination, our model leads to $\bar{E} = K'\omega$ (with $K' \propto K$), so that

$$C = 2K' \propto K. \tag{15}$$

2.1. The Excited Levels

If the system is immersed within a monochromatic radiation of frequence ω, it reaches a new equilibrium state which exhibits the same general properties as previously. In particular, it can be shown that relationship (9) remains valid, that $\delta\bar{E} = 0$, so that Eq. (14) is verified for this new state. Consequently, the average energy of the new state is equal to one of the eigenvalues of $\hat{H}$.

On the other hand, relation $m\tau\overline{\ddot{r}^2} = q\overline{\dot{r}E}$ applies to the new state, $\boldsymbol{r}$ being replaced by $\boldsymbol{r}^*$ corresponding to the new trajectory Γ^*. Taking the electric radiation $\boldsymbol{E}_{\text{rad}}$ into account, it is possible to show that, when the radiation is switched off, the radiated power on the trajectory Γ^* is greater than the absorbed one, so that the system returns again to its initial state. Consequently, in the absence of radiation, the average energy of the system is given by the lowest eigenvalue of $\hat{H}$, the others corresponding to excited levels.

2.2. The Electron Spin

At the relativistic level, Eq. (1) becomes

$$-\boldsymbol{f} + \dot{\boldsymbol{p}} = \boldsymbol{F}(\boldsymbol{r}) + q\boldsymbol{E}\left(t - \frac{\boldsymbol{n}\cdot\boldsymbol{r}}{c}\right) + q\frac{\dot{\boldsymbol{r}}}{c} \wedge \boldsymbol{H}\left(t - \frac{\boldsymbol{n}\cdot\boldsymbol{r}}{c}\right), \qquad (16)$$

with $f = \tau\left(1 - \frac{\dot{r}.n}{c}\right)\left(1-\beta^2\right)^{-1/2}\left(\ddot{p} - q\frac{\ddot{r}}{c} \wedge H\right) + \text{terms in } q^3(\boldsymbol{E}^2, \boldsymbol{H}^2, \boldsymbol{EH})$,

$$\ddot{p} = m\left(1-\beta^2\right)^{-3/2}\left(\dddot{r} + 3\frac{\dot{\boldsymbol{r}}}{c^2}\ddot{r}^2\right), \quad \beta^2 = \frac{\dot{\boldsymbol{r}}^2}{c^2}.$$

If we put $\boldsymbol{r} = R + \tilde{\boldsymbol{R}}$, $\boldsymbol{R}$ corresponding to the above defined trajectory Γ, $\tilde{\boldsymbol{R}}$ appears as a "trembling" around this latter (*Zitterbewegung*). To this motion corresponds a kinetic momentum, independent of the system, the *spin-momentum* σ. This momentum, relativistic in origin, must be added to the orbital momentum arising from the motion on Γ.

Owing to the properties of the universe field, this spin-momentum can be written as

$$\sigma = \boldsymbol{N} f(t), \qquad (17)$$

$\boldsymbol{N}$ being a unitary vector whose direction varies randomly versus time, and $f(t)$, a positive function of time. Consequently

$$\begin{aligned} \overline{\sigma_z} &= \overline{f(t)}.\overline{\cos\theta} = 0, \\ \overline{\sigma_z^2} &= \frac{1}{3}\overline{\sigma^2} = \frac{1}{3}\overline{f^2}. \end{aligned} \qquad (18)$$

We find again the quantum results, but σ_z^2 et σ^2 appear as average values, not as eigenvalues.

Moreover, the random fluctuation of σ allows us to explain that the spin orients very quickly within a magnetic field, parallel or anti-parallel with respect to this latter (Stern-Gerlach experiment).

Another important consequence of the fluctuation of σ is that the famous Bell inequality $|P_{ab} - P_{ac}| \leq 1 + P_{bc}$ is replaced by a weaker one which is never violated:

$$|P_{ab} - P_{ac}| \leq 3 + P_{bc}. \tag{19}$$

The non-local character assigned to quantum mechanics appears as being the simple consequence of a too restrictive measurement axiom, rather than reflecting any physical reality.

The term $q^3\left(\boldsymbol{E}^2, \boldsymbol{H}^2, \boldsymbol{EH}\right)$ has been neglected in the expression for the relativistic damping force (18), owing to its smallness. Its presence, nevertheless, explains that the magnetic moment of the positron is slightly different from that of electron.

3. COMPARISON WITH QUANTUM MECHANICS

The model we propose leads to results which are both consistent and inconsistent with the orthodox quantum theory.

On the one hand, indeed, the system exhibits a behavior quite analogous to that which quantum mechanics foresees for a particle subjected to a deterministic force. Among the convergence points we recall the position probability density, the existence of discrete states, and an equation that formally coincides with that of Schrödinger. Finally, the spin appears directly as an *eigen* kinetic momentum attached to the particle.

But, on the other hand, a fundamental difference between quantum mechanics and our model appears concerning the meaning of the values obtained for the dynamical properties of the system. In our model, the values of the various properties fluctuate with time, so that we can obtain their average values only. This results from the fact that the system is unceasingly exchanging energy with the rest of the universe, while quantum mechanics considers the system as being isolated. Consequently, strictly speaking, our model is not equivalent to quantum mechanics, or, at the very least, to *conventional* quantum mechanics. It is, indeed, well known that, if we completely symmetrize the operators with respect to q and p, instead of utilizing the rule $op(A^2) = (opA)^2$ for the squares of the various dynamical properties, the average quadratic dispersions of these latter quantities are different from zero, which signifies that their values fluctuate. In other words, our model corresponds to the completely symmetrized quantum formalism [2].

An important remark, moreover, has to be made. A measurement will give the value forseen by the operator formalism only if the duration of the measurement is greater than the ergodicity time.

If this is not the case, successive measurements will yield results different from one another. In order to obtain the quantum value, it is then necessary to form the average of the successive results for a sufficiently long time. In electronic systems, this time is of about $10^{-17}s$; in the case of the inversion of pyramidal molecules: $10^{-12}s$ for amines, one year for arsines; and *ca.*10^5 years for the inversion of an asymmetrical carbon [3].

To conclude, we will say that the introduction of an unceasing energy exchange between all the systems of the universe is a plausible hypothesis which permits us to construct a classical-like interpretation of quantum mechanics, although, in the practice, the quantum formalism remains an irreplaceable mathematical tool—at the present time, at least.

REFERENCES

1. A. Julg "A model for a quasi-ergodic interpretation of quantum mechanics," *Ann. Fond. L. de Broglie* **16**, 321–342 (1991).
2. A. Julg, "The complete symmetrization of quantum operators: new thoughts on an old problem," *Theor. Chim. Acta* **74**, 323–330 (1988).
3. A. Julg, "The problem of enantiomers: support for a new interpretation of quantum mechanics," *Croatica Chem. Acta.* **57**, 1497–1507 (1984).

DE BROGLIE WAVES AND NATURAL UNITS

Ludwik Kostro

Institute of Experimental Physics
University of Gdańsk
80-952 Gdańsk, Poland

De Broglie waves will be considered in relation to Stoney's and Planck's natural units that are determined by universal constants.

Key words: Universal constants, De Broglie waves, natural units.

1. DE BROGLIE WAVES

In 1905 A. Einstein [1] introduced his hypothesis of the particle-wave duality of light. As is well known, in 1923 and 1924 [2] L. de Broglie extended this duality to all kinds of elementary particles. He arrived at this result from studying Niels Bohr's model of hydrogen atom. According to this model, the physical quantity called action characterizing the motion of the rotating electron around the nucleus is quantized in integral multiples of Planck's constant h:

$$m v 2 \pi r_n = h n,$$

where mv is the momentum of the electron, $2\pi r_n$ is the circumference of its orbit, and $n=1, 2, 3 \ldots$.

De Broglie looked for the reason for the integers (n=1, 2, 3,...) in Bohr's model. Since integers also characterize standing waves (see fig.) De Broglie arrived at the idea that we may associate standing waves with the orbits of the rotating electron:

$$2 \pi r_n = \lambda n ,$$

where λ is the wavelength of the wave, and so he rewrote Bohr's equation $m v 2 \pi r_n = h n$ by replacing the circumference ($2 \pi r_n$) by the wavelength λ:

Waves and Particles in Light and Matter, Edited by
A. van der Merwe and A. Garuccio, Plenum Press, New York, 1994

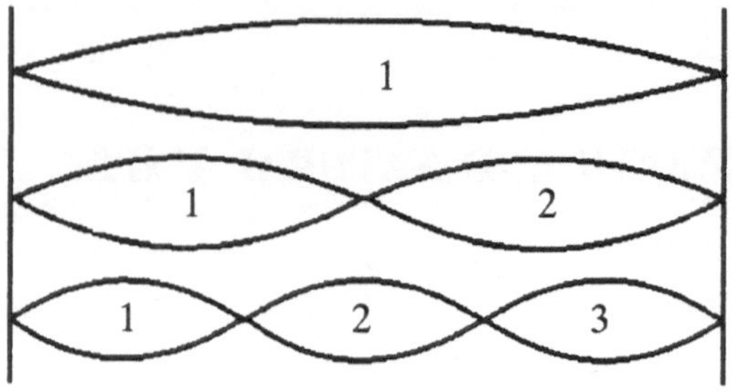

Fig. 1

$$m v \lambda n = h n.$$

In this way he discovered the well-known equation

$$\lambda = h / m v$$

and arrived at the conclusion that a wave is associated with the orbiting electron.

However, in his doctoral thesis he presented his discovery deductively, starting from the equation

$$m c^2 = h \nu$$

which became his basic assumption and basic premise. The opposite relativistivic transformations of the frequences

$$\nu = \nu_0 \left(1 - v^2/c^2\right)^{\pm 1/2}$$

of a clock (+sign) and of a wave (- sign), respectively, was his second basic premise.

On the basis of these premises, he assumed that we are dealing with an intrinsic clock-like vibration inside every particle, the frequency of which (in the rest frame) is given by

$$\nu_0 = m_0 c^2/h ,$$

and with an extrinsic wave-like vibration, which he called "wave-field" (*champ ondulatoire*), the frequency of which transforms according to the equation connected with the transformation of the particle's energy

$$E = E_0 \left(1 - v^2/c^2\right)^{-1/2} = h \nu = h \nu_0 \left(1 - v^2/c^2\right)^{-1/2}$$

As we can see, we are dealing here with a wave-like transformation of frequency $\nu = \nu_0 \left(1 - v^2/c^2\right)^{-1/2}$. De Broglie showed that the two vibrations (intrinsic and extrinsic) are in phase where the particle is present.

From his assumptions de Broglie deduced all properties of his wave, including its length $\lambda = h / mv$ and its superluminous velocity $u = c^2/ v$.

2. NATURAL UNITS

The physical world is quantitatively determined and therefore science of physics, by constructing models of physical phenomena, is able to introduce physical quantities that can be measured. Several physical quantities are related to each other. Some relations between them are always constant. In the models of physical phenomena constructed by physics, a particular part is played by the so-called universal constants. The most important of them are the following ones:

gravitational constant	$G = 6{,}673 \times 10^{-8}$ dyne cm^2 g^{-2}
velocity of light in vacuo	$c = 2{,}998 \times 10^{10}$ cm s^{-1}
elementary electric charge	$e = 4{,}803 \times 10^{-10}$ ES units
Planck's constant	$h = 6{,}625 \times 10^{-27}$ erg s
Boltzmann's constant	$k = 1{,}381 \times 10^{-16}$ erg K^{-1}

The relations between universal constants are also very important and should not be neglected. Stoney and Planck were convinced that some relations between these fundamental constants (raised to appropriate powers) form the natural units of length, time, mass (and temperature). According to them, these units are free of every conventionality, because they are determined by the nature itself through the universal constants.

2.1. Stoney's Natural Units

The first introduction of natural units (determined by universal constants) into physics is universally attributed to Max Planck (1858-1947). However, similar units (determined also by universal constants) were introduced over a quarter of a century before Planck by an Irish physicist George Johnston Stoney (1826-1911) [3]. They arose out of investigations that led to Stoney's successful prediction of electron charge. Stoney (a fellow of the Royal Society and Vice-president of the Royal Dublin Society from 1881 until his death in 1911) is famous for his introduction of the term "electron" to describe the elementary unit of electricity and for his calculation of its value from Faraday's law of electrolysis in 1874, long before its direct experimental measurement [3].

In 1874 Stoney delivered a lecture before Section A of the British Association at the Belfast Meeting. It was subsequently published in 1881 in the *Philosophical Magazine* [4]. His article, like his lecture, was entitled "On the physical units of Nature." In this article he drew attention to the fact that, although we have chosen all manner of arbitrary units, there are always three independent units that can be regarded as fundamental. He derived these units from three quantities he regarded as universal and independent of any anthropocentric bias [3-4]: Newton's gravitational constant G, the velocity of light c, and the basic unit of electricity e. In fact, on this occasion, he also gave the first calculation of its value by reformulating Faraday's law of electrolysis in appropriate form. Using also contemporary determinations of $c = 3\times10^{10}\,cm\,s^{-1}$ and $G = 6{,}67\times10^{-8}\,cm^3 g^{-1} s^{-2}$, he proceeded to determine, by dimensional analysis, the:

> fundamental units offered to us by Nature, upon which may be built an entire series of physical units deserving of the title of a truly Natural Series of Physical Units [4].

He obtained, in such a way, the following units:

length unit $$l_s = \left(G e^2/c^4\right)^{1/2} \approx 10^{-35}\,cm,$$

time unit $$t_s = \left(G e^2/c^6\right)^{1/2} \approx 0{,}33^{-45}\,s,$$

mass unit $$m_s = \left(e^2/G\right)^{1/2} \approx 10^{-7}\,g.$$

Using present-day values, Stoney's units of length, time, and mass are calculated to be $l_s = 1{,}380\times10^{-34}\,cm$, $t_s = 4{,}605\times10^{-45}\,s$, $m_s = 1{,}859\times10^{-6}\,g$.

The existence of Stoney's article has recently been indicated by J. D. Barrow [3]. This section of the present paper is based on Barrow's article.

Note that using Boltzmann's constant k, one is able to introduce also a temperature unit [5]

$$T_s = \left(e^2 c^4/ G k^2\right)^{1/2} = 1{,}210\times10^{31}\,K^\circ$$

2.2 Planck's Natural Units

Max Planck introduced his four natural units [6] in 1899 when he discovered the elementary quantum of action h. He inserted the same section concerning these units into several of his papers and textbooks [7-9].

Planck's units are determined by his new constant h and three of constants previously used by Stoney G, c and k. Here they are:

length unit $\quad l_p = \left(h\,G/c^3\right)^{1/2} = 3{,}99\times10^{-33}$ cm,

time unit $\quad t_p = \left(h\,G/c^5\right)^{1/2} = 1{,}33\times10^{-43}$ s,

mass unit $\quad m_p = \left(h\,c\,/\,G\right)^{1/2} = 5{,}37\times10^{-5}$ g,

temperature unit $\quad T_p = \left(h\,c^5\,/\,G\,k^2\right)^{1/2} = 3{,}6\times10^{32}$ K°.

Using present-day values of h, G, c and k we obtain $l_p = 4{,}0498\times10^{-33}$ cm, $t_p = 1{,}3509\times10^{-43}$ s, $m_p = 5{,}4573\times10^{-5}$ g, and $T_p = 3{,}5518\times10^{32}$ K°.

Planck, like Stoney, was convinced that he was introducing units that are independent of any conventionality and valid for all times. To quote Planck:

> ...they are independent of special bodies or substances and necessarily retain their significance for all times and all environments, terrestrial, human or otherwise ... so long as the law of gravitation and that of velocity of light in vacuum and two basic laws of thermodynamics remain valid. [10]

2.3. Relationship Between Stoney's and Planck's Units

Stoney's units differ from Planck's quantities by the constant factor $\left(e^2/h\,c\right)^{1/2} = 0{,}034$, which is the square root of Sommerfeld fine structure constant $\alpha = e^2/\hbar\,c$ divided by 2π [3–5]:

$$l_s = \left[\left(h\,G/c^3\right)\left(e^2/h\,c\right)\right]^{1/2} = l_p\left(\alpha/2\pi\right)^{1/2} = \left(Ge^2/c^4\right)^{1/2},$$

$$t_s = \left[\left(h\,G/c^5\right)\left(e^2/h\,c\right)\right]^{1/2} = t_p\left(\alpha/2\pi\right)^{1/2} = \left(Ge^2/c^6\right)^{1/2},$$

$$m_s = \left[\left(h\,c/G\right)\left(e^2/h\,c\right)\right]^{1/2} = m_p\left(\alpha/2\pi\right)^{1/2} = \left(e^2/G\right)^{1/2},$$

$$T_s = \left[\left(h\,c^5/G\,k^2\right)\left(e^2/h\,c\right)\right]^{1/2} = T_p\left(\alpha/2\pi\right)^{1/2} = \left(e^2c^4/G\,k^2\right)^{1/2}.$$

2.4. Physical Value of Stoney's and Planck's Units

Note that Stoney's and Planck's idea of natural units determined by universal constants is no longer tenable because, using the same constants (i.e., G, c, e, h, and k), one is able to introduce infinite number of lengths, times, masses and temperature determined by these constants. We obtain these sets using Planck's units and the Sommerfeld fine structure constant in the following way:

$$l_{n_{(m)}} = \left[\left(hG/c^3\right)^m \left(e^2/hc\right)^{-n}\right]^{1/2m},$$

$$t_{n_{(m)}} = \left[\left(hG/c^3\right)^m \left(e^2/hc\right)^{-n}\right]^{1/2m},$$

$$m_{n_{(m)}} = \left[\left(hc/G\right)^m \left(e^2/hc\right)^{n}\right]^{1/2m},$$

$$T_{n_{(m)}} = \left[\left(hc^5/Gk^2\right)^m \left(e^2/hc\right)^{n}\right]^{1/2m}.$$

where m=1, 2, 3 ... is the serial number of the sequences of the units and n = 0, ±1, ±2, ±3... is the number of unit in a sequence of the units. The above equations can be converted as follows:

$$l_{n_{(m)}} = \left(h^{n+m} c^{n-3m} e^{-2n} G^m\right)^{1/2m},$$

$$t_{n_{(m)}} = \left(h^{n+m} c^{n-5m} e^{-2n} G^m\right)^{1/2m},$$

$$m_{n_{(m)}} = \left(h^{m-n} c^{m-n} e^{2n} G^{-m}\right)^{1/2m},$$

$$T_{n_{(m)}} = \left(h^{m-n} c^{5m-n} e^{2n} G^{-m} k^{-2m}\right)^{1/2m}.$$

If m=1, then we are dealing with the first sequence of units. Here is the first sequence of lengths:

$$l_{0_{(1)}} = \left(hG/c^3\right)^{1/2},$$

$$l_{1_{(1)}} = \left(h^2G/e^2c^2\right)^{1/2},$$

$$l_{2_{(1)}} = \left(h^3G/e^4c\right)^{1/2},$$

. . . .

$$l_{n_{(1)}} = \left(h^{n+1} c^{n-3} e^{-2n} G\right)^{1/2}.$$

Note that, e.g.,

$$l_{23_{(1)}} = \left(h^{24} c^{20} e^{-46} G\right)^{1/2} = 22{,}94 \text{ cm},$$

$$l_{24_{(1)}} = \left(h^{25} c^{21} e^{-46} G\right)^{1/2} = 6{,}73 \text{ cm}$$

already have macroscopic dimensions and that

$$l_{34_{(1)}} = \left(h^{35} c^{31} e^{-68} G\right)^{1/2} = 0{,}103 \text{ parsecs},$$

$$l_{35_{(1)}} = \left(h^{36} c^{32} e^{-70} G\right)^{1/2} = 3{,}031 \text{ parsecs}$$

already have astronomical dimensions.

Note that the components of every set of four units $\left(l_{n_{(m)}}, t_{n_{(m)}}, m_{n_{(m)}}, T_{n_{(m)}}\right)$ are related to each other in such a way that, if the wavelength λ of a wave associated with photon is equal to $l_{n_{(m)}}$, then its period T is equal to $t_{n_{(m)}}$, the relativistivic mass m of the photon is equal to $m_{n_{(m)}}$, and the temperature of a gas of such photons is equal to $T_{n_{(m)}}$.

Although Stoney's and Planck's units can no longer be considered to be unique natural units, nevertheless, they are relevant in some branches of physics. Planck's units serve, e.g., in the attempts to quantize the gravitational field (see, e.g., H.-J. Treder [11]). In cosmology the first period of the cosmic evolution is called Planck's epoch, because Planck's units apparently played a fundamental role in it.

Note also that Planck's units are so related to each other that, if we assume the wavelength λ of a wave associated with a photon is equal to Planck's length

$$\lambda = l_p = \left(h G / c^3\right)^{1/2},$$

then, consequently, its period T is equal to Planck's time

$$T = \frac{\lambda}{c} = \frac{\left(h G / c^3\right)^{1/2}}{c} = \left(h G / c^5\right)^{1/2} = t_p,$$

the relativistivic mass m of the photon is equal to Planck's mass

$$m = h / c\lambda = \left(h / c\right)\left(h G / c^3\right)^{-1/2} = \left(h c / G\right)^{1/2} = m_p,$$

and the temperature of a gas of such photons is equal to Planck's temperature

$$T = mc^2/k = \left(hc/G\right)^{1/2}\left(c^2/k\right) = \left(hc^5/Gk^2\right)^{1/2} = T_p$$

And so we can assume that in Planck's epoch the electromagnetic radiation was made up of photon–maximons [12-13]. (Particles characterized by Planck's units are called "maximons" by Markow[14] and "planckons" by Staniukiewicz [14]).

In high–energy laboratories it is now possible to create elementary particles from photons, having high energy therefore in several cosmological models of the universe: e.g., in the Klein–Alfven model [15], the Charon model [16], and often also in the standard model [17] it is assumed that in the beginning there was only a gas of high–energy photons, and that all other particles and antiparticles have a photonic origin.

If we assume that in Planck's epoch there were only photon–planckons from which all other elementary particles and antiparticles (protons and antiprotons, neutrons and antineutrons, electrons and positions, etc.) were created. Thus the created particles and antiparticles were also maximons, because their velocities were only slightly less than the velocity of light. An elementary particle possessing the rest mass m_0 has Planck's characteristics (Compton wavelength equal to l_p, period equal to t_p, and relativistivic mass equal to m_p) when it moves at the velocity given by

$$V = c\left(1-m_0^2/m_p^2\right)^{1/2} = c\left(1-Gm_0^2/hc\right)^{1/2}.$$

This velocity is almost equal to the velocity of light because $m_0 <<< m_p$. It is so slightly less than c that we can even say it approaches the velocity c, it is practically equal to c.

Note that the temperature of a gas of rest mass particles moving at velocity V is only slightly less than Planck's temperature T_p and is given by

$$T = \left(hc^5/Gk^2\right)^{1/2}\left\{1-\left[1-\left(1-Gm_0^2/hc\right)\right]^{1/2}\right\}.$$

It is slightly less than T_p because $\left[1-\left(1-Gm_0^2/hc\right)\right]^{1/2}$ is almost equal to zero for $m_0 <<< m_p$.

Note also that density of matter in Planck's epoch is given by Planck's units

$$\rho_p = m_p/l_p^3 = \left(c^5/G^2h\right) = 0{,}821\times10^{96}\ \mathrm{kg\ m}^{-3}$$

All these data indicate that Planck's units will probably serve in future theories to eliminate singularities of the relativity theory; i.e. the infinite density of matter, the infinite temperature, the infinite mass, etc.

2.5. Stoney's and Planck's Units and the Fundamental Lengths of Quantum Mechanics, Electromagnetism and General Relativity

Quantum mechanics, electromagnetism and general relativity have introduced their theoretical fundamental lengths, which are already well established in present-day physics [18]:

(1) The Compton wavelength of a particle,

$$\lambda_C = h/ m c,$$

constitutes the fundamental length of quantum mechanics.

(2) The so-called "classical radius" or "electromagnetic radius" [19] of charged particles constitutes the fundamental length of electromagnetism,

$$R_{em} = e^2/ m c^2$$

(3) The fundamental gravitational length [18], called the Einstein-Schwarzschild gravitational radius [11],

$$R_G = G m/ c^2$$

has a well established place in general relativity.

Note that the mass of a particle m (recognized as a fundamental particle's physical quantity by all three theories) is related to its three fundamental lengths with the help of coefficients that are composed of the typical constant of the respective theory and of the velocity of light c, which constitutes the limiting velocity:

$$m = \left(h/ c\right) \lambda_C^{-1}, \tag{1}$$

$$m = \left(e^2/ c^2\right) R_{em}^{-1}, \tag{2}$$

$$m = \left(c^2/ G\right) R_G, \tag{3}$$

Combining Eq. (1) and (3) and also (2) and (3) we obtain respectively

$$\lambda_C R_G = \left(h G/ c^3\right) = l_p^2,$$

$$R_{em} R_G = \left(G e^2/ c^4\right) = l_s^2.$$

As we can see,

(a) the product of the particle's QM fundamental length λ_C and the fundamental gravitational length is always constant and equal to Planck's length unit raised to the second power; and

(b) the product of the particle's electromagnetic radius and particle's gravitational radius is also always constant and equal to Stoney's length unit raised to the second power.

This fact indicates that Planck's and Stoney's lengths are more fundamental than λ_C, R_{em} and, R_G. Planck's length appears here as fundamental both in quantum mechanics and general relativity. Stoney's length appears here as fundamental both in electromagnetism and general relativity.

And what about Planck's and Stoney's mass and time units? When the mass of particle increases because of the particle motion, $m = m_0\left(1-v^2/c^2\right)^{-1/2}$ or because of the particle's presence in a gravitational field, $m = m_0\left(1+\varphi/c^2\right)^{-1}$, where φ is the pseudo-Newtonan scalar gravitational potential, which determines the g_{00} component of the metrical tensor $g_{\mu\nu}$ describing the gravitational potentials in general relativity, $g_{00} = \left(1+\varphi/c^2\right)^{-1}$ [17] then the particle's fundamental gravitational length R_G elongates, whereas the particle's Compton wavelength reaches Planck's length and its electromagnetic radius reaches Stoney's length:

$$\lambda_C = R_G = \left(h\,G/c^3\right)^{1/2},$$

$$R_{em} = R_G = \left(G\,e^2/c^4\right)^{1/2}.$$

In this case the mass of the particle is equal, respectively, to Planck's mass unit,

$$m = h/\lambda_C c = \left(h/c\right)\left(h\,G/c^3\right)^{1/2} = \left(h\,c/G\right)^{1/2} = m_p,$$

or to Stoney's mass unit,

$$m = e^2/R_{em}c^2 = \left(e^2/c^2\right)\left(G\,e^2/c^4\right)^{-1/2} = \left(e^2/G\right)^{1/2} = m_s.$$

Planck's and Stoney's mass and length units appear here as limiting quantities.

An elementary particle possessing a rest mass m_0 acquires Planck's mass at a velocity

$$V = c\left(1 - G m_0^2 / h c\right)^{1/2}$$

or in the presence of a gravitational potential

$$\Phi = c^2\left[\left(G m_0^2 / h c\right)^{1/2} - 1\right],$$

i.e., when $g_{00} = \left(G m_0^2 / h c\right)^{1/2}$.

A charged particle acquires Stoney's mass when moving at speed

$$U = c\left(1 - G m_0^2 / e^2\right)^{1/2}$$

or in the presence of gravitational potential

$$\Phi' = c^2\left[\left(G m_0^2 / e^2\right)^{1/2} - 1\right],$$

i.e., when $g_{00} = \left(G m_0^2 / e^2\right)^{1/2}$.

As already mentioned, the velocities V and U are so slightly less than the velocity of light c that they can be considered practically equal to it. Note that also the scalar gravitational potential Φ and Φ' are slightly less than $\left|-c^2\right|$. The fact that at these velocities and potentials an elementary particle is characterized respectively by Planck's and Stoney's units, indicates that these units will help, in future, unified theories to eliminate singularities of relativity theory mentioned above.

Since $\lambda_C = c T_C$, $R_{em} = c t_{em}$, and $R_G = c t_G$ (where T_C is the Compton period, t_{em} is the time needed to transmit the electromagnetic action along the distance R_{em}, and t_G is the time needed to transmit gravitational action along the distance R_G), the equations

$$\lambda_C R_G = \left(h G / c^3\right), \quad R_{em} R_G = \left(G e^2 / c^4\right)$$

can be written as follows

$$c T_C c R_G = \left(h G / c^3\right), \quad c t_{em} R_G = \left(G e^2 / c^4\right).$$

After a simple conversion, we obtain

$$T_C t_G = \left(h G / c^5\right) = t_p^2, \quad t_{em} t_G = \left(G e^2 / c^6\right) = t_s^2.$$

Planck's and Stoney's time units appear here, like their length units, as units connecting general relativity with quantum mechanics and electromagnetism; Planck's and Stoney's length and time units relate the fundamental lengths and times of QM, EM and GR to each other.

The above considerations and data show that Planck's and Stoney's units, although they can no longer be considered as natural units, will probably serve as constants of a future unification between quantum mechanics, general relativity, and electromagnetism.

3. DE BROGLIE WAVES AND NATURAL UNITS

Let's note first that at the velocity $V = c\left(1 - G m_0^2/hc\right)^{1/2}$ (slightly less than c) or in the presence of the gravitational potential $\Phi = c^2\left[\left(G m_0^2/hc\right)^{1/2} - 1\right]$ (slightly less than $\left|-c^2\right|$), i.e., when the relativistic mass of the particle is equal to Planck's mass $m_p = \left(hc/G\right)^{1/2}$, the Compton wavelength is equal to Planck's length

$$\lambda_C = h/m_p c = h/c\left(hc/G\right)^{1/2} = \left(Gh/c^3\right)^{1/2}$$

and the Compton frequency

$$\nu_C = m_p c^2/h = c^2\left(hc/G\right)^{1/2}/h = \left(c^5/Gh\right)^{1/2}$$

is equal to Planck's frequency $\nu_p = \left(c^5/Gh\right)^{1/2}$

But at velocity $U = c\left(1 - G m_0^2/e^2\right)^{1/2}$ or in the presence of the gravitational potential $\Phi' = c^2\left[\left(G m_0^2/e^2\right)^{1/2} - 1\right]$, i.e., when the relativistivic mass of the particle is equal to Stoney's mass $m_s = \left(e^2/G\right)^{1/2}$, the Compton wavelength is not equal to Stoney's frequency. They are equal, respectively, to the length and to the frequency of the first sequence of units m=1 and n=1, units determined by the constants h, c, G, e:

$$\lambda_C = h / m_s c = h/c \cdot \left(e^2/G\right)^{-1/2} = \left(G h^2/e^2 c^2\right)^{1/2},$$

$$\nu_C = m_s c^2/h = c^2\left(e^2/G\right)^{1/2} / h = \left(e^2 c^4/G h^2\right)^{1/2}.$$

Thus not only Planck's and Stoney's units play a part here, but also other fundamental units. We are dealing with an analogical thing with respect to de Broglie waves.

At the velocity $V = c\left(1 - G m_0^2 / hc\right)^{1/2}$ the frequency of the de Broglie wave is equal to Planck's frequency

$$\nu_B = m_p c^2/h = c^2\left(hc/G\right)^{1/2}/h = \left(c^5 / G h\right)^{1/2}$$

and the length of the de Broglie wave is slightly longer than Planck's length

$$\lambda_B = h/m_p V = h \cdot 1/c \cdot \left[\left(hc / G\right) - m_0^2\right]^{-1/2}.$$

But at the velocity $U = c\left(1 - G m_0^2 / e^2\right)^{1/2}$ the frequency of the de Broglie wave is not equal to Stoney's frequency $\nu_s = 1/t_s = \left(c^6 / G e^2\right)^{1/2}$; it equal to the frequency of the first sequence units (m=1 and n=1),

$$\nu_B = m_s c^2/h = c^2\left(e^2 / G\right)^{\frac{1}{2}}/h = \left(e^2 c^4 / G h^2\right)^{\frac{1}{2}}$$

The de Broglie wavelength at velocity U is not slightly longer that Stoney's length but slightly longer than the length of the first sequence of units (m=1 and n=1),

$$\lambda_B = h / m_s U = h/c \left[\left(e^2 / G\right) - m_0^2\right]^{\frac{1}{2}}$$

REFERENCES

1. Einstein, A., *Ann. Phys.*, Leipzig **17**, 132-148 (1905).
2. de Broglie, L., Recherches sur la théorie des quanta, Doctoral thesis, University of Paris 1924.
3. Barrow, J. D., *Quart. F. R. Astr. Soc.* **24**, 24-26 (1983).
4. Stoney, G. J., *Phil. Mag.*, Ser. **5**, 381-390 (1881).

5. Kostro, L., "A deduction of natural units," unpublished paper, 1986.
6. Wadlinger, R. L., and Hunter G., "Max Planck's natural units," *Phys. Teacher,* 528-529, November 1988.
7. Planck, M., *Sitzungsber. Preus. Akad. Wiss.* (Mitteilung) **5**, 444 (1899).
8. Planck, M., *Ann. Phys.* Leipzig **1**, 69 (1900).
9. Planck, M., *Theorie der Wärmestrahlung* (Barth, Leipzig), (1906, 1912).
10. Planck, M., *Theory of Heat Radiation* (Dover, New York, 1959), pp. 174-175; quoted from Ref. 4.
11. Treder, H.-J., "Sul significato fisico delle Unita naturali di Planck, in: *Centenario di Einstein 1879-1979* (Giunti Barbera, Firenze, 1979), p. 905.
12. Kostro, L., *Spraw. Gdańsk. Tow. Nauk.* **8**, 107 (1981).
13. Kostro, L., *Phys. Lett.* 112A, 283 (1985).
14. Kuchowicz, B., and Szymczak, J. T., *Dzieje Materii Przez Fizyków Odczytane* (W.P., Warszawa 1978), p. 490.
15. Alfven, H., *Kosmologia i Antymateria*, (P.W.N., Warszawa, 1973).
16. Charon, J., *Théorie de la relativité complexe* (Albin Michel, Paris, 1977).
17. Weinberg, S., *Gravitation and Cosmology* (Wiley, New York, 1972).
18. Kittel, Ch., Knight, W. D. and Ruderman, M. A., *Mechanics, Berkeley Physics Course - Volume 1* (McGraw-Hill, New York, 1965), p. 302.
19. Treder, H.-J. and Bleyer, U., "Quantum gravodynamics and pregeometry," in *Problems in Quantum Physics, Gdańsk '87*, Kostro, L., Posiewnik, A., Pykacz, J., and Żukowski, M., eds. (World Scientific, Singapore, 1988).

A CLASSICAL MODEL FOR WAVE-PARTICLE DUALITY

A. F. Kracklauer

Belveterer Allee 23
99425 Weimar, Germany

A model of de Broglie waves based on the hypothesized existence of zeropoint electromagnetic background radiation is presented. Based on this model, a brief review of a paradigm for quantum mechanics is presented. Bell's theorem is examined and reinterpretated so as to preclude its negative conseqeunces. Finally, a possibly observable effect suggested by this model pertaining to the coherence length of particle beams is considered.

Key words: wave-particle duality, zeropoint radiation, foundations of quantum mechanics.

1. INTRODUCTION

Wave-particle duality, and its attendant philosophical principal of complementarity, is perhaps the most well know, counterintuitive concept introduced by the discipline of physics. Unlike virtually every other basic notion originating with the physical sciences, this principle is distinguished by its ambiguity and preternatural character. As such it has had a particular attraction to those put off by the stolid nature of conventional science. Nevertheless, because of its vague logic, duality is also a source of discomfort for those abhorring imprecision or "spookiness" in physical science.

The purpose of this presentation is to describe a model using concepts from classical physics for wave-particle duality [1]. The basis of this model is the electromagnetic zeropoint background, whose existence is the foundation of the theory of stochastic electrodynamics (SED) [2]. It will be argued below that the existence of such a zeropoint background engenders the properties observed for waves and particles that gives each characteristics of the behavior of the other in certain limiting circumstances.

Although a unified concept, wave-particles clearly can be distinguished into two categories: entities whose classical, large scale limit is wavelike (e.g., photons) and entities whose classical limit is particle-like (e.g., electrons, atoms, etc.). In both cases the zeropoint field, it will be argued herein, has the effect of introducing the surprising aspect heretofore attributed to ontological duality.

2. PARTICULATE BEHAVIOR OF ELECTROMAGNETIC WAVES

From the view point of SED, electrodynamic signals are classical waves whose fluctuations obey what is often called "photon statistics." It is the contention herein that the literal concept of a photon is superfluous, because observable manifestations attributed to their existence can be accounted for by waves which obey photon statistics.

The essential feature of zeropoint background radiation is determined by the requirement that its spectral energy density $E(\omega)$ be invariant under Lorentz transformations, in the sense that the total energy between two fixed numerical values of ω, a, and b be identical in all inertial frames; *i.e.*,

$$\int_a^b E(\omega)d^3k = \int_a^b E'(\omega)\gamma(1-v/c)d^3k, \quad \gamma := (1-v^2/c^2)^{-1/2}. \tag{2.1}$$

As physics, this stipulation is tantamount to the requirement that there be no distinguishable frames; and, it is based on the fact that, were it not true, the background would engender certain anisotropisms that in fact are not observed.

Equation (2.1) is satisfied by a linear spectral energy density $E(\omega)=$ (constant)ω where the constant scale factor is determined empirically to be Planck's constant $2\pi\hbar$[2,3].

The essence of photon statistics and the Planck blackbody spectrum can be found by manipulation of the following four equations involving the mean energy density $\overline{E_i}$, the mean square energy density $\overline{E_i^2}$, and the mean square deviation of the energy density $\overline{(\delta E_i)^2}$ of any two mutually incoherent radiation fields:

$$\overline{E_{\rm sum}} = \overline{E_1} + \overline{E_2}, \tag{2.2}$$

$$\overline{E_1 E_2} = \overline{E_1}\,\overline{E_2}, \tag{2.3}$$

$$\overline{(\delta E)^2_{\rm sum}} = \overline{(\delta E_1)^2} + \overline{(\delta E_2)^2} \tag{2.4}$$

and

$$\overline{(\delta E_i)^2} = \overline{E_i^2} - \overline{E_i}^2 - \overline{E_i}^2, \quad i = 1,2, \tag{2.5}$$

to obtain

$$\overline{(\delta E_{\rm T})^2} = \overline{E_{\rm T}^2} + 2\overline{E_{\rm T}}\,\overline{E_{\rm B}}. \tag{2.6}$$

where, in the case at hand, $\overline{E_{\rm B}}$ is the mean energy density solely of the background radiation, and $\overline{E_{\rm T}}$ is the same for a temperature-dependent radiation field. This latter field coexists with the background and is modified by it via the mutual interference terms which are included in $\overline{E_{\rm T}}$. Invoking the fluctuation theorem

$$\partial\overline{E_{\rm T}}/\partial\mu = \overline{(\delta E_{\rm T})^2}, \quad \mu := -1/kT, \tag{2.7}$$

for the thermal field at temperature T, where k is Boltzmann's constant, yields the differential equation

$$\partial\overline{E_{\rm T}}/\partial\mu = \overline{E_{\rm T}}^2 + 2\overline{E_{\rm B}}\,\overline{E_{\rm T}}, \tag{2.8}$$

whose solution, with $\overline{E_{\rm B}} = \hbar\omega/2$, is the Planck blackbody spectral energy distribution

$$\overline{E_{\rm T}} = \frac{\hbar\omega}{e^{\hbar\omega/kT} - 1}. \tag{2.9}$$

Eq. (2.6), written in the form

$$\frac{\overline{(\delta E_{\rm T})^2}}{\overline{E_{\rm T}}^2} = 1 + 2\frac{\overline{E_{\rm B}}}{\overline{E_{\rm T}}}, \tag{2.10}$$

elucidates the source of the dualistic nature of radiation [3]. Were the first term on the right to stand alone, Eq. (2.10) would characterize intensity fluctuations of a classical radiation field, while the second term alone, being proportional to $1/\overline{E_{\rm T}}$, gives an equation characterizing density fluctuations of a particle ensemble. Together they capture the essence of photon statistics and thus, without assuming the existence of discreet quanta, characterize the "quantized eletromagnetic field." [3]

The point here is that the fluctuation statistics observed at very low densities in anticorrelation experiments, and taken to demonstrate the discreetness of the photon, can be explained as well as a consequence of the existence of the zeropoint field.

3. WAVE BEHAVIOR OF PARTICLES

All particles with charge structure are considered to be in thermodynamic equilibrium with the zeropoint background via interaction with signals at frequences characteristic of their structure. The model for wave-like behavior of particles is based on the *Ansatz* that, in frames in which they are moving, particles will be deflected by diffraction patterns in the background signals to which they are tuned. Background waves, being conventional electromagnetic radiation, are diffracted by physical boundaries according to the usual principles of optics. For example, a particle moving towards a slit would equilibrate with a signal that is a standing wave in its own frame but which is a traveling wave in the slit's frame where it diffracts at boundaries such as those of the slit. On passage through the slit, the particle is subject to the lateral energy flux attendant to the diffraction pattern of the background signal to which it is tuned. In other words, it is envisioned that a particle will tend to be jostled into the energy nodes of the diffraction pattern of the "standing wave" to which it is tuned in its own inertia frame but which is a translating wave in the frame of the slit. This effect is similar to the way froth and debris tend to track the nodes of standing waves in rivers or sand tends to settle on the nodes of a vibrating membrane. An ensemble of similar particles in identical circumstances—*e.g.*, a beam of particles impinging on a slit—upon accumulation at the detector, discloses the diffraction pattern of the composite wave comprised of components with which the individual particles are in equilibrium.

Consider, for example, a neutral particle or system consisting of a dipole of opposite charges held apart by some internal structure modeled to first order by a simple spring with resonant frequency ω_0. According to the basic SED assumption of thermodynamic equilibrium with the background, the rest energy of this sytem constituting the particle will equal the energy in the background mode ω_0, which is also the resonant frequency of the system at which it is exchanging energy with the background; that is,

$$m_0 c^2 = \hbar\omega_0, \tag{3.1}$$

where a contribution of $\hbar\omega_0/2$ is made to the right side by both polarization states of the background mode. (For systems with more complex internal structure, ω_0 stands for the sum of the frequencies corresponding to the various possible interactions.) In its rest frame, with respect to each independent spatial direction, on the average a particle will tune into an antinode at its location, which, if the particle is located at $x = 0$, has an intensity proportional to the expression $2\cos(k_0 x)\sin(\omega_0 t)$. When projected onto a coordinate frame translating at velocity v with respect to the particle, that of a slit for example, this standing wave has the form of the modulated translating wave and is proportional to

$$2\cos\left(k_0\gamma(x - c\beta t)\right)\sin\left(\omega_0\gamma(t - c^{-1}\beta x)\right), \quad \beta := v/c. \tag{3.2}$$

This wave consists of a short wavelength carrier modulated at a wavelength $\lambda = (\gamma\beta k_0)^{-1}$, inversely proportional to the relative velocity of the particle with respect to the slit. The modulation on this wave is a relativistic kinematic effect. It arises from the difference in the Lorentz transformations of the oppositely translating components of a standing wave.

The modulated wave, upon propagation through a slit, for example, is diffracted according to Huygens' principle, such that the modulation diffraction pattern is imposed on the carrier's diffraction pattern. A particle bathed in this diffracted wave will experience a gross energy flux with a spatial pattern proportional to the square of the modulation intensity imposed on the fine-scale background wave driving the *Zitterbewegung.* In other words, according to this interpretation, boundary conditions on background zero-point waves modify the stochastic effects of *Zitterbewegung* on the orbits of material particles. The actual detailed motion of a particle, while it reflects the relatively large scale effects of the modulation, is very complex and jitters in consort with spatially modulated *Zitterbewegung.*

Now, a Lorentz transformation into the translating frame applied to both sides of the statement of energy equilibrium, Eq. (3.1), yields both

$$\gamma m_0 c^2 = \gamma \hbar \omega_0 \tag{3.3}$$

and

$$p' = \gamma m_0 c = \hbar \gamma \beta k_0. \tag{3.4}$$

From Eq. (3.4), $\gamma\beta k_0$ can be identified as the de Broglie wave vector from conventional QM.

Within this conceptual framework, the interpretational contradictions or ambiguities that arise in conventional QM with respect to duality are simply precluded from the start. Here particles are particles and waves are waves, although each is induced by the zeropoint background to exhibit certain behavior in the micro-domain reminiscent of the other. Particles are the primitive elements of this theory, whose existence is assumed axiomatically. Waves are mathematical constructs representing Fourier components of a decomposition or interactions occurring via action at a distance on the light cone (i.e., free electromagnetic fields without sources are regarded as artifacts of convenience). The purpose of the theory is to describe the motion of particles given *a priori* interacting via electromagnetism in the presence of zeropoint radiation. The zeropoint radiation itself is assumed to originate from those remaining charges in the universe excluded from immediate interest [1].

4. QUANTUM THEORY

The physical model of a background wave so affecting a particle's trajectory that it assumes a raylike character also provides a novel understanding of the physical significance of the Schrödinger equation and its solutions. The basic idea here is that the energy density of the zeropoint wave with which a particle is in equilibrium is a wave with half the wavelength of its supporting zeropoint wave, and this energy density pattern is effectively an agent inducing raylike characteristics in particle trajectories. Thus, if one considers a Gibbsian ensemble of similar particles with density $\rho(x,p,t)$, then the ensemble wave will be the Fourier composite over the variation in momentum of the ensemble.

Consider a particle subject to a force $\boldsymbol{F}$ and for which the density of trajectories on phase space is $\rho(x,p,t)$, where $\rho(x,p,t=0)$ can be regarded either as the distribution of initial conditions for similarly prepared particles, or, equivalently, as the *a priori* probability distribution of initial conditions for a single particle. Time evolution of $\rho(x,p,t)$ is governed by the Liouville equation

$$\frac{\partial\rho}{\partial t} = -\nabla\rho\cdot\frac{p}{m} + (\nabla_p\rho)\cdot\boldsymbol{F}, \quad \nabla_p := \sum_{i=x,y,z}\frac{\partial}{\partial p_i}. \tag{4.1}$$

By virtue of Eq. (3.4), each value of p in $\rho(x,p,t)$ is correlated with a particular wavelength of kinematical modulation. As proposed above, boundaries and geometrical constraints on the waves to which the particle is tuned cause these waves to diffract and interfere, so that gradients in their energy densities are induced. These gradients, in turn, result in spatial variations in the magnitude of the *Zitterbewegung*, the average effect of which is to systematically modify particle trajectories. Because the energy density of a wave is proportional to the square of its intensity, the wavelength of energy density oscillations caused by the modulation will be half that of the modulation itself. That is, the wavelength of the physical agent modifying trajectories, an energy gradient, is half that of the modulation. Further, an ensemble consisting of multiple particles, either conceptual or extant, will be guided by an ensemble of energy density waves derived from an ensemble of kinematical modulations. The spatial structure of this ensemble wave is found by taking the Fourier transform of $\rho(x,p,t)$ with respect to $2p/\hbar$, the wave vector for the phyiscal agent; *i.e.*,

$$\hat{\rho}(x,x',t) = \int e^{\frac{2ipx'}{\hbar}}\rho(x,p,t)dp, \tag{4.2}$$

for which the similarly transformed Liouville equation is

$$\frac{\partial\hat{\rho}}{\partial t} = \left(\frac{\hbar}{i2m}\right)\nabla'\nabla\hat{\rho} - \left(\frac{i2}{\hbar}\right)(x'\cdot\boldsymbol{F})\hat{\rho}. \tag{4.3}$$

Solutions for equations of this form are sought by first separating variables using a transformation of the form

$$r = x + x', \quad r' = x - x', \tag{4.4}$$

which yields

$$\frac{\partial \hat{\rho}}{\partial t} = \left(\frac{\hbar}{i2m}\right)\left(\nabla^2 - (\nabla')^2\right)\hat{\rho} - \left(\frac{i}{\hbar}\right)(r - r')\cdot \boldsymbol{F}\left(\frac{r+r'}{2}\right)\hat{\rho}. \quad (4.5)$$

In general, for an arbitrary form of the force $\boldsymbol{F}((r + r')/2)$, this equation still is not separable. However, for potentials having the form of a quadratic polynomial, the extraction the Schrödinger equation follows directly. Non quadratic potentials, which include the important case of the coulomb potential, may still be includable with novel analysis [1].

The logic of the extraction of the Schrödinger equation, given above, requires that physically realizable solutions be those for which the resulting phase space density is everywhere positive and such that $\rho(x, p, t = 0)$ is the appropriate initial condition. The relationship between solutions satisfying these physics requirements and the eigenfunctions of the Schrödinger equation is a complex question and is left for future study. However, it seems clearly auspicious in this regard, that thermal states, *i.e.*, mixed states with Boltzmann weighting factors, and coherent states give phyiscally interpretable, everywhere positive densities as well as other desirable traits [4].

5. BELL'S THEOREM

The above interpretation seems to achieve exactly what Bell's theorem is believed to preclude, namely an interpretation of QM with a classical underpinning with implied "hidden variable." [5] Thus, a reconciliation is needed. To this end, note that Bell's analysis is comprised of two separate elements: a theorem and a question.

The theorem is stated as follows: *If $A(a, \lambda)$ and $B(b, \lambda)$ are dichotomic functions such that each is independent of the other and if $\rho(\lambda)$ is a normalized density over a parameter space represented by λ, then A and B satisfy Bell-type inequalities, in particular*

$$1 + P(b, c) \geq |P(a, b) - P(a, c)|, \quad (5.1)$$

where

$$P(a, b) := \int d\lambda \rho(\lambda) A(a, \lambda) B(b, \lambda). \quad (5.2)$$

So much is mathematics. It is a rigours result beyond any contest.

However, physics in Bell's analysis is embedded in the following question: For spin correlations in the Bohm version of Einstein-Podolsky-Rosen (EPR) experiments, are expressions of the sort given by Eq. (5.2) identical to the QM calculation expressed as

$$< \sigma_1 \cdot a, \sigma_2 \cdot b >= -a \cdot b, \quad (5.3)$$

where $\boldsymbol{a}$ and $\boldsymbol{b}$ are direction unit vectors with respect to which the spin measurements are made? [5]

The answer on the surface to this question is clearly "no," as expression (5.3) does not satisfy the inequality (5.1). The reason behind this, however, is buried in the structure of quantum mechanics (QM) rather than mathematics. But given the vague foundations of QM, the significance of this answer is open to any amount of contest.

For example, ordinarily Eq. (5.3) is interpreted to be the cross-correlation of the results of individual spin measurements on the electrons in an EPR experiment. But, by definition, this correlation requires specific values from individual repetitions of the experiment; however this information is in principle and fact not existent in QM. That is, the results or individual events can not be predicted with QM, only "expectation values" can be computed. Thus, logic would seem to require that expression (5.3) can be actually only the correlation of expectation values rather than the correlation of individual events.

Because QM has neither an intuitive interpretation nor an axiomatic foundation, fixing precisely the meaning of expression (5.3) becomes a matter for argument. Those who find it easier to accept that mortals misconstrue things than that nature is non-local might elect heretically to view (5.3) as simply a variant expression of Malus's law describing the intensity of polarized light with respect to an arbitrary axis. Such an interpretation eerily conforms with the fact that virtually all tests of Bell's inequalities have been optical experiments wherein polarization is treated analogously to spin [6]. Moreover, actual dichotomic data fitting EPR conditions, i.e., raw sequences of plus- and minus-ones in equal quantities, can not be made to yield the sinusodial correlation evidenced by (5.3) even if they are credited with nonlocality, the inclusion of which is generally thought to be QM's way around Bell's result. (This observation is a variation of Bell's theorem; those who doubt it are challenged to find a counter-example.) The point is, the "no" answer to Bell's question might be utterly natural because the QM side of the equation has been misinterpreted. In this case, Bell's arguments pose no fundamental impediment to constructing a hidden variable interpretation for QM [7].

6. CONCLUSION

A new interpretation for QM, even if it has philosophical and pedagogical advantages, fails to meet the contemporary standard of science if it fails to predict an otherwise unexpected observable result. While such a prediction has yet to be made for the "zeropoint" interpretation of QM, perhaps the following preliminary notions will lead to a suitable experiment.

The coherence length of the de Broglie waves associated with particle beams is related to the velocity spread. According to the above model, *Zitterbewegung* will induce an irreducible coherence length for a given type of particle beam. Now, if *Zitterbewegung* is induced, as supposed above, by a real electromagnetic zeropoint field, then is seems natural that the zeropoint field would be more effective; that is, impel larger *Zitter* excursions for ionzied atomic beams than for neutral atomic beams which couple only by multipole interactions whose impulse is smaller. In other words, the intrinsic, irreducible coherence length for ionized beams should be larger than for neutral beams. In some manifestation this effect ought to be observable.

As other interpretations for the nature of de Broglie waves do not imply a dependence on the ionization state, the observance of such a dependence should constitute support for the interpretation espoused above.

REFERENCES

1. A. F. Kracklauer, *Phys. Essays* **5**, 226 (1992).
2. T. H. Boyer, "A brief survey of stochastic electrodynamics," in *Foundations of Radiation Theory and Quantum Electrodynamics*, A. O. Barut, ed. (Plenum, New York, 1980), p. 49; the references contained therein fully discuss the foundations of SED as presented herein. For contemporary work on SED, see, *e.g.*: T. W. Marshall and E. Santos, *Found. Phys.* **18**, 185 (1988); *Phys. Rev. A* **39**, 6271 (1989); T. H. Boyer, *Found. Phys.* **19**, 1371 (1989); D. C. Cole, *Phys. Rev. A* **42**, 1847 (1990); *Phys. Rev. A* **42**, 7006 (1990), and the references contained therein.
3. O. Theimer, *Phys. Rev. D* **4**, 1597 (1971).
4. P. H. E. Meijer, *Quantum Statistical Mechanics* (Gordon & Breach, New York, 1966), p. 17.
5. J. S. Bell, *Speakable and Unspeakable in Quantum Mechanics* (Cambridge University Press, Cambridge, 1987).
6. J. F. Clauser and M. A. Horne, *Phys. Rev. D* **10**, 526 (1974); J. F. Clauser, M. A. Horne, A. Shimony, and R. A. Holt, *Phys. Rev. Lett.* **23**, 880 (1969). A Aspect, P. Grangier, and G. Roger, *Phys. Rev. Lett.* **47**, 460 (1981); **49**, 91 (1982); A. Aspect, J. Dalibard, and G. Roger, *Phys. Rev. Lett.* **49**, 1804 (1982).
7. Further details are to be found in: A. F. Kracklauer, unpublished manuscript (as yet).

TOWARDS A PHILOSOPHY OF OBJECTS

Thomas Krüger

Friedrich-Engels-Strasse 18
51371 Leverkusen, Germany

We begin with a discussion of the ontic-ontologic difference between things and objects as entities *a priori* and *a posteriori*, respectively. It will be shown that only objects in this sense of the word are the matter of contemporary physics. This causes severe problems, since the attempt to apply any notion of object to the elements of any micro-world leads to the well-known quantum paradoxes. This consequence is unavoidable. Therefore, until physics starts to re-think its own fundaments, said paradoxes must remain unsolved. The thoughts of Martin Heidegger may serve as a token to overcome these difficulties, because they implicate a foundation of physics in terms of holistic performances rather than interacting but still isolated objects.

Key words: objects, realism, comprehensive theory, quantum mechanics and chemistry, holism.

On occasion of the Erice conference in 1989, John Bell presented a list of so-called *bad words* "which, however legitimate and necessary in application, have no place in a formulation with any pretension to physical precision" [1]. Hereto belong among others words like *system, micro-/macroscopic, observable* and *measurement.* But as we all know from our early schooldays, exactly these bad words exert an almost irresistable compulsion to use them always and on every convenient and inconvenient occasion. But for what? Just for enjoying the childlike pleasure to shock the adults! Nowadays, one would think, we have overcome this desire—but nevertheless there is no textbook about fundamental questions of quantum mechanics in which these bad words don't appear permanently or even represent the center of reflection. Why?

John Bell concedes that these *bad words* are "legitimate and necessary in application", i.e., we need them. But in how far and why? Is there

Waves and Particles in Light and Matter, Edited by
A. van der Merwe and A. Garuccio, Plenum Press, New York, 1994

any LEX, any law which would oblige us to make use of them? Hardly! But **while** 'doing' physics we make use of them. And therefore they are a mirror of **how** we 'do' physics. The question remains why these words have been called **bad** words. Couldn't this somehow imply that the manner how we 'do' physics is bad?

What does it mean to 'do' physics? We perform experiments. In doing so we deal with 'things' which we can touch and grasp. Beyond this mere touching and grasping we make some'thing' with them, i.e. we **handle** them. For the moment we don't pay attention to the question about our motive. We don't yet want to know whether we handle 'things' because we want to know what holds together the heart of the world or because this handling is nothing else than an interesting job to earn more or less money. For now it is only important that we make some'thing' with the 'things'. We handle them. So it comes to be *at hand* what has merely been present before. Therefore, these special kind of 'things' we are handling constitutes the set of what Martin Heidegger named *Zuhandenes* (those what is at hand).

Not only in physics but also in natural science as a whole, even in our daily life the word *objective* is highly appreciated. We say, statements can be objectively true or false, and we admonish ourselves to stay objective and set aside all subjective influences. But since my language is the mirror of just **my** world, the permanent use of the word *objective* points out that the existence of objects and therefore of subjects too—because without a SUB-IECTUM there are no OB-IECTA—is a characteristic feature of my world.

The question is now whether we 'do' physics with *Zuhandenem* or with objects. These two words denote distinct subgroups of what we called 'thing' But these subgroups are etymologically so different that complementary views of world are revealed by them.

As a human being we **explore** the primordial, mythic world of the beginning, i.e., the whole what surrounds us and where we belong to. We transcend the embedding in this whole by interrogating it as the whole what it is. Aristotle clothed this in the famous words: The astonishment is the beginning of all philosophy. The release mechanism of said exploration is just the own experience of being basically different with respect to all the other as a whole. Therefore we cannot experience compactness within the whole but rather the possibility of being distant, i.e., our world is voluminous, is porous, is structured like a sponge. Since our world permits the possibility of *distancity*—I crave your indulgence for introducing this curious Germanism—it has to be considered as being intrinsically open.

The experience of distance starts, as has been said above, with our own experience of being different from the 'rest' of the world. This original exploitation of the possibility of distancity, characterizing every human being, makes it possible to view our complement of the whole or, at least, some facets of it because the feature of distancity is the prerequisite for viewing any'thing' at all. So what we view is the 'remainder' of the world.

But since just we as upholders of the Occidental culture in contrast to a lot of other people as, for example, the Zen Buddhists cannot remain seated, we don't leave this contemplative viewing as it is. **We** turn from silence to action. Moreover, we both take hold of and **dissect** the 'rest' of the world! This we call the *technical* access [2].

We are used to dissect world and to split up into pieces what has been a lightly woven oneness before. So we **create** pieces by this dissection, and we call these pieces *objects* since their existence is being bound to the OB and the IACERE. The OB-IECTUM is what has been thrown against. Therefore the existence of objects depends on the **aggressive** exploitation of the possibility of distancity.

What we named *Zuhandenes* **arose** by handling from what has simply been present before, from the 'things'. The set of objects, however, has been **created by us** being urged by the internal dynamics of our Occidental, technical view of world. Therefore it is obvious that physics deals with objects rather than *Zuhandenem*, but we should always keep in mind that these 'objects' of physics are purely synthetic and not genuine entities. Objects **don't arise** in the analysis of the world. They don't exist *a priori*. They are **made** by just **our** way of behaving within the whole. So one could say: Objects are the self-made toys of physics.

But—for what at all do we then pursue science, especially physics? Superficially considered it is self-evident that we pursue science because we want to build up knowledge. Knowledge, however, is different from a mere accumulation of perceptions and observations, since it demands for **ordering** the perceived and observed events to unveil associations to generate **relations**. By creating knowledge through the generation of relations we proceed from perceiving and observing to pure cognition. Impetus for this is our **need for coherence**. The more we dissect world and thereby disintegrate its original oneness to a loose multipleness, the more a fundamental discomfort prompts us to look again for unifying structures within this artificially generated multifarious amount of parts. So we tend to revoke the dissection on a kind of sub-level. Within the dissected we strain after remains of oneness, we aim at unifying tendencies in the features of the parts. We dissect, and at the **same moment** we want to know what holds together the heart of the world. Somehow it seems that we only do physics to eliminate physics.

Howsoever, we experiment with objects, and all we want is to understand the outcome of our experiments, i.e., to cognize some sort of coherence. To this end we build a so-called *theory*, a structure consisting of mental images of the kind "we visualize ...so as if ..." and of relations combining these images. Said relations are especial in so far as they are translatable into the sign language of mathematics, what means that the theory's content is formalizable. So we can work with them by use of mathematics.

In our experiments we always deal with a **group** of at least partially different objects. For investigating the free descent, e.g., we need amongst others a solid body, a stop-watch, a surveyor's rod and the leaning tower. These are 'things' which we can touch and grasp. Usually we associate the outcome of an experiment unambiguously with either one or a small amount of the involved objects, i. e., we divide the group of all objects involved into two sub-groups, the one of the acting and the other of the re-acting objects. To the latter one belong, amongst others, our mesuring instruments. It is evident that a 'good' theory must consist of mental images and relations for **both** sub-groups.

But now there are experiments **also** involving objects we **cannot** touch. The most simple example for this is light. Since decades we know that a lot of all those experiments in which objects of the kind we cannot touch are acting or re-acting, yield an outcome the corresponding numerical value of which can be calculated in advance by means of traditional quantum mechanics (TQM). But—does this lead to any *understanding*?

Every mental image represents a certain description of a certain object, every relation a coherence between images. If said description, i.e., if the statement about some object is not clear and unambiguous, then a corresponding relation hardly can possess any content. In such a case the relation tends to deteriorate to a mere formalistic play, to an agglomeration of vague symbols. What the images mean determines what the relations mean. Without saying what, e.g., a wavefunction Ψ shall denote, a relation like the Schrödinger equation remains empty. It won't elucidate any 'thing'. Therefore it is, strictly spoken, physically meaningless. In such a case we certainly still can calculate but not understand.

The description again depends on our physical ideology which, based on the Occidental, technical view of world, is not only objectifying, i.e., dissecting, but in addition 'realistic'. The root of this frequently discussed term is the Latin substantive RES. RES denotes 'things' which are especial in so far as they are explicitly applicable and useful, and which accordingly have been manufactured to a certain purpose. Moreover, also matters and affairs are kinds of what the Romans called RES as, for example, their state, the RES PUBLICA, but in any case it is typical for every'thing' bearing the name RES that it is in **immediate and direct** contact with us. It is a concern of us; we have to do with it. So to be 'realistic' denotes no'thing' else than a certain way of behaving with regard to objects. It has no other meaning than to **handle** objects. The common utilization of RES to build up terms like 'real' or 'reality' distorts, strictly spoken, the original meaning and therefore leads to a couple of serious but home-made problems like the search for a 'realistic' interpretation of TQM or, moreover, the question about the 'reality' of a micro-world. We must confess that we often speak about a micro-world as if its possible constituents would be 'real' in the original sense, i.e., as if they could be handled like objects of our daily experience too. But we should always keep in mind that our physical ideology is 'realistic' and therefore contains the danger that we try to extend the 'realm' of 'realism' or 'reality' to the set of objects which can neither be touched nor grasped. We can deal with neutrons, e.g., but we cannot handle them!

According to the original meaning which we shall adopt for our further considerations, light is certainly an object of our daily experience, but despite of this we cannot say that it's 'real'. We perform experiments involving also those non-'real' objects but nevertheless try to understand the outcome on the basis of our physical, 'realistic' view of world. This must cause severe problems.

For objects which can be touched and grasped we have a 'good' theory consisting of Hamiltonian mechanics, thermo- and Maxwell's electrodynamics. Also for the said non-'real' objects which we shall call *quantum objects* we have a kind of theory named TQM. We know very well that TQM has a lot of hooks and eyes if we want to come to an **understanding** of what we assume to be the micro-world, so that we have to admit that TQM is first of all not very much more than a mere tool to **calculate** outcomes of experiments the so-called classical theory (theories) cannot explain. Let it be like it is: What we need is a comprehensive theory of the experiment as a whole, i.e., a theory for all the different objects involved. The classical theory is not able to play this part, but what's about TQM?

To the quantum objects created by us we allot a couple of potential properties. To formalize the theory we denote our quantum objects equipped with said potential properties by the code Ψ. If we now proceed from the prerequisite that

$$\Psi \in \mathcal{H} = \mathcal{L}_2\left(\boldsymbol{R}^{3n}\right), \quad n \in \boldsymbol{N},$$

then the potential properties can be formulated as self-adjoint operators $\hat{E}$ on the Hilbert space $\mathcal{H}$. These operators possess a unique spectral decomposition and the corresponding set of projection operators constitutes a Boolean algebra. This fact is of particular importance because it guarantees that every experiment actually leads to a result instead of only coming to an end. So we can be sure that the outcome in principle fits to our physical view of world. But as a further consequence the prerequisite given above leads to the following theorem: If Ψ_1 and $\Psi_2 \in \mathcal{H}$, then **every** *superposition* $(\alpha\Psi_1 + \beta\Psi_2)$ too is an element of this Hilbert space:

$$(\alpha\Psi_1 + \beta\Psi_2) \in \mathcal{H} \, \forall \, \alpha, \beta \in \boldsymbol{C},$$

and this means nothing else except that **every** superposition of the codes of two (or more) quantum objects could again be code of a quantum object. Since there is no reason to assume the domain of our theory not to be dense, we are compelled by the demand for intra-theoretical, logical consistency to accept that the new code generated by superpositioning corresponds to a **new** quantum object. So we have the possibility to create an arbitrary number of new quantum objects from a limited set of only two elements. Therefore our quantum objects satisfy neither the requirements of Kant's notion of object nor even those of his notion of substance. Moreover, even the meaning of *identity* has to be re-questioned.

Now the attempt to apply TQM to the experiment as a whole leads to a serious problem just because of the superposition principle or the new notion of object connected with it, respectively: First we construct a code Ξ covering all the objects taking part in the experiment. In the most simple case of only two objects involved (one re-acting and touchable object with the code Φ, the so-called measuring instrument, and a quantum object with the code Ψ) this can be done according to

$$\Xi = \Psi \otimes \Phi, \in \mathcal{H}_\Psi, \Phi \in \mathcal{H}_\Phi, \Xi \in \mathcal{H}_\Psi \otimes \mathcal{H}_\Phi \ ,$$

so that Ξ results as the tensor product of the two codes. In said experiment we are interested in the actualization of a potential property E of the quantum object. To describe this within the formal part of our theory we expand both Ψ and Φ in the series of corresponding eigenfunctions of $\hat{E}, \{\lambda_i\}$ or $\{\Lambda_j\}$, respectively. Since performing any experiment needs time, we want to know in how far Ξ changes from $t = 0$ to $t = t'$. The time dependency of Ξ is given by the Schrödinger equation, so that we finally arrive, as is generally known, at

$$\Xi(t) = \Sigma_n \gamma_n(t) \cdot \lambda_n \otimes \Lambda_n \qquad (1)$$

as code of the outcome of our experiment. [A discussion of the conceptual difficulties which would arise, e.g., from the idea of 'preparing' an 'original state' is not relevant for our way of thinking and therefore has been omitted, especially since it only would burden us with additional *bad words*.] What is the meaning of this equation? Above we have found out that **every** superposition of the codes of two quantum objects again is code of a **new** quantum object. But in this case the time development of Ξ leads to a superposition of tensor products, the factors of which are elements of different Hilbert spaces. We know that Λ has some'thing' to do with the measuring instrument and λ with the quantum object, and we also know that Ξ at a certain time t' represents the outcome of the experiment. But Ξ does **not** reflect it **unambiguously** despite the fact that the pointer of our measuring instrument has a definite position. It is obvious that the range of statements and predictions TQM allows is limited as soon as we attempt to apply this theory also to objects which are 'real' in the sense given above. This, however, is not a miracle since we describe the interaction of two different kinds of objects within **one** system of mental images and relations. Yet there is a pragmatic way out: We introduce the so-called *reduction postulate* and thereby guarantee that the final superposition **1** decomposes in a 'reasonable' manner, i.e., we thereby guarantee that both the measuring instrument remains touchable and the pointer allows a reading.

Meanwhile we know that every scientific theory satisfying the requirements of logical consistency contains both *notions* which are *not explainable* by means of the theory itself and *propositions* which can*not* be *proven* using the arithmetics of the corresponding formal system alone. But it is questionable, how many of these *notions* and *propositions* are necessary to make a theory applicable. In our context we must ask above all whether the reduction postulate is 'really' necessary. [Here I want to point out that until now we could discuss fundamentals of TQM without stressing the notion of *probability*. In connection with the reduction postulate, however, we would have to introduce it as an additional notion.]

But: Can we solve **all** problems which arise from the attempts to explain the meaning of superpositions by introducing the reduction postulate and the notion (and concept) of probability? Surely not! Primas adduces the example of the optically active amino acid alanine [3]: In a pharmacy we can buy both D- and L-alanine and furthermore the cheaper, racemic mixture of these two, i.e., we can hold objects in our hands which to describe within the framework of TQM is at least not absurd. About molecules and the relation of chemistry to TQM we will speak later on. For now we only assign a code to each of the different 'kinds' of alanine. Denoting the dextrorotatory form by $|\Psi_D\rangle$ and the other one by $|\Psi_L\rangle$ we obtain the projection operator

$$|\Psi_D\rangle\langle\Psi_D| + |\Psi_L\rangle\langle\Psi_L|$$

as code of the racemic mixture. But if we now should try to buy a 'kind' of alanine which corresponds to one of the two superpositions $|\Psi_D\rangle + |\Psi_L\rangle$ or $|\Psi_D\rangle - |\Psi_L\rangle$, respectively, both admitted by TQM, we would get into severe troubles because a corresponding molecule *in praxi* does not 'exist'.

On the other hand we had pointed out that the demand for intratheoretical, logical consistency boils down to the demand for an unambiguous correspondence between code and quantum object. In the best case we can only sail round this reef if we assume that molecules are no quantum objects. But this assumption would simply lead us to another facet of the basic problem in discussion.

Now we arrive at the question how chemistry and TQM are related to one another. Already in the early times of TQM Dirac has claimed that it covers the whole area of chemistry too, but meanwhile we know that TQM does not permit one of the basics of chemistry, that is the notion of molecule (see, among others, [4] and [5])! We '*make*' molecules by conscious, mental elimination of certain interactions. We 'make' molecules by sifting groups of elementary particles from a texture of interactions the most important of which is the coupling of electronic and nuclear motion. Basis of this process is, of course, a quite concrete idea of how a molecule as such a one shall be. According to this idea which is based on vivid images taken from classical physics, we eliminate some couplings in order to create new objects from an already 'existing' one. This process differs not basically from the objectification of the world we had mentioned in the beginning of this paper. Put it into plain words: We **want** to have molecules; therefore we **make** them. But in contradistinction to the 'general' objectification we don't dissect the whole, *τὰ πάντα*, *das Seiende im Ganzen und an sich*, i.e., every'thing' what **is**, taken together and considered as a whole. We only dissect an already created quantum object. But thereby we obtain a set of sub-objects which are essentially different from the original quantum object because they show up a certain touchability despite still belonging to a kind of micro-world. This touchability becomes apparent in what the chemists call *substance* and, e.g., in the fact that we at least can '*visualize*' the tertiary structure of DNA helices by means of electron microscopy.

Now molecular crystals, liquids and gases form the bridge of a **continual** transition from the single molecules, created by dissecting quantum objects, up to the touchable substances of our daily experience. Just because this transition, this increase in dimension, is continually, it again demonstrates the central difficulty of TQM: It makes sense, without any doubt, to require that TQM fades into classical mechanics in the asymptotic case of large' objects. But the simple statement that effects of the order of magnitude of h can be neglected for 'macroscopic' objects is erroneous. Planck's constant h is a **constant** with an unambiguous numerical value, not an expansion parameter for which the consideration of such an asymptotic behaviour would be meaningful. In addition, there are also non-exotic 'systems' for which the $h \rightarrow 0$ limes of the Schrödinger equation does not yield the corresponding equation of classical mechanics as has recently been shown [6].

Summing up, it looks like as if we could obtain a comprehensive theory of the experiment as a whole **neither** by use of TQM alone **nor** by combining TQM with so-called classical theories. At first glance the idea of Ghirardi, Rimini, and Weber (GRW) [7] offers a smart way out of this dilemma. They pay regard to the content of the reduction postulate, i.e., the guarantee that 'macroscopic' objects turn out to be localized in TQM too, by endowing the formalism with two additional parameters which control a simple mechanism of localization and therefore could be considered as natural constants. By this way the formalism of TQM is extended to cover also some aspects of the behaviour of 'macroscopic' objects. This **would** allow for a comprehensive theory of the experiment as a whole, but formulas can only be the expression of a theory and not its content. So the attempt to solve a problem of a theory by simply changing its form, the only function of which is to provide us with numerical values, is similar to the idea to prove the pope's dogma of infallibility by using Chanel No. 5 instead of incense—even if the odour of Chanel **would** seem to be more pleasant! Moreover, our choice of the numerical values of said parameters decides about what in our world shall be called 'large' or 'small', respectively, and surely this is a debatable point. Therefore the idea of GRW has to be considered as a purely formalistic modification of TQM which is inspired by experience so that it could be a useful tool for practical purposes. A theory in the strict sense of the word, however, it is not, because it does not lead to a deepened understanding.

A lot of authors have shown that, strictly speaking, TQM may be applied in the case of perfectly isolated quantum objects only. This isolation manifests itself in the time symmetry of the ordinary Schrödinger equation as well as in the bidirectional deterministic time evolution of $\Psi(t)$ described by the famous theorem of Stone [8]. In the case of a coupling to any environment, however, *dissipative* forces play an important part. Hereto among others also the coupling to those 'macroscopic' objects could belong which we had called measuring instruments. Including these dissipative forces in the formalism of TQM leads to **nonlinear** Schrödinger equations with a broken time symmetry [9]! Moreover, it seems to turn out that the superposition principle is no more valid to its full extent. This means no'thing' less than that the explicit consideration of couplings of formerly isolated quantum objects to objects with infinitely many degrees of freedom, to somehow *macroscopic* objects, could diminish the difficulties we had in building up a comprehensive theory of the experiment as a whole. We cannot avoid to become aware of a fascinating analogy: The 'world' of non-linear, symmetry broken **formulas** which has been unveiled by taking account of the action of dissipative forces coupling different **kinds** of objects seems to re-evoke the image of the original, mythic world of the beginning. But nevertheless also in this way we would put the cart before the horse since despite of all modifications of the formalism the problems of the notion of object remain unsolved.

Over and over again we have to notice that the essential hindrance of all attempts to understand experiments lies in the notion of object—and all 'realistic' interpretations of it have still more aggravating consequences. But the point is **not** that our common notion(s) of object lead(s) to difficulties which could be avoided by simply generating another notion! The snag is that we create objects at all! Together with the objects in themselves also the complex fabric of the relations of all these objects arises by our coercive grabbing, by our dissecting and separating of what we had called *world.* Physics nowadays is mixed up in the multidimensional hierarchy of said fabric. It is evident that this hierarchy can only be penetrated from **outside** of the underlying, object creating thinking. So a comprehensive scientific theory must come from there, from beyond the 'idea' to create objects. Its root must be the certitude that *τὰ πάντα*, ***das Seiende*** *im Ganzen und an sich*, is not some'thing' so different and far away from us that we simply can go to investigate it without being intrinsically involved. Man, as Martin Heidegger named it, is *der Hüter* (the guardian) *des Seins*—but often we are too big for these boots.

Physics led us to the point where we begin to understand that our world is a self-made one. Physics itself can only be understood as soon as we begin to understand **how** we have made just this world. So the first step is to cognize the difference between being *der Hüter des Seins* and being the maker of the world. Only this cognition will lead us to the comprehensive theory often mentioned above, and in this theory the word *theory* will again bear its original meaning: *respectful viewing of the present in its unveiledness* [10].

REFERENCES

1. John Bell, "Against 'Measurement'", in Arthur I. Miller, ed., *Sixty-Two Years of Uncertainty* (NATO ASI Series B, Vol. 226, Plenum, New York, 1990).
2. See, e.g., Martin Heidegger, "Die Frage nach der Technik" and "Wissenschaft und Besinnung," in Martin Heidegger, *Vorträge und Aufsätze* (Neske, Pfullingen, 1978). As an easy-to-read introduction, the following book can be recommended: Silvio Vietta, *Heideggers Kritik am Nationalsozialismus und an der Technik* (Niemeyer, Tübingen, 1989). Heidegger expressed his thoughts very plainly in his letter of 2 September, 1963 to Takehiko Kojima, published in Hartmut Buchner, ed., *Japan und Heidegger* (Thorbecke, Sigmaringen, 1989), p. 222 ff.
3. Hans Primas, *Chemistry, Quantum Mechanics and Reductionism* (Springer, Berlin, 1983), p. 12.
4. R. G. Woolley, *Adv. Phys.* **25**, 27 (1976).
5. Hans Primas, *Chimia* **36**, 293 (1982).
6. Song Ling, *J. Chem. Phys.* **96**, 7869 (1992).
7. Gian-Carlo Ghirardi and Alberto Rimini, "Old and new ideas in the theory of quantum measurement," introductory review, in Arthur I. Miller, ed., *loc. cit.*
8. Dieter Schuch, Habilitation Thesis, Johann Wolfgang Goethe-University, Frankfurt am Main, 1990.

9. M. H. Stone, *Proc. Natl. Acad. Sci. U. S.* **16**, 172 (1930).
10. Thomas Krüger, "Quantenmechanik und eigentliches Denken," unpublished paper.

THE ENERGY-MOMENTUM TRANSPORT WAVE FUNCTION

Miroslaw J. Kubiak

Planetarium and Astronomical Observatory
ul. J. Krasickiego 1-7
PL-86-300 Grudziadz, Poland

In this paper a very simple model of the *energy-momentum transport wave function* $\psi(\boldsymbol{r},t)$ is presented. In the frame of this simple model, we discuss how the information (energy and momentum) "*from the inside of an electron to the outside*" is transported.

Key words: transfer of information in nature, quanta of action, diffusion equation of the EMTWF.

1. TRANSFER OF INFORMATION IN NATURE: A VERY SIMPLE MODEL

Let us consider a body with mass m moving with a constant velocity v ($v << c$, where c is the speed of light).

According to the simple formula $E = p^2/2m$, where p is the momentum, we can calculate the energy E for the body. But we do not know how the energy and momentum are transported in the spacetime with the body during interactions.

Let us assume that the energy and momentum are transported with the body along path x during time t by means of the *energy-momentum transport wave function* (EMTWF),

$$\psi(x,t) = \exp\left[(Et - px)/q\right] \tag{1}$$

where q is the *quantum of action.*

We differentiate Eq. (1) with respect to t:

$$\frac{\partial\psi(x,t)}{\partial t} = (E/q)\psi(x,t) \tag{2}$$

and we differentiate Eq. (1) twice with respect to x:

$$\frac{\partial^2\psi(x,t)}{\partial x^2} = (p/q)^2\psi(x,t) \tag{3}$$

Equations (2) and (3) become equivalent if we multiply Eq. (3) by the coefficient $\mu = q/2m$. Thus

$$\frac{\partial\psi(x,t)}{\partial t} = \mu\,\frac{\partial^2\psi(x,t)}{\partial x^2}\,. \tag{4}$$

This is the *diffusion equation of the energy-momentum transport wave function* (μ is the coefficient of diffusion).

In our very simple model, transfer of information (energy and momentum) takes place through the *diffusion of the EMTWF*. The coefficient μ describes *the speed of diffusion of the EMTWF* for the four foundamental interactions. The coefficient of diffusion μ for electromagnetic (EM) and gravitational interaction we have discussed in paper [1]. *The above consideration we used only to obtain the diffusion equation of the EMTWF.* This model we used to explain how the information (energy and momentum) "*from the inside of an electron to the outside*" is transported.

For the EM interaction [1] we have

$$\frac{\partial\psi(\mathbf{r},t)}{\partial t} = \mu_e\nabla^2\psi(\mathbf{r},t), \tag{5}$$

where $\mu_e = \left(ke^2\right)/\left(2m_e c\right)$, $k = 1/4\pi\varepsilon_0$ is the constant of Coulomb's law in the SI system of units, e the elementary charge of EM interactions, m_e the mass of the electron. For the simplicity of our considerations, we omit in the following the constant of 1/2 in the coefficient of diffusion.

Equation (4) describes the diffusion of the EMTWF without the source. Let us try to give an answer to the question: what should be the source of the *energy-momentum transport wave function* $\psi(\mathbf{r},t)$?

Let us assume, that the source of the $\psi(\mathbf{r},t)$ is the energy and momentum of the particle $E^2 = (pc)^2 + \left(m_e c^2\right)^2$, in the particular case, the rest-energy of the electron $E = m_e c^2$, which is trasported by means of quanta of action connected with the EM interactions h_e where $h_e = ke^2/c$ [2].

Therefore, we modify Eq. (5) through the addition of the term $m_e c^2/h_e$. We find:

$$\frac{\partial\psi(\mathbf{r},t)}{\partial t} = \mu_e\nabla^2\psi(\mathbf{r},t) + m_e c^2/h_e. \tag{6}$$

Note that the $(m_e c^2/h_e) = c/r_e = 1.05\times10^{23}[1/s]$, where r_e is the *classical radius of the electron.*

According to Eq. (6), we can say, that as a result of the diffusion phenomenon, the $\psi(\boldsymbol{r},t)$ is disappearing in time, which should lead to decay of the electron. But an electron is a stable particle and is not decaying. This signifies that inside of electron there exist forces which are responsible for the stability of the electron and prevent the decay of the electron impossible. Let us postulate that these forces are described by the term of the form: $-\gamma(\boldsymbol{r},t)\psi(\boldsymbol{r},t)$, where $\gamma(\boldsymbol{r},t)$ is any function. In this paper we assume $\gamma(\boldsymbol{r},t) = \gamma = \text{const}$. Our equation has the form:

$$\frac{\partial\psi(\boldsymbol{r},t)}{\partial t} = \mu_e\nabla^2\psi(\boldsymbol{r},t) - \gamma\psi(\boldsymbol{r},t) + m_e c^2/h_e \tag{7}$$

or

$$\mu_e\nabla^2\psi(\boldsymbol{r},t) - \frac{\partial\psi(\boldsymbol{r},t)}{\partial t} + \frac{m_e c^2}{h_e} = \gamma\psi(\boldsymbol{r},t), \tag{7a}$$

where the left side of Eq. (7a) describes *the diffusion with the source*; the right one compensate *the diffusion process as a result of forces which make decay of the electron impossible.*

For the static case, Eq. (7) has the form

$$\mu_e\nabla^2\psi(\boldsymbol{r}) - \gamma\psi(\boldsymbol{r}) + m_e c^2/h_e = 0, \tag{8}$$

or, dividing by μ_e,

$$\nabla^2\psi(\boldsymbol{r}) - \beta^2\psi(\boldsymbol{r}) = -r_e^{-2}, \tag{9}$$

where $\beta^2 = (\gamma/\mu_e)$.

If the electron is at rest at the origin of coordinate system, then the solution (for the central symmetry) has the form

$$\psi(\boldsymbol{r}) = (r_e\beta)^{-2} - (\beta r)^{-1}\exp(-\beta r).$$

Omission in the above equation the constant of $(r_e\beta)^{-2}$ yields

$$\psi(\boldsymbol{r}) = -(\beta r)^{-1}\exp(-\beta r). \tag{10}$$

This equation has a form similar to the well-known Yukawa equation for the potential. The *energy-momentum transport wave function* $\psi(\boldsymbol{r},t)$ describes the transport of information (energy and momentum) "*from the inside of the particle to the outside*". We can calculate the energy E and momentum p according to the following formulae:

$$p(\boldsymbol{r},t) = -h_i\ \nabla\psi(\boldsymbol{r},t) \tag{11a}$$

$$E(\boldsymbol{r},t) = h_i\ \frac{\partial\psi(\boldsymbol{r},t)}{\partial t}, \tag{11b}$$

where h_i is the "*elementary quantum of action for the four fundamental interactions*" [2].

Assume, that inside of electron there exist forces which are responsible for the stability of the electron and that the information from the inside of the electron to its outside is transported by means of the h_w, where h_w is the "*elementary quantum of weak action at short-range distances*" [2]. We have:

$$p(r,t) = -h_w \, \nabla\psi(r,t) \tag{12a}$$

$$E(r,t) = h_w \, \frac{\partial\psi(r,t)}{\partial t}, \tag{12b}$$

The constant h_w depend on the type of interacting particles and on the type of bosons (W^+, W^-) that carry weak interactions [2]. In our case, for $\psi(r) = -(\beta r)^{-1}\exp(-\beta r)$, we have

$$\begin{aligned} \text{momentum}: \; p(r) &= -(h_w/\beta r^2)\exp(-\beta r)\left(\frac{h_w}{r}\right)\exp(-\beta r), \\ \text{energy}: \; E(t) &= 0. \end{aligned} \tag{13}$$

As we assumed, in the inside of the electron exists forces which are responsible for the stability of the electron. Now let us assume also that these forces are transported by means of the particles whose speed is close to the speed of light c. Multiplying both sides of Eq. (12) by the c and omitting the term r^{-2}, we arrive at the energy E_w of weak interactions as a function of the r:

$$E_w(r) \sim pc = -(h_w c/r)\exp(-\beta r).$$

This energy is decreasing exponentially and is transported to the distance $R = \beta^{-1}$, where $\beta^{-1} = (\mu_e/\gamma)^{1/2} = \lambda_{Cw}, \lambda_{Cw} = h/(m_w c)$, is the Compton wavelength of the bosons W, and m_W is the mass of the bosons W, which are the carriers of weak interactions.

For the distances greater than the Compton wavelength of the bosons (W^+ and W^-), the weak interactions vanish quickly.

According to the formula $\beta^{-1} = (\mu_e/\gamma)^{1/2} = \lambda_{Cw}$ we can determine $\gamma = \mu_e(\lambda_{Cw})^{-2}$.

Summing up, if our very simple model is correct, then the transport of information (energy and momentum) "*from the inside of the electron to the outside*" is described, for the static case, by means of the following equation:

$$\nabla^2\psi(r) - (\lambda_{Cw})^{-2}\psi(r) = -r_e^{-2}, \tag{14}$$

whose solution (for the central symmetry) is

$$\psi(r) = (\lambda_{Cw}/r_e)^2 - (\lambda_{Cw}/r)\exp(-r/\lambda_{Cw}) \tag{15}$$

The energy of weak interaction is expressed by the formula:

$$E_w(r) = -h_w \nu_{Cw}(\lambda_{Cw}/r)\exp(-r/\lambda_{Cw})$$

where ν_{Cw} is the Compton frequency [2].

The wavelength λ_{Cw} is the distance at which the EM interaction is equal to the weak interaction, i.e., $\mu_e = \mu_w$ or $\alpha_e = \alpha_w$, where μ_w is the coefficient of diffusion for the weak interaction, $\alpha_e = h_e/h$ is the the coupling constant for the EM interaction, and $\alpha_w = h_w/h$ is the coupling constant for the weak interaction.

If we would like to obtain full information on how energy and momentum are transported "*from the inside of the electron to the outside,*" we would have to solve the equation

$$\frac{\partial\psi(r,t)}{\partial t} = \mu_e\nabla^2\psi(r,t) - \mu_w\eta^2\psi(r,t) + (E/h_e) \qquad (16)$$

as a soliton wave, where $\eta = \eta(r,t)$ is any function.

2. CONCLUSION

In this paper I have presented the simple model of the *energy-momentum transport wave function* $\psi(r,t)$.

Assuming, that the weak forces are responsbile for the permanence of the electron, in the frame of this simple model, we came to the conclusion that the information "*from the inside electron to the outside,*" for the static case, is transported by means of the *energy-momentum transport wave function*

$$\psi(r) = (\lambda_{Cw}/r_e)^2 - (\lambda_{Cw}/r)\exp(-r/\lambda_{Cw}).$$

If we would like to obtain full information on how energy and momentum are transported "*from the inside of the electron to the outside,*" we would have to solve the nonlinear equation

$$\frac{\partial\psi(r,t)}{\partial t} = \mu_e\nabla^2\psi(r,t) - \mu_w\eta^2\psi(r,t) + (E/h_e)$$

as a soliton wave, where $\eta = \eta(r,t)$ is any function.

ACKNOWLEDGMENT

I am grateful to the sponsors: Delegatura Kuratorium in Grudziadz (Poland), Instituto Nationale di Fisica Nucleare, Commissione delle Comunita Europee and Universita di Bari, Diparitimento di Fisica (Italy) for the financial support, which made possible my participation in the workshop "Waves and Particles in Light and Matter," on the occasion of Louis de Broglie's 100th anniversary, in Trani, Italy, 24–30 September 1992.

REFERENCES

1. M. J. Kubiak, *Physics Essays* **5**, 422 (1992).
2. L. Kostro, *Physics Essays* **1**, 64 (1988).

THE SPACETIME STRUCTURE OF QUANTUM OBJECTS

Vito Luigi Lepore

Dipartimento di Fisica, Università di Bari and
Istituto Nazionale di Fisica Nucleare, Sezione di Bari
173 Via G. Amendola,
I-70126 Bari, Italy

According to de Broglie, a quantum object like a neutron is composed of a localized corpuscle surrounded by a guiding wave, both assumed to be objectively evolving in the spacetime. The wave guides the corpuscle towards the zones of maximum intensity. In the following we will discuss the mathematical formulation of these ideas and show that spectral filtering is fully comprehensible on adopting this approach.

Key words: de Broglie waves, coherence, neutron interference.

1. INTRODUCTION

In a recent paper [1] the result of a neutron interferometry experiment has been reported. A neutron beam is allowed to interfere in a silicon perfect crystal skew-symmetric interferometer (Fig. 1). The beam is split into two rays on the first plate. After having traversed the interferometer following different routes, these rays are recombined on the fourth plate. An aluminum slab intersects both routes. On rotating this slab, a phase difference can be induced between the two paths. Denoting by α the angle of rotation and by λ the neutron wavelength, the phase difference is

$$\phi_\lambda = N_p b_p \lambda d \frac{\sin^2\theta_I \sin\alpha}{\cos^2\theta_I - \sin^2\alpha}, \tag{1.1}$$

where θ_I is the Bragg angle of the interferometer; N_p, b_p and d are, respectively, the atom density, nuclear scattering length, and the thickness of the aluminum slab. On one of the two routes, there is also a sample of bismuth placed perpendicular to the beam. This sample changes the optical path length. The corresponding phase shift is given by

$$\chi_\lambda = -NbD\lambda, \tag{1.2}$$

where N, b, and D are, respectively, the atom density, the neutron-nuclear scattering length, and the thickness of the sample. Note that the purpose of the Al slab is to induce a small phase shift, whereas the purpose of the Bi sample is to induce a large change in the optical length of one of the two paths of the interferometer. Several interferograms are collected on using samples of increasing thickness. The contrast of the interference pattern falls rapidly to the minimum 2.2% of the contrast without the sample when a sample 8 mm long or more is inserted in the interferometer. With this condition, the difference between the optical lengths of the two paths of the interferometer is greater than the coherence length of the incident wave packet, and the interference pattern is wiped out. However, even in the extremely unfavorable case of a sample 20 mm long inserted in the interferometer, and a perfect crystal analyzer placed on the outgoing incoherent beam, it is possible to restore coherence with a contrast equal to the 95.2% of the contrast observed when both the bismuth sample and the analyzer crystal are removed. The difficulty of giving an interpretation in terms of real waves and real corpuscles lead the authors to conclude:

> We know the neutron is a particle when it is emitted, and again when it is detected, but between these two times, the physical connection between the neutron particle and the wave packet remains hidden,...."

According to de Broglie [2], a quantum object like a neutron is composed of a localized corpuscle surrounded by a guiding wave, both assumed to be objectively evolving in the spacetime. The wave guides the corpuscle towards the zones of maximum intensity. For example, in an interferometer a corpuscle is transmitted along the arm where the constructive interference takes place. However, if we try to understand the previous experiment on using this model, then immediately a question arise: Since the wave packets are superimposed out of the coherence length, the corpuscles arriving on the fourth plate are transmitted or reflected independently of the phase difference between the two routes; how is it then possible that the beam becomes sensitive to the phase difference on placing an analyzer crystal after the interferometer? In the following we will address this problem.

2. THE DE BROGLIE MODEL

Let us state explicitly the basic assumptions of the de Broglie model for neutrons [3]: (i) A neutron is composed of a localized corpuscle sur-

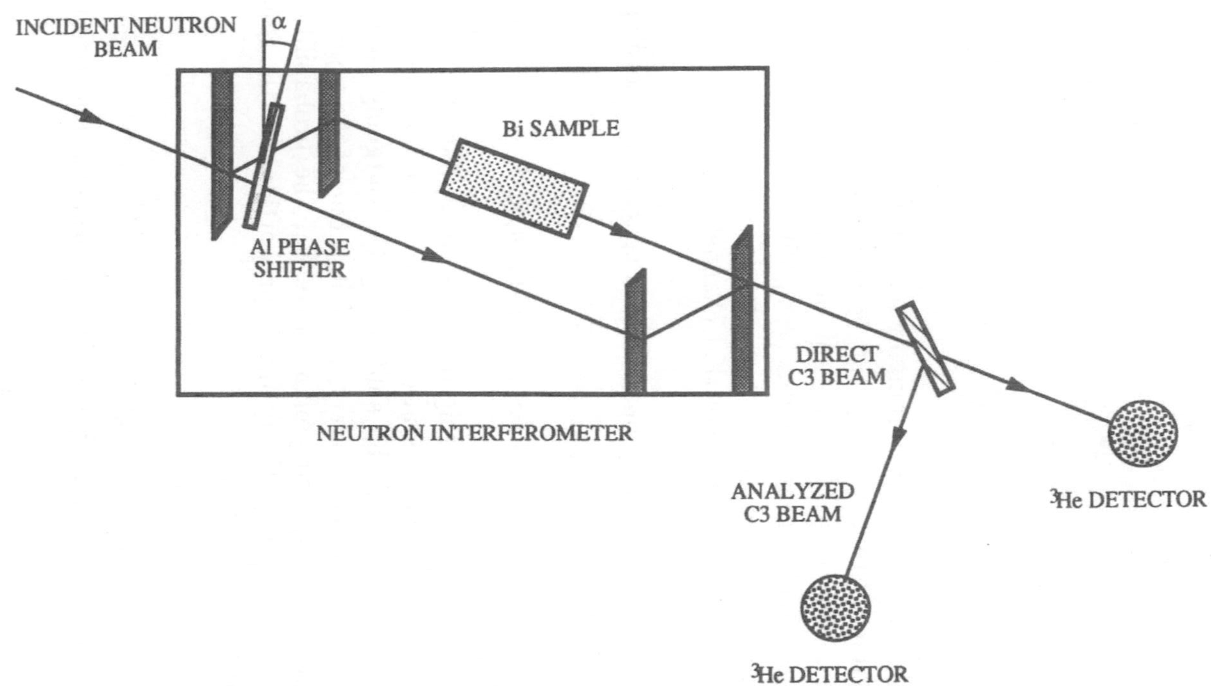

Fig. 1. Outline of the experiment on spectral filtering of a neutron beam. An incident neutron beam is split in two beams on the first plate of a skew-symmetric interferometer. The split beams are recombined on the fourth plate of the interferometer. If the analyzer crystal is removed, the overall outgoing beam impinges on a ^{3}He detector. When an analyzer crystal is placed in the direct C3 beam, a filtered out part of the beam is reflected towards another ^{3}He detector.

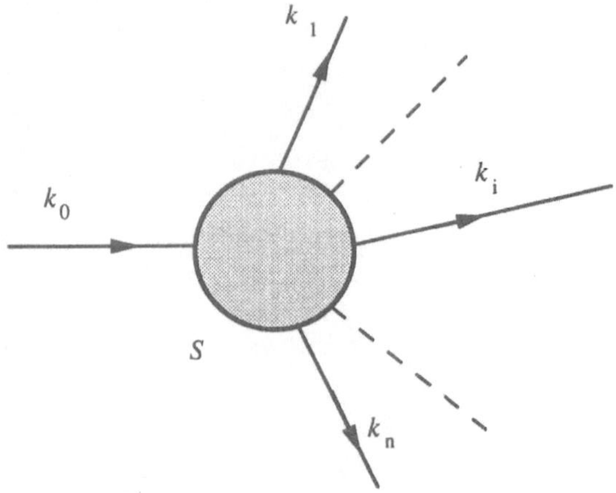

Fig. 2. A corpuscle impinging on a scattering system S along the k_0 channel is transmitted along one of the possible outgoing channels $k_1, \ldots, k_n$ with a probability proportional to the square modulus of the wave in that channel.

rounded by a real wave $\Phi(x, y, z, t)$ propagating in spacetime according to the Schrödinger equation

$$i\hbar \frac{\partial \Phi}{\partial t} = -\frac{\hbar^2}{2m} \nabla^2 \Phi + V(x, y, z, t) \Phi(x, y, z, t). \tag{2.1}$$

(ii) Let us consider a neutron scattering on a system S, like the plate of an interferometer or an analyzer crystal (Fig. 2), impinging along the k_0 channel. The incoming wave will be split along the outgoing channels $k_1, \ldots, k_n$. The corpuscle, on the contrary, will follow one of the n possible paths. If one assumes an impinging plane wave Φ_0, the probability P_j that the corpuscle will follow the jth channel is proportional to the square modulus of the outgoing wave Φ_j [4],

$$P_j = \frac{|\Phi_j|^2}{|\Phi_1|^2 + \ldots + |\Phi_n|^2}, \quad j = 1, \ldots, n. \tag{2.2}$$

Notice that the probabilities $P_1, \ldots, P_n$ satisfy the normalization condition

$$\sum_{j=1}^{n} P_j = 1, \tag{2.3}$$

and that the intensity of the incoming wave is equal to the sum of the intensities of the outgoing waves,

$$|\Phi_0|^2 = |\Phi_1|^2 + \ldots + |\Phi_n|^2. \tag{2.4}$$

The generalization of (2.2) to the wave packet case is straightforward. The scattering system divides the incoming wave packet $\Phi(x, y, z, t)$ along the n outgoing channels. Assuming that the outgoing wave packets occupy the volumes $V_1, \ldots, V_n$, the probability P_j becomes

$$P_j = \frac{\int_{V_j} |\Phi_j(x, y, z, t)|^2 d^3x}{\int_{V_1} |\Phi_1(x, y, z, t)|^2 d^3x + \ldots + \int_{V_n} |\Phi_n(x, y, z, t)|^2 d^3x}, \quad j = 1, \ldots, n. \tag{2.5}$$

Naturally, even in this case, the normalization condition (2.3) for the probabilities P_j is fulfilled and Eq. (2.4) for the conservation of the intensities becomes

$$\int_V |\Phi_0(x, y, z, t)|^2 d^3x = \int_{V_1} |\Phi_1(x, y, z, t)|^2 d^3x + \ldots + \int_{V_n} |\Phi_n(x, y, z, t)|^2 d^3x, \tag{2.6}$$

where V is the volume occupied by the incoming wave packet. Of course, if there is more than one incident channel, as happens on the fourth plate of the interferometer in Fig. 1, then one should write the sum of the incident intensities on the left-hand side of (2.4) and (2.6). It should be stressed that we are attributing two physical properties to the guiding wave $\Phi(x, y, z, t)$: First, it is the objective wave field in spacetime for every single event and, secondly, it gives the probabilities of scattering in repeated events, stated in (2.2) or (2.5).

3. THE DESCRIPTION OF THE EXPERIMENT

According to the authors of Ref. [1], the normalized spectrum of the source can be assumed to be Gaussian

$$g(\lambda) = \frac{1}{\sqrt{2\pi}\sigma_0} \exp\left(-\frac{(\lambda - \lambda_0)^2}{2\sigma_0^2}\right), \tag{3.1}$$

with $\lambda_0 = 2.3456\mathring{A}$ and $\sigma_0 = 0.0120\mathring{A}$. A more accurate description of the spectrum can be obtained on using a double-Gaussian function. However, this would introduce unessential complications for the main points under discussion.

Let us consider the neutron interferometer in Fig. 1. If we consider an incoming wave packet with Fourier components given by $A(k)$, the outgoing wave packet after the interferometer and before the analyzer crystal is

$$\Phi_f(z, t) = \int_{-\infty}^{+\infty} \frac{dk}{\sqrt{2\pi}} A(k) \left(-\frac{1}{2\sqrt{2}}\right) [1 + \exp i(\phi_\lambda + \chi_\lambda)] \exp i(kz - \omega t), \tag{3.2}$$

where we have assumed that the reflection and transmission coefficients are $\frac{i}{\sqrt{2}}$ and $\frac{1}{\sqrt{2}}$, respectively, and ϕ_λ and χ_λ are given by (1.2) and (1.2). The effect of the analyzer crystal is described by the associated transmission function $T(\lambda)$, with the square modulus given by [1]

$$W(\lambda) = |T(\lambda)|^2 = \exp\left(-\frac{(\lambda-\lambda_0)^2}{2\sigma_A^2}\right), \tag{3.3}$$

with σ_A^2 proportional to the mosaic width of the crystal. Therefore, for the incoming wave packet $A(k)$, the amplitude of the wave after the analyzer crystal becomes

$$\Phi_a(z,t) = \int_{-\infty}^{+\infty} \frac{dk}{\sqrt{2\pi}} A(k)T(\lambda)\left(-\frac{1}{2\sqrt{2}}\right)[1+\exp i(\phi_\lambda+\chi_\lambda)]\exp i(kz-\omega t). \tag{3.4}$$

The probability P_f^W that a neutron traverses the interferometer, which is obtained on inserting (3.2) in (2.5), is

$$P_f^W = \frac{\int_{-\infty}^{+\infty} G(\lambda)\frac{1}{4}\left[1+\cos\left(\phi_\lambda+\chi_\lambda\right)\right]d\lambda}{\int_{-\infty}^{+\infty} G(\lambda)d\lambda}, \tag{3.5}$$

where $G(\lambda)$ is the spectrum of the incident wave packet.

On inserting in (2.5) the expressions of the wave packets before and after the analyzer crystal given by the Eqs. (3.2) and (3.4), respectively, it is possible to calculate the probability P_a^W that a corpuscle present in the outgoing direct C3 beam of the interferometer is reflected by the analyzer crystal:

$$P_a^W = \frac{\int_{-\infty}^{+\infty} G(\lambda)W(\lambda)\frac{1}{4}\left[1+\cos\left(\phi_\lambda+\chi_\lambda\right)\right]d\lambda}{\int_{-\infty}^{+\infty} G(\lambda)\frac{1}{4}\left[1+\cos\left(\phi_\lambda+\chi_\lambda\right)\right]d\lambda}. \tag{3.6}$$

The probability P_{af}^W that a neutron represented by a wave packet both crosses the interferometer along the C3 direct beam and is reflected by the analyzer crystal is then

$$P_{af}^W = P_f^W P_a^W. \tag{3.7}$$

The explicit calculation in (3.7) leads to

$$P_{af}^W = \frac{\sigma_T}{4\sigma_0}\left[1+\exp\left(-\frac{N^2b^2\sigma_T^2D^2}{2}\right)\cos\left(\phi_{\lambda_0}+\chi_{\lambda_0}\right)\right], \tag{3.8}$$

with $\sigma_T = \sigma_0\sigma_A/\sqrt{\sigma_0^2+\sigma_A^2}$. If the analyzer crystal is removed, then $P_a^W = 1$ and the probability of crossing the interferometer, according to (3.5), is

$$P_f^W = \frac{1}{4}\left[1+\exp\left(-\frac{N^2b^2\sigma_0^2D^2}{2}\right)\cos\left(\phi_{\lambda_0}+\chi_{\lambda_0}\right)\right]. \tag{3.9}$$

Let us consider the most interesting case

$$\frac{1}{Nb\sigma_0} \ll D < \frac{1}{Nb\sigma_T}, \tag{3.10}$$

reached by the authors of Ref. [1] with bismuth samples of different lengths between 8 mm and 20 mm. In this condition the probability that a neutron crosses the interferometer is independent of the phase difference of the two routes (see Eqs. (3.9)). Thus, if one measures the counting rate when the analyzer crystal is removed, no interference pattern appears. By contrast, when the analyzer crystal is placed between the interferometer and the detector, the probability that a neutron crosses the interferometer and is reflected by the analyzer crystal which is given by the Eq. (3.8) depends on the position of the Al phase shifter and the interference becomes visible. On viewing the neutrons as described by a wave packet, the probability P_f^W that a corpuscle crosses the interferometer along the direct beam is given by (3.9). In the condition (3.10) the probability P_f^W does not depend upon the position of the Al phase shifter because the superposition of the wave packets on the fourth plate takes place out of the coherence length. However, the outgoing beam carries the dependence upon the phase difference as shown in Eq. (3.2). After the analyzer crystal the expression for the wave packet is given by Eq. (3.4) The probability P_a^W that the corpuscle is reflected by the analyzer crystal, calculated from (2.5) does depend upon the phase difference between the paths of the interferometer and the interference shows up when the condition (3.10) is satisfied. It is then clear that the interference for the wave packet case is produced on the analyzer crystal.

4. CONCLUSION

We have explicitly stated the assumptions underlying a realistic approach based on the de Broglie idea of a wave "piloting" the corpuscle, both with plane waves and wave packets. We have also shown that within this scheme it is possible to rederive the quantum mechanical formulae tested in the experiment under discussion. Spectral filtering in neutron interferometry, then, is fully understandable if we assume that a neutron is made up of a corpuscle guided by a wave, both existing objectively in spacetime.

ACKNOWLEDGMENTS

The author is particularly grateful to Professor H. Rauch for many stimulating discussions and for the friendly hospitality during a visit to the Atominstitut der Österreichischen Universitäten.

REFERENCES

1. H. Kaiser, R. Clothier, S. A. Werner, H. Rauch, and H. Wölwitsch, *Phys. Rev.* **A45**, 31 (1992).
2. L. de Broglie, *Non-Linear Wave Mechanics* (Elsevier, Amsterdam, 1960).
3. V.L. Lepore, *Phys. Rev. A*, **28**, 111 (1993).
4. M. Born, *Z. Phys.* **37**, 863 (1926); English translation in *Quantum Theory and Measurement*, J. A. Wheeler and W. H. Zurek, eds. (Princeton University Press, Princeton, 1983), p.52.

HØFFDING AND BOHR: WAVES OR PARTICLES

R.N. Moreira

Departamento de Física
Fac. de Ciências, Universidade de Lisboa
Campo Grande, Ed. C1, 1700 Lisboa, Portugal

After Faye has published his book in 1991 concerning the relationship between Høffding and Bohr, Favrholdt answers once again (in his book of 1992) challenging everyone who intends to defend, following Faye, the influence of Høffding on Bohr to prove any kind of resemblance between their thoughts. Though I consider that Faye has already answered him, I intend with this small work to call the attention to one point that so far, inclusive in Faye's works, has not been sufficiently stressed. The main problem posed to quantum physics, the need to use either a continuous or a discontinuous descriptions of quantum systems, and the way Bohr solved the problem introducing the concept of complementarity between these descriptions, had already been pointed out by Høffding as early as in 1921 in his book *Relation as a Category*, a book that Bohr is likely to have read.

Key words: complementarity, continuity, discontinuity, psychology, philosophy, physics.

1. INTRODUCTION

Since 1966, when Max Jammer in his book *The Conceptual Development of Quantum Mechanics*[1] first addressed the problem of the various philosophical influences which Bohr had been subject to, the problem was no longer left unstudied. In 1970, Holton explored in his article "The Roots of Complementarity"[2] and in 1974, Feuer,[3] in his book *Einstein and the Generation of Science* did the very same thing. The analysis of this problem has continued giving scope to a considerable number of works, the most outstanding ones being those of two Danes, David Favrholdt[4] and Jan Faye.[5]

The position of Favrholdt can be summed up by saying that he defends the arrival of complementarity as a logical and inevitable consequence of the quantum formalism and regarding which, Bohr, in a work without any philosophical orientation, showed his originality and genius by removing the profoundness, and I would dare say, the "definitive" consequence underlying in it. Favrholdt thus adopts a radical internalist vision of the history of science in approaching this topic.

On the other hand, in various works which he has published on this topic, Faye tries to demonstrate that the Danish philosopher Harald Høffding, a friend of Niels Bohr's father, his philosophy professor and personal friend, would have reached a principle of complementarity in psychology, a field in which he was an internationally renowned expert and which Bohr might have used by resorting to physics.

This is at least the stand that he takes in his article of 1979,[6] although in the book that he published in 1991,[7] he goes further stating that Høffding had formed this principle of complementarity in the field of psychology on a general epistemological principle. He considers this principle as linked to the philosophical attitude of Høffding, to which he refers as objective anti-realist, a position that later would be shared by Bohr.

The historical analysis that these two Danish scholars have undertaken has been almost exhaustive. Both scholars, because they share the same mother tongue with Bohr and Høffding, are in a privileged position to carry out such a task.

My analysis of the topic has definite common points with Faye's, regarding the influence of Høffding on Bohr, even though I do not share his same theory concerning the objective anti-realism of Høffding and Bohr. I think that what Faye calls objectivity must be considered as inter-subjectivity, and this with lead us to the conclusion that Bohr's observer should be considered as a psychological one. Faye arrives at such a conclusion stating that the subject-object relation would be the fundamental relation for Høffding and also for Bohr. In my opinion, this relation alone, though fundamental for Høffding, and for almost all philosophers, would not be enough to arrive at the concept of complementarity in how it would be applied to physics. The complementarity between two of the categories considered as fundamental for Høffding, namely continuity and discontinuity, and their relation with cognitive constraints of Man, imposed by the phenomenology of the spirit that, according to Høffding, would be psychology, would in my opinion, reveal itself as being much more strictly applicable to the fundamental problems raised by quantum physics.

2. THE RELATIONSHIP BETWEEN HØFFDING AND BOHR

The philosophical perceptions of Høffding, which influenced Bohr, just as Faye defend, and I agree with him, can be considered as a kind of a synthesis between Kant's, Spinoza's and Kierkegaard's thoughts. This will be enough to understand his way of thinking and consequently Bohr's one.

In the book which Favrholdt has already published in 1992, after attempting to take apart the arguments which Faye had used to backup his thesis, he once again launched a challenge (the quotation marks are mine):

> Those supporters of the "Høffding-myth" who do not intend to give it up, really have only two choices. Either must maintain that Bohr was influenced by Høffding without being aware of it, which presupposes the "false" premise that the similarity between Bohr and Høffding's view was great enough to make it possible to speak of an influence. Or they must maintain that some new evidence will appear in support of their hypothesis.[8]

Regarding the similarity between the philosophical perceptions of Høffding and Bohr I consider that Faye's works are already an answer to Favrholdt. However, I believe that it is possible to bring it out even more, if we concentrate on Høffding's attitude regarding the two categories that he considered to be included in the group of fundamental ones: continuity and discontinuity. Faye mentions it in his book, particularly when he studies the relation between Bohr and Kierkegaard thoughts, but in the remainder of his book he didn't explore this problem as we would hope.

Contrarily to Favrholdt, I consider that the known facts, most of them compiled by Faye in his article of 1988 and in his book of 1991, are enough to show Høffding's influence on Bohr. It will be worthless to mention everything that we can find in the works just mentioned. From all this information, I wish to quote only what Bohr said in the posthumous homage in honour of Høffding in 1931 at the Royal Academy of Sciences and Letters[9] that sums up their relationship. Bohr said that he had (italics mine)

> ... the privilege of having been *in close contact* with Høffding *since* my early youth, because my father was an intimate friend of his and I had, *at all stages of my life*, the possibility of benefiting from his true *scientific and philosophical spirit*, that you all know from Høffding's work, but which took on special meaning through *direct contact*.

This makes clear that they had been in close contact during all Bohr's life, since his early youth until the Høffding's death. After this, it seems very difficult to defend the opposite.

But I think that there is one fact that, until now, has not been sufficiently explored. In a letter to Meyerson dated the 12th of February of 1924, Høffding informs Meyerson that Bohr had congratulated him for the use of the word "relation" instead of "relativity", in the title of his essay written in Danish[10] about the concept of relation. It would be useful to refer to the terms in which this letter was written (italics mine):

> *My friend Niels Bohr congratulated me for having used the word "relation" in the title of my Danish essay.*[11]

This statement of Høffding compels us to admit two things: in the first place, the fact that also at this time there was a close relationship between the two of them, and, in second place, that Bohr was aware of what this book said, the contents of which he had, after all, personally discussed with him. It would be hard to believe that Bohr would have given an opinion on the title of a book without basing himself on what he read in that book.[12] We can still add that in the course of several pages of this book, Høffding expounds on Physics and its concepts. Certainly this would do not leave Bohr indifferent. Thus, we are led to conclude that Bohr was not only interested in the writings of Høffding during the time in which he had been his philosophy teacher but also that, at a later stage, when he had even been involved in the crisis of his first atomic theory.

I think this is something that so far has not been sufficiently looked into. In fact, if we go back to the works that, in one way or another, deal with the relationship between Bohr and Høffding, only two refer to this fact.[13] However, neither of them explore it as we would hope. Faye never mentions it, and Favrholdt, in his most recent book, quotes it for the very first time, though he forgets it immediately afterwards, since he does not come to any conclusion or refers to it again.

Everything points to the fact that the relationship between Bohr and Høffding went on for several years. Related to this, we must not forget to mention a letter from Høffding to Meyerson on the 30th of December 1926[14] (and this date is of particular importance because it was the time when there was a full-blown upheaval of ideas in the field of physics), a letter in which he states (italics mine):

> ...Some weeks ago, (M.) Niels Bohr gave a speech of utmost interest in our Academy. I will try to give you an outline of the most relevant points.
>
> In a previous publication, (M) Niels Bohr (as I demonstrated in my essay on the concept of analogy, translated into German, p. 80) had already stated that the relation between "what the spectrum denotes and what the theory of electrons allows us to understand" is a relation of analogy. In his recent conference on "the theory of atoms and the undulatory movement" he maintains that *we cannot decide whether the electron is an undulatory movement (in this case we can avoid dis-*

> *continuity) or a mere particle (with discontinuity between the particles). Certain equations point us to the former perception, while others to the latter. No single image or word can give an answer to all the equations - In a conversation he had with me after the conference, (M.) Bohr told me that he is more and more convinced of the need to resort to symbolism if we want to express the latest results of Physics.*

This information that is given to us by Høffding is of great importance as, apart from clearly showing that there were "behind the scenes" discussions between them at the Academy, also shows that Bohr and Høffding discussed the problems confronting physics during this crucial period of its history, in this century (Bohr's conference took place on the 17th of December 1926). A time when there was a sharp conflict between the concepts of continuity and discontinuity that, as we shall see later, had always been studied by Høffding, which even led him to introduce the concept of complementarity in between the categories of continuity and discontinuity, expressed exactly in the book mentioned before.

In 1928, the relationship between both of them became more intense. In that same year, Bohr becomes a constant visitor at the Høffding's home at a time when Høffding had fallen ill. This fact is reported to us in letters written by Høffding to Meyerson. In one of such letters dated the 30th of March 1928,[15] he shows once more that the problem of the complementarity between these two concepts have been discussed by them when he says (italics mine):

> Last night (M.) Bohr visited me and told me that his latest investigations on the concept of the electron would soon be published in *Nature* (the English magazine). In this article he seeks to overcome the difficulty created by the fact that the electron should be considered as a particle which is found in a specific place, as well as a source of energy. *Behold, an old problem which is still found in the limits of the science of nature.*

In yet another letter,[16] Høffding refers to these visits of Bohr in a very interesting way (italics mine):

> What always gives me great pleasure are the visits of (M.) Niels Bohr. He tells me of his work and *we go on to discuss its philosophical consequences.*

It is important to point out the following: during this very important period Bohr discussed this topic with Høffding. Bohr himself acknowledges the important role played by Høffding in his attempts to solve the problems that had come about with the study of the atom. This revelation appeared in a newspaper article written by Bohr in 1928, on the occasion of the 85th birthday of Høffding[17] (italics mine):

> In the study of the phenomena of nature we often find ourselves confronted with problems which require a reassessment of the concepts subjacent to the understanding of our observations. Whenever there has been a crisis caused by external circumstances, stemming from an apparent conflict between old and new experiences which threatens to create an obstacle to the attempts of human thought towards delving into the secrets of nature, that crisis *has had an incalculable value for scientists who were able to get support and find starting points towards new advances, in the philosophers' attempts to clarify the foundations and limits of human intellectual activity.* Without feeling disturbed by external conditions, the philosophers, only due to a need for harmony of thought, have broadened our knowledge and *have created a world-wide attitude toward emerging difficulties and a broad understanding of relativity and complementarity which affects all human concepts... We are all in debt to Professor Høffding because, through his personal but very objective description of the conquests of philosophy in the struggle to determine the conditions of our cognition and emotions, he thus contributed to a greater understanding of the basis of our work, while wholeheartedly supporting us in our endeavours.*

Favrholdt belittles this confession basing himself on what he calls his own generality. Bohr, in fact, speaks about the philosophers in general, but also makes specific mention to Høffding; it is necessary to emphasise that Bohr was never one to dedicate himself to in-depth study of the thoughts of various philosophers or of any given philosophical topic. This is recognised by Favrholdt himself. Indeed, if Bohr knew anything about philosophy and about *"the attempts of philosophers to clarify the foundations and limits of human intellectual activity"* that, in turn, would have *"created a general attitude towards emerging difficulties and a broad understanding of relativity and complementarity in all human concepts"*, he fundamentally knew it through Høffding. Thus, when Bohr speaks about the *"personal"* presentation by Høffding on the struggle of philosophers to understand *"the conditions of our cognition and emotions"*, we can genuinely consider that he was truly familiar with this *"personal"* presentation and that indeed it had a decisive influence upon him.

Høffding's reaction to this speech can be observed in a letter he wrote to Meyerson:[18]

> Among the praises published by the press on that day, an article of (M.) Niels Bohr caused me great happiness. (M.) Bohr declared that he discovered ideas in my books that helped scientists in their "understanding" of his work and thus, were of great help to them. It is a great

> satisfaction for me to frequently be aware of my lack of specialised preparation in natural science.

I think that Høffding fully understood that Bohr, even speaking in general terms, was in fact also referring to him. In virtue of the fact that this article was published in a year when both men had maintained assiduous contact, Høffding was probably more informed regarding what Bohr meant than Favrholdt is.

In another written reply to a request for clarification from Kalle Sorainen, a Finnish scholar who followed Høffding's philosophy, Bohr, after once again recalling all the contacts he had with Høffding since his youth, states that (italics mine)

> ...As can be seen from Høffding's latest work I have in the years before his death many deep and searching discussions with Høffding on recent developments in physics...As can be seen from Høffding's latest work I have in the years before his death many deep and searching discussions with Høffding on recent developments in physics...but what I think is particularly admirable is his lack of bias and the very serious effort he made to see the new developments in relation to the *general epistemological position* which he himself had formed as a result of extensive studies of psychological problems over a period of many years... [19]

As we can see from this letter, Bohr was aware of "the general epistemological position" of Høffding that reinforces my belief that it was essentially that knowledge, which carried with it unavoidable consequences to all fields of knowledge, the factor that made him the founder of the interpretation of quantum formalism, adopting as its foundation, the Principle of Complementarity.

In fact, Bohr understood Høffding's way of thinking rather well. His posthumous homage speech in honour of Høffding, which he delivered in the Royal Academy of Sciences is proof of this (italics mine):

> That which we see (in Høffding) is a sober and yet enthusiastic report of, in the best sense of the word, a scientific attitude regarding the psychological stages. A prominent aspect is the struggle to keep the balance between analysis and synthesis where it is never forgotten that *the whole is made up of individual parts which, in turn, always appear in the perspective of the whole.* This peculiar objectivity of Høffding's report of psychology was certainly much more important to many of those present at his lectures and perhaps even more so for the readers of his book than any one of us is able to explain...

Bohr is referring to Høffding's book *Outline of a psychology based on experience* and this is one of the sentences that forces Faye to defend that Bohr made use of the perceptions of Høffding in psychology and applied them to physics. However, with these words Bohr could possibly be referring to other works of Høffding such as *The Human Thought*, *Totality as a Category* or *Relation as a Category*, where Høffding deals with that problem, that is, that "the whole is made up of individual parts which, in turn, always appear in the perspective of the whole."

The sentence that we underlined above is an important proof of the deep knowledge that Bohr had of Høffding's way of thinking. I agree with Faye when he says that the concept of the whole or totality plays in Høffding a very important role,[20] and Bohr shows here that he was perfect aware of it. Thus, this is another statement that shows that Bohr knew what Høffding defends on these books written between 1910 and 1921.

3. PSYCHOLOGY, PHILOSOPHY AND SCIENCE

Høffding was a prolific author. His long life was filled with the writing of an enormous amount of articles and books about the most various subjects. Starting with theology, going on to Ethics, Psychology, the History of philosophical and scientific thought, analysis of various philosopher's thoughts, even reaching the theory of knowledge, which we can consider the peak of his work; all of the above caught his attention.

If indeed psychology was the subject that most contributed to his being recognised relatively early on outside his native country, we cannot but emphasise that his monumental work *History of Modern Philosophy* also became internationally renowned, to the point that even as late as the 1950's this work was still being recommended to Parisian students.[21] In that work, apart from the analysis of the development of European thought after the scientific revolution of the 17th century, Høffding shows to have reflected intensely on the development of scientific thought and how it could be connected to philosophy, per se.

However, it was in 1910, upon his writing of *The Human Thought*, that Høffding showed that he had begun to reveal in a more consistent way his own philosophical perceptions.

From Høffding's correspondence with Meyerson we know that, after admitting the influence that Kierkegaard had over him at the beginning of his career, he decided to place a great importance on the theory of scientific knowledge, emphasising the theory of categories. The mentioned correspondence leads us to the conclusion that the works that most clearly reveal his thoughts in a more consistent manner would be *The Human Thought*,[22] *Totality as a Category*,[23] *Relation as a Category*,[24] *The Concept of Analogy*[25] and *Epistemology and Life Perceptions*,[26] the last four considered as an addendum to the first. This correspondence with Meyerson, published in 1939, is another extremely useful document to explain the exact thought of Høffding because it corresponds to a very important period falling between 1918 and 1931. In this correspondence Høffding clarifies many issues.

This is not the place to make a detailed analysis of Høffding's way of thinking, therefore here I will only give some notes in answer to the challenge launched by Favrholdt in his book published this year.

Favrholdt defends that "...Høffding throughout his life thought that continuity and causality were essential clues."[27] In fact, following Høffding, continuity was part of the fundamental categories. However, the same happened with discontinuity as well. In that which concerns causality, Faye has already shown that Høffding did not defend a principle of causality but rather a concept of causality. The difference resides in the fact that the former exists in an ontological level while the latter exists only at an epistemological one. This means that, for Høffding, causality would only have a heuristic value, or rather, it would only be a concept that man would use to try to understand the world, and not something that would characterise such a world. Evidently, this is nothing new. Kant categories means precisely this. However, Høffding stresses Kant's approach to the problem in the *Prolegomena to Any Future Metaphysic*, were we can reach categories through psychology. However, Høffding, categories can "born" and "died". On this point even Favrholdt seems to agree when he states that for "... Høffding, a category also seems to be something like a temporary guideline or a heuristically good idea."[28] In this point he was not kantian. Already in his work *Outline of a psychology based on experience*[29] when studying the problem of will, Høffding states that (italics mine):

> Here, as in general, *the causal law is nothing but a hypothesis or a postulate...*[30]

Man would try to circumscribe phenomena within causality as a way to understand them. Meanwhile, even if up until then science would have been successful as well as its implicit concepts, as that which we have just mentioned,[31] that would not mean that this need could always be met. As Høffding states in another letter to Meyerson: "As Leibniz said: Success does not demonstrate the truth behind an hypothesis."[32] The concept of causality was not free from this problem.

This position of Høffding is based on his conviction, just referred, that psychology would play a primordial role in the theory of scientific knowledge. Let us hear what he has to say (italics mine):

> The independence of psychology in view of the theory of knowledge results from the fact that *scientific thought is not able to separate itself from the general laws of the life of consciousness...* Both discovery and demonstration are a result of the psychic work. *In terms of psychology they must be possible... Thus psychology is an introduction, a type of Phenomenology of the Spirit.*[33]

Psychological laws could not therefore be violated in scientific activity and further on. Continuing, he states (italics mine):

> Meyerson... through the analysis of the conjectures of modern science (in the French translation "Suppositions") returns to the opposition between *identity* and *succession*. Causality, in a strict sense, involves from the onset a reciprocal equivalence of this type, the possibility of a substitution and therefore, identity. However, according to him, legality can exist without reciprocal equivalence; here substitution is not possible and this is the reason why Meyerson does not want, in this case, to resort to the term causality. The principle of Carnot (or entropy) reveals the importance of this difference. Throughout the history of science Meyerson discovers a continuous and invincible tendency of the human spirit to form the concept of anything that subsists in spite of all succession. But, considering that time cannot be eliminated, which proves that the principle of Carnot which sets a limit to the substitution of the stages of nature, we remain before the opposition between *identity* and *succession* and that to which it is connected, between *quantity* and *quality* which transposes us towards the relation between *thought* and *sensation* which is the very aim of the study of psychology. In this way and once again, we go back through the *categories* from the *(scientific) principles* to the *psychic functions*.[34]

We see that Høffding uses Meyerson's way of thinking to emphasise the primacy of psychology when face to face with the theory of knowledge and, therefore, inevitably to that of science. However, he does not only quote Meyerson (italics mine):

> We also find in the works of Henri Poincaré the triplicity of what we have just spoken of. His analysis of the theories of modern mathematics and physics transposed him to the opposition between *continuity* and *discontinuity* and to its continual struggle towards the development of science. The aim of psychology is, according to him, to study this opposition because it relates to the opposition which is ever being reborn between intuition and thought, being this the reason why Henri Poincaré... contests that there can be a theory of knowledge independent from all psychology: it would be as impossible as a science without scientists.[35]

Thus, for Høffding, the laws or the constraints discovered in psychology would be insurmountable, and all intellectual activity that inevitably would result from it, would have to be subjected to it. All of it would have to be "psychologically possible".

In the book *Relation as a Category*, a title which Bohr vividly approved, we find the most pronounced statement of complementarity by Høffding, and in close connection with the problems raised by quantum mechanics. Once again, referring to the *continuity-discontinuity dualism* he states (italics mine):

> In physics, the recent doctrine of Quanta according to which energy is not expelled in a continuous pattern but rather, in leaps,[36] is based on mathematical laws in which such liberation by leaps appears to be necessary. In all these cases,[37] we search for theoretical points of view that may help us to understand discontinuity. Given the fact that the leaps can be arranged in series, the work that consists in finding out a new continuity has already began *even if this continuity is more formal than real*.
>
> These problems have already been exposed by ancient Greek philosophy. The "One" by Parmenides, simultaneously meaning continuity and identity, meant that as far as he could see, there was no difference between the two things. On the contrary, Heraclitus and Democritus insisted on discontinuity. The struggle went on until modern times and it is precisely this struggle that motivated platonism and its opponents. Kant considered this opposition as a rational necessity. His doctrine of antinomies attempts to proof that there is an absolute limit to reason. The great founder of critical philosophy *did not see that discontinuity cannot ever be the last word of thought, unless it relates to a problem*; thus, his "antithesis" exploring that point of view are correct in opposing his "thesis" which defend that there are indeed absolute discontinuities.[38]

This last sentence must not be misinterpreted. Høffding is speaking on the level of thought and, therefore, at the level of the language that expresses it. Høffding is talking at an epistemological level and not at an ontological one, which, as we have seen before, seemed to him to be inappropriate. And Høffding goes on to say (italics mine):

> *Continuity and discontinuity are correlatives which supply one another (why not even say complement one another as we will see just immediately? - RNM). They represent different points of view and different operations*; The history of science show that in these categories both one and the other assume primacy, but in such a way that the struggle between them is permanently rekindled. No one shed a more clarifying light upon them than Henri Poincaré when he says: "This

> struggle will go on as long as there is science, as long as mankind thinks, because it is due to the result of *the two irreconcilable needs of the human spirit from which that same spirit cannot strip itself unless it ceases to exist,* the need *to understand* and we are not able to understand but that which is finite, as well as the need *to see* and we are not able to see except an extension which is infinite."[39]

Even if Høffding makes some amends in some of Poincaré's statements but without offending that which is essential, what is truly important to emphasise here is that he accepts the existence of an incompatibility between the categories of continuity and discontinuity that have their origin in the constraints of human thought, which psychology brings out, (let us not forget the subordination of the categories of the already mentioned psychological functions). However, although incompatible, they are also inevitable. When Høffding says that discontinuity can never be the last word this does not mean that the discovery of the quantum of action would be a rebuke of the Høffding's way of thinking. As I have already pointed out, Høffding is talking at an epistemological level. In fact, at this level, discontinuity does not represent the last word. In the interpretation of quantum formalism introduced by Bohr, both categories mentioned above are necessary. We cannot rescind either from the perception of continuity or of discontinuity. The complementarity principle represents precisely that. It would not be possible to consider either a Hegelian way of overcoming this contradiction (as defended by both Broglie and Einstein) or to opt for only one of the two (the particle theory initially defended by Heisenberg) or another one (the undulatory theory upheld by Schrödinger). In his book *Totality as a Category*, Høffding had already stated:

> The absolute multiplicity [discontinuity - RNM] is as irrational as absolute continuity. Both continuity and discontinuity bring about problems, the former being the one of finding more connections and the latter more differences.[40]

The complementarity principle that Bohr upholds may well have been found in Høffding's writings, as has been pointed out by Faye. Let us not forget that Bohr admitted, in accordance with the interpretation that Høffding gave to his words,[41] that "ideas which helped scientists in the *understanding* of their work, thus having truly helped them"[42] had been found in his books.

Clearly, in itself, this is not enough for Bohr to reach his interpretation of the quantum formalism, but the philosophical foundation is clearly found here. This is my answer to the first part of Favrholdt's challenge in which he refutes any similarity between Bohr's and Høffding's thoughts.

From a historical point of view, which is the second part of Favrholdt's challenge, it would be good to recall that this passage is probably found in one of Høffding's books which Bohr is likely to have read. As mentioned before, Bohr congratulates Høffding on the choice of the title This would only be acceptable if Bohr had read and then also discussed its contents with him Besides that, we cannot but recall that Bohr spoke at a conference in the Danish Royal Academy of Sciences and Letters on the 17th of December 1926, in which he approached the problem of continuity and discontinuity in quantum mechanics, having stated that (italics mine)

> *we cannot decide whether the electron is an undulatory movement (in this case we could avoid discontinuity) or a particle (with discontinuity among the particles) Certain equations take us to the former perception, while others take us to the latter No image no word can answer all the equations* [43]

And Høffding goes on to say

> In a conversation he had with me after the conference (M) Bohr told me that he was more and more convinced of the necessity of symbolisation if we want to display the latest results in Physics [44]

Knowing Høffding's opinion about the matter, there were probably no major conflicts of ideas between them By this I do not mean that this conversation by itself would have been essential for Bohr in finding his interpretation of quantum formalism, but nobody can deny that it would, in any possible way, have gone against Bohr's convictions It will not be over stressing to say that the previous sentence of Poincaré, so praised by Høffding, could have been challenged, given that, according to him, nobody more than Poincaré himself had given a more enlightening explanation on the relation between continuity and discontinuity Of course we could say that complementarity was something common to several spirits at that period, but no one can deny that it was the touch stone of Høffding's thought

Høffding's perceptions do not clash with Bohr's profound perceptions Thus, it would neither have been only the reading of this book nor this conversation that could have helped Bohr find an interpretation (his own) of quantum formalism (this only proves he enormous similarity between their views) A possible reason for this is that they matched his oldest convictions Heisenberg reports a fact that sheds light on this On a boat trip in which he took part together with Bohr and some of Bohr's friends he [Bohr] "was full of the new interpretation of quantum theory, and as the boat took us full sail southwards, in sunshine, there was plenty of time to tell us of this scientific event and to reflect philosophically on the nature of atomic theory Bohr began by talking *of the difficulties of language*, of *the limitations of all our means of expressing ourselves*, which one had to take into account from the very beginning if one wants to practice science, *and he explained how satisfying it was that this limitation had already been expressed in the foundation of atomic theory in a mathematically lucid way.* Finally, one of his friends remarked drily, "But, Niels, this is not really new, you said exactly the same thing, ten years ago".[45] In an interview given by Heisenberg to Kuhn the time span mentioned by Bohr's friend changed from ten to twenty years. If the latter is correct, then we are actually in the period when Bohr admitted having had close contact with Høffding.

ACKNOWLEDGEMENTS

I would like to thank Professor F. Selleri for giving me the opportunity to discuss with him this talk. However this does not mean that he would agree with the ideas expressed in it.

REFERENCES

Bohr, N. [1985], *Collected Works*, L. Rosenfeld ed. (North-Holland, Amsterdam, 1972 onwards) (presently unfinished).

Bohr, N. [1928], "Ved Harald Høffding's 85 Aars-Dag" (On the occasion of Harald Høffding's 85th Birthday), Berlingske Tidende (March 10, 1928).

Bohr, N. [1931], "Mindeord over Harald Hoffding" (In memoir of Harald Høffding) Oversigt over Det kgl. Danske Videnskabernes Selskabes förhandlinger 1931-1932, pp. 131-136.

Bohr, N. [1985], *Atomic Physics and Human Knowledge* (Wiley, New York, 1985).

Favrholdt, D. [1976], "Niels Bohr and Danish philosophy," *Danish Yearbook of Philosophy*, Vol. 13, 1976, Munksgaard, Copenhagen.

Favrholdt, D. [1979], "On Høffding and Bohr. A reply to Jan Faye.," *Danish Yearbook of Philosophy* 16 (1979), pp. 73-77.

Favrholdt, D. [1985], "The cultural background of the young Niels Bohr", *Rivista di Storia della Scienza*, Vol. 2, No. 3, Nov. 1985.

Favrholdt, D. [1991], "Remarks on the Bohr-Høffding relationship", *Stud. Hist. Phil. Sci.*, Vol. 22, No. 3, pp. 399-414, 1991.

Favrholdt, D. [1992], *Niels Bohr's philosophical background* (Munksgaard, Copenhagen 1992).

Faye, J. [1979], "The influence of Harald Høffding's philosophy on Niels Bohr's interpretation of quantum mechanics", *Danish Yearbook of Philosophy* 16, (1979), pp. 37-72.

Faye, J. [1988], "The Bohr-Høffding relationship reconsidered", *Stud. Hist. Phil. Sci.*, Vol. 19, No. 3, pp. 321-346, 1988.

Faye, J. [1991], "Niels Bohr: His Heritage and Legacy. An Anti-Realist View of Quantum Mechanics" (Kluwer Academic, Dordrecht, 1991).

Feuer, L. S. [1982], *Einstein and the generation of science* (Transaction Publishers, New Brunswick, 1989).
Folse, H. J. [1985], *The Philosophy of Niels Bohr* (North-Holland, Amsterdam, 1985).
Høffding, H. [1908], *Histoire de la philosophie moderne* (Alcan, Paris, 1908).
Høffding, H. [1909], *Esquisse d'une psychologie fondée sur l'expérience* (Alcan, Paris, 1909).
Høffding, H. [1910], *La pensée humaine* (Alcan, Paris, 1910).
Høffding, H. [1925], *La relativité philosophique* (Alcan, Paris, 1925). This French translation contains two books of Høffding in danish: *Totalitatet som kategori* (Totality as a category), (Det kgl. D. Vid. S. Skrifter. 6. række, hist. filosofisk afd. 2., Copenhagen 1917), and *Relation som kategori* (Relation as a category) (Det kgl. D. Vid. S. Filosofisk Meddelelser, I, 3, Copenhagen, 1921).
Høffding, H. [1927], *Les conceptions de la vie* (Alcan, Paris, 1927).
Høffding, H. [1931], *Le concept d'analogie* (Vrin, Paris, 1931).
Høffding, H. e Meyerson, É. [1939], *Correspondence entre Harald Høffding et Emile Meyerson*, Frithiof Brandt, Hans Høffding and Jean Adigard des Gautries, eds. (Munksgaard, Copenhagen, 1939).
Høffding, H. [1962], *Os problemas da filosofia (Problems of philosophy)* (Lux, Lisbon, 1962).
Holton, G. [1970], "The roots of complementarity," *Dædalus* 99, Fall 1970.
Jammer, M. [1966], *The conceptual development of quantum mechanics* (McGraw-Hill, New York, 1966).
Rozental, S. (ed.) [1968], *Niels Bohr. His Life and Work as Seen by his Friends and Colleagues* (North-Holland, Amsterdam, 1968). First Danish edition of 1964.

NOTES

1 Jammer [1966].
2 Holton [1970].
3 Feuer [1974].
4 Favrholdt [1976], [1979], [1985], [1991], [1992].
5 Faye [1979], [1988], [1991].
6 See note 5.
7 See note 5.
8 Favrholdt [1992], p. 118.
9 Bohr [1931].
10 Høffding [1925]. Published in Danish in 1921.
11 Høffding [1939], p. 70.
12 Bohr would have read and discussed this book with Høffding betwween 1921 and February of 1924, when Høffding tells it to Meyerson, but it seems to me more likely that it would happen closer to the last date.

[13] Up to recently, only Feuer had referred to it (Feuer [1982]). However, Favrholdt also makes reference to it in the book he published this year.
[14] Høffding [1939], p. 131.
[15] Hoffding (1939), p. 150.
[16] Høffding (1939), p. 169.
[17] Bohr [1928].
[18] Letter of the 30th of March 1928, Hoffding [1939]. pp. 149-150.
[19] Extracted from Faye [1991]. p. 71.
[20] Faye [1991], p. 90.
[21] Information given to me by Prof. Andrade e Silva.
[22] Published in 1910.
[23] Published in 1917.
[24] Published in 1921.
[25] Published in 1923.
[26] Published in 1925.
[27] Favrholdt (1992), p. 107
[28] Favrholdt [1992], p. 81.
[29] Høffding [1903]. The first Danish edition dates back to 1882. I consulted the French translation *Esquisse d' une psychologie fondée sur l' expérience*, (Alcan, Paris, 1909).
[30] Op. cit. p. 434.
[31] Causality belonged to real categories, which are the special forms that fundamental categories took when applied by natural sciences.
[32] Høffding [1939], p. 9. Letter dated the 17th of July 1921.
[33] Høffding [1924], pp. 23-24. This French translation includes two separate works in the Danish editions: *Totality as a Category* (1917) and *Relation as a Category* (1921). The sentences that I quoted belong to his book *Totality as a Category*.
[34] Op. cit., pp. 25-26.
[35] Op. cit. p. 27.
[36] Concerning this point, Høffding quotes Poincaré, *L'hypothése des quanta, Dernières pensées, Chap. VI.*
[37] Høffding had just spoken about the discontinuity-continuity dualism in psychology ("The psychological discontinuity does not constitutes, generally speaking, nothing but a problem; it is never a solution") and in biology, where the same thing happens with mutations.
[38] Høffding [1924]. pp. 197-198. This quotation is already part of this work that corresponds to *Relativity as a Category*, separately published in Danish.
[39] Henri Poincaré, "Les conceptions nouvelles de la matière," in *Le Matérialisme actuel*, p. 67. Quoted from Høffding's *La relativité philosophique*, pp. 198-199.
[40] Op. cit. p. 57. Is part of the book *Totality as a Category*.
[41] As I have already stated, the interpretation of Høffding seems to me to be much more trustworthy than that of Favrholdt's.
[42] Høffding [1939], p. 149.
[43] Høffding [1939], p. 131.
[44] Høffding [1939], p. 131. Letter dated the 30th of December 1926.
[45] Heisenberg, in S. Rozental, ed., Niels Bohr, *His Life and Work as Seen by his Friends and Colleagues*, (North-Holland, Amsterdam, 1968), pp. 106-107. The first English edition is of the previous year, and the first Danish edition dates of 1964.

PHOTONIC TUNNELING EXPERIMENTS: SUPERLUMINAL TUNNELING

G. Nimtz, A. Enders and H. Spieker

II. Physikalisches Institut, Universität zu Köln
Zülpicher Str. 77, D-50937 Köln, Germany

We report on microwave experiments with evanescent modes. The propagation of these "tunneling" modes appear to be instantaneous inside the barrier region independent of barrier length. Data were obtained from various potential barrier configurations and with two basically different measuring procedures. All the results point to a superluminal propagation of evanescent modes through opaque barriers. We question the analogy between evanescent mode propagation and quantum mechanical tunneling.

Key words: evanescent modes, tunneling time, superluminal velocity, quantum theory.

1. INTRODUCTION

The tunneling time of wave packets has been studied for half a century in innumerable theoretical papers [1,2]. In 1962 Hartman [3] and more recently Low and Mende [4] have pointed out that according to quantum theory tunneling should take place in zero time. Tunneling time experiments with electrons are rather difficult to perform and even more difficult to interprete [5], on the other hand it was demonstrated that the equivalent tunneling time of classical wave packets can be measured and interpreted easier with evanescent modes [6-8]. Evanescent mode propagation in a waveguide is even equal to the one-dimensional particle tunneling problem even formally [10,11].

The propagation time of evanescent modes have not been studied much before, neither theoretically nor experimentally. Feynman mentioned that Newton has introduced a kind of medium accompanying his light particles, which informs the light particle instantaneously about the thickness of the photonic barrier due to partial reflection from two surfaces [12].

We shall introduce microwave experiments which were performed recently by Enders and Nimtz [6-8,11]. The experiments have shown that the propagation time through opaque barriers becomes independent of barrier length and that the apparent delay time is essentially caused by the phase change at the barrier boundaries. In addition we present experimental data of a double-barrier configuration which have confirmed the instantaneous propagation in any evanescent region. The double-barrier experiment has revealed that at non-resonant tunneling conditions the delay time of the transmitted evanescent narrow mode packet is independent of the distance between the two barriers. The instantaneous tunneling with microwaves was quite recently reproduced in an analogous optical experiment [9].

All the microwave experiments were carried out at frequencies between 5 and 10 GHz (vaccum wavelengths between 6 and 3 cm) in a waveguide which is a rectangular metal tube. The eigenvalues for the lowest mode of such a waveguide with cross section a·b, where a<b holds, is given by

$$k^2 = (2\pi/c)^2(\nu^2 - \nu_c^2),$$

with the cut-off frequency $\nu_c = c/2b$ and c the vacuum velocity of light. For $\nu < \nu_c$ the wave number k is imaginary and we are dealing with an evanescent mode, which corresponds to the quantum mechanical tunneling solution [10,11].

2. EXPERIMENTAL RESULTS

The microwave tunneling set-up is sketched on top of Fig. 1. The narrow part in the center of the waveguide is operated at frequencies below its cut-off value in the evanescent regime. The corresponding frequency (energy) diagram is shown in the same figure, where we illustrate also the two different methods of time measurement, the stationary and the amplitude modulated procedure. The stationary measurement, i.e. the experiment is carried out in the frequency domain, delivers the propagation time of a distinct wave packet via Fourier transform [6]. The amplitude modulated signal allows a direct time measurement of the signal front [7]. Both experimental procedures resulted in the same tunneling time. (The measured time is essentially asymptotic in character, since it is derived as an asymptotic characteristic for completed scattering events involving wave packets

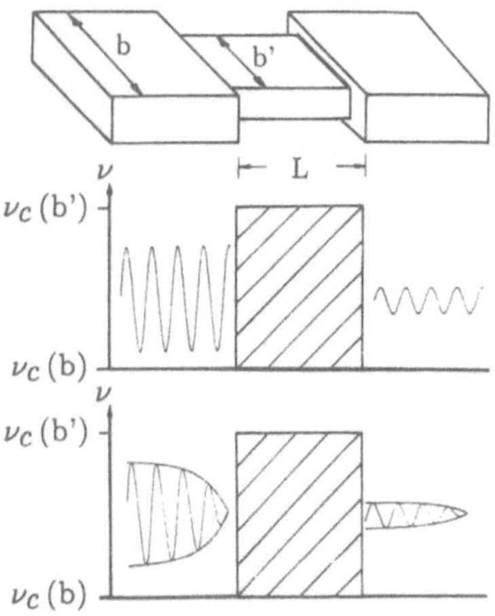

Figure 1: Experimental set-up with transitions from two larger to one smaller waveguide (top), illustrations of stationary measurement (middle) [6], and of the amplitude modulated measurement (bottom) [8].

narrow in k-space [1,7]. The data correspond in this sense to Hartman's calculated phase time [3]).

In Fig. 2 we present amplitude modulated signals with two slightly different carrier frequencies in order to illustrate the dispersion effects of the tunneling process. In this time domain experiment one signal was sent through a classical step attenuator (power attenuation by 40dB, its delay time uncertainty is smaller than 25 ps independent of attenuation) whereas the second wave packet tunneled through an undersized waveguide (ν_c=8 GHz) of 60 mm length. The length corresponds to a vacuum travelling time of 200 ps. Such a delay time was not observed, so that the tunneling appeared to be instantaneous.

The experimental tunneling time is displayed in Fig. 3 as a function of barrier length. For short barriers $a \cdot \epsilon < 1$, where a is the barrier length and ϵ is the wave number equivalent to the barrier height of the evanescent region with $\epsilon = \pi(1/b'-1/b)$,the barrier transmission time or tunneling time is longer than the equal time. The latter time corresponds to the same distance in vacuum.

However, for $a \cdot \epsilon > 1$ (the opaque regime) the tunneling time is constant and independent of barrier length. With increasing thickness the constant tunneling time, which is determined by interference effects at the barrier entrance, yields a superluminal barrier transition velocity eventually. Obviously the experimental data of of evanescent electromagnetic modes (dots) do agree with the quantum mechanical calculation [11].

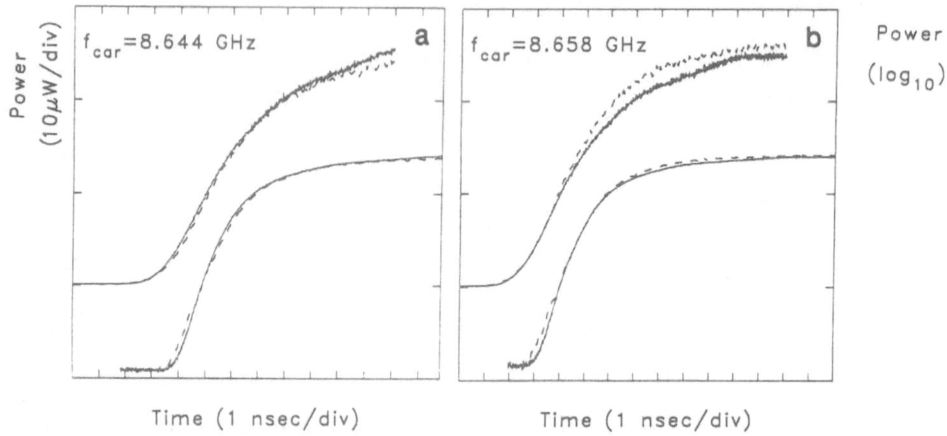

Figure 2: Power envelope of carrier frequency a) 8.644 GHz and b) 8.658 GHz for the rising edge of the reference pulse (zero length of the evanescent waveguide section, 40 dB attenuation of step attenuator, solid line) and the pulse behind the evanescent region (length 60 mm, 0 dB attenuation of step attenuator, dashed line); a) 800 and b) 1000 equidistant points were taken for each trace [8].

In the following experiment we have studied the tunneling time of a double barrier structure which is sketched in Fig. 4. The experimental arrangement was composed of two barriers performed by two undersized waveguides as described in [7] with interbarrier distances either L = 0 and L = 57 mm. At the finite interbarrier distance two resonant transmission frequencies at 7.6 and 6.9 GHz occured as shown in Fig. 5. In this experimental arrangement we have determined the tunneling time of a narrow wave packet (width 1%) at the non-resonant frequency 7.3 GHz. We have found the propagation time of the signal front and of its maximum to be equal for both interbarrier lengths L = 0 and L = 57 mm. Accordingly the tunneling time of the non-resonant mode is the same for the two interbarrier lengths. At the choosen carrier frequency the tunneld signal had nearly the same transmission loss for both barrier configurations. In 57 mm free space the propagation time is 190 ps, in the waveguide which is operated above cut-off we have measured 400 ps as shown in Fig. 6, the value is in agreement with the group velocity in the waveguide. The double barrier experiment is confirming the previous results on tunneling time with single barriers, subtracting the interference effects from the incident and reflected waves in general the tunneling of evanescent modes appears to do so in zero time.

This very double barrier experiment inspired the following Gedanken-experiment. We perform the same double barrier experiment but with an interbarrier distance as long as 1 km. There are much more resonant transitions lines having a very narrow frequency distance from each other. Reproducing the former experiment of non-resonant tunneling at a transition minimum with a mode packet, this has to be very narrow, with only

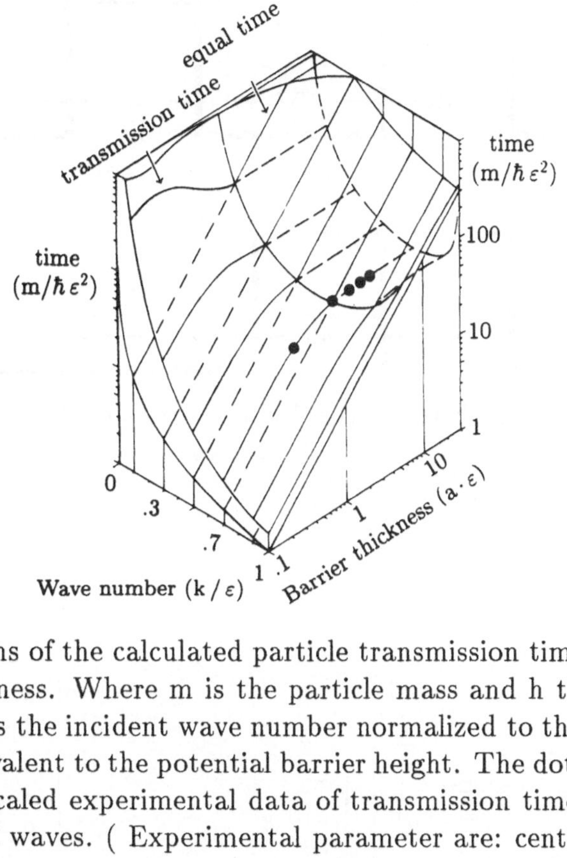

Figure 3: Graphs of the calculated particle transmission time as a function of barrier thickness. Where m is the particle mass and h the Planck constant, and k/ϵ is the incident wave number normalized to the wave number ϵ, which is equivalent to the potential barrier height. The dots represent the appropriately scaled experimental data of transmission time of evanescent electromagnetic waves. (Experimental parameter are: center frequency of the Gaussian-like wave packet ν = 8.7 GHz, ν_{c1} = 6.56 GHz, ν_{c2} = 9.49 GHz, a = 10, 40, 60, 80, 100 mm. For more details see Refs.[6,11].)

a frequency width of 10^{-6} times the carrier frequency in order to avoid significant dispersion effects. For a microwave carrier frequency of 10 GHz we can only transmit a signal of 10 kHz. This frequency width would be enough to modulate and to transmit Mozart's Forty for 1 km without a time delay compared with the same signal sent through the double barrier with zero interbarrier distance. Is that compatible with special relativity ?

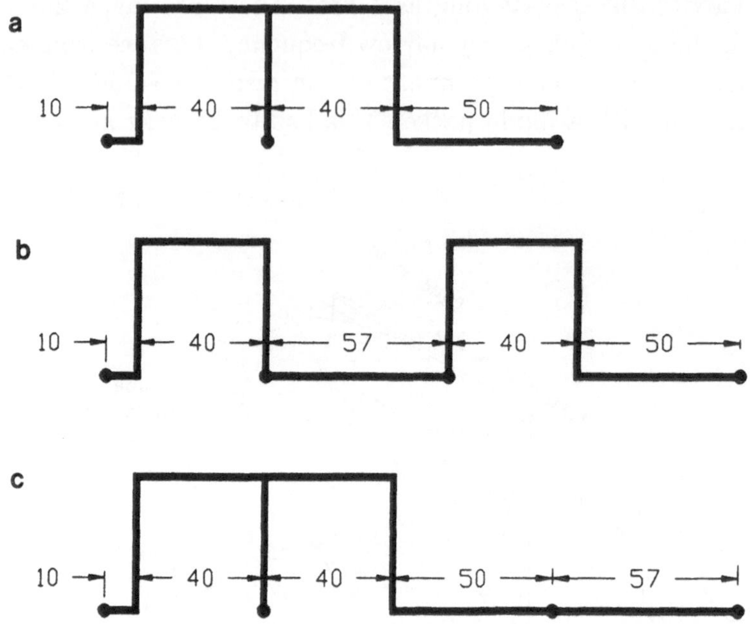

Figure 4: Sketch of the double-barrier configurations.

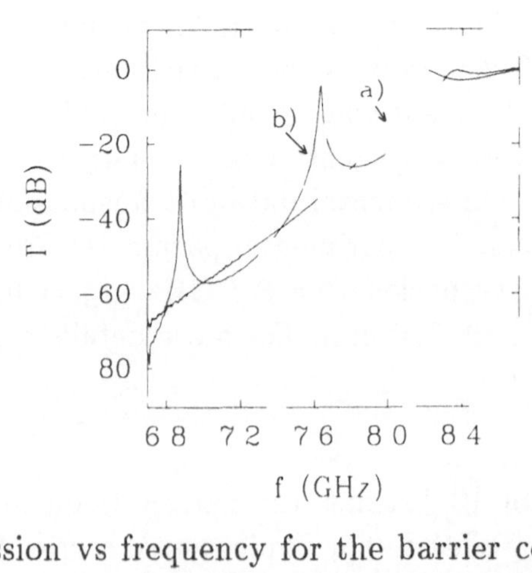

Figure 5: Transmission vs frequency for the barrier configurations a) and b) of Fig. 4.

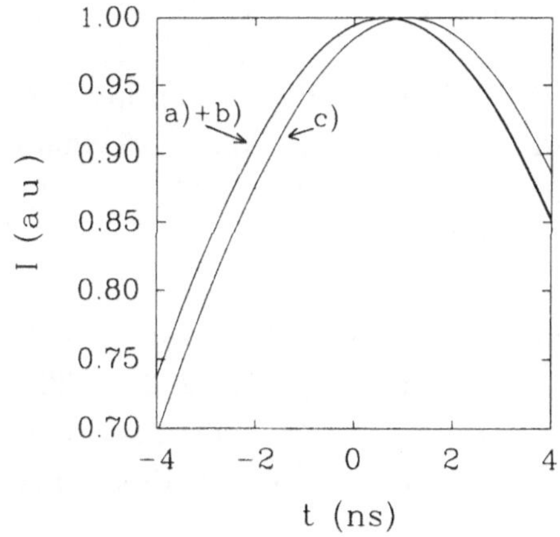

Figure 6: Transmitted intensity of a Gaussian-like wave packet vs time for the three barrier configurations of Fig. 4. Note that a) and b) appear at the same time. The value t = 0 corresponds to the time the center of the wave packet enters the barrier structure.

3. SUMMING UP

The classical Helmholtz equation yields field solutions for waveguides which are mathematically equal to the probability solutions of the Schrödinger equation [10,11]. Our classical tunneling experiments with electromagnetic evanescent modes have revealed that the tunneling proceeds instantaneously inside the barrier. In the case of opaque barriers this results in a superluminal group velocity of wave packets. We expect that this behaviour is representative fore every tunneling process of classical fields as well as of quantum mechanical wave packets - in agreement with some theoretical studies [2-4,11].

ACKNOWLEDGMENTS

We gratefully acknowledge discussions with P.Busch, F.Hehl, E.Mielke, P.Mittelstaedt, and D.Kreimer and the generous technical support by H.Aichmann and W.Strasser from Hewlett-Packard/Germany.

REFERENCES

[1] E.H.Hauge and J.A.Stovneng, *Rev.Mod.Phys.* **61**, 917 (1989).
[2] V.S.Olkhovsky and E.Recami, *Phys.Rep.* **214**, 339 (1992).
[3] Th.E.Hartman, *J.Appl.Phys.* **33**, 3427 (1962).

[4] F.E.Low and P.F.Mende, *Ann.Phys. (New York)* **210**, 380 (1991).
[5] D.Esteve et al., *Physica Scripta* **T29**, 121 (1989).
[6] A.Enders and G.Nimtz, *J.Phys.I (France)* **2**, 1693 (1992).
[7] A.Enders and G.Nimtz, *Phys.Rev.B* **15**, 9605 (1993).
[8] A.Enders and G.Nimtz, *J.Phys.I (France)* **3**, 1089 (1993).
[9] A.M.Steinberg, P.G.Kwiat and R.Y.Chiao, *Phys.Rev.Lett.* **71**, 708 (1993).
[10] Th.Martin and R.Landauer, *Phys.Rev.A* **45**, 2611 (1992).
[11] A.Enders and G.Nimtz, *Phys.Rev.E* **48**, 632 (1993).
[12] R.P.Feynman, *QED: The Strange Theory of Light and Matter*, 7th printing (Princeton University Press, Princeton, New Jersey, 1988), p.23.

POSSIBLE TESTS OF NONLINEAR QUANTUM MECHANICS

Miroslav Pardy

Department of Theoretical Physics
and Astrophysics
Masaryk University
Kotlářská 2, 611 37 Brno
Czech Republic

The logical support is given for the competence of the nonlinear Schrödinger equation with term $-b(\ln|\Psi|^2)\Psi$, b being positive. The stationary states of a particle in a box are determined and the measurement methods for determination of the constant b are proposed.

Key words: nonlinear theory, stationary states, box.

Many authors have suggested that the quantum mechanics based on the Schrödinger equation is only an approximation of some nonlinear theory with the nonlinear Schrödinger equation.The motivation for introduction of the nonlinear theory was in expecting of the better understanding of the synergism of waves and particles. In year 1976 Białynicky-Birula and Mycielski [1] considered the generalized Schrödinger equation with the additional term $F(|\Psi|^2)\Psi$, where F was some arbitrary function which they later specified to be $-b(\ln|\Psi|^2)$ as a consequence of the factorization of the wave function for the composed systems.The same equation was also obtained by Lemos [2] within the context of the stochastic quantum mechanics.In such a way the basic equation of the more realistic quantum mechanics was postulated as follows [1]:

$$i\hbar\frac{\partial\Psi}{\partial t} = -\frac{\hbar^2}{2m}\Delta\Psi + V\Psi - b(\ln|\Psi|^2)\Psi, \tag{1}$$

where b is positive and the nonlinear term contents the unit constant with the dimensionality of (lenght)^3.

It is easy to show that Eq.(1) follows from the hydrodynamical model of quantum mechanics [3-6] after adding into the Euler hydrodynamical equation the pressure $-b|\Psi|^2$ and by the inverse procedure to the derivation of the hydrodynamical model.

The solution of Eq. (1) can be assumed in the soliton-wave form (in the one-dimensional case with $V = 0$):

$$\Psi(x,t) = cG(x - vt)e^{i(kx-\omega t)}, \tag{2}$$

where c, v, k and ω are real numbers. After insertion of function (2) into Eq. (1), we get

$$\Psi(x,t) = ce^{a/B} exp - \left[\frac{1}{4}B(x - vt + d)^2\right] e^{i(kx-\omega t)} \tag{3}$$

where

$$v = \frac{\hbar k}{m}, \tag{4}$$

$$B = \frac{4mb}{\hbar^2}, \tag{5}$$

and

$$a = \frac{B}{2} - \frac{2m}{\hbar}\omega + k^2 - \frac{2m}{\hbar^2}b\ln c^2, \tag{6}$$

with d being some constant. The constant c can be determined from the normalization condition

$$\int_{-\infty}^{\infty} \Psi^*\Psi \, dx = 1. \tag{7}$$

The probability density is $\delta_m(\xi) = \Psi^*\Psi$, and from Eq. (7) we get (for $d = 0$)

$$\delta_m(\xi) = \sqrt{\frac{m\alpha}{\pi}}e^{-\alpha m\xi^2}, \quad \xi = x - vt, \quad \alpha = \frac{2b}{\hbar^2}. \tag{8}$$

It is obvious that $\delta_m(\xi)$ is the delta-generation function, which tends for $m \to \infty$ just to the Dirac delta function. The advantage of the δ_m function is in its behaviour for the sufficiently large mass m, because in this case it describes the strongly localized motion of a particle. In other words, function δ_m describes the classical motion of a particle with large mass m in contrast to the standard quantum mechanics. Such behaviour obviously gives the logical support for the competence of the logarithmic quantum mechanics.

The natural application of the nonlinear Schrödinger equation is the problem of the stationary states of the particle in the box. While we have used the specific iteration procedure for the stationary states problem in the case of the Gross-Pitaevskii equation [7], here we use some simplifications. After insertion of the function

$$\Psi(x,t) = exp - [i(E/\hbar)t]\phi(x) \tag{9}$$

into Eq. (1), we get, in the one-dimensional case,

$$\phi'' + k^2\phi = D\phi\ln|\phi|, \tag{10}$$

with

$$k^2 = \frac{2mE}{\hbar^2}, \quad D = -\frac{4mb}{\hbar^2}. \tag{11}$$

We suppose that the approximate solution of Eq.(10) is of the form

$$\phi(x) = A \sin \kappa x, \tag{12}$$

where A and κ are to be determined. For $|\sin \kappa x \cdot \ln |A|| \gg |\sin \kappa x \cdot \ln |\sin \kappa x||$, we get

$$(k^2 - \kappa^2) = D \ln |A|; \tag{13}$$

and, from the boumdary conditions $\phi(0) = \phi(L) = 0$ and from the normalization condition of the wave function on the space interval $(0, L)$, we get

$$\phi(x) = \sqrt{\frac{2}{L}} \sin \frac{n\pi}{L} x. \tag{14}$$

The corresponding energies of the stationary states then are

$$E_n = \frac{\hbar^2 \pi^2}{2mL^2} n^2 - b \lg \frac{2}{L}; \quad n = 1, 2, 3, \ldots \quad . \tag{15}$$

The last formula indicates that it is not possible to determine the constant b spectroscopically, because relation $\hbar\omega = E_n - E_m$ does not contain the constant b. On the other hand, the force between boundaries, due to the the existence of the particle in the box, is $F = -\partial E/\partial L$, i.e.,

$$F_n = \frac{\hbar^2 \pi^2}{mL^3} n^2 - \frac{b}{L} \tag{16}$$

for one particle in the box, and NF_n for the N noninteracting particles in the box. It can in principal be measured by the same methods that were used in the case of the Casimir effect [8,9]. Of course, the difficulties will be greater than in the case of the Casimir effect.

The other possibility for the measurement of E_n, one which is here considered for the first time, is to consider the following experiment: Two rods with square cross section, are given, the near ends being at the distance L apart and forming a potential box of width L. Suppose particles are impinging on the rods in the locality of the boundary of the gap and are reflected. It is evident that the resonance absorption of particles by the gap occurs for a velocity that is determined by the equation

$$\frac{1}{2} m v_n^2 = E_n \tag{17}$$

where v_n is the velocity perpendicular to the plane of the gap. If the source of particles is fixed to the rotating disk, then the velocity of the emitted particles can be continuously changed in order to get the resonance velocity, just as in case of the Mössbauer experiment, and in such a way that it gives the possibility for determination of the constant b.

For $\hbar$= 1.05 x 10^{-34} J, $L = 10^{-7}$ m, m = 1,67 x 10^{-27}kg, b = 3.3 x 10^{-15}eV = 3.3 x 1.6 x 10^{-34}J, $n = 10$, , we get E_{10} = 3,25 x 10^{-25}J - 8.87 x 10^{-33}J.

In this calculation we have used the results of the interferometric search for the nonlinear term in the Schrödinger equation of Shull et al. [11] and Gähler et al. [12], who got the upper limit of b of $\leq$ 3.3 x 10^{-15}eV.

REFERENCES

[1] I. Białynicky-Birula and J. Mycielski, *Ann. Phys.* (*N.Y.*) **100** (1976) 62.
[2] N. A Lemos, *Phys. Lett.* **94A** (1) (1983) 20.
[3] E. Madelung, *Z. Phys.* **40** (1926) 322.
[4] D. Bohm, *Phys. Rev.* **85** (1952) 166; *ibid.* 180.
[5] D. Bohm and J. Vigier, *Phys. Rev.* **96** (1954) 208.
[6] N. Rosen, *Nuovo Cimento* **19B** (1) (1974) 90.
[7] M. Pardy, *Phys. Lett.* **A 140** (1) (2) (1989) 51.
[8] M. J. Sparnaay, *Physica* **24** (1958) 751.
[9] D. Tabor and R. M. S. Winterton, *Proc. Roy Soc. A* **312** (1969) 435.
[10] R. L. Mössbauer, *Z. Phys.* **151** (1958) 124.
[11] C. G. Shull, D. K. Atwood, J. Arthur, and M. A.Horne, *Phys. Rev. Lett.* **44** (1980) 765.
[12] R. Gähler, A. G. Klein, and A. Zeilinger, *Phys. Rev. A* **23** (1981) 1611.

DETECTION OF EMPTY WAVES CONTRADICTS EITHER SPECIAL RELATIVITY OR QUANTUM MECHANICS

Jarosław Pykacz

Institute of Mathematics, University of Gdańsk
ul. Wita Stwosza 57, 80-952 Gdańsk, Poland

It is shown that the very possibility of detecting hypothetical empty waves, combined with very general and apparently innocent assumption about their physical behaviour, contradicts either special relativity or orthodox quantum-mechanical description of EPR-type experiments.

Key words: empty waves, EPR state, faster-than-light communication.

The existence of empty waves is a straightforward consequence of de Broglie-Bohm [1,2] interpretation of quantum mechanics. According to this interpretation a microobject consists of a "material" part and its guiding "pilot" wave which is usually assumed not to carry energy or momentum. Whenever a microobject, e.g., in an interference experiment, faces a choice of different trajectories, its "material" part travels through one path only while its guiding wave fills all possible paths being, in all but one path, an "empty wave." Although properties of empty waves were never defined precisely, they could not be used to explain quantum phenomena without assuming that they interact somehow with experimental devices and that the influence of such interactions on the motion of empty waves is identical or at least very similar to the influence of respective interactions on motion of microobjects predicted by orthodox quantum mechanics. Actually, even quick glimpse through numerous papers on empty waves (see, for example, [3–7]; Chapter 4 of Selleri's book [8] reviews the most important papers) shows that it is taken for granted that the impact of all experimental devices (beamsplitters, shutters, mirrors, phase shifters, etc.) on empty waves is the same as predicted for their respective "material" objects by already existing non-empty-wave theories able to describe the same phenomena. Therefore, the following minimal assumption about hypothetical empty waves seems to be unavoidable.

Minimal Assumption (MA). In no experimental device empty waves associated with a "material" object being in the state ρ can enter regions of space in which the presence of this object is forbidden by existing "working" theories.

To illustrate this assumption let us consider spin-1/2 particles with definite orientation of spins, say in the direction x, which can be achieved by sending the particles through the suitably oriented Stern-Gerlach apparatus. If a particle appearing in the "spin-x up" outgoing channel is sent through the second, identically oriented Stern-Gerlach apparatus, it is sure to appear again in the "spin-x up" channel and quantum mechanics says that its presence in the "spin-x down" outgoing channel of the second apparatus is impossible. According to our MA, no empty wave of any spin-1/2 particle emerging from the "spin-x up" outgoing channel of the first apparatus enters the "spin-x down" channel of the second one. However, if the second apparatus is oriented in the direction y, the previously mentioned particle can be detected with equal probability in both outgoing channels of the second apparatus, therefore our MA allows its hypothetical guiding wave to enter both of them. If the MA is not adopted, then any advocate of the pilot-wave hypothesis would meet serious difficulties in explaining why in the simplest two-slit experiment interference vanishes when one path is made inaccessible to particles.

The considerations presented above yield a clear contradiction between quantum-mechanical description of correlated microobjects (EPR-type experiments) and non-faster-than-light communication hypothesis if we assume that detection of empty waves is at all possible.

Let us consider the EPR experiment in the Bohm version [9] involving spins. A source produces pairs of correlated spin-1/2 particles which fly in opposite directions. Particles going to the left enter the Stern-Gerlach apparatus with two "ordinary" (i.e. non-sensitive to empty waves) detectors placed in outgoing channels. Particles going to the right fly through the identical Stern-Gerlach apparatus and then enter either the channel which ends with an ordinary detector, or the channel which ends with a hypothetical empty-wave detector able to detect not only particles but also the very presence of particle's guiding wave even in the absence of its piloted particle. We assume that the intensity of the source is so low that emissions of subsequent pairs are well separated in time. This implies that simultaneous firing of both detectors in the right arm of the device means that only one of them detected a particle (possibly accompanied by its wave) while the other one detected the genuine empty wave. We assume also for simplicity that the experimental device is "ideal," in particular that all detectors work with 100% efficiency.

Let us start our analysis with considering the case when the left arm of the device is shorter than its right arm. Such configuration makes the arrival of the left-hand side particle to its Stern-Gerlach apparatus and detectors prior to the arrival of its correlated twin-brother to the Stern-Gerlach apparatus placed in the right arm of the device. If Stern-Gerlach apparata placed in both arms of the device are identically oriented, say in the direction x, detection of a particle which goes to the left in the "spin-x up" or "spin-x down" channel fixes, by the well-known EPR correlations, the result of the same measurement performed on its twin-brother which goes to the right. Such measurement is actually performed, since either the ordinary right-hand side detector fires, or it does not fire but the hypothetical detector sensitive both to particles and empty waves fires. Simultaneous firing of both right-hand side detectors is, according to our preliminary discussion, impossible if our MA works.

The situation is quite different if the left-hand side of the device is designed to measure spin of particles in the directon y. Now their respective twin-brothers can be detected with equal probability in both , "spin-x up" and "spin-x down" outgoing channels of the right-hand side Stern-Gerlach apparatus. This, again according to our preliminary discussion, means that in at least some cases (actually in 50% of them since the whole device was assumed to be "ideal") both right-hand side detectors fire simultaneously.

The contradiction with special relativity becomes evident when the distance between two Stern-Gerlach apparata is so big that we are able to change the left-hand side of the configuration from measuring spin-x to measuring spin-y before the light signal passes from the one side to the other. Since such change leads to physically distinguishable behaviour of right-hand side detectors (they fire simultaneously or not) we see that the very possibility of detecting empty waves combined with the apparently innocent MA about their behaviour makes faster-than light signalling possible, provided that ordinary quantum mechanics correctly predicts behaviour of EPR-correlated particles.

Still another difficulty is met when we try to establish how the considered experiment will be seen from different moving frames. The situation when both right-hand side detectors fire is definitely different from the situation when only ordinary detector fires and according to the very idea of special relativity this should not depend on the reference frame in which an observer is placed. However, if the events of passing of the correlated particles through their respective Stern-Gerlach apparata are spatially-separated, then by choosing suitable reference frames we can make any of the two events prior to the other which, according to the previous discussion, should lead to physically distinguishable consequences. Such danger does not exist according to the ordinary quantum-mechanical description of EPR-like experiments since it is only the possilibity of simultaneous detecting of a particle and its empty wave by two different detectors that creates the problem.

The same problems caused by the relativity of temporal order of spatially-separated events are bound to emerge in Szczepański's [10] (see also Chapter 4 of [8]) proposal of experimental exhibition of wave-packet reduction for an ordinary non-entangled state of a single photon. Szczepański's proposal is based on a hypothetical amplifier which produces "material" output when triggered by an empty wave (cf. "Laser Gain Tube" discussed in details in the Chapter 4 of [8]). Since such a device followed by a detector is actually an empty-wave detector [5,8], we are not surprised that its assumed existence leads to the contradiction of orthodox quantum-mechanical concepts (instantanous wave-packet reduction in Szczepański's paper) with relativistic description of the phenomenon.

The next difficulty caused by the existence of hypothetical empty-wave detectors arises when we take into account that according to the orthodox quantum mechanics the state of single EPR-particle is a mixed state which can be equally well described as 50%-50% mixture of "up" and "down" states in any chosen direction. Therefore, our MA should in no case exclude the possibility of simultaneous firing of both detectors in the right-hand side of the device. This indicates that the very idea of detecting empty waves is either incompatible or at least very hard to be combined with the idea of a mixed state of the orthodox quantum mechanics.

In order to make our analysis complete we have to mention that according to the "Paris School" of followers of de Broglie [11,12] each "pilot" wave acts exclusively on its own guided particle, so the detection of empty waves through their action on other independent particles or particle systems is regarded as impossible. However, such position leads to experimental indistinguishability of proposed interpretation from the orthodox one. Therefore, despite the appealing pictorial character of de Broglie interpretation, the orthodox one could be preferred on the basis of Ockham's razor argument.

Finally, let us mention that it is possible to obtain the same contradiction between quantum-mechanical description of EPR-type experiments and special relativity also when the hypothetical empty-wave detector is not sensitive to the "material" part of a microobject. Such situation was considered in Ref. [13] for hypothetical empty waves associated with photons in a device designed to measure either linear or circular polarization. We are convinced that similar demonstration is possible for empty waves associated with other microobjects as well. The "ideal" character of the studied device is of course also not crucial since we can make the distance between two arms of the device arbitrary long in order to allow enough many microobjects to come to "our" arm to make the phenomenon visible before any light signal from the "remote" arm arrives. This feature is important since generally proposals for building empty-wave detectors are not based on assumed possibility of detecting an empty wave of a single particle, but on the expected influence of empty waves on probabilistic features of other microobjects which can be exhibited only through statistical behaviour of these other microobjects. Therefore, our demonstration does not depend on any specific feature of a hypothetical empty-wave dectector but only on very possibility of detection of empty waves combined with assumed validity of ordinary description of EPR-type experiments and very weak assumption about the expected "reasonable" behaviour of empty waves.

ACKNOWLEDGMENTS

The author is greatly indebted to the organizers of the Trani Workshop for the invitation, to the Batory Foundation for a travel grant, and to the Corpus Christi College for the hospitatlity extended to him in September 1993 under the Oxford Colleges Hospitality Scheme for East European Scholars (financially supported by the British Council and the Open Society Fund) when the final version of the paper was written.

The work was also supported by the University of Gdańsk grant BW/5100-5-0117-3.

REFERENCES

1. L. de Broglie, *Non-Linear Wave Mechanics: A Causal Interpretation* (Elsevier, Amsterdam, 1960).
2. D. Bohm *Phys. Rev.* **85** 166, 180 (1952).
3. J. Andrade e Silva and M. Andrade e Silva, *C. R. Acad. Sci.(Paris)* **290**, 501 (1980).
4. J. R. Croca, *Found. Phys.* **17** 971 (1987); *Phys. Lett. A* **124** 22 (1987).
5. G. Tarozzi, *Lett. Nuovo Cimento* **42** 438 (1985).
6. J. R. Croca, A. Garuccio, V. L. Lepore, and R. N. Moreira, *Found. Phys. Lett.* **3** 557 (1990).
7. M. Schmidt, and F. Selleri, *Found. Phys. Lett.* **4** 1 (1990).
8. F. Selleri *Quantum Paradoxes and Physical Reality*, A. van der Merwe, ed. (Kluwer, Dordrecht, 1990).
9. D. Bohm, *Quantum Theory* (Prentice Hall, Engelwood Cliffs, 1951).
10. A. Szczepański, *Found. Phys.* **6** 427 (1976).
11. P. R. Holland, and J. P. Vigier, *Phys. Rev. Lett.* **67**, 402 (1991).
12. J. P. Vigier, private communication.
13. J. Pykacz, *Phys. Lett. A* **171**, 141 (1992).

INTERFEROMETRY WITH VERY COLD NEUTRONS

E.M.Rasel[1], K.Eder[1], J.Felber[2], R.Gähler[2]
R.Golub[3], W.Mampe[4], A.Zeilinger[1]

[1] *Institut für Experimentalphysik, Universität Innsbruck, Austria*
[2] *Fakultät für Physik, Technische Universität München, Germany*
[3] *Hahn Meitner Institut, Berlin, Germany*
[4] *Institut Laue Langevin, Grenoble, France*

We describe the features and the performance of a Mach-Zehnder interferometer for very cold neutrons installed at the HFR-Grenoble. Different applications in fundamental physics are discussed in detail and a new type of a neutron interferometer is presented.

Key words: interferometer, very cold neutrons, Aharonov-Bohm effects, nonlinear wave mechanics, electric charge of the neutron, sideband interferometer.

1. THE VCN INTERFEROMETER

1.1. Principle of Operation

The interferometer is of the Mach-Zehnder type with three transmission phase gratings for beam-splitting, -recombining and detection (Fig.1). For a neutron wavelength of 100 Å ($v = 40m/s$) and a total length of $4m$ ($0.5m$ at present), a beam separation of $\simeq 1cm$ is possible – enough to place sophisticated equipment in one arm of the interferometer. The interference pattern is achromatic and independent from the input divergence (general properties of a Mach-Zehnder with gratings), and a wide beam with 10% spread in wavelength can be used. The n-intensity for the full length of $4m$ will be about $1\, n/s$.

Waves and Particles in Light and Matter, Edited by
A. van der Merwe and A. Garuccio, Plenum Press, New York, 1994

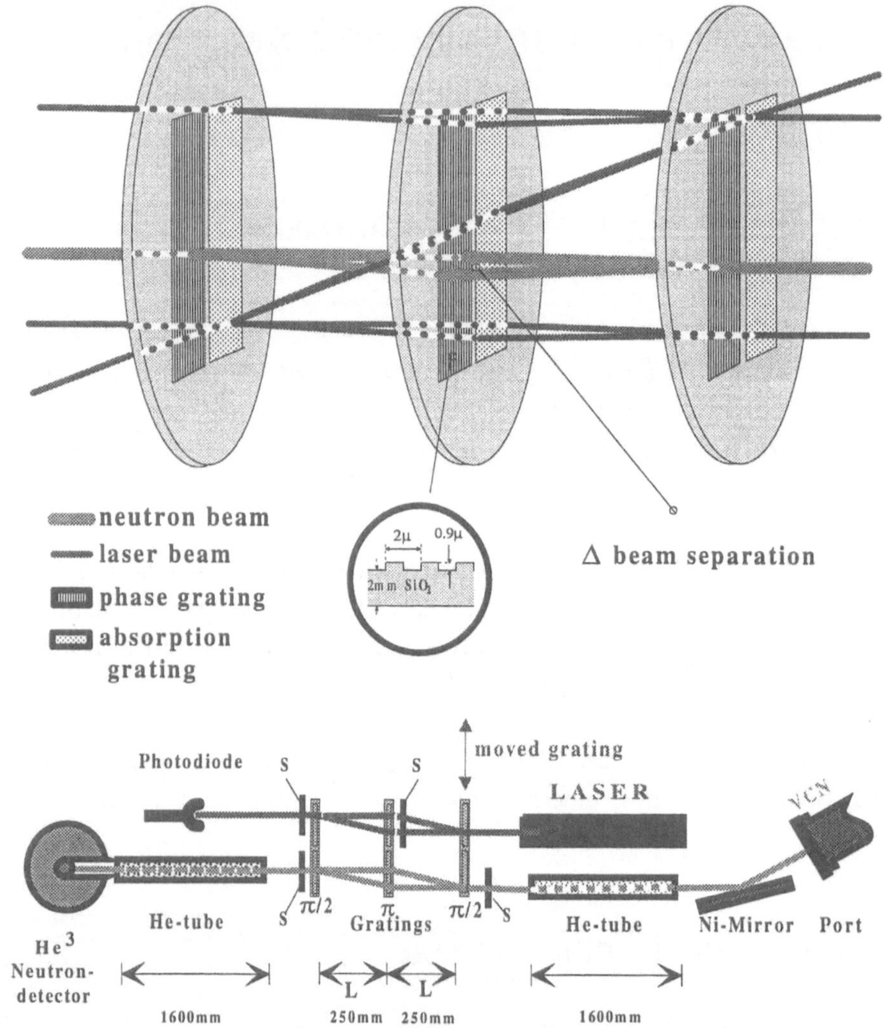

Fig. 1: Setup of the Interferometer

1.1.1. The Gratings

Quartz plates with sputter-etched profiles of $2\mu m$ lattice constant on an area of $1cm$ width and $8cm$ height are used as phase gratings. With a profile depth of $0.5\mu m$ (this corresponds to a phase shift of $\pi/2$ and gives optimum intensities for both beams) and $1\mu m$ wide steps and grooves (1:1 grating), the aspect ratio is well within conservative limits and the diffraction patterns are close to the ideal shape and intensity. From the 1. and 3. grating we exploit the 0. and 1. order diffracted beam. From the 2. grating only the 1. order diffracted beam is used and the 0. order is suppressed by choosing a profile depth of $\simeq 1\mu m$ (phase shift of π). We loose hardly any intensity from diffraction into higher orders because of the 1 : 1 ratio mentioned above.

1.1.2. The Optical Bench

The interferometer together with a beam collimation system and the detectors are mounted on an optical bench of $6m$ length. It is carefully isolated against vibrations (see below) and equipped with an electronic level with a long time stability of better than 0.1 microradians. The variation in fall height of about 2 cm for the neutrons of different velocities requires an alignment of the lattice vectors of all three gratings vertical to the local g-vector within microradian precision.

1.1.3. Alignment of the Interferometer

The gratings can be rotated with a precision better than 0.1 microradians around the direction of the n-beam, in order to align them parallel to each other within parts of the lattice constant. For this purpose three optical Mach-Zehnder-interferometers with their gratings on the same substrates as the neutron ones are positioned such that neutrons and light travel parallel to each other in a close distance. The difference in wavelength (factor 60) is just compensated by an increase of the lattice constant by the same factor. These three light interferometers (two of them parallel and one tilt to the n-beam axis) are operated in such a way that any misalignment of the light gratings (and consequently the n-gratings) is automatically detected and compensated within the required precision.

1.1.4. Vibration Isolation

Much effort was put into the vibration isolation. Any acceleration parallel to the lattice vectors of the gratings (lateral movements or rotations) causes a shift of the interference pattern. Due to the long n-flight time in the interferometer ($\simeq 0.1s$), even low frequency vibrations ($\simeq 1Hz$) will reduce the contrast of the interference pattern. An active control system has been built to lower the vibrational level of the optical bench to values well below $1\mu m$ even for these low frequencies.

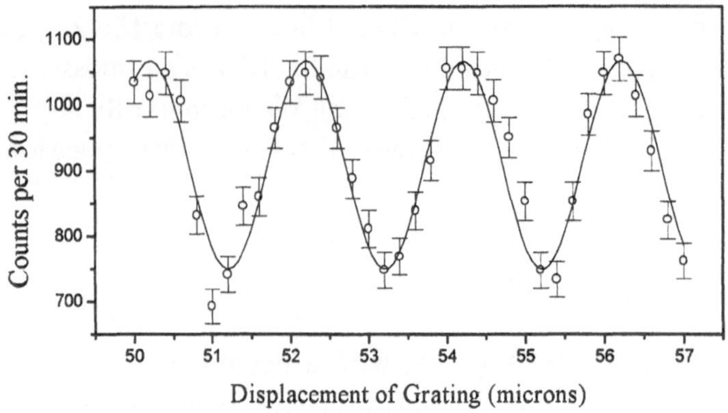

Fig. 2: Interference Pattern

1.2. Experiments for the VCN-interferometer

A first test of the interferometer has been done (Fig.2). We now give a brief description of some experiments, which will be performed, once the HFR-Grenoble is back in operation again.

1.2.1. Neutron Analogue of the Bohm-Aharonov-Effect

In their original paper [1], Bohm and Aharonov proposed two different experiments for electron interferometry. Both should demonstrate phase shifts without any classical effect on either beam separately, which can be considered as a basic feature of generalized BA-effects. One of the two experiments was never performed because of serious technical problems with the time scales involved: In both arms of an electron interferometer there should be conductive tubes, with electostatic potentials Φ applied for definite times when the electron wave packets are clearly inside the tubes. As the phase shift $\Delta\varphi_e$ is proportional to $\int q\Phi dt$, no dependence of an wave vector $\boldsymbol{k}$ and consequently no classically observable shift of the envelope of the wave packet occurs. On the other hand, $\Delta\varphi_e$ clearly gives rise to a shift of the interference pattern.

A nice neutron analogue of this experiment can be performed without major problems, if we replace the electrostatic potentials by magnetic fields in a VCN-interferometer [2] (Fig.3). The phase shift $\Delta\varphi_m$ for neutrons inside the field regions (made by coils) is proportional to $\int \mu \cdot B dt$ (with $\boldsymbol{\mu}$ the magnetic moment of the neutron). The phase shift is once again independent of $\boldsymbol{k}$, if the field is only switched on for times, when the wave packets are well inside the homogeneous field regions. Because of the low neutron velocity

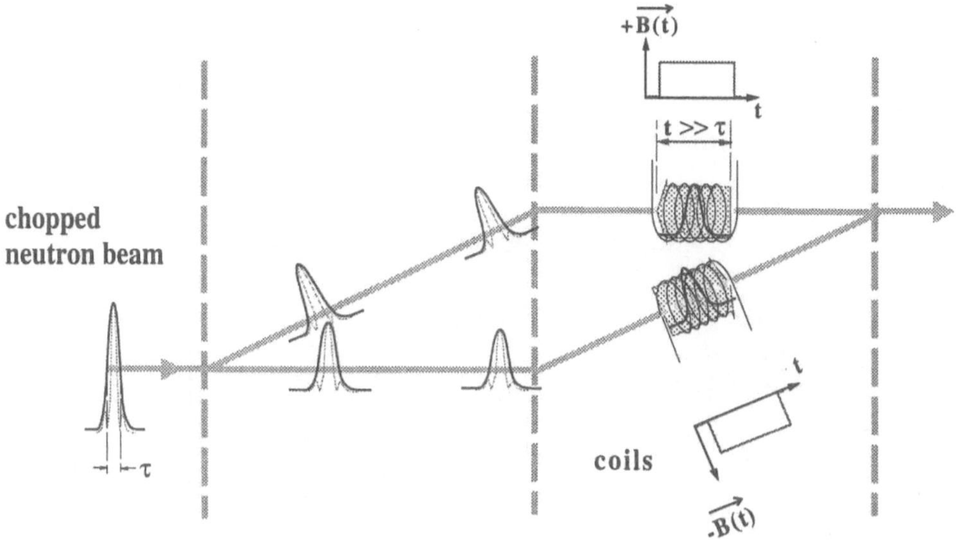

Fig. 3: Aharonov-Bohm Effect

($\simeq$ 40m/s) and the long length of $\simeq$ 1m for the field regions, the experiment can be performed with high precision. It should be possible to shift the interference pattern by thousands of fringes without any loss of contrast.

1.2.2. The Electric Charge of the Neutron

Despite its commonly accepted neutrality, an improvement on the present limit of the neutron charge (10^{-21} e-charges [3]) is of interest in theoretical physics:

Superstring theory allows charge fluctuations Δq of about 10^{-22} to 10^{-21} elementary charges for the neutron. The time scale of this effect is unknown and an experimental verification is certainly a challenging task.

Any nonzero neutron charge is of importance for some "'classical" conservation laws (baryon-, leptonnumber, charge). As shown by Feinberg and Goldhaber [4] , these would become dependent from each other in the case of a nonzero neutron charge, no matter how small this charge might be. Furthermore a measurement of the n-charge probes the apparent symmetry of charges, which is not based on fundamental grounds but just on experience.

All former experiments on the n-charge q_n look for a beam deflection in a static transverse electric field. The VCN-interferometer offers an alternative scheme: We apply different electrostatic potentials $\pm\Phi_e$ along both

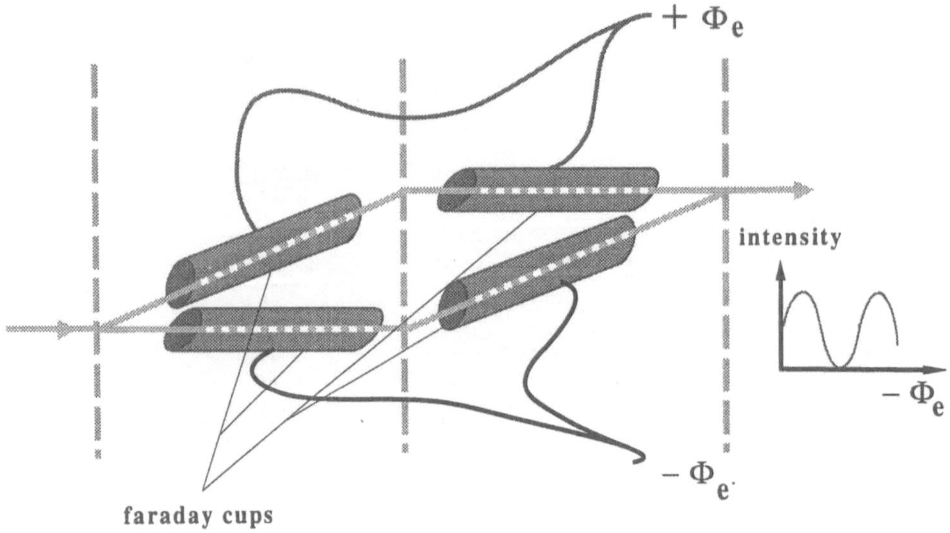

Fig. 4: Electric Charge of the Neutron

pathes L_1 and L_2 and the potential difference will drop in an insulator between them (Fig.4). The phase difference $\Delta\phi_e$ accumulated along both pathes is given by $\Delta\phi_e \sim 2 \cdot \int q_n \Phi_e dt$, leading to a lateral shift of the interference pattern.

To some extent this version avoids the $\boldsymbol{E} \times \boldsymbol{v}$ effect and allows higher sensitivity for a charge measurement because the maximum fields inside insulators are higher than in vacuum, which limited the maximum field strength in the former experiments. We expect a sensitivity of about 10^{-22} electron charges, which is a factor of 10 below the present value.

1.2.3. Nonlinear Wave Mechanics

In the Schrödinger equation any term dependent on $|\Psi|^2$ added to the Hamiltonian, gives rise to a change in potential energy, when a wave packet spreads out in space. Consequently the introduction of such a nonlinear term (with the right sign) limits the spreading and gives solutions closer to classical mechanics with more point like particles instead of wave packets which will, for the linear equation, extend to infinity. A special typ of the nonlinear term $(-b \cdot \ln(a^n \cdot |\Psi|^2))$ had been chosen by Bialynicki-Birula and Mycielski, to allow for separability of noninteracting subsystems [5].

Neutron optics offers an excellent possibility to test this class of nonli-

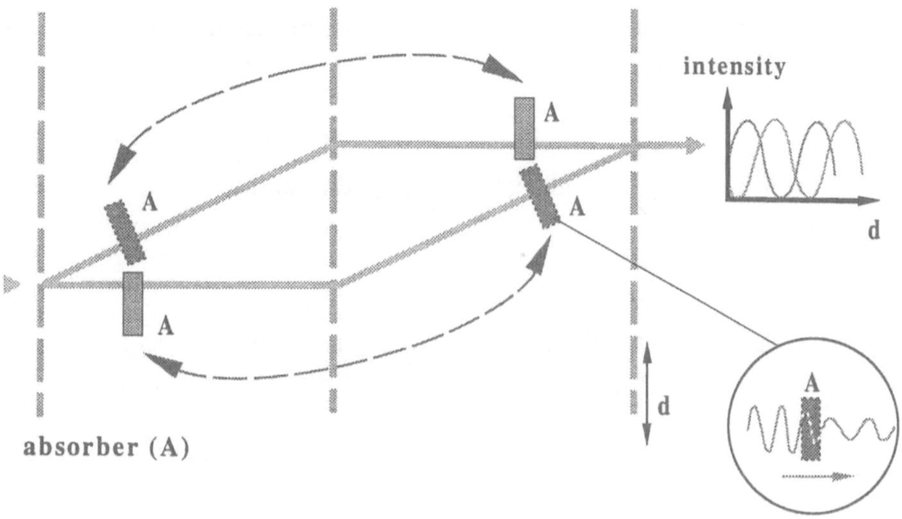

Fig. 5: Non-Linear Wave Mechanics

near Schrödinger equations: In one of these experiments [6] , partial absorbers were put in the two pathes of a single crystal neutron interferometer (Fig.5). The attenuation of the amplitude changes the value of the nonlinear term and consequently leads to a change in wave vector $\vec{k}$. As a consequence the interference pattern should be shifted, if we place the absorbers at different positions of the flight paths. No shift was found and a value of $b \leq 10^{-13}$ eV was deduced from the data .

In a second experiment, a strong lateral gradient of $|\Psi|^2$ was created in a highly collimated neutron beam and a possible deflection due to the nonlinear term was sensed, leading to $b \leq 10^{-15}$ eV [7].

With the VCN-interferometer the first experiment can be repeated with a much higher precision: From the bigger length and mainly from the much smaller value of $\boldsymbol{k}$ we will gain a factor of about 5000 in sensitivity (i.e. b in the order of 10^{-16} eV). This might lower the nonlinear term to irrelevant values.

2. A NEUTRON SIDEBAND INTERFEROMETER

When a neutron is reflected by a moving potential step the incoming wave function is phase modulated causing a characteristic energy spread of the outgoing state. If we take a plane monochromatic wave as the incoming state and let the potential step oscillate harmonically with frequency ω_v and amplitude a_0 (Fig.6), the reflected spectrum shows discrete lines, which are spaced by integers of $\hbar\omega_v$. An approximate solution of the reflected wave can be given by

$$\Psi_{refl} \simeq \sum_n J_n(\varepsilon) \cdot e^{i(k_n x - \omega_n t)} \quad , \tag{1}$$

with

$$\begin{aligned} \Psi_{inc} &= e^{i(k_0 x - \omega_0 t)} \quad , & a(t) &= a_0 \sin(\omega_v t) \\ \omega_n &= \omega_0 + n \cdot \omega_v \quad , & n &= 0, 1, \ldots, \pm\infty \\ k_n &= \sqrt{2m\omega_n/\hbar} \quad , & \varepsilon &= 2a_0 k_0 \end{aligned} \tag{2}$$

and $J_n(x)$ is a Besselfunction of first kind and n-th order.

A similar kind of spectrum is known from frequency modulation of electro-magnetic waves. If there is an additional wave vector component ($k_{||}$) parallel to the surface (this component is unaltered by the vibrating system) the reflected wave is diffracted into discrete angles

$$\tan\phi_n = \frac{k_n}{k_{||}} \quad , \tag{3}$$

with $n = 0$ meaning specular reflection.

As the phases of the outgoing partial waves Ψ_n (Ψ_n refers to ϕ_n) are mutually coherent, such a device may be used as a "sideband" beam splitter in a matter wave interferometer.

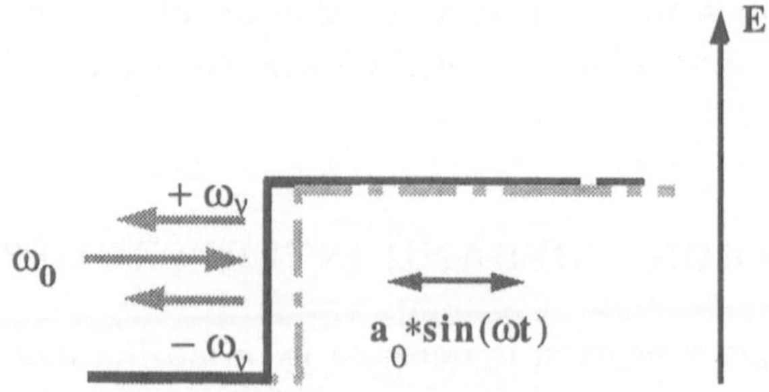

Fig. 6: Vibrating Potential Step

In a normal (i.e. stationary) interferometer only interferences between wave components with equal energies are observable, so the spread in energy caused by one vibrating beam splitter has to be '"reversed" by (at least) one second. If this reversal is not made, it is possible to test the incoming beam on coherence between partial waves with different energies making it possible to determine the longitudinal coherence of the neutron beam. This interesting experiment was suggested in [8].

Figure 7 shows a possible assembly of an interferometer using vibrating devices. Two mirrors (S_1 and S_3) oscillate with ω_v, one (S_2) with $2 \cdot \omega_v$. This makes the two arms of the interferometer symmetrical with respect to path lengths and energy shifts and therefore prevents sensitivity of the interferometer to both dispersive and '"Sagnac" effects.

The relative energies units of $\hbar\omega_v$(with respect to the incoming wave) of the different flight paths are indicated by integer numbers.

By proper choice of the amplitude a_0 the amplitudes of the different reflected modes can be adjusted. For $\varepsilon \simeq 1.84$ Eq.(1) gives $J_0(\varepsilon) \simeq 0.31$, $J_{\pm1}(\varepsilon) \simeq \pm0.58$ and $J_{\pm2}(\varepsilon) \simeq 0.31$, all other modes being negliable. In fig.7 only dominating modes are shown.

The possible path deviation $\Delta\phi = \phi_{+1} - \phi_{-1}$ for an experiment with typical (conservative) parameters (λ = 100 Å, ϕ_0 = 10°, $\omega_v = 2\pi$ MHz, a_0 = 9.25 nm) is about 1.3°, giving rise to a beam separation of 2 cm per 1 m flight path.

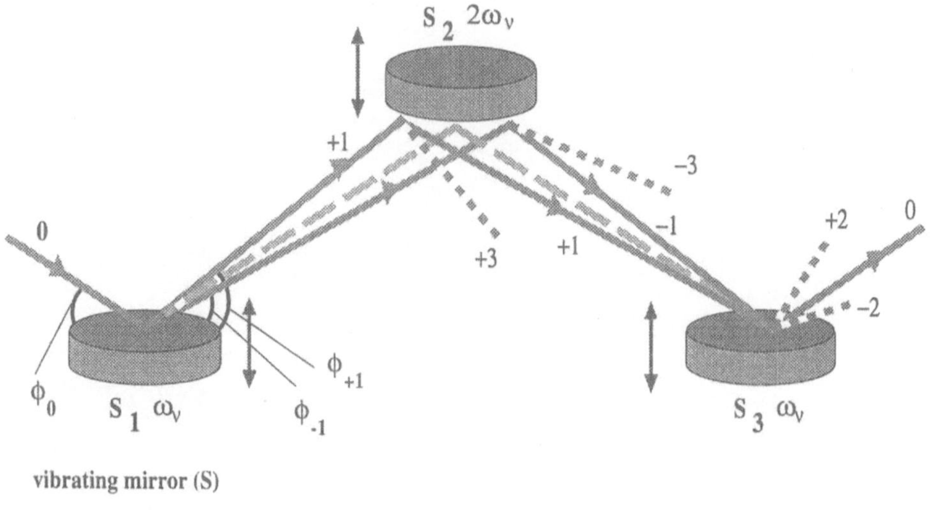

Fig. 7: Neutron Sideband Interferometer

A technichal realization of such oscillating devices is in progress using resonantly vibrating piezoelectric samples, which are coated with nickel mirrors.

REFERENCES

[1] Y.Aharonov and D.Bohm. *Phys.Rev.* **115** (1959) 485.
[2] A.Zeilinger. *J.Phys.* (Paris) **45**, suppl. C3 (1984) 213.
[3] J.Baumann, R.Gähler, J.Kalus, and W.Mampe. *Phys. Rev. D* **37** (1988) 3107.
[4] G.Feinberg and M.Goldhaber. *Proc. Natl. Acad. Sci. USA* **45** (1959) 1301.
[5] I.Bialynicki-Barula and J.Mycielski. *Ann. Phys.* (Leipzig) **100** (1976) 62.
[6] C.G.Shull, D.K.Atwood, J.Arthur, and M.A.Horne. *Phys. Rev. Lett.* **44** (1980) 765.
[7] R.Gähler, A.G.Klein, and A.Zeilinger. *Phys. Rev. A* **23** (1981) 1611.
[8] R.Golub and S.K.Lamoreaux. *Phys. Lett. A* **162** (1992) 122.

VELOCITY-SYMMETRIZING SYNCHRONIZATION AND CONVENTIONAL ASPECTS OF RELATIVITY

F. Selleri

Dipartimento di Fisica, Universitá di Bari
INFN-Sezione di Bari
I-70126 Bari, Italy

We show that a very natural synchronization procedure is expressed by the requirement that the velocity of S_0 relative to S be equal and opposite to that of S relative to S_0, if S_0 is the "absolute" inertial frame and S any other inertial frame. This new procedure is compatible with Einstein synchronization and with slow-transport synchronization only within the special theory of relativity. The practical equivalence of the Lorentz and Tangherlini transformations is discussed.

Key words: synchronization, relativity theory.

1. VELOCITY-SYMMETRIZING SYNCHRONIZATION

In a well-known series of papers [1–3] Mansouri and Sexl (MS) presented and developed the very interesting program of looking for possible deviations from special relativity in the outcome of various experiments. For this purpose they defined a class of rival theories, introduced through the following spacetime transformations between inertial systems

$$x = b(x_0 - vt_0), \tag{1a}$$

$$t = a\, t_0 - Ex, \tag{1b}$$

where $S_0(x_0, t_0)$ is the privileged (ether) system, $S(x, t)$ is an inertial system moving with velocity v parallel to the positive x axis with respect to S_0, and a, b, E are functions of the absolute velocity v. In (1) we modified slightly the MS notation, in particular by putting $E = -\varepsilon$, but maintained the original meaning of a and b. In their whole analysis, MS treated a, b, E as three independent functions to be determined in different experiments and/or by suitable clock synchronization procedures.

A similar approach has been developed in [4] by studying in more detail the problem of setting up the most general transformation laws between inertial systems in ether theories. The result was that spacetime transformations are given by

$$x = b(x_0 - vt_0), \tag{2a}$$
$$t = b(t_0 - Ex_0), \tag{2b}$$

The only essential difference between (1) and (2) is that the latter contain *only two* functions of absolute velocity, a, and E.

In order to understand the source of this difference, consider the motion of the origin of S (defined by $x = 0$) with respect to S_0. From the first equation (1) one has immediately

$$x_0 = vt_0, \tag{3}$$

as it should be. Consider instead the origin of S_0 (defined by $x_0 = 0$) as seen from S: A simple calculation based on (1) gives

$$x = -v\frac{b}{a + bvE}t. \tag{4}$$

Therefore, the velocity of S_0 as seen from S is equal and opposite to that of S seen from S_0 if and only if the fraction in the right-hand side of (4) is unity, that is, if

$$E = E_V \equiv \frac{1}{v}\left(1 - \frac{a}{b}\right), \tag{5}$$

which is precisely the relation found in [4] between these coefficients. The previous determination of E corresponds to a new synchronization procedure which can be called "velocity-symmetrizing synchronization." It amounts to the following: A measuring rod is produced in S_0 and taken there as unit of length. This allows one to measure x_0. Many identical measuring rods are produced in S_0 and given to the moving systems $S, S', \ldots$. This "giving" implies an acceleration of the rod which is brought to rest in the new system. One does not need to worry about the eventual deformations induced by the acquired absolute velocity. Whatever happens, the rod will be used as unit of length in the moving system. In this way x can be measured in S, x' in S', and so on.

Coming to time intervals, one can proceed as follows: Many identical clocks are produced in S_0, such that one can correct their rates, if needed. Some of them are used in S_0 to measure t_0 and synchronized with the Einstein procedure, since the privileged system S_0 is by definition that inertial frame in which the velocity of light is isotropic.

The remaining clocks are given to $S, S', \ldots$, and their rates are regulated in each frame in order to meet the requirement of equal and opposite velocities. S_0 has now space and time units and can thus measure the absolute velocities $v, v', \ldots$ of $S, S', \ldots$, respectively. The observer in S_0 will inform those in $S, S', \ldots$ of the exact values of their absolute velocities. At this point the rate of the clocks in the moving frames will be adjusted in such a way that S_0 will be seen from all points of $S, S', \ldots$ to move with velocity $-v, -v', \ldots$, respectively.

At this point, space and time are completely fixed in all inertial frames and there is nothing left that can be fixed in order to make sure that, if S' is seen to move with velocity $\tilde{v}$ in S then S is seen to move with velocity $-\tilde{v}$ in S'. In fact it can be shown that, for any two systems S and S' moving wtih absolute velocity v and v', respectively, one has [4]

$$\text{velocity of } S' \text{ with respect to } S = \frac{v' - v}{1 - v'E(v)},$$

$$\text{velocity of } S \text{ with respect to } S' = \frac{v - v'}{1 - vE(v')},$$

which are equal and opposite only if

$$\frac{E(v')}{v'} = \frac{E(v)}{v},$$

that is if

$$E(v) = v/c^2, \tag{6}$$

where c is a constant with dimensions of velocity, given that E has obviously the dimensions of inverse velocity. We thus conclude that only if E has its well known relativistic expression can the symmetry relation be satisfied for all inertial frames. In general it does not hold for *two moving* frames in an ether theory, even if the velocity-symmetrizing synchronization is adopted between the absolute system and all moving systems.

2. COMPARISON OF SYNCHRONIZATION PROCEDURES

Two classes of synchronizing procedures can be considered [1]:

(i) *External synchronization*, where the clocks of any moving system S are set by some type of comparison with the clocks of the absolute system S_0. To this class belongs the velocity-symmetrizing synchronization considered in the previous section since every inertial observer regulates his clocks after having been informed by the observer in S_0 of the numerical value of his absolute velocity. A different synchronization procedure can be carried out by adjusting the clocks of S to $t = 0$ whenever they fly past a clock in S_0, which shows $t_0 = 0$. This leads to

$$E = 0$$

and therefore to *absolute simultaneity*, since (1) becomes

$$x = b(x_0 - vt_0), \tag{7a}$$

$$t = a\, t_0, \tag{7b}$$

and therefore two events taking place at the same time t_0 in two different points x_{01} and x_{02} are judged also in S to take place at the same time $t = a\, t_0$.

(ii) *Internal synchronization*, which does not require S_0 to be known to the observer in S who makes use only of operations defined in S. The two most important procedures belonging to this class are *Einstein synchronization* and *slow-transport synchronization.* The former convention is based on the postulate that the velocity of light is constant and equal to c in all inertial frames. As shown in [1], it leads to

$$E = E_E \equiv \frac{av/c^2}{b(1 - v^2/c^2)}\,. \tag{8}$$

Synchronizing clocks by slow transport leads instead to

$$E = E_T \equiv -\frac{1}{b}\frac{da}{dv}\,. \tag{9}$$

Of course only one procedure can be adopted, but one can investigate the compatibility of E_V given by (5) with E_E and E_T. By equating E_V and E_E, one gets

$$\frac{1}{v}\left(1 - \frac{a}{b}\right) = \frac{a}{b}\,\frac{v/c^2}{1 - v^2/c^2}, \tag{10}$$

which gives

$$b = \frac{a}{1 - \beta^2}, \tag{11}$$

with $\beta = v/c$. Furthermore by equating E_V and E_T, one gets

$$\frac{1}{v}\left(1 - \frac{a}{b}\right) = -\frac{1}{b}\frac{da}{dv}; \tag{12}$$

which, on using (11), becomes

$$\frac{da}{a} = \frac{-\beta d\beta}{1 - \beta^2}, \tag{13}$$

an equation from which one obtains immediately

$$a = \sqrt{1-\beta^2}, \tag{14}$$

since the condition $a \to 1$ if $v \to 0$ must be imposed. From (11) and (5), one now gets

$$b = \frac{1}{\sqrt{1-\beta^2}}, \tag{15}$$

$$E = \beta/c. \tag{16}$$

From (14)–(16), it follows that the transformations (1) become identical with the Lorentz transformations. Therefore the three considered procedures of clock synchronization are compatible with one another only within the special theory of relativity.

3. RELATIVITY OF RELATIVITY

Mansouri and Sexl [1] argued that a theory maintaining absolute simultaneity is equivalent to special relativity by considering the Tangherlini transformations [5]

$$x = \frac{x_0 - vt_0}{\sqrt{1-v^2/c^2}}, \tag{17a}$$

$$t = \sqrt{1-v^2/c^2}\, t_0. \tag{17b}$$

Obviously (17) implies absolute simultaneity, since $\Delta t_0 = 0$ leads to $\Delta t = 0$. These transformations can be obtained from the well-known Lorentz transformations by applying a different convention about clock synchronization, more precisely by applying the absolute synchronization briefly discussed in Sec. 2. Since the difference between Tangherlini and Lorentz is conventional, MS argued that the two theories are completely equivalent as far as the explanation of experimental results is concerned. This conclusion is truly extraordinary, since it empties the principle of relativity (and its negation) of any physical role: A theory explicitly based on ether, in which a privileged frame exists and is recognized as such by all observers, leads to all the well-known predictions of special relativity! The main argument of MS in favour of their thesis is the clock paradox, and we repeat their reasoning in the following in order to stress a neglected but important point. The spacetime diagram corresponding to the transformation (17) is shown in Fig. 1. Since $t_0 = 0$ implies $t = 0$, the two space axes x_0 and x

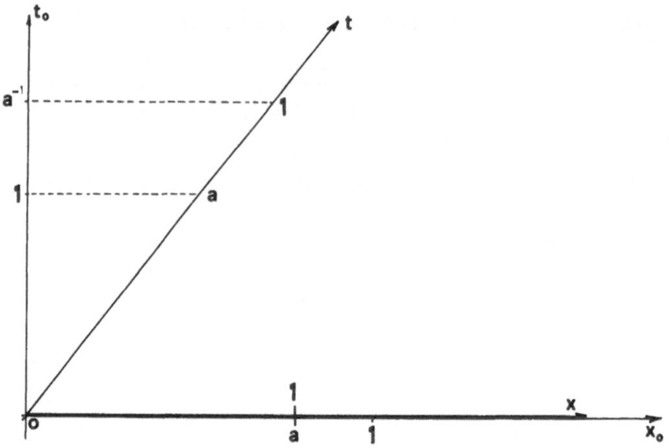

Fig. 1. The spacetime diagram corresponding to the Tangherlini transformations.

are coincident. The t_0 axis is vertical, and the t axis is given by $x = 0$. Thus, by (17), it obeys the equation

$$t_0 = \frac{1}{v} x_0. \tag{18}$$

The time t of the t axis is related to t_0 through (17b):

$$t = a\, t_0, \tag{19}$$

so that it is scaled by a factor $a = \sqrt{1-\beta^2}$ with respect to the absolute time t_0.

Consider now the clock-paradox as represented in Fig. 2: There are two oservers, A and B, and two possible travels. In Fig. 2a the observer A remains all the time at rest in S_0, while B travels away from A for a fraction a of a year (corresponding exactly to 1 year of the t_0 time). After this, B turns back his spaceship and travels in the opposite direction (from P to H), reaching the point where A is at rest exactly at time $t_0 = 2$ years. The point P has coordinates $t_0 = 1, x_0 = v$, and the B time for the back journey, given by the pseudo-Euclidean length of PH in Fig. 2a, is $\sqrt{1 - v^2/c^2} = a$. Therefore, the total time spent by B in the travel OPH is $2\sqrt{1-\beta^2}$ and B arrives in H younger than A.

In Fig. 2b the twin B travels away from S_0 at constant speed v, while A remains at rest in S_0 for a fraction a of a year, after which he jumps on a spaceship and travels in the same direction as B, but faster, in such a way to meet B after 2 years of t time, that is, in the point Q of

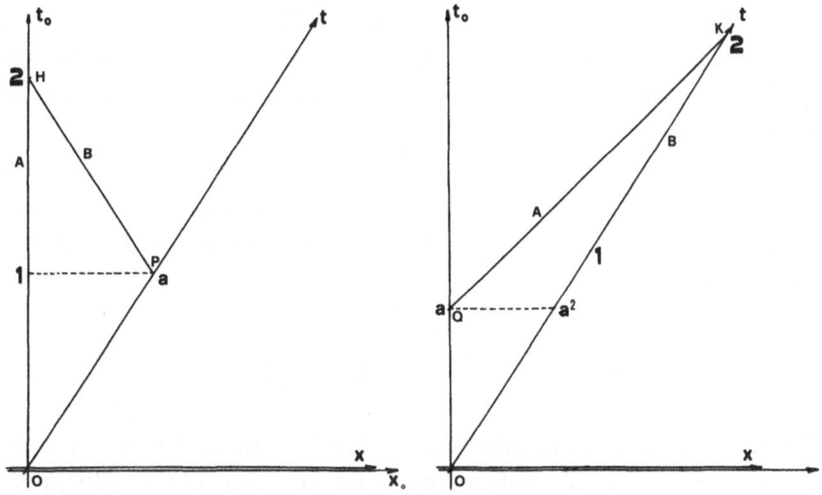

Fig. 2. The twin paradox considered from the point of view of the ether theory of Fig. 1.

coordinates $\frac{2}{a}, \frac{2v}{a}$. In the t_0, x_0 plane the straight line QK has the equation

$$t_0 = \frac{2 - a^2}{2v} x_0 + a, \tag{20}$$

as one can easily check. This means that the speed of A in QK is

$$w = \frac{2v}{2 - a^2}. \tag{21}$$

The total time spent by A in his rest system during his journey QK is given by the pseudo-Euclidean length of QK, that is, by

$$\overline{KQ} = \left[\left(\frac{2}{a} - a\right)^2 - \frac{1}{c^2}\left(\frac{2v}{a}\right)^2\right]^{1/2} = a. \tag{22}$$

Therefore, the total time spent by A in his travel OQK is 2a and the situation is completely symmetrical between Fig. 2a and Fig. 2b *as far as the traveling twin is concerned.* For the twin remaining always in the same system (A in Fig. 2a, B in Fig. 2b), the total time observed is again the same (2 years), but the time at which the traveling twin is seen to turn around is different. In the case of Fig. 2a, such a time is judged to be $t_0 = 1$ year; in the case of Fig. 2b it is $t = a^2$ years. We can conclude that the Tangherlini description of the twin paradox is exactly the same as that of special relativity only if all observers look only inside their respective inertial systems, but that differences arise (in the turn-around time) if they look at the behaviour of other systems. For example, A knows to be in the privileged frame in Fig. 2a, since only in such a case can the turn-around time of B be 1 year!

The previous considerations can easily be extended to a different point P of Fig. 2a: If the time spent by B traveling away from A before turning around is α (instead of a), then the total time measured by B for the travel OPH is

$$\Delta t_B = \alpha + \left[\alpha^2 + 4\left(1 - \frac{\alpha}{a}\right)\right]^{1/2}.$$

Exactly the same time is required by A for the trip OQK if it remains the time $t_0 = \alpha$ at rest in S_0 before pursuing B. One can easily check that the twin "paradox" remains perfectly symmetrical for the two observers A and B if initially they are at rest in two different *moving* systems, instead than in S_0 and S as considered above, of course when the Tangherlini transformations are properly generalized.

REFERENCES

1. R. Mansouri and R. Sexl, *Gen. Rel. Grav.* **8**, 497 (1977).
2. R. Mansouri and R. Sexl, *Gen. Rel. Grav.* **8**, 515 (1977).
3. R. Mansouri and R. Sexl, *Gen. Rel. Grav.* **8**, 525 (1977).
4. F. Selleri, *Z. Naturforsch.* **46a**, 419 (1990).
5. F. Tangherlini, *Suppl. Nuovo Cimento* **20**, 1 (1961).

TRAJECTORIES OF PARTICLES INTERACTING WITH ENVIRONMENTS

Timothy P. Spiller

School of Mathematical and Physical Sciences
University of Sussex, Falmer, Brighton
Sussex, BN1 9QH United Kingdom

We discuss the behaviour of open quantum systems, quantum objects coupled to environments, in the Schrödinger picture. The object state evolves in a non-linear and stochastic manner. We illustrate this for a frictional environment by calculating the de Broglie-Bohm trajectories for some simple examples.

Key words: quantum mechanics, Bohm theory, trajectories, environment, dissipation, quantum friction.

1. INTRODUCTION

The purpose of this work is to discuss the behaviour of open quantum systems in the Schrödinger picture, and with reference to the de Broglie-Bohm quantum potential viewpoint [1–3]. We take some particular quantum object as the focus of attention. The reason we refer to an open quantum system is that this object is considered to be coupled to many other degrees of freedom, its environment. If this environmental coupling is non-negligible then the object cannot be described by the usual Schrödinger evolution of a wavefunction of its configuration space co-ordinates alone, leaving aside the adiabatic approximation.

There is, of course, great interest in the concept of non-Schrödinger evolution of quantum states [4–14], whether this be an intrinsic aspect of the state evolution, or due to the extrinsic effect of an environment. The measurement problem may not be a problem after all, if wavefunctions in fact collapse (albeit infrequently for individual fundamental objects) as a natural part of their evolution. Here we shall focus on extrinsic non-Schrödinger evolution cased by an environment; our examples will make this quite apparent. We thus assume that any intrinsic non-Schrödinger evolution is small by comparison. However it is worth pausing to contemplate the distinction between intrinsic and extrinsic non-Schrödinger state evolution, for it is not clear (at least to the author) that this distinction can be made at a physical, rather than a philosophical, level. We know from experiment that any intrinsic components of non-Schrödinger evolution, (for individual fundamental i.e. microscopic object states,) if such components exist, are constrained to be very weak, or occur very infrequently, or both. It is therefore not clear that experiments designed to identify such effects could distinguish between them and those due to the environment. One could attempt to amplify the non-Schrödinger behaviour being sought by using a large number of microscopic objects. However this will be met by a corresponding increase in the degree of difficulty in isolating such a system from its environment. This is particularly so, for example, in the case of gravitational coupling, which cannot simply be turned off, or screened, as is possible with electromagnetic interactions.

We now turn to our open quantum system, object plus environment. The textbook approach for treating an object in such a situation is to use a reduced density operator. The reduction is made by integrating over the degrees of freedom of the environment, which are not of direct interest. If the system is open, so the environment contains an infinite number of degrees of freedom, then the evolution of the object density operator is in general irreversible. Pure states evolve to mixed states.

We note the following points about the density operator approach:

(i) In models of real systems, the microscopic description of the environment may not be known. It may be the case that only the average effect of the environment on the object is known. If so, it may well be preferable to simply choose an appropriate density evolution for the object which incorporates this effect. Rather than taking the trouble to construct a microscopic model environment purely on the grounds that integrating it out generates a desired effect on the object, one may as well omit this procedure and instead start by directly modelling the density evolution equation.

(ii) The density operator equation, being a matrix equation, may be difficult to solve. When attacked using a computer, a large amount of memory may be required [15].

(iii) The density operator only gives statistical predictions. It gives no feel at all for the behaviour of an individual quantum object in a single run of an experiment, such as an excited atom in a trap undergoing a cascade decay. Whilst one clearly cannot expect to *predict* the outcome of one such run of an experiment, it is clearly of interest to investigate models which can *mimic* such events, and which give the same predictions as the density approach when averaged over many trials [15].

It is this last point, (iii), which concerns us in this paper. Schrödinger picture models which mimic the individual runs of an experiment on a quantum object coupled to some form of environment involve specifying a non-linear stochastic equation for the evolution of the quantum state of the object. The equation should be consistent with the appropriate density operator evolution equation [8–10]. Clearly it must also reduce to the appropriate Schrödinger equation for the isolated object if the coupling to the environment is taken to zero. We note that Heisenberg picture models, which describe the evolution of the appropriate operators for the object whilst it interacts with its environment, can also be given [16,17]. However we stick to the Schrödinger picture here as we want to consider object trajectories using the de Broglie-Bohm interpretation [1–3].

2. BOHM APPROACH

This interpretation involves breaking the non-relativistic Schrödinger equation for a single particle of mass m,

$$H\psi = \left[\frac{-\hbar^2}{2m}\nabla^2 + V(\boldsymbol{x})\right]\psi = i\hbar\frac{\partial\psi}{\partial t}, \tag{1}$$

into two parts. We restrict ourselves to the non-relativistic case in the discussion here. By defining $\psi = R\exp[iS/\hbar]$, for R and S real, one obtains

$$\frac{(\nabla S)^2}{2m} + V + Q + \frac{\partial S}{\partial t} = 0, \tag{2a}$$

$$\nabla\cdot\left[\frac{P\nabla S}{m}\right] + \frac{\partial P}{\partial t} = 0, \tag{2b}$$

having defined

$$Q \equiv \frac{-\hbar^2\nabla^2 R}{2mR}, \quad P \equiv R^2. \tag{3a) (3b}$$

Clearly (2a) is the Hamilton-Jacobi equation apart from the additional so-called quantum potential Q, and (2b) expresses conservation of probability density P, bearing in mind the guiding equation

$$\boldsymbol{p} = m\dot{\boldsymbol{x}} = \nabla S. \tag{4}$$

In this interpretation of quantum mechanics both the particle and the wave ψ are objectively real. The particle follows a trajectory which is a solution of (4); this will be non-classical if the quantum potential Q, given by (3a), plays a significant role in the solution of (2a). For a particular solution to the Schrödinger equation (1), $\psi(\boldsymbol{x},t)$, an ensemble of possible trajectories exists; each member of the ensemble being specified by its initial ($t = 0$) point $\boldsymbol{x}(0)$. Quantum uncertainty enters because such an initial particle position cannot be experimentally controlled. Averaging over the ensemble of trajectories (with a weight $|\psi(\boldsymbol{x}(0),0)|^2$) yields the conventional expectation values [2,3]. It is of particular relevance to the discussion here to note that any attempt to experimentally determine a trajectory requires interactions which will modify the quantum potential [2,3]. In this sense, the trajectories computed from the solution to the Schrödinger equation for the particle ignoring such interactions are not amenable to experimental investigation. In essence, this is the motivation for our discussion here. Rather than consider the trajectories of quantum particles ignoring such interactions, we consider the trajectories of quantum objects which are undergoing interaction with their environment.

This clearly requires an additional assumption. We know that for systems of more than one particle, the trajectories in fact lie in the full configuration space of the whole system. However we are only interested in one object, part of an open quantum system, and how it moves in its own configuration space. We therefore need a prescription for finding these "reduced trajectories." This emerges quite naturally when we consider the non-linear stochastic equation which the state of the object under interaction with its envirnoment is taken to obey. If the interaction does not destroy, or absorb, the particle, and here we restrict ourselves to interactions of this kind, then we know that an equation of the form of (2b) must still emerge when we decompose the state evolution equation. We must retain a conserved probability density. It is thus natural to take the quantity which plays the role of the velocity in this conservation equation, and use it in the guiding equation (4).

3. PARTICLE TRAJECTORIES WITH A DISSIPATIVE ENVIRONMENT

The example we focus on is a dissipative environment, one which on average removes energy from the object of interest. Classically the concept of frictional loss arises because a choice is made to reduce the configuration space of the system under consideration down to that of one part. This reduction is accompanied by a modification to the equations of motion of that part; a frictional force is added in. Here we simply make this reduction for a quantum system. We consider two examples of non-linear stochastic state evolution equations.

The first of these is

$$\left[\frac{-\hbar^2\nabla^2}{2m} + V(\boldsymbol{x}) - \boldsymbol{x}\cdot\boldsymbol{F_R}(t) + \frac{\hbar\Gamma}{2i}\log\left[\frac{\psi}{\psi^*}\right] + W(t)\right]\psi = i\hbar\frac{\partial\psi}{\partial t}. \quad (5)$$

This replaces the Schrödinger equation (1) when the particle is interacting with a lossy environment. Equation (5) is due to Kostin [18,19]. The stochasticity is due to $\boldsymbol{F_R}(t)$, a random force (which is only a function of time) caused by the environment, and the non-linear log term generates the frictional loss. Γ is a conventionally defined frictional coefficient, with dimensions of $[\text{time}]^{-1}$. $W(t)$ is chosen to satisfy

$$W(t) = -\frac{\hbar\Gamma}{2i}\int d^3\boldsymbol{x}\,\psi^*\log\left[\frac{\psi}{\psi^*}\right]\psi, \quad (6)$$

purely so that the expectation value of the square bracket in (5) gives the average energy of the particle. However it can be absorbed into the phase of ψ, and does not play an important role. Kostin's is not the only Schrödinger picture quantum friction equation on the market. However in a recent paper [20] we have argued that it is to be preferred. This is because using the same decomposition as for Eqs. (2), one finds that the probability conservation equation (2b) is unchanged, and (2a) becomes

$$\frac{(\nabla S)^2}{2m} + V + Q + \frac{\partial S}{\partial t} + \Gamma S - \boldsymbol{x}\cdot\boldsymbol{F_R}(t) + W(t) - 0. \quad (7)$$

Equation (7) is nice because the ΓS term is exactly what one would choose in order to modify the classical Hamilton-Jacobi equation to allow for friction. This, along with the fact that (2b) is unchanged, means that ∇S still plays the role of the momentum, and so the guiding equation in the form (4) is still appropriate for the calculation of the particle trajectories.

In general stochastic non-linear differential equations require immediate resort to numerical computation. For illustrative purposes here, however, we choose to ignore the stochastic term. In the state evolution approach, the non-linear term represents the dissipative effect of the environment and the stochastic term represents the fluctuations. By the well known fluctuation-dissipation theorem [21–24] these are related, the mean square fluctuations being proportional to the dissipation present and the environmental temperature. Thus ignoring the stochastic term is tantamount to assuming that the quantum object has significant coupling to a dissipative environment which is at a low temperature, so that its thermal fluctuation effects can be neglected in comparison to its dissipative effects. We also implicitly ignore quantum fluctuations. Our choice has the advantage that one can make headway analytically. We consider two cases; both involve the damping of oscillations for a one-dimensional oscillator with a potential $V(x) = m\Omega^2 x^2/2$.

The first case is the damping of an oscillatory coherent state. The appropriate solution to (5) (without the stochastic term) is due to Kan and Griffin [25]

$$\psi(x,t) = \frac{\alpha^{\frac{1}{2}}}{\pi^{\frac{1}{4}}} \exp\left[\frac{-\alpha^2(x-X)^2}{2}\right] \exp\left[i\left[\frac{xP}{\hbar} - g\right]\right] \tag{8}$$

Here $\alpha^2 \equiv m\Omega/\hbar, X \equiv X_0 \exp(-\Gamma t/2)\cos\Omega_r t, P \equiv m\dot{X}$, and

$$g \equiv \frac{\Omega t}{2} + \frac{1}{\hbar}\int_0^t dt' \left[\frac{P^2}{2m} - \frac{m\Omega^2 X^2}{2} - \Gamma P X\right]. \tag{9}$$

We note that, as would be expected from the behaviour of a damped classical oscillator, the peak in the probability density P found from (8) oscillates at the shifted frequency $\Omega_r \equiv (\Omega^2 - \Gamma^2/4)^{\frac{1}{2}}$ as its amplitude decays. This is shown in Fig. 1. The particle trajectories which follow from (4) taking S from (8) are simply damped cosines. We show an ensemble of these in Fig. 2.

The second case is the damping of weak coherent quantum oscillations. If one takes the oscillator ground state u_0 and a small admixture of an excited state u_e, an approximate solution to (5) (without the stochastic term) is [20]

$$\begin{aligned}\psi(x,t) = \exp\left[\frac{-i}{\hbar}\int^t dt' W(t') - \frac{i\Omega}{2\Gamma}\left(1-e^{-\Gamma t}\right)\right] \\ \cdot\left\{\left(1-\alpha_0^2\, e^{-\Gamma t}\right)^{\frac{1}{2}} u_0 + e^{-i\Omega t}\alpha_0\, e^{-\Gamma t/2} u_e.\right\}\end{aligned} \tag{10}$$

In this case the coherent oscillations occur at the oscillator frequency Ω. With the definitions $y \equiv \alpha x, \tau \equiv \Omega t, \delta \equiv \Gamma/\Omega$ and with the choice $u_e = u_1$, the trajectory Eq. (4) reads

$$\frac{dy}{d\tau} = \frac{-2^{\frac{1}{2}}\alpha_0\, e^{-\delta\tau/2}\left(1-\alpha_0^2\, e^{-\delta\tau}\right)^{\frac{1}{2}}\sin\tau}{\left[1-\alpha_0^2\, e^{-\delta\tau} + 2\alpha_0^2\, e^{-\delta\tau}y^2 + 2^{3/2}\alpha_0\, e^{-\delta\tau/2}\left(1-\alpha_0^2\, e^{-\delta\tau}\right)^{\frac{1}{2}} y\cos\tau\right]}. \tag{11}$$

With the alternative choice $u_e = u_2$, we obtain

$$\begin{aligned}\frac{dy}{d\tau} = -2^{3/2}\alpha_0\, e^{-\delta/2}\left(1-\alpha_0^2\, e^{-\delta\tau}\right)^{\frac{1}{2}} y\sin 2\tau\big[1-\alpha_0^2\, e^{-\delta\tau} \\ + \alpha_0^2\, e^{=\delta\tau}\left(2y^2-1\right)^2/2 + 2^{\frac{1}{2}}\alpha_0\, e^{-\delta\tau/2}\left(1-\alpha_0^2\, e^{-\delta\tau}\right)^{\frac{1}{2}}\left(2y^2-1\right)\cos 2\tau\big]^{-1}.\end{aligned} \tag{12}$$

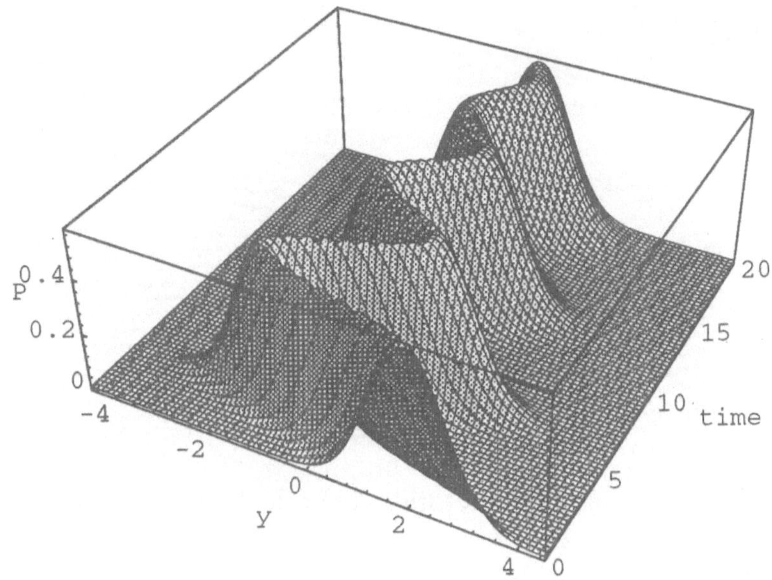

Fig. 1. The probability density $P = \psi^* \psi$, plotted as a function of the dimensionless quantities $y \equiv \alpha x$ and $\tau \equiv \Omega_r t$. We have chosen $\alpha X_0 = 2$ and $\Gamma / \Omega_r = 0.2$.

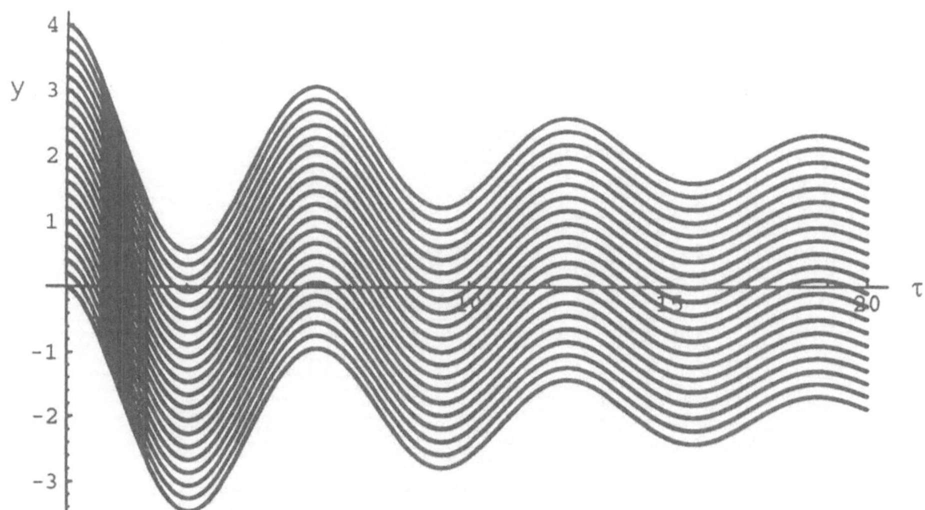

Fig. 2. An ensemble of trajectories $x(t) = x_0 + X_0 \exp(-\Gamma t/2) \cos \Omega_r t$, plotted in terms of $y = \alpha x$ as a function of $\tau = \Omega_r t$. We have chosen $\alpha X_0 = 2, \Gamma / \Omega_r = 0.2$, and a range of αx_0 from -2 to 2.

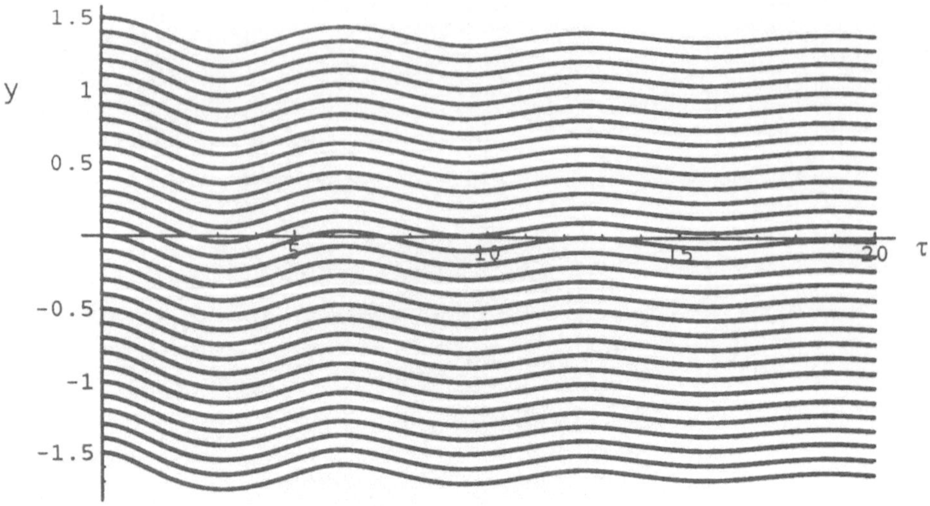

Fig. 3. An ensemble of trajectories $y(\tau)$ calculated from Eq. (11), choosing $\alpha_0 = 0.1$ and $\delta = 0.2$.

We show ensembles of trajectories calculated from these equations in Figs. 3 and 4. In the case $u_e = u_1$, the trajectories all follow the form of the

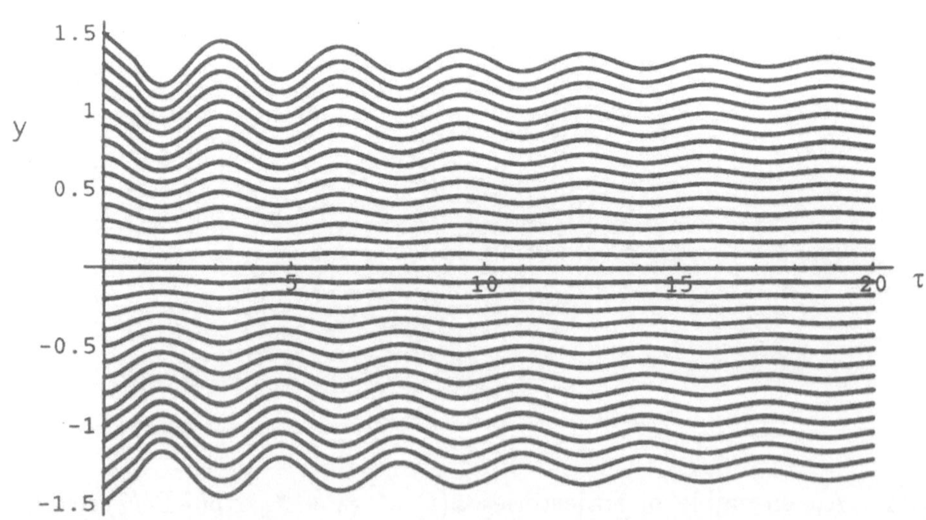

Fig. 4. An ensemble of trajectories $y(\tau)$ calculated from Eq. (12), choosing $\alpha_0 = 0.1$ and $\delta = 0.2$.

expectation value

$$<y>= 2^{\frac{1}{2}}\alpha_0\, e^{-\delta\tau/2}\left(1-\alpha_0^2\, e^{-\delta\tau}\right)^{\frac{1}{2}}\cos\tau \qquad (13)$$

The appropriately weighted average over the ensemble indeed yields this result [20]. In the case $u_e = u_2$, then of course $< y >= 0$. However individual trajectories still exhibit oscillations (at twice the frequency of the $u_e = u_1$ case). Clearly though, from Fig. 4, the trajectories either side of the origin mirror each other, and the average is zero at all times.

Our second example equation is

$$\left[(1-ik)\left[\frac{-\hbar^2}{2m}\nabla^2+V(\boldsymbol{x})\right]+ik<H>_\psi\right]\psi=i\hbar\frac{\partial\psi}{\partial t}. \qquad (14)$$

This equation is due to Gisin [26]. As we again ignore the stochastic effects of the environment, we have omitted any such terms in (14). The expectation value of the Hamiltonian is evaluated for the state ψ, and so is a function of time, and k is the damping parameter. It is easy to prove [26] that the evolution of ψ preserves the integrated probability, and that

$$\frac{d}{dt}<H>_\psi= 2k\left[<H>_\psi^2 - <H^2>_\psi\right]<0, \qquad (15)$$

so energy is removed by the environment as k is positive. Many other physically reasonable properties also follow [26]. However, despite the conservation of integrated probability, the probability density flow equation is not simply (2b). Instead one finds

$$\nabla\cdot\left[\frac{P(1+k)^2\nabla S}{m}\right]-\frac{2kP}{\hbar}\left[\frac{\partial S}{\partial t}+<H>\right]+\frac{\partial P}{\partial t}=0. \qquad (16)$$

This means that the particle velocity in this model cannot simply be given by $\nabla S/m$; the second term in (16) has to be absorbed into the first, and a new definition of the velocity selected accordingly.

This can be achieved. We illustrate how using the example of a quantum decay. Once again, despite the non-linear nature of (14), analytic solutions exist [26]. We take the simple one

$$\psi=\left[\cos\eta e^{-k\omega_0 t}\psi_0+\sin\eta e^{-k\omega_1 t}\psi_1\right]N, \qquad (17)$$

where ψ_0 and ψ_1 are solutions to the usual Schrödinger equation ($k = 0$ in (14)), with $E_0 < E_1$, and

$$N\equiv\cos^2\eta e^{-2k\omega_0 t}+\sin^2\eta e^{-2k\omega_1 t} \qquad (18)$$

Here η is a parameter. Choosing it close to $\pi/2$ means that the $t = 0$ state is $\psi \simeq \psi_1$ and the state at large times is $\psi = \psi_0$. The quantum object undergoes a decay, and gives up the energy $\sim E_1 - E_0$ to its environment. Taking the simple harmonic oscillator $V(x) = m\Omega^2 x^2/2$, and ψ_0 and ψ_1 as the ground and first excited states respectively, we illustrate this decay in Figs. 5–8. These show the time dependence of $< H >$, the $t = 0$ and large time probability densities, and the time evolution of the density. Clearly the expectation value of position (we use $y \equiv \alpha x$) vanishes for any oscillator energy eigenstate. However it exhibits time dependence as the decay occurs. This is shown in Fig. 9. Taking due account of (16), the trajectory equation for our example (in dimensionless form) reads

$$\frac{dy}{d\tau} = \frac{-2^{\frac{1}{2}} \sin 2\eta \left[(1+k^2)\sin\tau + ke^{-2k\tau}N^{-1}F\right]}{\left[e^{k\tau}\cos^2\eta + 2y^2 e^{-k\tau}\sin^2\eta + 2^{\frac{1}{2}}\sin 2\eta\cos\tau\right]}, \tag{19a}$$

where F is defined to be

$$F \equiv 2^{-\frac{1}{2}} y \sin 2\eta - ke^{2k\tau}N \ \sin\tau + \left[e^{k\tau}\cos^2\eta - e^{-k\tau}\sin^2\eta\right]\cos\tau \tag{19b}$$

and N is given in (18). An ensemble of trajectory solutions to (19a) is shown in Fig. 10. Two bunches of initial y-values are chosen, corresponding to the initial peaks in the probability density. As well as oscillating for the period of the decay, the trajectories clearly converge towards the position of the final peak in the density. This provides a graphic illustration of the process of a simple quantum decay.

Clearly for our example, as we have ignored the stochastic effects of the environment, the decay will occur at the same time interval for repeated 'runs' of the calculation. In order to achieve a distribution of intervals for when the decay occurs, the stochasticity must be included. It is interesting to note that then one can begin (at $\tau = 0$) with a pure first excited state for the oscillator; in our example here we had to include a small piece of the ground state to initiate the decay. Stochastic contributions to the state evolution would automatically produce such a piece sometime in the future evolution of an intially pure state. This will "seed" the decay, and thus it will not occur at the same interval in repeated runs of the calculation[14], in keeping with repeated runs of an experiment observing a decay.

The model discussed here can be contrasted with discussions of a decay made purely within the Bohm interpretation of Schrödinger evolution [27,28]. Here only one mode, or degree of freedom, of the environment is retained, so the full configuration space trajectories of the particle and environment system can be calculated. In our example we have, implicitly,

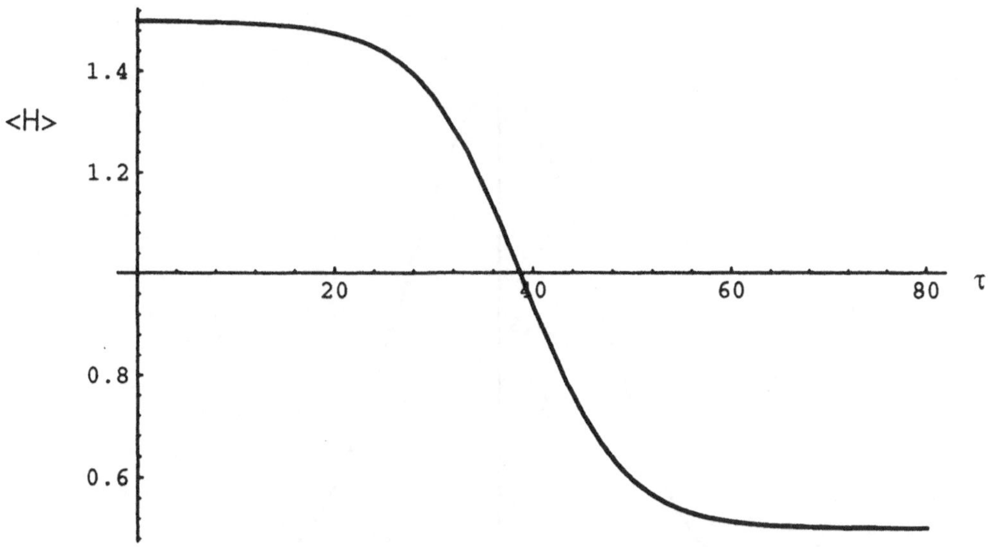

Fig. 5. The energy expectation value $< H >$ (in units of $\hbar\Omega$) as a function of time for the solution (17). We have defined $\tau \equiv \Omega t$ and chosen $\eta = 1.55$ and $k = 0.1$ for this and the subsequent figures.

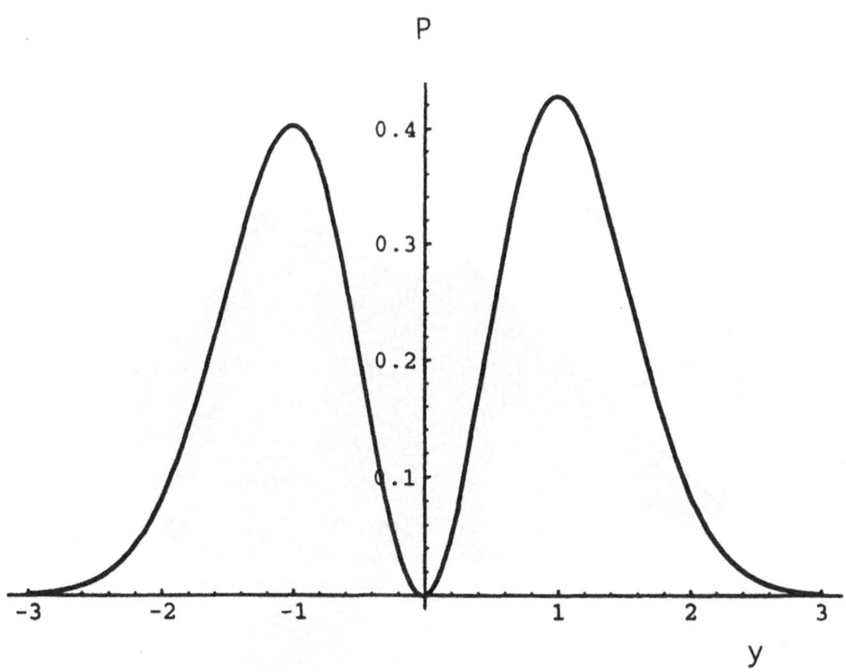

Fig. 6. The $\tau = 0$ probability density $P = \psi^*\psi$ for the solution (17), shown as a function of $y = \alpha x$.

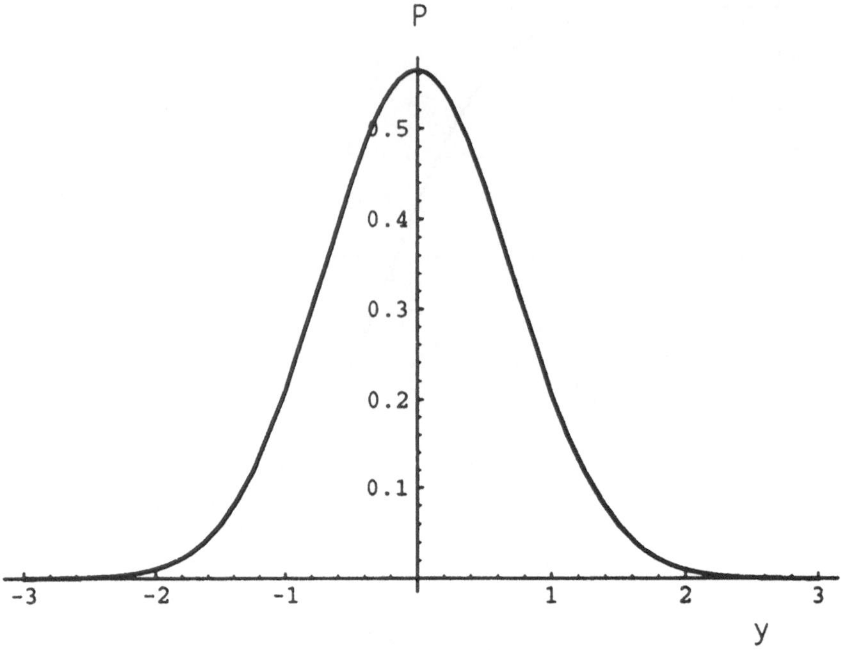

Fig. 7. The $\tau = 80$ probability density $P = \psi^*\psi$ for the solution (17), shown as a function of $y \equiv \alpha x$.

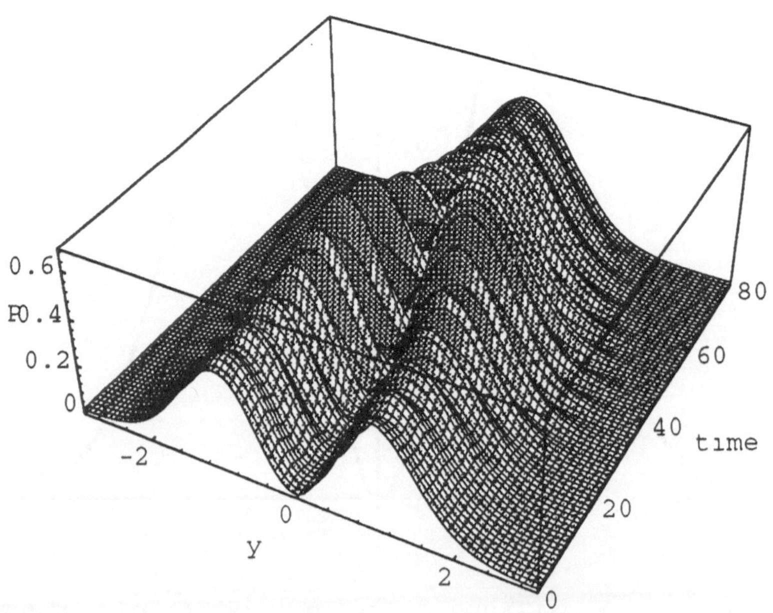

Fig. 8. The time evolution of the probability density $P = \psi^*\psi$ for the solution (17).

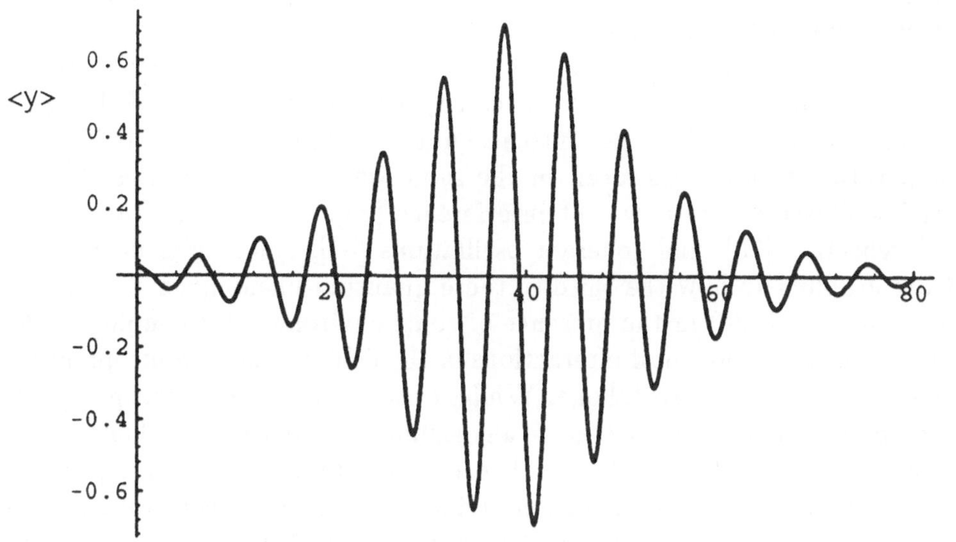

Fig. 9. The expectation value of position $<y>$ for the solution (17), plotted as a function of τ.

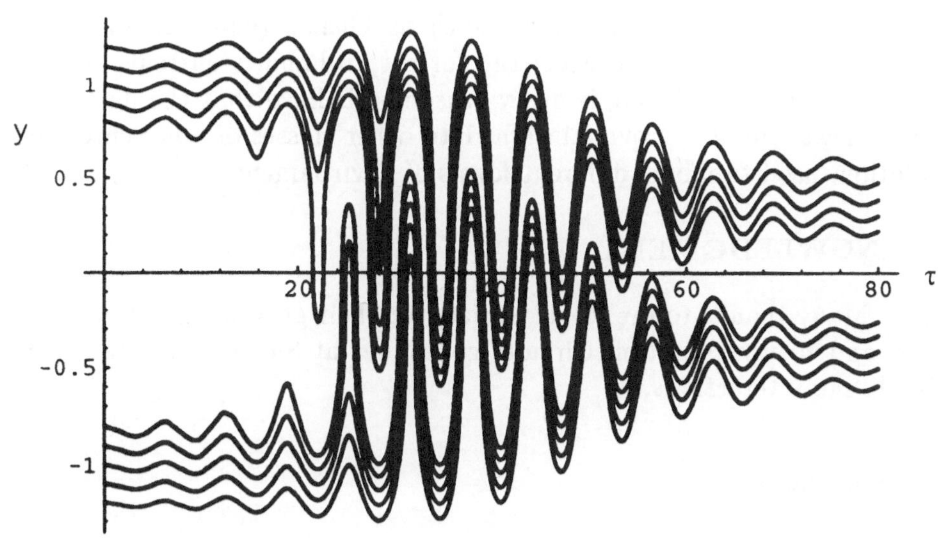

Fig. 10. An ensemble of trajectories $y(\tau)$, calculated from (19) for a range of initial values $y(0)$.

retained all the environmental degrees of freedom. Thus the full configuration space trajectories clearly cannot be calculated. That is why we have computed the 'reduced' particle trajectories directly, using a modified evolution equation for the particle state.

Obviously all the results presented here are calculated from an overly simplified approach. As we have stressed, the stochastic effect of the environment should also be taken into account to produce more realistic finite temperature results. However, in the same way that trajectories provide striking illustrations of two-slit interference [29], barrier penetration and well reflection [30], and coherent oscillations [31], our results do demonstrate graphically how the third of these quantum phenomena can persist (at least for a time) in the presence of some environmental coupling. The role of such environmental interactions in the field of macroscopic quantum phenomena [32] is a privotal one. Whilst experiments on microscopic quantum objects can frequently remove any significant environmental effects (so pure Schrödinger state evolution then provides an accurate description of the behaviour between preparation and measurement), this is not so when the quantum degrees of freedom under investigation are macroscopic collective co-ordinates. It is very hard to decouple these from environmental effects, so the persistence of quantum phenomena under such conditions [24,33,34] is of great interest. For our example of quantum friction, the de Broglie-Bohm view of the Schrödinger picture provides helpful insight, and leads one naturally to the choice of Kostin's equation, (5), out of the many alternatives. For other equations, such as Gisin's (equation (14)), this view highlights the need to select the correct probability current, and thus trajectory equation. There is every reason to believe that the de Broglie-Bohm views will also provide insight into other classes of non-Schrödinger evolution, used to model different forms of environment.

ACKNOWLEDGMENT

Many thanks to the Royal Society for their generous funding of this work, and to my Quantum Circuits colleagues at Sussex for innumerable stimulating interactions.

REFERENCES

1. L. de Broglie, in *Rapport au Ve Congres de Physique Solvay* (Gauthier-Villars, Paris, 1930).
2. D. Bohm, *Phys. Rev.* **85**, 166; 180 (1952).
3. D. Bohm, B. J. Hiley, and P. N. Kaloyerou, *Phys. Rep.* **144**, 321 (1987).
4. D. Bohm, J. Bub, *Rev. Mod. Phys.* **38**, 453 (1966).

5. P. Pearle, *Phys. Rev. D* **13**, 857 (1976).
6. P. Pearle, *Int. J. Theor. Phys.* **18** 489 (1979).
7. P. Pearle, *J. Stat. Phys.* **41**, 719 (1985).
8. N. Gisin, *Phys. Rev. Lett.* **52**, 1657 (1984).
9. N. Gisin, *Helv. Phys. Acta* **62**, 363 (1989).
10. L. Diósi, *J. Phys. A* **21**, 2885 (1988).
11. L. Diósi, *Phys. Lett. A* **129**, 419 (1988).
12. G.-C. Ghirardi, A. Rimini, and T. Weber *Phys. Rev. D* **34**, 470 (1986).
13. G.-C. Ghirardi, P. Pearle, and A. Rimini, *Phys. Rev. A* **42**, 78 (1990).
14. N. Gisin, I. C. Percival, *J. Phys. A* **25**, 5677 (1992).
15. For an interesting discussion of this point, see Ref. 14.
16. G. W. Ford, M. Kac, and P. Mazur, *J. Math. Phys.* **6**, 504 (1965).
17. G. W. Ford, J. T. Lewis, and R. F. O'Connell, *Phys. Lett. A* **128**, 29 (1988).
18. M. D. Kostin, *J. Chem. Phys.* **57**, 3589 (1972).
19. M. D. Kostin, *J. Stat. Phys.* **12**, 145 (1975).
20. T. P. Spiller, P. S. Spencer, T. D. Clark, J. F. Ralph, H. Prance, R. J. Prance, and A. J. Clippingdale, *Found. Phys. Lett.* **4**, 507 (1991).
21. H. Nyquist, *Phys. Rev.* **32**, 110 (1928).
22. H. B. Callen, J. H. Welton, *Phys. Rev.* **83**, 34 (1951).
23. R. Kubo, *J. Phys. Soc. Japan* **12**, 570 (1957).
24. R. P. Feynman, F. L. Vernon Jr., *Ann. Phys.* **24**, 118 (1963).
25. K. K. Kan, J. J. Griffin, *Phys. Lett.* **50B**, 241 (1974).
26. N. Gisin, *J. Phys. A* **14**, 2259 (1981).
27. C. Dewdney, and M. M. Lam, in *Information Dynamics*, H. Atmanspacher and H. Scheingraber, eds. (NATO ASI Series B **256**) (Plenum, New York, 1991), p. 329.
28. C. Dewdney, G. Horton, M. M. Lam, Z. Malik, and M. Schmidt, *Found. Phys.* **22**, 1217 (1992).
29. C. Philippidis, C. Dewdney, and B. J. Hiley, *Nuovo Cimento* **52B**, 15 (1979).
30. C. Dewdney, and B. J. Hiley, *Found. Phys.* **12**, 27 (1982).
31. T. P. Spiller, T. D. Clark, R. J. Prance, H. Prance, and D. A. Poulton *Found. Phys. Lett.* **4**, 19 (1991).
32. For a recent review, see T. P. Spiller, T. D. Clark, R. J. Prance, and A. Widom, *Prog. Low Temp. Phys.* **XIII**, 219 (1992).
33. A. O. Caldeira, and A. J. Leggett, *Phys. Rev. Lett.* **46**, 211 (1981).
34. A. O. Caldeira, and A. J. Leggett, *Ann. Phys.* **149**, 374 (1983).

ON THE WAVE SYSTEM THEORY OF THE EPR EXPERIMENT

Thomas B. Andrews

3828 Atlantic Avenue
Brooklyn, New York 11224

The EPR polarization experiment is interpreted in terms of a new basic theory of physics called the "Wave system theory." This theory was developed to explain why elementary particles occur but is also applicable to the EPR experiment. The wave system theory assumes the universe is a pure wave system and hypothesizes that the elementary particles are the constructive interference peaks of the wave modes. Using this theory, it may be shown that the wave amplitudes associated with each photon exist physically at each calcite crystal detector. Consequently, the quantum mechanics or QED calculations may be shown to have a physical basis. This enables the results of the EPR experiment to be logically understood without invoking hidden variables or non-local effects.

Key words: EPR experiment, photon polarization, QED, hidden variables, non-local effects, wave amplitudes.

1. INTRODUCTION

In a photon polarization EPR experiment, an atom emits two photons with the same polarization in opposite directions along the z axis. The polarization of each photon is observed by separate calcite crystals. When the crystals are orientated the same, the same polarizations are measured at each crystal. Bell [1] then argued that, because we can predict the result of measuring the polarization of either photon by first measuring the polarization of the other photon, the result of any measurement must be predetermined. This implies that "hidden values" are carried by each photon.

When the calcite crystals are not orientated the same, the probability of measuring the polarizations the same is $\cos^2 v$ where v is the angle between the two crystals. For this case, Bell developed Bell's inequality. Bell's inequality is based on two explicit assumptions: (1) The EPR results are predetermined by unknown or hidden parameters; (2) the polarization measured at one calcite does not depend in any way on the orientation of the other calcite. The second assumption is referred to as the locality assumption.

Bell's inequality predicts results that differ from quantum mechanics. However, the EPR experiments agreed with predictions of quantum mechanics.

The most likely conclusion then is that hidden values do not exist. Nevertheless, some physicists question the locality assumption in Bell's inequality and speculate that superluminal signaling is responsible for the EPR experimental results. Unfortunately, neither quantum mechanics nor Bell's inequality offer a rationale for deciding which assumption is incorrect.

The EPR problem originated over 50 year ago. Bell's inequality made it possible to test EPR experimentally and further deepened the problem. Despite an enormous amount of effort to solve the problem, EPR continues to puzzle physicists. Consequently, I believe a solution to the EPR problem does not exist within either the formalism of quantum mechanics or the current theories of the physical world. Furthermore, I suggest that resolving the EPR problem is vital to the advancement of basic physical theory.

Consequently, I propose a new basic theory of the physical world, called the "Wave system theory" [2]. This theory was developed primarily to explain why elementary particles occur but has much wider physical applications. Through its use, many basic laws of physics have been derived, for example, Newton's law $F = ma$ and the time-independent Schrödinger equation.

2. WAVE SYSTEM THEORY

H. Giorgi [3] has proved that waves must exist given an infinite linear system in which the laws of nature exhibit both translational symmetry and time symmetry. The proof is as follows: The modes of oscillation of any system with these symmetries are governed by the representations of both the space and time translation groups. The solution of each representation is a complex exponential in space and time, respectively. Combining these solutions results in left and right progressive waves, given by $\exp i(\omega t \pm kx)$, which form a standing wave mode.

Assuming Giorgi's proof applies and the physical universe is physically a pure wave system, it is hypothesized that the elementary particles such as the proton and electron are the constructive interference peaks of the wave modes of the wave system. These peaks are extremely localized and widely separated by destructive interference regions and thus correspond closely to our picture of elementary particles. This represents a new paradigm in physics.

2.1. Stability of Wave System

Since the electron and the proton are stable particles, it is first necessary to show that the constructive interference results in a stable system and stable particles. The stability of the wave system is based on the following two principles: (1) The wave system frequency is reduced if the mass density and tension are increased at the wave peaks; (2) as the number of particles interacting increase, the system frequency increases.

To apply the first principle to the wave system, the mass density and tension are assumed proportional to the local energy density of the wave system. Then, the frequency of the wave modes will decrease when constructive interference produces high energy concentrations at the peaks. Conversely, if the constructive interference is reduced, the frequency will increase. But this requires a global increase in the energy of the system which cannot occur. Consequently, once formed, the constructive interference peaks must continue to interfere constructively.

The energy density at a peak is proportional to the square of the number of wave modes constructively interfering. Consequently, the decrease in the frequency of the system is very large. For example, if the natural frequency of the wave system without constructive interference is f_o, complete constructive interference reduces the frequency [2] to

$$f_N = f_o/N. \tag{1}$$

Since the number of wave modes increases with the distance between particles, the frequency could go to zero as N increases. However, by the second principle, there is a lower bound on the system frequency determined by the number of interacting particles in the system.

The minimum frequency [4] or lower bound is given exactly by the eigenvalue equation

$$f_p = \frac{1}{2\pi}\sqrt{n}\sqrt{k/m}, \tag{2}$$

where n is the effective number of particles interacting and k/m is the average interaction constant between any two particles.

The process which determines the lower bound on the frequency is as follows: First, the increase in the number of wave modes is driven by the decrease in the eigenvalue f_N. Then, as the number of wave modes constructively interfering increases, the distance between particles in the universe increases. This decreases the wave energy reflected from the particles back into the local system since the particles are further apart. In turn, this increases the mean interaction radius of the system, the volume of the system and thus the number of interacting particles.

A stable frequency of the wave system is attained when $f_N = f_p$. This stable frequency is about 10^{23} hertz. At this stable frequency, the mean interactive radius and mass density of the universe are also determined.

2.2. Local Effects of the Wave System Theory

The wave system constrains a mass particle or a photon to follow a path where the energy density is greatest in order to minimize the frequency of the wave system. This corresponds to a path (generally one of many) along which the particle remains in phase with the wave modes, since along this path the particle has the greatest energy density. Quantum mechanics depends on this effect.

For a mass particle, the in-phase condition is determined by the particle's de Broglie wavelength, $\lambda = h/(\gamma m v)$, and its spin state. For a photon, the in-phase condition is determined by the photon's wavelength and polarization state.

2.3. Quantum Electrodynamics

Quantum Electrodynamics (QED) [5] is based on a set of empirical rules for calculating interaction amplitudes. These rules are: (1) If an event can happen in alternative ways, the amplitudes for the different ways are added. (2) If an event occurs as a succession of steps or depends on independent sub-events, then the amplitudes of each step or sub-event are multiplied. In general, each amplitude is represented by the length of an arrow which rotates with distance as it moves. One complete rotation occurs for each wavelength a photon moves.

However, when considering the polarization of photons, rotation is not a factor. A photon polarized in the χ direction at one position will still be polarized in the χ direction at any other position.

From the wave-system viewpoint, the amplitudes of the waves are real physical effects. Thus, the square of the amplitude determines the physical intensity or local energy density. After normalization, the intensity corresponds to the quantum probability. This explains the mysterious use of wave amplitudes in QED. Thus, the quantum probability is clearly a logical result of the wave nature of the universe. To paraphrase this, "the wheels and gears" of nature are the physical interactions of a pure wave system.

3. POLARIZATION CALCULATIONS

In QED, the physical assumption that the two photons are emitted in the same state of polarization is represented by the singlet state, given by

$$|\,\psi\,\rangle \;=\; \frac{1}{\sqrt{2}}\,|\,(xx+yy)\,\rangle. \tag{3}$$

Formally, xx represents *two ordinary photons* polarized along the x axis and yy represents *two extraordinary photons* polarized along the y axis.

In this paper, the polarization calculations will be based on a generalization of the singlet state which assumes both photons are emitted either ordinary at the random angle χ with amplitudes x' or extraordinary at the angle $\chi + 90°$ with amplitudes y'.

3.1. Amplitude for Measuring Both Photons Ordinary

Only the case of measuring both photons ordinary will be considered. Other cases are similar. This calculation is based on adding the amplitudes for the two possible events which result in measuring both photons ordinary — both photons emitted ordinary or both photons emitted extraordinary.

The amplitude for measuring both photons ordinary, given that both photons are emitted ordinary at the angle χ, is the product

$$\langle\, x_\phi x_\theta \,|\, \psi_{x'x'} \,\rangle \;=\; \langle\, x_\phi \,|\, \psi_{x'} \,\rangle\langle\, x_\theta \,|\, \psi_{x'} \,\rangle. \tag{4}$$

This follows from a specific rule in QED for combining independent amplitudes. Here independence implies that the physical measurement of each photon is strictly a local interaction and that no signals are exchanged between the two calcite crystals. Then, using the coordinate transformations

$$|\,x'\,\rangle \;=\; \cos(\chi-\phi)\,|\,x_\phi\,\rangle \;+\; \sin(\chi-\phi)\,|\,y_\phi\,\rangle \tag{5}$$

and

$$|\,x'\,\rangle \;=\; \cos(\chi-\theta)\,|\,x_\theta\,\rangle \;+\; \sin(\chi-\theta)\,|\,y_\theta\,\rangle, \tag{6}$$

we find

$$\langle\, x_\phi x_\theta \,|\, \psi_{x'x'} \,\rangle \;=\; \frac{1}{\sqrt{2}}\cos(\chi-\phi)\cos(\chi-\theta). \tag{7}$$

Next, reference the polarization calculations to one of the calcite crystal axes, say ϕ. On substituting new variables, given by $z' = \chi - \phi$ and $z = \theta - \phi$, into Eq. (7), the equation in the new variables is seen to be

$$\langle\, x_\phi x_\theta \,|\, \psi_{x'x'} \,\rangle \;=\; \frac{1}{\sqrt{2}}\cos z'\cos\,(z'-z). \tag{8}$$

Then, using the identity

$$\cos(z' - z) = \cos z' \cos z + \sin z' \sin z, \tag{9}$$

for the second factor, Eq. (8) reduces to

$$\langle\, x_\phi x_\theta \mid \psi_{x'x'} \,\rangle \;=\; \frac{1}{\sqrt{2}}(\cos^2 z' \cos z + \cos z' \sin z' \sin z). \tag{10}$$

Eq. (10) is the partial amplitude for measuring both photons ordinary, given that both were emitted ordinary at the angle χ.

Similarly, the amplitude for measuring both photons ordinary given that both are emitted extraordinary is

$$\langle\, x_\phi x_\theta \mid \psi_{y'y'} \,\rangle \;=\; \frac{1}{\sqrt{2}} \sin z' \sin(z' - z) \tag{11}$$

which reduces to

$$\langle\, x_\phi x_\theta \mid \psi_{y'y'} \,\rangle \;=\; \frac{1}{\sqrt{2}}(\sin^2 z' \cos z - \cos z' \sin z' \sin z). \tag{12}$$

Finally, the two amplitudes for the measurement of both photons ordinary, given by Eqs. (10) and (12), must be added together, since they result from indistinguishable events. Interference occurs at this point in the calculation since the amplitudes are added. The sum is

$$\langle\, x_\phi x_\theta \mid \psi_{x'x'} \,\rangle + \langle\, x_\phi x_\theta \mid \psi_{y'y'} \,\rangle \;=\; \frac{1}{\sqrt{2}} \cos z = \frac{1}{\sqrt{2}} \cos(\theta - \phi) \tag{13}$$

because the second terms of each equation add to zero and $\sin^2 z' + \cos^2 z' = 1$. The random angle χ does not appear in the final result, showing that Eq. (13) holds independently of the polarization angle of the emitted photons.

4. PHYSICAL INTERPRETATION OF THE EPR EXPERIMENT

Interpretation involves understanding physically why the EPR experimental results are predicted by quantum mechanics and why Bell's Inequality is violated by quantum mechanics.

The major difficulty in the interpretation is the physical meaning of

$$\langle\, x_\phi x_\theta \mid \psi_{x'x'} \,\rangle \;=\; \frac{1}{\sqrt{2}} \cos(\chi - \phi) \cos(\chi - \theta). \tag{14}$$

This equation represents the partial amplitude to measure an ordinary photon at each crystal given that two ordinary photons are emitted.

The multiplication of amplitudes in Eq. (14) follows from the independence of the polarization measurements at each crystal. This is an empirical rule from QED. However, the mathematics is indifferent as to whether both amplitudes appear at each crystal or only one amplitude appears at each crystal. But, from the wave-system point of view, the two amplitudes multiplied together represent the physical amplitude of a wave at each crystal. This, I believe, is where the current interpretation of the EPR experiment fails.

It is, therefore, necessary to assume there are two alternative ways for the wave amplitudes associated with each of the emitted photons to interact with the two crystals. Label by A the polarized amplitude associated with one of the emitted photons and B the polarized amplitude associated with the other photon. In the wave system theory, the polarized wave amplitudes are emitted equally in both directions. Thus, the amplitude A occurs at crystal 1 and the amplitude B at crystal 2. The amplitude A also occurs at crystal 2 and B at crystal 1. The important point is that both amplitudes A and B exist physically at each crystal, although only one photon is detected at each crystal. In the following, the amplitudes corresponding to the two alternatives will be referred to as the "AB amplitudes."

Without the above two physical alternatives, multiplying and then adding amplitudes at each crystal would correspond to purely mathematical operations and would be devoid of any physical meaning. This would not be consistent with a logical physics.

4.1. Crystals Orientated at Same Angle

Consider first the case where the crystals are orientated the same. Given that two ordinary photons are emitted, the partial amplitude for two ordinary photons to be measured when $\theta = \phi$ is

$$\langle\, x_\phi x_\phi \mid \psi_{x'x'} \,\rangle \;=\; \frac{1}{\sqrt{2}} \cos^2(\chi - \phi). \tag{15}$$

Then, based on the AB amplitudes discussion in the previous section, the same partial amplitude occurs at each crystal.

Similarly, given that two extraordinary photons are emitted at the angle χ, the partial amplitude to measure two ordinary photons is

$$\langle\, x_\phi x_\phi \mid \psi_{y'y'} \,\rangle \;=\; \frac{1}{\sqrt{2}} \sin^2(\chi - \phi). \tag{16}$$

The final amplitude for measuring two ordinary photons is the sum of the partial amplitudes of Eqs. (15) and (16):

$$\langle\, x_\phi x_\phi \mid \psi_{x'x'}\,\rangle + \langle\, x_\phi x_\phi \mid \psi_{y'y'}\,\rangle = \frac{1}{\sqrt{2}}(\cos^2 z' + \sin^2 z') = \frac{1}{\sqrt{2}}, \qquad (17)$$

where $z' = \chi - \phi$.

But the same argument can be made in terms of the angle θ at crystal 2 since the configuration of the two polarizers is completely symmetric. Then, we have

$$\langle\, x_\theta x_\theta \mid \psi_{x'x'}\,\rangle + \langle\, x_\theta x_\theta \mid \psi_{y'y'}\,\rangle = \frac{1}{\sqrt{2}}(\cos^2 w' + \sin^2 w') = \frac{1}{\sqrt{2}}, \qquad (18)$$

where $w' = \chi - \theta$. Eq. (18) shows that the amplitude AB at crystal 2 for both photons to be ordinary is also $1/\sqrt{2}$. The intensity or probability to measure two ordinary photons is then the square of the amplitude and equal to 1/2.

In the same way, it may be shown that if an extraordinary photon is measured at one crystal, an extraordinary photon must be measured at the other crystal. What cannot occur is an ordinary photon measured at one crystal and an extraordinary photon measured at the other crystal, since the total amplitude for these events is zero.

4.2. Crystals Orientated at Different Angles

If calcite 2 is orientated at the same angle as crystal 1, the photon at crystal 2 is polarized the same as the photon at crystal 1. This was shown in the preceding section. Therefore, when crystal 2 is orientated at the angle $v = \theta - \phi \neq 0$, the amplitude of the photon at crystal 2 to be polarized the same as the photon at crystal 1 will be equal to the component, $\cos v$. Note that both the physical orientation of crystal 2 relative to crystal 1 and the physics of the experiment determines the amplitude to be polarized the same, not knowledge at either crystal of the orientation of the other crystal.

This two-step process is analogous to a polarization experiment consisting of two calcite crystals arranged in series. Consider a photon which passes through the first crystal polarized ordinary. Assuming the second crystal is at an angle v with the first, the amplitude to measure an ordinary polarized photon at the second crystal is $\cos v$.

5. CONCLUSIONS

The violation of Bell's inequality by the EPR experiments proves to many physicists that hidden variables do not exist and to other physicists that non-local effects occur. However, application of the the wave-system theory shows that the EPR problem is logically understandable without hidden variables or non-local effects.

Furthermore, this explanation of the EPR experiment shows that interference is essential to obtaining the EPR effects. Since interference effects do not occur in the derivation of Bell's inequality, Bell's inequality must be inapplicable to the EPR experimental results and by extension to quantum mechanics.

More generally, the fact that the EPR experimental results can be understood logically using the wave system theory is additional evidence that the world is quantum mechanical in principle.

REFERENCES

[1] J.S. Bell, "On the Einstein-Podolsky-Rosen paradox," *Physics* **1**, 195-200 (1965).

[2] T.B. Andrews, "A wave system theory of quantum mechanics," *Physica* B **151**, 351-354 (1988).

[3] H. Giorgi, "An overview of symmetry groups in physics," talk presented at 1990 fall meeting of the New England Section of the American Physical Society, Yale University; Harvard preprint number HUTP-90/A065, dated 10/90.

[4] F.Y. Chen, "Similarity transformations and the eigenvibration problem of certain far-coupled systems," *Am. J. Phys.* **38** 1036 (1970).

[5] R.P. Feynman, *QED: The Strange Theory of Light and Matter* (Princeton University Press, Princeton, 1985).

CFD OR NOT CFD? THAT IS THE QUESTION

Donald Bedford

Department of Physics
University of Natal
King George V Avenue
Durban 4001, South Africa

The question as to whether counterfactual definiteness, in some form, and however disguised, is needed to formulate any version of Bell's theorem is crucial to our understanding of the nature of quantum non-locality. If the answer can be shown to be "yes," then the formulation of Bell's theorem in an indeterministic universe (such as this one may be) is not possible. It is argued here that the answer depends on one's approach to the measurement problem.

Key words: non-locality, Bell's theorem, measurement problem, counterfactual definiteness.

Is counterfactual definiteness (introduced by Henry Stapp to the quantum world a couple of decades ago) needed to formulate Bell's inequality? Equivalently, is Bell's inequality formulatable in an indeterministic universe? Since the success of quantum theory strongly implies that the universe we are living in is indeterministic, it is essential that these questions be answered if we are to know whether we are dealing with non-locality or not.

Mars 1. Let us be clear on the notion of CFD. You and I are on Earth, and a colleague is on Mars, which happens to be 10 light-minutes away. I give you good odds that a particular volcano is not erupting now, 1200 UT. You decline to bet. A few minutes later, word arrives from our colleague that the volcano started erupting at 1159 UT. I say: "If you had bet, you would have won." You agree. Your choosing to bet or not bet could not influence a volcano on Mars, particularly since the eruption began before you made your choice.

Mars 2. Suppose I offer you the same odds that a particular isolated neutron our colleague has in a bottle has not decayed by now, 1200 UT. You decline. Word soon arrives from our colleague that it decayed at 1159 UT. I say: "If you had bet you would have won." You hesitantly agree, arguing that your choice could not have influenced even a quantum event in the past. Of course, for some observers, the decay was in your future, but in any case it was spacelike separated from the event of your choosing not to bet and hence was uninfluenceable by it.

Mars 3. Our colleague places a half-silvered mirror on top of the (no longer erupting) volcano and returns to base camp. We send a single photon towards the mirror and ten minutes later (when the photon arrives at the mirror) I offer you odds that it will be transmitted. You decline. Ten minutes later you observe the photon's return. I say: "If you had bet you would have won." You disagree, saying that this amounts to a kind of relativistic delayed choice experiment: "It is not at all clear that, had I chosen to bet, the experimental conditions would have been essentially the same. In fact, there can be little doubt that the two experiments are different (my retina, which, after all, played a critical role in the experiment, would have been physically different had I said "Yes" instead of "No") and could have had different outcomes." I am forced to agree with you.

GHZ [1] 1. Three spin-half particles which are in state

$$S = 2^{-1/2}\left(Z_1^{+}Z_2^{+}Z_3^{+} - Z_1^{-}Z_2^{-}Z_3^{-}\right)$$

(Z_1^{+} is the z spin = +1/2 eigenstate for particle 1, etc.) leave Earth in directions 120 degrees to one another heading for three planets (named 1, 2 and 3), 10 light minutes away. On each planet a colleague chooses, at 1159 UT, just before the particles arrive, to measure its x or y spin. The results are radioed back to us on earth. *Quantum theory predicts* that under the different choices shown in the left column, the outcomes (times 2) will satisfy the equations shown in the right column:

$$\begin{array}{ll} X_1X_2X_3 & x_1x_2x_3 = -1 \\ X_2Y_2Y_3 & x_1y_2y_3 = +1 \\ Y_1X_2Y_3 & y_1x_2y_3 = +1 \\ Y_1Y_2X_3 & y_1y_2x_3 = +1 \end{array}$$

I say: "But this is not possible! Clearly, if X_1 is chosen, $x_1(X_2X_3) = x_1(Y_2Y_3)$, since in some reference frame, x_1 occurs before our colleagues on 2 and 3 make their (*free*) *choices*, and thus cannot be influenced by these choices (*locality*). The same holds for all the other possibilities. Therefore, the product of the left hand members of the above four equations is +1 since each x_1, etc. appears twice. The product of the right hand members is −1. There is thus an incompatibility between the three *underlined* notions above:

Free choices, the predictions of quantum theory, locality [2]."

You say: "But the way you have formulated the locality condition involves counterfactual definiteness, which itself is clearly incompatible with quantum theory, which tells us, loud and clear, that unperformed experiments don't have results. In saying that x_1 has the same value in the first two equations above, even if Y_1 were chosen, you are saying just that." I counter: "But there is no other way to formulate the locality condition, and moreover, it does *not* involve CFD. I am saying only that if it *were* measured, the result would be independent of what happened elsewhere. That's not CFD." [3, 4]

GHZ 2. As for GHZ 1, but with human free choice of X or Y replaced by (quantum) random choice. I claim then that quantum theory and locality are incompatible with indeterminacy rather than free choice. You aver that, now that the "choices" are quantum random events, it is *a fortiori* true that the value of x_1 will not necessarily be the same in two different experiments, and the incompatibility cannot be established.

Mars 3'. Just before 1200 UT our colleague takes a neutron and Schrödinger's cat to the top of the quiescent volcano, sets up the famous experiment with neutron decay as trigger, and returns to base camp. At 1159 UT I give you odds that Schrödinger's cat will be alive at 1200 UT. You decline. At 1209 UT you detect the neutrino from the decay. I say: "If you had bet you would have won." You disagree, arguing that*clearly* now we are talking about two crucially different experiments. In detecting the neutrino you participated directly in the decay process. Presumably, before the detection event, the neutron is in a state of |having decayed > + |not having decayed >, and the cat is consequently |dead > + |alive >.

CFD? What is apparent here is that our responses to the measurement "problem" determine our interpretations of x_1 having the same value in two different experiments. I believe that the dying of the cat qualifies as an irreversible phenomenon, a permanent record, capable of "collapsing the wave packet." CFD is not needed to know that the cat would have been dead whether you had decided to bet or not. You maintain that the measurement "problem" has not been "solved," and that, in any case, we should be looking at what quantum theory (in the form of John Wheeler, anyway) tells us: No phenomenon is a phenomenon until it is an observed phenomenon. For you, knowing that the cat would have been dead, whether you had decided to bet or not, is CFD.

GHZ 3'. Use a quantum random event on each planet to chose X or Y, and replace the human observers with video transmitters. You receive the data. I argue that the video transmitters collapsed the wave packet, and you disagree. For you, the collapse occurred when you received the data, and therefore no non-locality is indicated. Furthermore, you insist that CFD is thus again needed to formulate my "locality" condition. I suggest that if they were quantum video transmitters (which consist of at most a few atoms and make no permanent record) I might be forced to agree, but as things stand in the real-world laboratories, where conditions are or will be something like GHZ 3', we disagree regarding the measurement process, and that accounts for our disagreement as to whether we need CFD or do not need CFD to formulate the non-locality condition.

For those whose views on the measurement process correspond, however roughly, to those of the character played by "I" in the dialogue above, the incompatibilities referred to in GHZ 1 and GHZ 2 must be taken seriously. The formation of a permanent record, e.g., in a human brain, or for that matter, in any irreversible process, cannot be allowed to be influenced by events in the future without a serious breakdown in causality, and there is no need for CFD for this to be so. Those whose views correspond to "you" above can legitimately cry "CFD!", or, at least in terms of experiments to date, argue as "you" did in GHZ 3' that non-locality is not indicated anyway. It is likely to be a long time before there is any consensus on measurement, and it certainly will be a long time before spacelike separated human choices of the kind we have been talking about are possible. The debate is far from over.

REFERENCES

1. D.M. Greenberger, M. Horne, and A. Zeilinger, *in Bell's Theorem, Quantum Theory and Conceptions of the Universe*, M. Katatos, ed. (Kluwer Academic, Dordrecht, 1989).
2. H.P Stapp, "Significance of an experiment of the Greenberger-Horne-Zeilinger kind," Lawrence Berkeley Laboratory preprint No. 29377. Stapp augments the trio of incompatible conditions to a quartet by the inclusion of Unique Results (i.e., the rejection of the Everett many-worlds interpretation).
3. D. Bedford and H.P. Stapp, "Non-locality of the Rastall model," *Found. Phys.* **19** 397 (1989); "Bell's theorem in an indeterministic universe," Lawrence Berkeley Laboratory preprint No. 29836.
4. R.K. Clifton, J.N. Butterfield, M.J. Redhead, "Nonlocal influences and possible worlds," *Brit. J. Phil. Sci.* **41** 5 (1990).

QUANTUM PARTICLE AS SEEN IN LIGHT SCATTERING

Lajos Diósi

KFKI Research Institute for Particle and Nuclear Physics
H-1525 Budapest 114, POB 49, Hungary

A possible mathematical model is proposed for motion of illuminated quantum particles seen by eyes or similar devices mapping the scattered light.

Key words: wave function, localization, ito-stochastic equation.

Spatial motion of free quantum particles is described by the Schrödinger wave equation

$$\frac{d}{dt}\psi(x) = \frac{i}{2m}\Delta\psi(x). \tag{1}$$

Typical solutions of this equation show an unlimited growth of the wave packet width σ.

If the particle is illuminated, then the scattered light will show the trajectory of the quantum particle. Of course, the free particle Schrödinger equation (1) will no longer be valid. If a single photon of incoming wave number k_i has scattered with the final wave number k_f, then the scatterer's wave function obtains a unitary factor

$$\psi_i(x) \rightarrow \psi_f(x|\theta) := exp\,(ik_{fi}x)\,\psi_i(x), \tag{2a}$$

where the scattering angle θ between k_i and k_f, resp., is distributed according to the modulus square of the scattering amplitude f:

$$p(\theta) = \frac{1}{4\pi}|f(\theta)|^2. \tag{2b}$$

The processes (2ab) interrupts the otherwise deterministic evolution (1) of the wave function.

We shall assume a certain diffuse light of wave number k. Then the scattering processes will occur randomly at a given rate denoted by ν. To

Waves and Particles in Light and Matter, Edited by
A van der Merwe and A Garuccio, Plenum Press, New York, 1994

get an easy insight into the resulting stochastic process, we shall assume that $\lambda = 2\pi/k$ is much bigger than σ and, furthermore, that the repetition frequency ν of scatterings is big as compared to the time scale of the free dynamics (1) of the particle. The following approximate equation will then describe the dynamics of the illuminated quantum particle:

$$\frac{d}{dt}\psi(x) = \frac{i}{2m}\Delta\psi(x) - iF(t)x\psi(x), \tag{3}$$

where $F(t)$ is a certain stationary white noise of correlation $\gamma\delta(t)$. The dispersion takes the form

$$\gamma = \nu k^2 \int 4sin^2(\theta/2)p(\theta)\frac{d\Omega}{4\pi}. \tag{4}$$

Equation (3) is "almost" an ordinary Schrödinger equation. The particle's motion is influenced by the random force $F(t)$. This force imparts fuzzy phases to the wavepacket but does not prevent its delocalization. Hence, the linear stochastic Eq. (3) is not suitable to represent the experienced trajectory of the quantum particle.

What is wrong with our attempt? Where have we lost the well shaped positions and trajectories of our particle clearly *seen* when illuminated?

We have to go back to the elementary scattering process (2ab). Observe that the jumps (2a) of the wave function assumes *a particular* classification of the photon states. Actually, the final states (as well as the initial ones) have been classified according to their momenta. Obviously, trajectories can not be observed via the scattering *angles* of the photons. They are only observed by identifying the position where the scattered light has emerged. One has to think of a lens inserted on the path of each scattered photon, making an optical map of the scatterer particle.

Let us present the mathematical model of the above set-up. Introduce the special Fourier transform of the scattering amplitude:

$$\ell(x|\xi) \equiv \frac{1}{4\pi}\int e^{ik_{f,i}(x-\xi)}|f(\theta)|^2 d\Omega. \tag{5}$$

The influence of a single scattering process on the particle's wave function can thus be written as

$$\psi_i(x) \rightarrow \psi_f(x|\xi) \equiv \mathcal{N}^{-1}(\xi)\ell(x|\xi)\psi_i(x). \tag{6a}$$

This jump differs very much from the previous one (2a). The jump (6a) is nonlinear and *localizes* the wave function in the neighbourhood of the random position ξ. The probablity distribution of ξ is simply related to the normalization factor of the "reduced" wave function:

$$p(\xi) = \mathcal{N}^2(\xi). \tag{6b}$$

In the same approximation that we assumed for the process (2ab), the nonlinear process (6ab) leads to the following nonlinear counterpart of the phenomenological Eq. (3):

$$\frac{d}{dt}\psi(x) = \frac{i}{2m}\Delta\psi(x) - F(t)(x - < x >)\psi(x) - \frac{\gamma}{2}(x - < x >)^2\psi(x). \tag{7}$$

The derivation of this Ito-stochastic differential equation can be learnt from Ref.[1].

Let us make a formal comparison of Eq. (7) to the "almost" ordinary Schrödinger Eq. (3). The main difference on the rhs is the lack of the factor i in the second term as if the random force had become pure imaginary. (The third term then comes in just to restore normalization of the wave function.) To the authors knowledge, this *imaginary* random force has no direct physical interpretation than that it has come out from Eqs. (1) and (6ab) in the given approximation.

In a search for stochastic equation of state reduction, Gisin [2] discovered in fact the related couple of Eqs. (3) and (7) and showed that Eq. (3) did not reduce, while Eq. (7) did. *Mutatis mutandis*, we mentioned above that Eq. (3) possesses no trajectory-like solutions while, by now not very surprisingly, the nonlinear counterpart (7) has shown to have them.

In Ref.[3] we calculated the stationary solutions of the Eq. (7). These solutions are squeezed Gaussian wave packets,

$$\psi(x) = \left(2\pi\sigma_\infty^2\right)^{-3/4} exp\left(ix < p > -\frac{1-i}{4\sigma_\infty^2}(x - < x >)^2\right), \tag{8}$$

of stationary width $\sigma_\infty = (2/m\gamma)^{1/4}$. The center of the wave packet follows the classical straightline trajectories with $d < x > /dt = < p > /m = const.$, apart from a tiny random walk around it. Exact results [3,4] mimic as if the particle momentum $< p >$ were influenced by the *real* random force $F(t)$, which was not the case at the level of the Schrödinger Eq. (7). Nevertheless, the stationary random walk of the averaged momentum $< p >$ turns out to satisfy the Newton equation

$$\frac{d}{dt} < p > = F(t). \tag{9a}$$

The quantum average of the position satisfies the following equation:

$$m\frac{d}{dt} < x > = < p > + 2m\sigma_\infty^2 F(t), \tag{9b}$$

containing the anomalous second term on its rhs. It would, of course, be important to discuss how the quality of trajectories depends on the intensity and on the spectrum of the light.

In summary, we claim that the processes (6ab) and, especially, the fenomenological Eq. (7) have turned out to be a suitable mathematical model for the seen/detected motion of an illuminated quantum particle. It is, furthermore, worthwhile to mention a few further applications [5–8] of nonlinear stochastic Schrödinger equations, related to the quantum measurement problem.

ACKNOWLEDGMENT

This work was supported by the Hungarian Scientific Research Fund under Grant No 1822/1991.

REFERENCES

1. L. Diósi, *Phys. Lett.* **129A**, 419 (1988).
2. N. Gisin, *Phys. Rev. Lett.* **52**, 1657 (1984).
3. L. Diósi, *Phys. Lett.* **132A**, 233 (1988).
4. D. Gatarek and N. Gisin, *J. Math. Phys.* **32**, 2152 (1991).
5. N. Gisin, *Helv. Phys. Acta* **63**, 929 (1989).
6. L. Diósi, *Phys. Rev.* **A40**, 1165 (1989).
7. G. C. Ghirardi, P. Pearle, and A. Rimini, *Phys. Rev. A* **42**, 78 (1990).
8. A. Barchielli and V. P. Belavkin, *J. Phys. A* **24**, 1495 (1991).

$e^2 = \alpha h$, THE ONLY PHYSICALLY JUSTIFIED FORMULATION OF ELECTRON CHARGE, AND THE RESULTING ELECTRON ENERGY PARADIGM

Gerhard Dorda

Siemens AG
Corporate Research and Development
Otto Hahn-Ring 6
D-81730 Munich, Germany

It is shown that the formulation for the square of the electric charge given by $e^2 = \alpha h$ has priority over other formulations. It allows reference values to be defined for quantized representations of all electromagnetic quantities. A consequence of $e^2 = \alpha h$ is the formulation of the electron energy paradigm. It links five fundamental reference energies by means of the coupling constant α raised to different powers. As shown, this paradigm results from the experimental findings of atomic light spectra, Ohm's law, $1/f$ noise, the quantum Hall effect, the Aharonov-Bohm effect, and electron spin. An attempt to interpret the electron energy paradigm is presented.

Key words: electron charge, units, space and time, quantum effects, Aharonov Bohm effect, noise.

1. INTRODUCTION

In recent years, investigations of transport phenomena in solids have shown universal characteristic features with increasing frequency. Examples of such phenomena are:

(a) The quantum Hall effect (QHE) [1].
(b) Quantum transport in GaAs heterostructures with a point contact [2,3].
(c) Superconductivity in thin metal films [4].
(d) Temperature dependence of the magnetoresistance [5].
(e) Quantum transport through thin amorphous Si films [6].

Waves and Particles in Light and Matter, Edited by
A. van der Merwe and A. Garuccio, Plenum Press, New York, 1994

Typical of all these current transport phenomena is the (quantized) relation to the factor e^2/h. The task that we had set ourselves in these investigations was to determine the significance of this factor. It will be shown that a physical understanding of e^2/h and its application to known experimental phenomena leads to the paradigm of the electron energy. For the purpose of elucidating this paradigm, all aspects required for its derivation will be examined in the following sections.

2. THE FORMULATION OF e^2

Planck's constant h is unambiguously formulated on the basis of the energy-time relationship and thus presents no problem. In contrast, different formulations exist for the square of the electric charge e^2. Depending on the system of electromagnetic units used, the various e^2's differ not only in their value but also in their dimensions. In quantum electrodynamics (QED), e^2 is given by the following dimensionless number [7]

$$e^2 = 4\pi\alpha \tag{1}$$

in conjunction with the normalization condition

$$\hbar = c = 1,$$

where $\alpha \approx 1/137$ is the coupling constant and c the speed of light. The correctness of this formulation has been sufficiently confirmed by the agreement of the theoretical results of QED with the experimental data. It is therefore justified to start with Eq. (1) when changing over to the customary length-mass(energy)-time system of units.

In the SI system of units, e^2 is defined in the following way [8]:

$$(e^2)_{SI} = 4\pi\varepsilon_0\alpha\hbar c, \tag{2}$$

where ε_0 is the dielectric constant and c the speed of light. The quantity e has the dimension of Coulomb.

In the CGS system of units, e^2 is defined as [8]

$$(e^2)_{CGS} = \alpha\hbar c \tag{3}$$

and has the dimension energy $\times$ length.

It is evident that Eq. (1) also permits the formulation

$$e^2 = \alpha h, \tag{4}$$

as an additional variant of (2) and (3). The dimension of e^2 in (4) is energy $\times$ time.

Equations (2), (3), and (4) offer three different formulations for e^2. Experience shows that the electric charge can be defined by means of the basic mechanical units. Analogously to the formulation for the mass M, which is given solely by the formula $M\,c^2 = E$, thus allowing a single plausible representation by means of the basic mechanical units energy E, time, and length, it can be expected that *only one of the equations (3) and (4) represents the simplest and physically most plausible form.* Thus, in the following we will compare the dimensions of the electromagnetic quantities of the CGS system with those of the new electromagnetic system of units resulting from the formula $e^2 = \alpha\,h$. We will not consider the SI system because of its additional, free definition of the ampère, the permeability of vacuum μ_0, and the permittivity of vacuum ε_0.

3. THE MJS SYSTEM OF UNITS

As can be seen from Table I, a new electromagnetic system of units is derived from the definition $e^2 = \alpha\,h$. We call this system the MJS (meter-joule-second) system of units. For the sake of clarity, energy rather than mass has been chosen as the basic unit. The difference between the CGS and MJS systems is shown there by the dimensions of the electromagnetic quantities. The difference can be best seen in the dimensions of the resistance, voltage, capacitance, and inductance. Moreover, in the MJS system the dimensions of the electric field E differ from those of the electric displacement D, and the dimensions of the magnetic field H differ from those of the magnetic flux density B, which is not the case within the CGS system.

4. PRIORITY OF $e^2 = \alpha\,h$

In this section, four features of electricity will be selected which convincingly testify to the priority of $e^2 = \alpha\,h$:

(a) The reference value for the conductance e^2/h yields the speed $3.48 \cdot 10^7$ cm/s in the CGS system which has no physical relevance either in magnitude or dimension. In contrast, the reference factor e^2/h has in the MJS system the significance of the coupling constant α: it represents the coupling between different phase-coherent states of the electrons. In this way it justifies the generalization of Bohr's quantum model by its application to the description of transport phenomena in solids, as was hypothetically assumed in [9,10].

(b) The priority of $e^2 = \alpha\,h$ is also supported by the formula for the

Table 1. The dimensions of the electromagnetic quantities presented in the CGS and meter-joule-second (MJS) systems of units. The quantity of mass has been replaced by that of energy.

Symbol	Quantity	CGS dimensions	MJS dimensions
e	electric charge	$(\text{erg cm})^{1/2}$ (e.s.u.)	$(\text{Js})^{1/2}$
Φ	magnetic flux	$(\text{erg cm})^{1/2}$	$(\text{Js})^{1/2}$
I	electric current	$(\text{erg cm})^{1/2}$ / s	$(\text{Js})^{1/2}$
V	voltage	$(\text{erg cm})^{1/2}$ / cm	$(\text{Js})^{1/2}$ /s
R	resistance	s/cm	1
E	electric field	$(\text{erg cm})^{1/2}$ / cm^2	$(\text{Js})^{1/2}$ / ms
H	magnetic field intensity	$(\text{erg cm})^{1/2}$ / cm^2	$(\text{Js})^{1/2}$ / ms
D	electric displacement	$(\text{erg cm})^{1/2}$ / cm^2	$(\text{Js})^{1/2}$ / m^2
B	magnetic flux density	$(\text{erg cm})^{1/2}$ / cm^2	$(\text{Js})^{1/2}$ / m^2
C	capacitance	cm	s
L	inductance	cm	s
ε_0	permittivity of vacuum	1	s/m
μ_0	permeability of vacuum	1	s/m

reference voltage V_0, which is defined in the MJS system as

$$V_0 = \Phi\, f_e, \tag{5}$$

where Φ is the flux quantum. It is given by

$$\Phi = \frac{h}{e} = \frac{e}{\alpha}, \tag{6}$$

which represents a variant of $e^2 = \alpha\, h$. Equation (5) is obtained from the reference energy $e\, V_0$, which is also known as the atomic reference energy [11]

$$e\, V_0 = 2\, Ry = 2\, R_\infty\, h\, c = h\, f_e = 27.2\,\text{eV}, \tag{7}$$

where Ry is the Rydberg energy and R_∞ the Rydberg constant given by

$$R_\infty = \frac{1}{2}\,(2\pi\, a_e)^{-1}. \tag{8}$$

Here $a_e = 7.25$ nm is the reference length [9,10] and $f_e = c/2\pi\, a_e = 6.58 \cdot 10^{15}$ Hz the reference frequency. The formula for the electric voltage in terms of the flux quantum Φ as expressed by Eq. (5) can immediately seen to be in full agreement with the Josephson effect, whereas the CGS system does not show this agreement explicitely.

(c) The capacitance in the CGS system has the dimension length whose value increases with decreasing spacing between the plates of the capacitor. This is obviously nonsense. In the MJS system, on the other hand, the capacitance represents the integral coupling time as a measure for the number of $e-\Phi$ couplings which must understandably increase with decreasing spacing between the capacitor plates.

(d) The quantum Hall effect (QHE) is given by the relationship

$$N = \nu \frac{e}{h} B \,, \qquad \text{for} \qquad \nu = 1,2,3,\ldots, \tag{9}$$

or by

$$D = \nu \frac{e^2}{h} B, \tag{10}$$

where N is the capacitively induced charge density, B the magnetic flux density, D the electric displacement density, and ν stands for a quantum number. In the MJS system, the relationship

$$D = \alpha B \tag{11}$$

is obtained for the ground state $\nu = 1$. This relation demonstrates that, due to the coupling constant α, a clear difference between D and B is obtained in spite of their dimensions being identical. This same difference is also obtained between E and H [10]. It is thus expedient, also from the viewpoint of formulating E, D, H and B, to give priority to the formula $e^2 = \alpha h$ and consequently to the MJS system.

5. THE QUANTUM CHARACTER OF ELECTROMAGNETIC QUANTITIES

Analogous to the QHE, the current transport shows a quantitative character of the following form [2,3,6]

$$I = 2n \frac{e^2}{h} V, \quad n = 1,2,3,\ldots. \tag{12}$$

In view of the priority of $e^2 = \alpha h$ discussed here, this relationship can be rewritten in the new form

$$I = 2n \alpha V. \tag{13}$$

The fact that the coupling constant cannot be quantized necessarily leads to the conclusion that I and V, and also D and B in view of (11), and thus E and H must be quantizable. The reference value for the current I_0 is obtained by means of the reference voltage V_0 given in (5); i.e., for $2n = 1$, we have

$$I_0 = e\, f_e. \tag{14}$$

To define the reference values E_0, H_0, D_0, and B_0, we require a reference area S_0. As already shown elsewhere, this area is given by $S_0 = 2\pi\, a_e^2$, which again leads to the definition of a two-dimensional reference density of electrons $N_0 = (2\pi\, a_e^2)^{-1} = 3 \cdot 10^{15}\mathrm{m}^{-2}$. This N_0 and thus E_0 and D_0 are in agreement with experimental findings [9,10], especially for two-dimensional systems such as MOSFETs. Thus we assume, on the basis of (11), that the reference value of the magnetic flux density B_0 is also related to S_0 and is therefore given by

$$B_0 = \frac{\Phi}{2\pi\, a_e^2}. \tag{15}$$

6. EXPERIMENTAL VERIFICATION OF THE REFERENCE VALUE B_0

The frequency analysis of the Aharonov-Bohm effect (ABE) yields a support for the correctness of the derivation of the reference value for the magnetic flux density $B_0 = \Phi/2\pi\, a_e^2$. The ABE was discovered experimentally on small thin gold rings by Washburn, Webb, et al. [12-14]. Figure 1 presents the results of the Fourier analysis on the measured oscillations of a large ring and Fig. 2 shows the results for a small ring. We can see from Fig. 1 that a fine structure of the h/e peak occurs with increasing range of the magnetic field. We interpret this structure as an area quantization of the metal ring investigated according to the relationship

$$B = \frac{B_0}{n^2} = \frac{\Phi}{n^2\, 2\pi\, a_e^2}, \tag{16}$$

with n as a quantum number. The position of the structure for the quantum number $n = 39, 40, 41$, and 42 is given in Fig. 1 and for $n = 14, 15, 16$ and 17 in Fig. 2. The quantum numbers listed here correspond to the given width of the metal rings with good accuracy. As can be seen from Figs. 1 and 2, the agreement between the peak positions calculated by means of Eq. (16) and experimentally observed maxima in the fine structure is surprisingly good, so that these data may be regarded as an experimental confirmation of the reference value B_0.

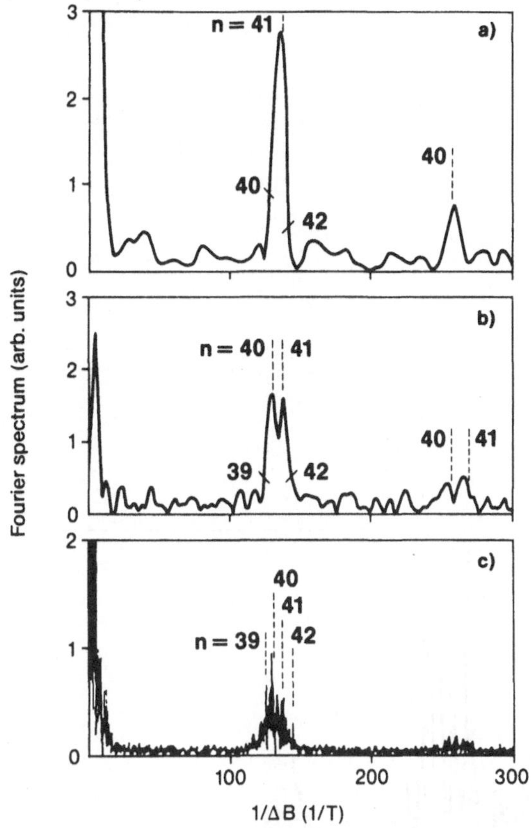

Fig. 1. Fourier spectrum (in arbitrary units) of the Aharonov-Bohm oscillations for a 820nm diameter loop, linewidth 40nm. Range of magnetic field: a) $\Delta B = 0.2\,\mathrm{T}$, b) $\Delta B = 0.4$ T, c) $\Delta B = 8$ T. (Redrawn after Washburn, Webb et al. [12,13])

7. $1/f$ NOISE

It will be demonstrated in this section that the empirical equation first formulated by Hooge for describing the $1/f$ noise in solids can be interpreted in a very interesting new way through the application of the MJS system. Using Hooge's equation, the strength of the noise S_V at constant current is defined as [14,15]

$$S_V = \alpha_H \frac{V^2}{N f}, \tag{17}$$

where V is the voltage, f the frequency, α_H a fitting parameter, and N is a number which is assumed to be the number of free electrons or atoms

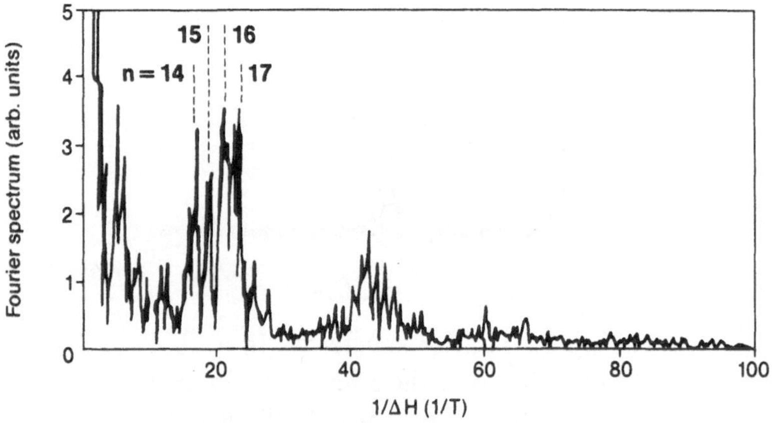

Fig. 2. Fourier spectrum (in arbitrary units) of the Aharonov-Bohm oscillations for a 325nm diameter loop, linewidth 40nm. Range of magnetic field: $\Delta B = 6$ T. (Redrawn after Washburn, Webb et al. [13])

of the specimen under test. The noise strength S_V, also known as the spectral power density, is an energy parameter, as the measurements show by means of thermoelements. As S_V must be invariant with regard to the system of units, it follows that, depending on the electromagnetic system selected, a transformulation constant must be applied to the voltage value V^2. Equation (17) is formulated within the SI system. If we wish to obtain the dimension energy for S_V in the SI system, α_H must take on the significance of conductance. This is, however, in contradiction to experimental findings, which show that α_H is almost universal in character [16,17]. In contrast, by using the MJS system, we directly obtain the dimension energy for V^2/f, which again supports the priority of $e^2 = \alpha\, h$. If we accept the priority of (4) and thus the MJS system as a fact of nature, and when using the SI system, the correction factor $2\varepsilon_0\, c = 5 \cdot 10^{-3}\, \Omega^{-1}$ must necessarily occur at V^2, due to the invariance of energy. The correction factor is obtained according to the transformulation equation

$$(V^2)_{MJS} = (2\varepsilon_0\, c\, V^2)_{SI}\,. \tag{18}$$

In this context, it is therefore highly remarkable that many authors have in fact found an empirical factor α_H in the order of magnitude of 10^{-3} (see the review papers [16] and [17]). This again means that the empirical Hooge equation is to be represented in the MJS system by

$$S_V = \frac{(V^2)_{MJS}}{N\, f}\,. \tag{19}$$

Evidently a new interpretation is required for the number N. According to the new model, N characterizes not only the number of phase-coherent states of the electrons [9,10], but also the strength of localization of these states. The background for this interpretation will become obvious in the following section, where the electron energy paradigm is formulated, and in which equation (19) plays an important role.

8. THE ELECTRON ENERGY PARADIGM

As shown in Secs. 5 and 7, the application of the relationship $e^2 = \alpha\, h$ and the MJS system allows a deeper understanding of a) Ohm's law as well as b) of Hooge's empirical equation for the description of $1/f$ noise and enables c) the formulation of the reference quantity of the magnetic flux density B_0, confirmed by the experimental data on the Aharonov-Bohm effect. These findings allow a new relationship to be formulated between five reference energies of the electron:

(A) The *umklapp* energy of the electron spin is given by

$$\mathbf{E}_{spin} = 2\mu_B\, B, \tag{20}$$

where μ_B is the Bohr magneton. If we now express μ_B and B in (20) within the MJS system and the reference value B_0 (15) is used for B, the *umklapp* reference energy $\mathbf{E}_{o,0}$ is obtained as follows:

$$\mathbf{E}_{o,0} = 2\mu_B\, B_0 = \frac{e\,\hbar\,\Phi}{m_0\, 2\pi\, a_e^2} = \alpha^{-2}\, eV_0 = \alpha^{-2}\, \mathbf{E}_{o,2} \tag{21}$$

Here Eq. (5) and Bohr's angular momentum equation have been applied, whose ground state is given by

$$m_0\, \alpha\, c\, a_0 = \alpha^2\, m_0\, c\, a_e = \hbar, \tag{22}$$

where m_0 is the rest mass of the electron and a_0 the Bohr radius. We have designated the reference energy for the electron potential eV_0 as $\mathbf{E}_{o,2}$.

Note: The index zero at all energies $\mathbf{E}$ denotes the reference value of the quantizable $\mathbf{E}$.

(B) Ohm's law in the ground state can be rewritten in the following form, using Eq. (5), (13), and (14):

$$I_0 = \alpha\, V_0 \tag{23}$$

By multiplying the current by the charge e, a relationship is obtained between the reference energies for the current $\mathbf{E}_{o,1}$ and for the electric potential $\mathbf{E}_{o,2}$:

$$\mathbf{E}_{o,1} = e\, I_0 = e^2\, f_e = \alpha\, eV_0 = \alpha\, e\, \Phi\, f_e = \alpha\, \mathbf{E}_{o,2} \tag{24}$$

(C) The variable N in the energy equation (19) for the $1/f$ noise was interpreted as a number characterizing the number of phase-coherent electron states and their degree of localization in the specimen under test. Thus N may be regarded as a quantum number assigned to the electron state. Using the reference frequency f_e for f and taking the ground state, i.e., $N = 1$, the following relationship is obtained for the reference state $\mathbf{E}_{o,3}$ of the $1/f$ noise:

$$\mathbf{E}_{o,3} = \frac{V_0^2}{f_e} = \Phi^2\, f_e = \alpha^{-1}\, eV_0 = \alpha^{-1}\, \mathbf{E}_{o,2}. \tag{25}$$

(D) The relationship between the rest mass energy of the electron $\mathbf{E}_{o,4}$ and the Rydberg energy (7) has been known for a long time but has been without any revealing interpretation so far. It can be expressed as

$$\mathbf{E}_{o,4} = m_0\, c^2 = \alpha^2\, 2Ry = \alpha^2\, eV_0 = \alpha^2\, \mathbf{E}_{o,2} \tag{26}$$

By combining Eqs. (21), (24), (25), and (26), we finally obtain *the electron energy paradigm*:

$$\boxed{\mathbf{E}_{o,0} = \alpha^1\,\mathbf{E}_{o,1} = \alpha^2\,\mathbf{E}_{o,2} = \alpha^3\,\mathbf{E}_{o,3} = \alpha^4\,\mathbf{E}_{o,4}}\,, \tag{27}$$

where the reference energies for quantum considerations are given by

$$\mathbf{E}_{o,0} = 2\mu_B\,B_0\,, \qquad \mathbf{E}_{o,1} = e\,I_0 = e^2\,f_e\,, \qquad \mathbf{E}_{o,2} = eV_0 = h\,f_e = e\,\Phi\,f_e\,, \tag{28}$$

$$\mathbf{E}_{o,3} = \frac{V_0^2}{f_e} = \Phi^2\,f_e\,, \qquad \mathbf{E}_{o,4} = m_0\,c^2\,.$$

Equation (27) is an empirical equation. It represents a summary of findings in the following fields of experimental physics: atomic line spectra, electric transport (Ohm's law), $1/f$ noise, the quantum Hall effect, the Aharanov-Bohm effect, and the spin effect of the electron. This equation is an energy equation. It sets up a relationship between five reference energies by means of the coupling constant α raised to different powers, so that it is invariant with regard to the electromagnetic system of units.

9. ATTEMPT TO INTERPRET THE PARADIGM

The basic energy $\mathbf{E}_{o,0}$ represents a yes-no statement and thus belongs to the category of information.

The energy $\mathbf{E}_{o,1}$ is an extension of $\mathbf{E}_{o,0}$ since, in addition to the yes-no content, it contains a frequency component, i.e., a statement with respect to time. Space does not play any role at this juncture, as clearly shown in the QHE by the phase-coherent electron state which is independent of size.

The energy $\mathbf{E}_{o,2}$ reveals an extension of $\mathbf{E}_{o,1}$ by the additional coupling of space, in particular one-dimensional space. This can be recognized from energy changes related to $\mathbf{E}_{o,2}$ as represented, for example, by the emission of light which propagates in space in a one-dimensional direction.

As can be deduced from the experience gained with the $1/f$ noise, the energy $\mathbf{E}_{o,3}$ expresses not only the behavior of the applied voltage, but also its modification by the imperfections in the material, such as ions. This means a modification of the electric flux resulting inevitably in a kind of localization, which in turn is a manifestation of the crossing of two one-dimensional spaces. We therefore assume that $\mathbf{E}_{o,3}$ represents the coupling of the yes-no statement to time and to the two-dimensional space.

The energy $\mathbf{E}_{o,4}$ is defined in terms of the rest mass of the electron and the speed of light. As experience suggests, the mass requires a three-dimensional space. We therefore envisage that $\mathbf{E}_{o,4}$, the highest reference

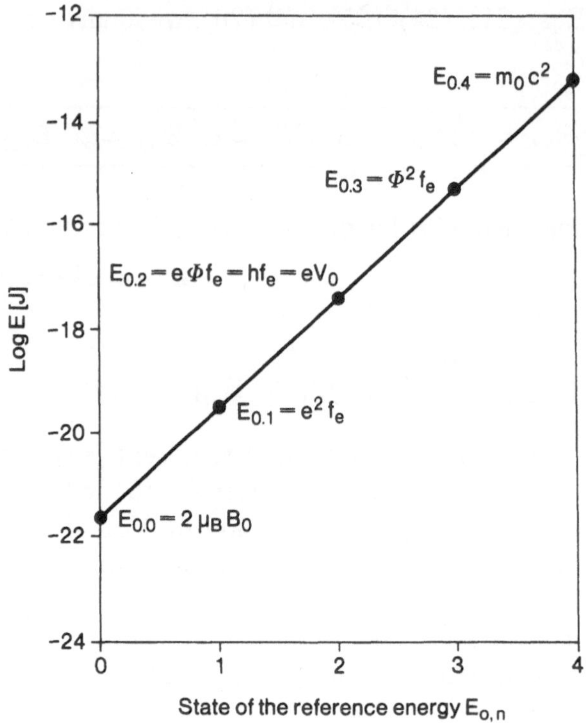

Fig. 3. Values for the five reference energies $\mathbf{E}_{o,n}, n = 0,\ldots,4$, of the electron energy paradigm

energy, reflects the coupling of the yes-no, i.e., the (+ −) statement to time and to the three-dimensional space.

10. CONCLUDING REMARKS

Values for the reference energies of the electron energy paradigm are shown in Fig. 3 on a logarithmic scale. This representation suggests new ways of considering quantized energy transformations.

The attempt at an interpretation of the paradigm presented in Sec. 9 can be taken as a starting point for space-time considerations. This new type of representation simultaneously questions the wave character assigned to the electron and suggests the replacement of the electron wave model by a digital, pulse-like yes-no, i.e. (+ −) form.

Finally, from these considerations we may conclude that underlying the electron dualism is the different relation of the electron energy state to space: i.e., the electron "wave" state (related to $\mathbf{E}_{o,2}$) is a specific expression of the coupling of the yes-no and time states to the one-dimensional space, whereas in the case of the electron mass ($\mathbf{E}_{o,4}$) the coupling is to the three-dimensional space.

REFERENCES

1. K.von Klitzing, G. Dorda, and M. Pepper, *Phys. Rev. Lett.* **45**, 494 (1980); R. E. Prange and S. M. Girvin, *The Quantum Hall Effect* (Springer, Berlin 1987).
2. B. J. van Wees, H. van Houten, C. W. J. Beenakker, J. G. Williamson, L. P. Kouwenhoven, D. van der Marel, and C. T. Foxon, *Phys. Rev. Lett.* **60**, 848 (1988).
3. D. A. Wharam, T. J. Thornton, R. Newbury, M. Pepper, H. Ahmed, J. E. F. Frost, D. G. Hasko, D. C. Peacock, D. A. Ritchie, and G. A. C. Jones, *J. Phys. C* **21**, L209 (1988).
4. R. C. Dynes, J. P Garno, and J. M. Rowell, *Phys. Rev. Lett.* **40**, 479 (1978); B. G. Orr, H. M. Jaeger and A. M. Goldman, *Phys. Rev. B* **32**, 7586 (1985); B. G. Orr, H. M. Jaeger, A. M. Goldman, and C. G. Kuper, *Phys. Rev. Lett.* **56**, 378 (1986); H. M. Jaeger, D. B. Haviland, A. M. Goldman, and B. G. Orr, *Phys. Rev. B* **34**, 4920 (1986).
5. R. G. Clark, S. R. Haynes, J. V. Branch, J. R. Mallet, A. M. Suckling, P. A. Wright, P. M. W. Oswald, J. J. Harris, and C. T. Foxon, *Surface Sci.* **229**, 25 (1990).
6. J. Hajto, A. E. Owen, S. M. Gage, A. J. Snell, P. G. LeComber, and M. J. Rose, *Phys. Rev. Lett.* **66**, 1918 (1991).
7. N. N. Bogoliubov, and D. V. Shirkov, *Introduction to the Theory of Quantized Fields* (Interscience Monographs in Physics and Astronomy **3**, R. E. Marshak, ed.) (Wiley, New York, 1959).
8. "Symbols, units and nomenclature in physics," Document U.I.P. 11 (1965) S.U.N. 65–3.
9. G. Dorda, *Superlattices and Microstructures* **7**, 103 (1990).
10. G. Dorda, *Proceedings of the 12th General Conference of Condensed Matter Division, EPS*, Prague, April 1992; *Physica Scripta T*, **45**, 297 (1992).
11. L. D. Landau and E. M. Lifschitz, *Quantenmechanik*, P. Ziesche, ed. (Akademie-Verlag, Berlin, 1979), p. 116.
12. R. A. Webb, S. Washburn, C. P. Umbach, and R. B. Laibowitz, *Phys. Rev. Lett.* **25**, 2696 (1985).
13. S. Washburn, C. P. Umbach, R. B. Laibowitz, and R. A. Webb, *Phys. Rev. B* **32**, 4789 (1985).
14. S. Washburn and R. A. Webb, *Adv. Phys.* **35**, 375 (1986).
15. F. N. Hooge, *Phys. Rev.* **29A**, 139 (1969); *Physica B* **83**, 14 (1976).
16. F. N. Hooge, *Physica B* **162**, 344 (1990).

17. L. K. J. Vandamme, *Noise in Physical Systems and 1/f Noise*, M. Savelli, G. Lecoy, and J.-P. Nougier, eds. (Elsevier, Amsterdam, 1983), pp. 183–192.

SPACETIME APPROACH TO WEINBERG-SALAM MODEL: WAVES AND PARTICLES IN SPACETIME STRUCTURE

Osamu Hara

Atomic Energy Research Institute
Nihon University, Tokyo, Japan

A spacetime approach to the Weinberg-Salam model is discussed based upon the extended-particle model, in which the lepton or the weak boson field appears as a part of the spacetime structure. A possibility is also discussed to check the validity of this scheme experimentally.

Key words: Weinberg-Salam model, extended particle model, tetrad field.

The Weinberg-Salam model has scored a great success, establishing it as a standard model. But it is still a phenomenological theory waiting for a more fundamental understanding. Here we would like to discuss a possibility to understand it in relation to the spacetime structure.

Our starting point is an assumption that the fundamental particles have an extended structure with the degree of freedom of intrinsic rotation around its center of mass, which is described by a tetrad the origin of which is placed at the center of mass (body frame). To be more precise, it is a spatial rotation in a spacelike plane perpendicular to the four-velocity of the center of mass v^{μ}, since, among the internal motion which is given by the succession of the Lorentz transformation of body frame in general, the boost cannot be distinguished from the external motion of the center of mass, and so must be excluded from the internal motion.

Thus the motion of the system is described by tracing the time development of the coordinate of the center of mass and that of the orientation of the body frame. We meet here, however, with a big problem when general relativity is taken into account. The point is that in this case the axiom of parallels is violated, which means that the relative orientation of the tetrads at different world points cannot be defined, and so we cannot trace the time development of the orientation of the body frame.

Fortunately, however, the clue to this difficulty is afforded by the tetrad formalism given by Weyl and Cartan [1]. In this formalism we introduce a tetrad $\{e_m^{\ \mu}\}$ at every world point x^μ. Here an important fact is that the basic unit vector $e_m^{\ \mu}$ satisfies

$$D_\mu e_m^{\ \nu} = 0, \tag{1}$$

where D_μ is the covariant derivative. This means that $e_m^{\ \mu}$ is *parallel* throughout all of spacetime. Therefore the orientation of the body frame can be specified referring to the tetrad $\{e_m^{\ \mu}\}$, the origin of which coincides with the center of mass of the particle at that moment. Thus the tetrad formalism is essential in order to describe the motion of our particle in conformity with the theory of general relativity.

The tetrad field has sixteen components. Among them, ten are related to the metric tensor $g_{\mu\nu}$ through the relation $e_{m\mu}e^m_{\ \nu} = g_{\mu\nu}$, and the remainder are angular variables describing its orientation. They are given, at every world point, by quantities like the relativistic Euler angles or, equivalently, a pair of two-component spinors u and v referring to some arbitrarily fixed local Lorentz frame. Denoting this as $x^\alpha_{(0)}$ frame, its unit vector $d_\alpha^{\ \mu}$ is given by $(\partial x^\mu/\partial x^\alpha_{(0)})$ at $x^\alpha_{(0)} = 0$, assuming that the point under consideration corresponds to the origin of the $x^\alpha_{(0)}$ frame. Under the local Lorentz transformation, they are transformed as

$$\chi \to \chi' = \begin{pmatrix} a & b \\ c & d \end{pmatrix} \chi, \quad ad - bc = 1, \tag{2}$$

where

$$\chi = \begin{pmatrix} u \\ v \end{pmatrix}, \tag{3}$$

and the direction cosine between the two tetrads $\{ d_\alpha^{\ \mu} \}$ and $\{ e_m^{\ \mu} \}$ is given by

$$h_m^{\ \alpha} = -(\chi^* S_m \sigma^\alpha \chi)/\sqrt{\rho\rho^*}, \tag{4}$$

where $\rho = ui\sigma_2 v$ and $S_m = (\boldsymbol{s}, 1/2)$, with $\boldsymbol{s}$ denoting $(\mu_2 \boldsymbol{\mu} \mu_2)/2$, where μ_i is the Pauli matrix in (u, v) space.

Thus the tetrad field leads to the introduction of an angular variable χ in addition to the metric tensor $g_{\mu\nu}$. But thus far no physical meaning has been given to it. This is because the Lagrangian of the tetrad field has been constructed to be invariant under the local Lorentz transformation, which involves six arbitrary functions. In that case, from Noether's theorem, the solution of the Euler equation contains six arbitrary functions. Therefore, the orientation of the tetrad is left completely arbitrary, which means that it cannot be physical. This is analogous to the case of electromagnetic field, in which its longitudinal component is not physical because of the invariance under the local gauge transformation.

The basis of this requirement is, however, not necessarily obvious when the tetrad $\{e_m{}^\mu\}$ is considered as a reference frame to describe the rotational motion of the extended particle. To see this, let us discuss in some detail how to describe the motion of our particle in the general relativistic case. Then we find that, since the internal motion is restricted to the spatial rotation in a spacelike plane perpendicular to v^μ, the body frame $\{ a_m{}^\mu \}$ can be written as $\{ (a_i{}^\mu), v^\mu \}$, where $(a_i{}^\mu)$ is a triad in a spacelike plane perpendicular to v^μ formed by three spacelike unit vectors $a_i{}^\mu (i = 1,2,3)$, and that at every world point it is always possible, through a suitable boost of the body frame, to bring its zeroth axis to the direction of the zeroth axis of the tetrad $\{ e_m{}^\mu \}$ at that point. With this specification of the body frame, the triad $(a_i{}^\mu)$ is in a spacelike plane spanned by the triad $(e_i{}^\mu)$, where $(e_i{}^\mu)$ is the spatial part of the tetrad $\{ e_m{}^\mu \}$ formed by three spacelike unit vectors $e_i{}^\mu (i = 1,2,3)$. Therefore, $(a_i{}^\mu)$ can be specified referring to $(e_i{}^\mu)$, and the orientation of $(a_i{}^\mu)$ thus specified gives directly the angle of rotation of the particle.

Therefore, the motion of our particle can be described by tracing the time development of the coordinate of the center of mass referring to some world reference frame and that of the orientation of $(a_i{}^\mu)$ referring to $(e_i{}^\mu)$. Clearly there is no preferred direction in the triad $(e_i{}^\mu)$. So its arbitrary rotation should be possible, and the Lagrangian of the tetrad field must be invariant under the local spatial rotation.

Contrary to this, however, there is no reason to require the invariance under the boost. This is because our prescription for defining the angle of rotation is essentially to define the angle between the two triads $(a_i{}^\mu)$ and $(e_i{}^\mu)$ under the condition that the body frame $\{ a_m{}^\mu \}$ is *at rest* relative to the reference frame $\{ e_m{}^\mu \}$. As the result the effect of any boost of $\{ e_m{}^\mu \}$ is compensated by the corresponding change of boost of $\{ a_m{}^\mu \}$ to bring it at rest relative to the boosted $\{ e_m{}^\mu \}$, leaving unchanged the angle of rotation of the particle. Therefore, when the tetrad $\{ e_m{}^\mu \}$ is considered as a reference frame to describe the rotational motion of the extended particle, the boost is not a meaningful operation, and so it is unnecessary to keep room for it in the equation of motion of $\{ e_m{}^\mu \}$. This means that it is redundant to require the invariance under the boost (contraposition of Noether's theorem). Because of this, we shall restrict in the following the transformation under which the Lagrangian of the tetrad field is required to be invariant from the local Lorentz transformation to the local spatial rotation. Even if the local Lorentz transformation is thus restricted to the local spatial rotation, the original tetrad formalism can be reproduced literally by the analytic continuation of the angle of rotation into complex [2].

Under this restriction, a non-trivial Lagrangian exists for χ, since the local spatial rotation contains only three arbitrary functions compared to six in the case of the local Lorentz transformation. This enables us to interpret it as expressing a real physical object.

To include the inversion, u and v must be replaced by four-component Dirac spinors ξ and η. They are related by $u^{\pm} = (1 \pm \gamma_5)/2 \cdot \xi$ and $v^{\pm} = (1 \pm \gamma_5)/2 \cdot \eta$ with $u^{+} \equiv u, v^{+} \equiv v$. Under the inversion they are transformed as $u^{+} \rightleftharpoons u^{-}, v^{+} \rightleftharpoons v^{-}$ and, in each reference frame, $h_m{}^{\alpha}$ is given by *either* of the helicity eigenstates (u^{+}, v^{+}) and (u^{-}, v^{-}), where, when u and v are replaced by ξ and $\eta, h_m{}^{\alpha}$ is given by

$$h_m{}^{\alpha} = -i \{\bar{\chi}(1 \pm \gamma_5)/2 \cdot S_m \gamma^{\alpha} \chi\} / \sqrt{\rho\rho^*}. \tag{5}$$

This means that S_m should be replaced by $S_m{}^{\pm} = (1 \pm \gamma_5)/2 \cdot S_m$. Therefore, if the right (left)-handed component has S-spin 1/2 in one reference frame, then the left (right)-handed component has S-spin 0.

Then we find that, in terms of these variables, the Lagrangian is given by

$$L = \sqrt{-g} \left\{ \bar{\chi} \Gamma^{\mu} D_{\mu}^{\pm} \chi + \frac{1}{2} g \bar{\chi} (1 \pm \gamma_5) \chi \cdot H \right\} + h.c., \tag{6}$$

$$D_{\mu}^{\pm} = \partial_{\mu} A_{\mu}^{i} S_i{}^{\pm} + B_{\mu}, \tag{7}$$

where A_{μ}^{i} and B_{μ} denote, respectively, the gauge field for the local spatial rotation and the local phase transformation. Explicitly A_{μ}^{i} is given by

$$A_{\mu}^{i} = e^{j\rho} e^{k\sigma} (\gamma_{\rho\sigma\mu} + K_{\rho\sigma\mu}), \quad ijk \text{ cyclic}, \tag{8}$$

where $\gamma_{\rho\sigma\mu}$ and $K_{\rho\sigma\mu}$ denote, respectively, the Ricci's rotation coefficient and the contorsion. $\Gamma^{\mu} = d_{\alpha}{}^{\mu} \gamma^{\alpha}$, and H is the Higg's field.

The Lagrangian (6) is invariant under the local spatial rotation and the local phase transformation, that is, $SU(2) \times U(1)$. If we identify, therefore, S with the weak isospin I, our S-spin assignment structure leads to the following two possibilities:

case I: right-handed $I = 1/2$ and left-handed $I = 0$,
case II: right-handed $I = 0$ and left-handed $I = 1/2$.

Nature chooses case II. This is nothing but the $W - S$ scheme. So we shall identify χ with the lepton (for example, electron and electron-netrino). Then we find that, in this scheme, the lepton appears as a part of the spacetime structure together with the gravitational field $g_{\mu\nu}$.

One of the characteristic features of this model is that the weak boson field W_μ is given by Eq. (8). If we consider the Minkowski limit for simpliticity, $\gamma_{\rho\sigma\mu}$ vanishes, and so W_μ is given essentially by contorsion. This means that W_μ is of a universal one. So, for example, the Maxwell equation $\Box A_\mu = j_\mu$ is supposed to be replaced by

$$\eta^{\alpha\beta} D_\alpha B_\beta A_\mu = j_\mu. \tag{9}$$

This introduces a new direct coupling between the electromagnetic field and W_μ. Therefore, it is expected that through this interaction it should be possible, at least in principle, to check the validity of this scheme experimentally.

REFERENCES

1. H. Weyl, *Z. Phys.* **56** (1929) 330; E. Cartan, *The Theory of Spinors* (MIT Press, Cambridge, Massachusetts, 1966).
2. D. Bohm, P. Hillion, T. Takabayasi, and J. P. Vigier, *Prog. Theor. Phys.* **23** (1960) 496.

COMPUTER PORTRAYALS OF THE SINE-GORDON BREATHER AS A MODEL OF THE DE BROGLIE DOUBLE SOLUTION

Albert J. Hatch[1]

Department of Physics, Roosevelt University
430 South Michigan Avenue
Chicago, Illinois 60605

Computer portrayals are presented of a one-dimensional model of fundamental particles as developed by U. Enz from the breather solution of the nonlinear sine-Gordon equation in the context of soliton physics. Particles portrayed are the stationary and the moving basic nucleon and the stationary electron. The portrayals illustrate the nonlinear characteristics of the u wave anticipated by de Broglie as part of his proposed "double solution" in wave mechanics. Linear plots of u vs x for a moving breather exhibit the familiar single hump profile which moves with group velocity $v_g = \beta c$. Plots of the signed logarithm of u vs x reveal both an apex corresponding to the single hump of the linear plot plus the de Broglie ψ waves which are strongly modulated by the exponential decrease ($\sim$ 43 decades per de Broglie wavelength) and which move with phase velocity $v_p = c/\beta$. The semilogarithmic plot makes possible a unique portrayal of the wave packet representing a moving particle which shows the *coexistence* of both particle *and* wave.

1. INTRODUCTION

An important feature of de Broglie's early work in the mid-1920's on the wave nature of matter was that he recognized the necessity for two wave functions. One was the familiar ψ wave of wave mechanics, which contained information on the energy and momentum of matter. The other was a u wave, which he envisioned as having additional information on the spatial and temporal structure of matter. The overwhelming advocacy of the statistical interpretation of the ψ wave by others starting in the late

Waves and Particles in Light and Matter, Edited by
A. van der Merwe and A. Garuccio, Plenum Press, New York, 1994

1920's led de Broglie to set aside his work on the u wave until the early 1950's when he returned to his original concept of the "double solution" and attempted to find an explicit formulation for the u wave. Although de Broglie saw clearly the nonlinear nature of the u-wave problem, he was never able to obtain a solution and subsequently summarized his efforts and views in his 1960 book *Nonlinear Wave Mechanics.* [1]

An apparent solution of this long-standing problem has been found recently by Enz [2], who has shown that de Broglie's postulates for a one-dimensional u wave are met by the properties of the breather solution of the sine-Gordon equation in the context of soliton physics. Since modern soliton physics originated in the mid-1960's it now appears in restrospect that de Broglie was ahead of his time in being able to identify the nonlinear mathematical nature of the u-wave problem long before the solutions to such problems had been developed.

Computer plots have long been an important tool in the study of soliton phenomena, starting with the famous studies of Zabusky and Kruskal. [3] Examples of computer plots of sine-Gordon breathers can be found in Lamb, [4] in Eilenberger, [5,6] and in Drazin and Johnson. [7] In all of these examples the nonessential constants have been suppressed so that the computations represent only the essential mathematical properties of the breathers. No open-literature examples of particle-specific computations are known to the author.

The primary purpose of this paper is to present the results of computer studies which illustrate the application of Enz's work to specific examples such as nucleons and electrons. A secondary purpose is to present a new semi-logarithmic portrayal of the u-wave which simultaneously exhibits both the wave and the particle properties of a moving particle.

2. THEORY

Enz invokes the one-dimensional sine-Gordon equation [2,8]

$$u_{xx} - \left(1/c^2\right) u_{tt} = \left(1/d^2\right) \sin u, \tag{1}$$

where c is the speed of light and d is a fundamental length defined below. This is a purely classical nonlinear wave equation. The breather solution of Eq. (1) is given by [2,9]

$$u = 4 \tan^{-1} \left(\frac{s \sin \left[(r/d) \left(c(t - t_0) - \beta x\right) \left(1 - \beta^2\right)^{-1/2} \right]}{r \cosh \left[(s/d) \left(x - x_0 - vt\right) \left(1 - \beta^2\right)^{-1/2} \right]} \right). \tag{2}$$

Here $s = \sin q$ and $r = \cos q$ depend on an arbitrarily chosen value $q =$ constant. The fundamental length d is related to the Compton wavelength λ_c of the "particle" by [2]

$$\lambdabar_c = \lambda_c/2\pi = d/r. \tag{3}$$

The "particle" speed is $v = \beta c$. The initial conditions are position x_0 (the "center" of the breather) and time $t_0 = 0$ in the present work.

Enz's description of the breather solution can be paraphrased as follows: The wave function u represents an extended oscillating soliton-antisoliton pair (a standing wave) either stationary or moving at the speed v along the x axis. It is given in closed form by elementary functions, has no singularities in either x or t, and exhibits the Lorentz invariance of a moving system. In this one-dimensional case u contains all the information in the ψ wave function *plus* additional information pertinent to the finite spatial distribution of energy, the internal motion, and the total energy of the breather corresponding to its mass.

The term "particle" is put in quotes here to distinguish it from the customary point particle concept of quantum mechanics where the location is given in probabilistic terms by the square of the absolute value of an associated ψ wave packet. In the breather there is no point particle. Instead, the "particle" is represented by the entire u-wave function in the appropriate space (restricted to the x axis for the one-dimensional sine-Gordon equation). According to Enz, "The breather as a whole can be considered as an extended particle distributed continuously along the x axis with a mass given by the total field energy of the u field...The major part of the total energy is contained in a space region of the order of the size b." (The quantity b is defined below.)

From Eq. (2), Enz obtains expressions for various parameters of interest. Thus the period of the breather oscillation is

$$\tau_b = (2\pi d/rc)\left(1 - \beta^2\right)^{1/2}. \tag{4}$$

Alternatively, in terms of λ_c from Eq. (3), the period becomes

$$\tau_b = (\lambda_c/c)\left(1 - \beta^2\right)^{1/2}. \tag{5}$$

The wavelength of the oscillation is

$$\lambda_b = (2\pi cd/rv)\left(1 - \beta^2\right)^{1/2}. \tag{6}$$

In terms of λ_c, this becomes

$$\lambda_b = (\lambda_c/\beta)\left(1 - \beta^2\right)^{1/2}. \tag{7}$$

The half-height width b of the breather profile (for $0 < r \ll 1$) is found from

$$r \approx 2\exp\left(-(b/2d)\left(1-\beta^2\right)^{-1/2}\right). \tag{8}$$

It is useful to recast this relation in the following form:

$$b/\lambda_c \approx -(\cos q/\pi)\ln(\cos q/2). \tag{9}$$

Enz has also shown that a "Planck's constant of the breather," $\hbar_b$, can be obtained from

$$\hbar_b = W_0 d/rc, \tag{10}$$

where W_0 is the rest mass energy of the breather. In terms of λ_c, this becomes

$$\hbar_b = W_0\lambda_c/2\pi c. \tag{11}$$

At this juncture it is useful to point out that Eqs. (4), (6), (8), and (10) are taken directly from Enz's work, whereas the corresponding Eqs. (5), (7), (9), and (11) are all recast in terms of λ_c and/or q, which either makes them more useful for computations or clarifies their dependence on the parameters λ_c and q, or both. Thus, for example, it is seen from Eqs. (5) and (7) that the period and wavelength of the breather are independent of q, whereas from Eq. (9) the "size" b is dependent on q.

Finally, Enz shows that Eq. (10) leads to expressions for the momentum of the breather:

$$p_b = \hbar_b k_b, \tag{12}$$

where k_b is the wave vector (propagation number) and for the breather wavelength

$$\lambda_b = h_b/m_b v. \tag{13}$$

Enz points out that Eqs. (12) and (13) "are a perfect analogy to the corresponding quantum mechanical relations; *they emerge, however, from the entirely classical field solutions* $u(x,t)$." (Italics by Enz.)

3. COMPUTATIONS

The computations can be organized under three headings: (A) Dependence of various parameters on the arbitrarily chosen value of q. (B) Typical breather profiles and anharmonic properties for specific particles both at rest and moving. (C) The use of both linear and logarithmic plots of u vs x to illustrate the simultaneous interdependent roles of breather u waves as particles having a group velocity $v_g = \beta c$ and as conventional de Broglie waves having a phase velocity $v_p = c/\beta$.

A. Dependence of Parameters on q

Breathers are portrayed as a sequence of profiles $u(x)$ computed from Eq. (2) at fixed time steps during a single cycle of the oscillation. For each profile there is a peak value u_p which occurs at $x = x_0$ for a stationary breather. As the time steps progress the values of u_p increase to a maximum value u_{pm} at 1/4th cycle, then to a negative maximum at 3/4th cycle, etc. Inspection of Eq. (2) shows that these peak maximum values occur at sin[ ] $\rightarrow \pm 1$ and cosh[ ] $\rightarrow 1$ which gives $u_{pm} = 4\tan^{-1}(s/r) = 4q$.

The half-height width b of each breather profile normalized to the Compton wavelength λ_c depends on q as given in Eq. (9). This is an approximation for $0 < r \ll 1$ and hence it fails for small values of q. This failure is shown in Fig. 1 wherein are plotted values of b/λ_c from Eq. (9) and values determined from direct division of b (as measured on breather profiles) by λ_c. Thus at $q = \frac{1}{2}\pi - 0.2$, Eq. (9) gives the approximate value $b/\lambda_c = 0.146$ whereas direct measurement of $b = 0.213\text{E} - 15$ m (see Fig. 2 for a nucleon) combined with $\lambda_c = 1.323\text{E}{-15}$ m (also for a nucleon) gives $b/\lambda_c = 0.161$, the correct value. The difference between the two values is $\sim 10\%$. For smaller values of q the difference increases rapidly as in Fig. 1.

Representative profiles of stationary breathers at three different values of q are shown in Fig. 2. The half-height width is shown only for the $\frac{1}{2}\pi - 0.2$ profile. Most of the breather energy resides in this central region. All other parameters correspond to a stationary nucleon, $\lambda_c = 1.323$ E-15 m. From Eq. (2) it can be shown that the general shape of the stationary breather at any specified time can be described as that of an arctan (sech$x$). It is instructive to plot the logarithm of u vs x as in Fig. 3 for the same values of q as in Fig. 2. Fig. 3 shows clearly the exponential character of $u(x)$ and in particular the asymptotic character at large $|x - x_0|$. The slope of this asymptote can be obtained by getting the linear slope of Eq. (2) at a time such that sin[ ] $\rightarrow 1$, using the exponential forms of the hyperbolic functions and keeping only those terms that are significant for $|x - x_0| > b$. The result is

$$du/dx = -8\left(s^2/rd\right)\exp\left(-\left(s/d\right)\left(x - x_0\right)\right). \tag{14}$$

Thus for a plot of $\log_{10}u$ the slope is $-0.43429(s/d)$ for the asymptotic wings of the plots in Fig. 3. Alternatively, it is useful to rewrite this slope as $-0.43429(\tan q/\lambda_c)$. This semilog plot reveals properties of the breather at low values of u which are lost in the linear plot of Fig. 2. Thus the central hump which is the prominent signature feature of the breather in Fig. 2 is reduced to a gently rounded apex in Fig. 3, whereas the asymptotic wings which are a prominent feature of Fig. 3 are not visible in Fig. 2. However, although the apex in Fig. 3 is much less prominent than the central hump

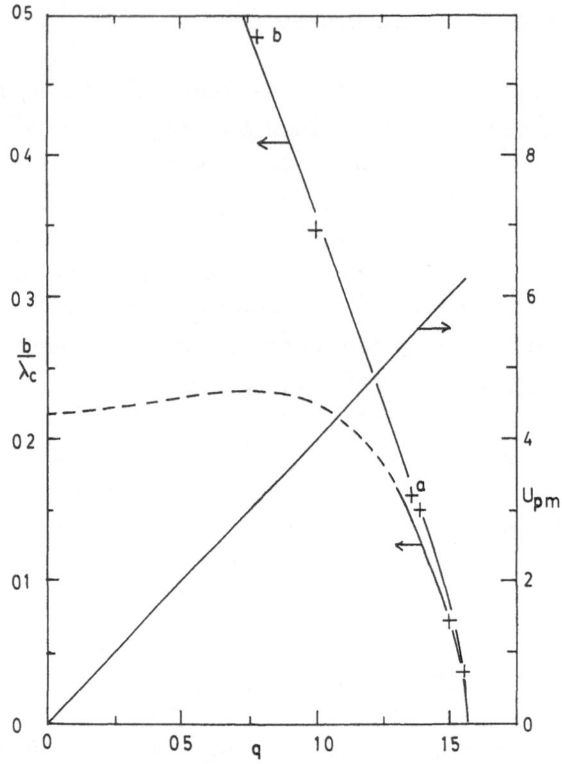

Fig. 1. Dependence of breather maximum potential u_{pm} and normalized breather half-height width b/λ_c on the arbitrary parameter q.

in Fig. 2, it must be remembered that the apex region is still the locus of the major part of the breather energy. This semilogarithmic plot turns out to be especially useful in portraying the associated de Broglie waves for a moving breather as will be shown further on.

From the above three figures and the related discussion it appears that there is no unique value of q which is optimum in any way and hence the value $q = \frac{1}{2}\pi - 0.2$ as used by Enz will be used throughout the rest of this work.

B. Typical Breather Profiles

Of the constant parameters s, r, d, v, x_0 and t_0, (or alternatively q, λ_c, β, x_0 and t_0) the only particle-specific parameter is $d = r\lambda_c/2\pi$ from Eq. (3), and this depends ultimately on the particle mass. From Eqs. (5)

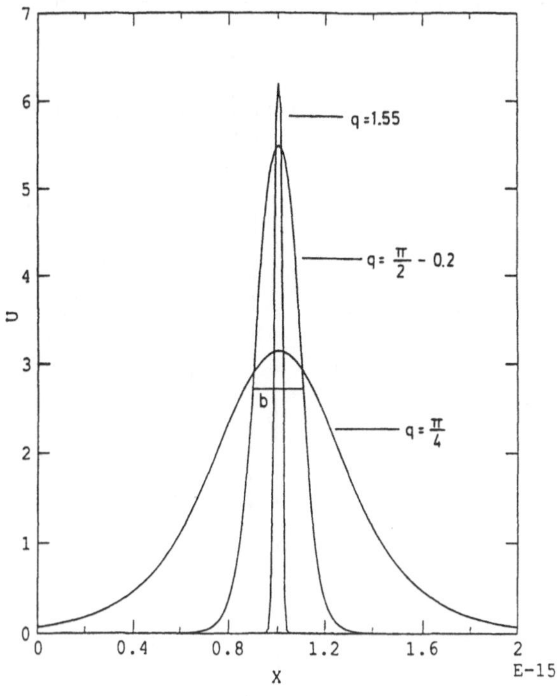

Fig. 2. Linear plots of breather profiles for three different values of q. See text for details.

and (7) it is clear that the breather period τ_b and wavelength λ_b are independent of q but dependent on λ_c and β. The time-step interval used in most computations corresponds to a basic phase angle step of 5°. The integer I shown on most computer generated figures designates the time steps, $I = 72$ corresponding to one cycle.

Figure 4 portrays typical breather profiles during one cycle for the free nucleon at rest. The pertinent parameters are λ_c = 1.323E−15 m, d = 4.1832E−17 m, $\beta = 0, x_0$ = 1 E−15 m and $t_0 = 0$. Phase angle steps are shown only for 0°, 15°, 30° and 90° in the first quarter cycle and for modular values in the other quarter cycles for clarity. The nucleon breather half-height width (at 90°) is b = 0.213E−15 m giving a normalized value $b/\lambda_c = 0.161$.

Figure 5 portrays typical breather profiles during one cycle for the free electron at rest. The pertinent parameters are λ_c = 2.429E−12 m, d = 7.6800E−14 m, $\beta = 0, x_0$ =E−12 m, and $t_0 = 0$. The electron breather half-height width is b = 0.390E−12 m, giving $b/\lambda_c = 0.161$, the same as for the free nucleon. Thus the only difference between Figs. 4 and 5 is a factor of 1836 in the horizontal (x) scale.

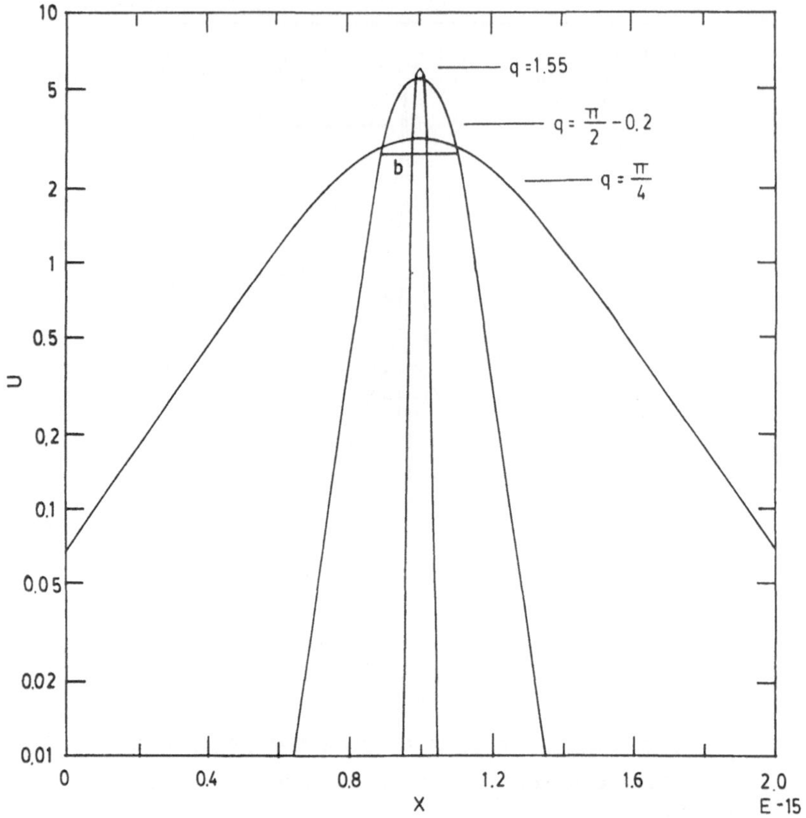

Fig. 3. Semilogarithmic plots of stationary breather profiles for the same parameters as in Fig. 2. These plots are hand drawn. Horizontal intervals are 0.01E−15 m. Details in text.

In addition to the obvious spatial anharmonic character of the breather profiles there is also a less obvious temporal anharmonic character. This is shown by the plot in Fig. 6 of the peak values u_p throughout the 72 steps of the computation for the stationary nucleon breather. Compared to the normal sine curve shown, the plot of u_p is highly anharmonic, approaching a square wave. A Fourier analysis of this plot gives the following result:

$$u_p(t) = 6.54 \sin \phi + 1.46 \sin 3\phi + 0.58 \sin 5\phi + 0.28 \sin 7\phi + 0.10 \sin 9\phi,$$

where $\phi = \omega_b \tau$. There is a residue in this sum corresponding to a standard deviation of $\sigma = 0.08$ or 1.5%. The temporal anharmonic character of u_p evidently comes from the circumstance that u is an arctan function of the inner sine function in Eq. (2).

A careful review of earlier descriptions of plots of breather oscillations reveals no explicit awareness of the temporal anharmonic character of sine-Gordon breathers. Thus Lamb [4] remarks that the solution is

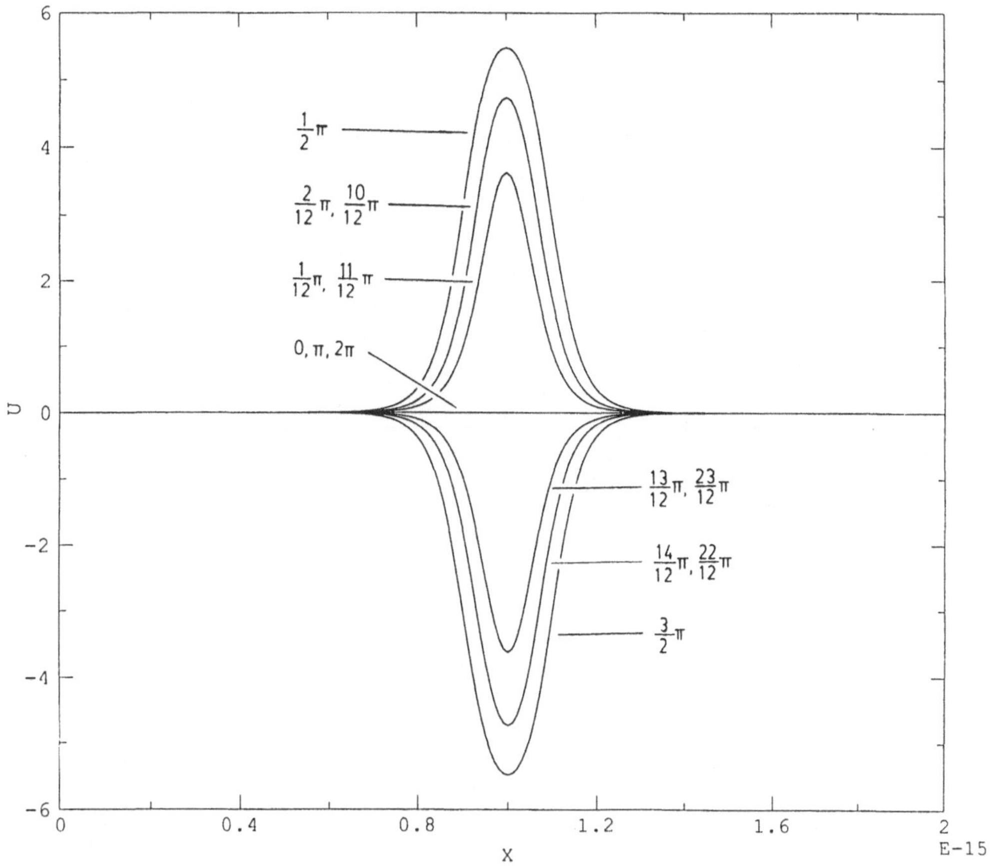

Fig. 4. Typical breather profiles for a stationary nucleon having a Compton wavelength 1.323E−15 m.

periodic and his Fig. 5.4 shows profiles only at phase angles in steps of $\frac{1}{2}\pi$, thereby seeming to imply that the solution is harmonic (as distinct from periodic). Eilenberger [5,6] has several figures showing interactions between breathers and solitons in which the breather oscillations (~ 20 profiles per cycle) appear to be harmonic. However, close inspection of the plots indicate that they may indeed have the anharmonic character found here. Enz [2] states that the breather oscillation is "highly anharmonic near the centre of the breather but harmonic far (distance $>> b$) from it," but makes no distinction between spatial and temporal anharmonicity.

Figure 7 is a 3-dimensional plot of the stationary nucleon breather. The only real spatial coordinate is x. The two virtual coordinates are the time t (one complete cycle) and the breather potential u as a function of x and t. There are 73 profiles of u vs x similar to the examples portrayed in Fig. 4 and encompassing 72 time steps each corresponding to 5° phase

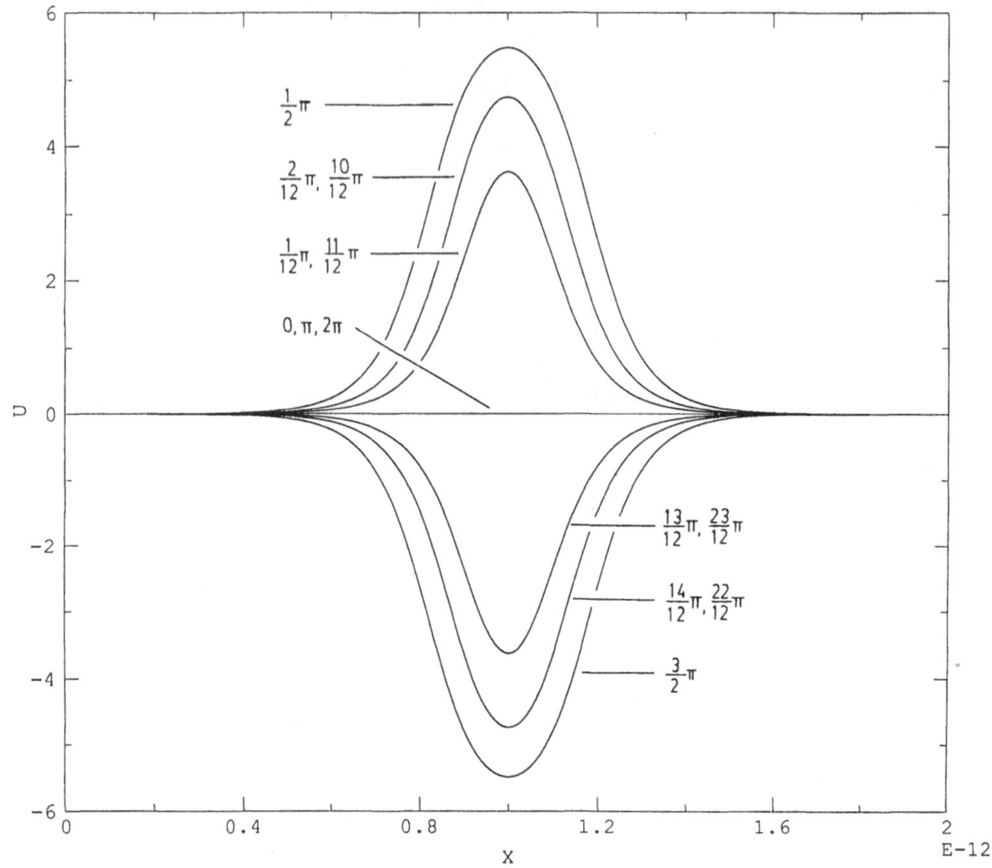

Fig. 5. Typical breather profiles for a stationary electron having a Compton wavelength 2.429E−12 m.

angle steps. There is an equal number (73) of values $u(x)$ in each profile, thus making a 73× 73 array in x and t. The range of x values in this portrayal is from 0.66E−15 to 1.38E− 15 m in steps of 0.01E−15 m. The "center" of the breather "particle" is at x_0 = 1.00E−15 m. However, this position is not to be interpreted in any way as corresponding to a point particle. The concept of "particle" here is that according to Enz as already discussed following Eq. (3). Furthermore, the extended "particle" is not limited to the visible portion portrayed in Fig. 7 but extends in highly attenuated form to $\pm\infty$ on the x axis.

Typical profiles for the moving nucleon breather ($\beta = 0.3$) are shown in Fig. 8. Nine of the ten profiles are for phase angles which are integral multiples of 45° and are designated by I = 0, 9, 18...72, where the phase angle is $\omega_b t = (I/72)360°$. Of the nine, all except that for $I = 18$ have the single-hump profiles characteristic of the stationary breather shown in

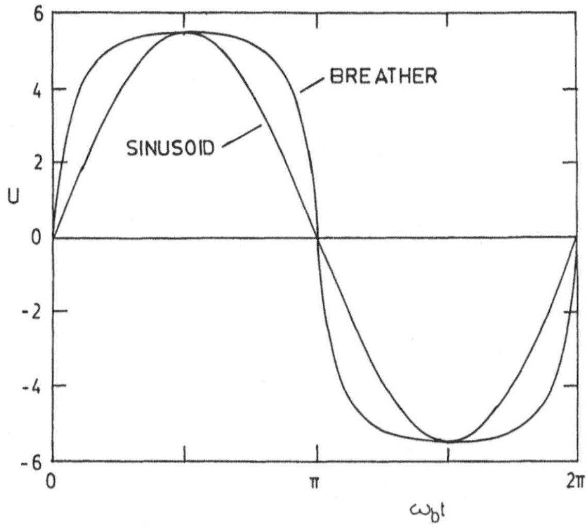

Fig. 6. Temporal character of peak ordinates u_p of stationary breather profiles through one cycle compared with a normal sine curve.

Fig. 4. However, instead of the profiles in Fig. 8 having a shape which is an exact arctan (sech$x$) function, they are slightly perturbed from this shape by the compound nature of the arguments of the sin[ ] term and the cosh[ ] term in Eq. (2). That is, both of these arguments are functions of both x and t for the moving breather rather than of just one. This circumstance is displayed most prominently by the sinuous profiles for $I = 18$ and $I = 58$ in Fig. 8. Neighboring profiles show the sinuous shape diminishing and becoming barely noticeable (visually) within $\sim \pm 20°$ of these profiles. An important overall property of the moving breather is that the distance traveled by the breather from the initial profile ($I = 0$, peak at $x = 1.00\text{E}{-15}$ m) to the corresponding profile one cycle later $I = 72$, peak at 1.38E−15 m) is 0.38E−15 m which agrees well with the calculated value $v_g \tau_b = 0.37862\text{E}{-15}$ m where v_g is the group velocity 9E7 m/s of the breather.

Another unique albeit minor characteristic of the moving breather profiles is that the amplitude u_{pm} of the initial profile (Fig. 8, $I = 0$) depends on the relative velocity of the breather. A detailed calculation shows that the (negative) maximum amplitude occurs for $\beta = 0.314$ rather than for $\beta = 0.3$ as used for Fig. 8. The temporal behavior of the profiles for the moving breather is essentially the same as for the stationary breather.

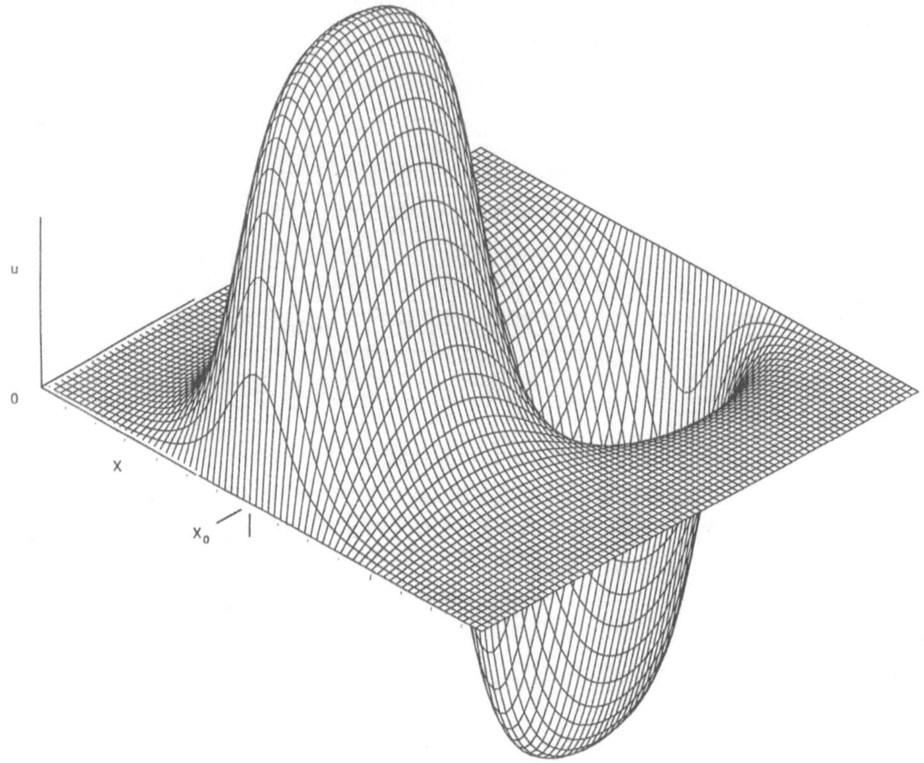

Fig. 7. A 3-D plot of a stationary breather potential $u(x,t)$. Detailed description in text.

C. De Broglie Waves

Associated with the moving breather portrayed in Fig. 8 there should be a de Broglie wave having a wavelength according to Eq. (7) of $\lambda_b =$ 4.207E−15 m. Such a wave is illustrated schematically in Fig. 2 of Enz's 1986 paper. [2] From Eqs. (5) and (7) one finds the phase velocity of these waves to be $v_p = c/\beta$. To look for evidence of these waves the x range of the computations was extended four-fold to 8E−15 m. The linear plot of $u(x)$ in Fig. 9 showed no evidence of such waves but the computer printout did. Accordingly, plots were made of $\log u$ vs x on the specially prepared graph shown in Fig. (10). This is a double semilogarithmic graph.[2] The upper half for positive u (labeled z) above the 0 ordinate has a logarithmic scale ranging nominally from 0 at the top (labeled 70) to −70 at the mid-point (labeled 0). The lower half for negative $u(z)$ below the 0 ordinate has a logarithmic scale ranging nominally from 0 at the bottom (labeled −70) to −70 at the midpoint (labeled 0). The upper and lower halves of this graph (but not the plotted results) are essentially mirror images of each other. When the linear results associated with Fig. 9 are transformed to the double semilogarithmic graph of Fig. 10, the de Broglie waves stand out prominently as sloping (nearly) straight line segments similar to Fig. 3 except that here each line segment corresponds to a half de Broglie wavelength. From the crossover points read from the computer printout the wavelength is found to be $\lambda_b = 4.21\text{E}{-15}$ m, in good agreement with the calculated value above. The major portion of each line segment has a slope given by $-0.43429(\tan q/\lambda_d)$ as in Fig. 3. Numerically, this is ~ -43 decades per de Broglie wavelength. Thus the double semilog plot can be characterized as a de Broglie wave strongly modulated by the exponential slope. The ends of each line segment exhibit slight curvature toward the central axis at $z = 0$ where crossover occurs. Inspection of computer output shows that the central breather peak of the linear plot in Fig. 9 corresponds to the center of the short flat-top apex of the semilogarithmic plot in Fig. 10. The flat top is due to the circumstance that the range of values of $\log u$ within the x range of $\sim (\frac{1}{2}b)$ is less than the resolution of the laser printer used to make the figure.[3] Although the apex in Fig. 10 is not nearly as prominent visually as the linear peaks in Fig. 9, it is nevertheless the region which contains most of the energy of the breather.

It is insightful to express the slope of the straight line envelope of the semilogarithmic plot of Fig. 10 in terms of a decrement $\Delta \log_{10} u$ per de Broglie wavelength λ_b. It can be shown that this is

$$\Delta \log_{10} u = -(\log_{10} e)(2\pi \tan q/\beta)\left(1-\beta^2\right)^{1/2}. \qquad (15)$$

For the parameters $q = \frac{1}{2}\pi - 0.2$ and $\beta = 0.3$ used here one obtains $\Delta \log_{10} u = -42.8 \cong -43$, independent of the mass m.

Movement of the de Broglie waves is portrayed in the sequence of profiles shown in Fig. 11. Of these ten semilogarithmic profiles, nine are for phase angles which are integral multiples of 45° and are designated by $I = 0, 9, 18 \ldots 72$, while the tenth is for the $I = 58$ profile, just as for the sequence of linear profiles already shown in Fig. 8. Comparison of the profiles for $I = 18$ and $I = 58$ in Figs. 8 and 11 show that the sinuous profiles in Fig. 8 are the result of a split in the apex in Fig. 11 at the appropriate vertical crossover lines, part of each split apex occurring in the positive portion of each plot and part in the negative portion.

Visual indication of the de Broglie waves is most obvious from the motion toward the right of the vertical crossover lines which pass through $z = 0$ and which therefore correspond to the null points at intervals of a half de Broglie wavelength. From these figures and the computer printout

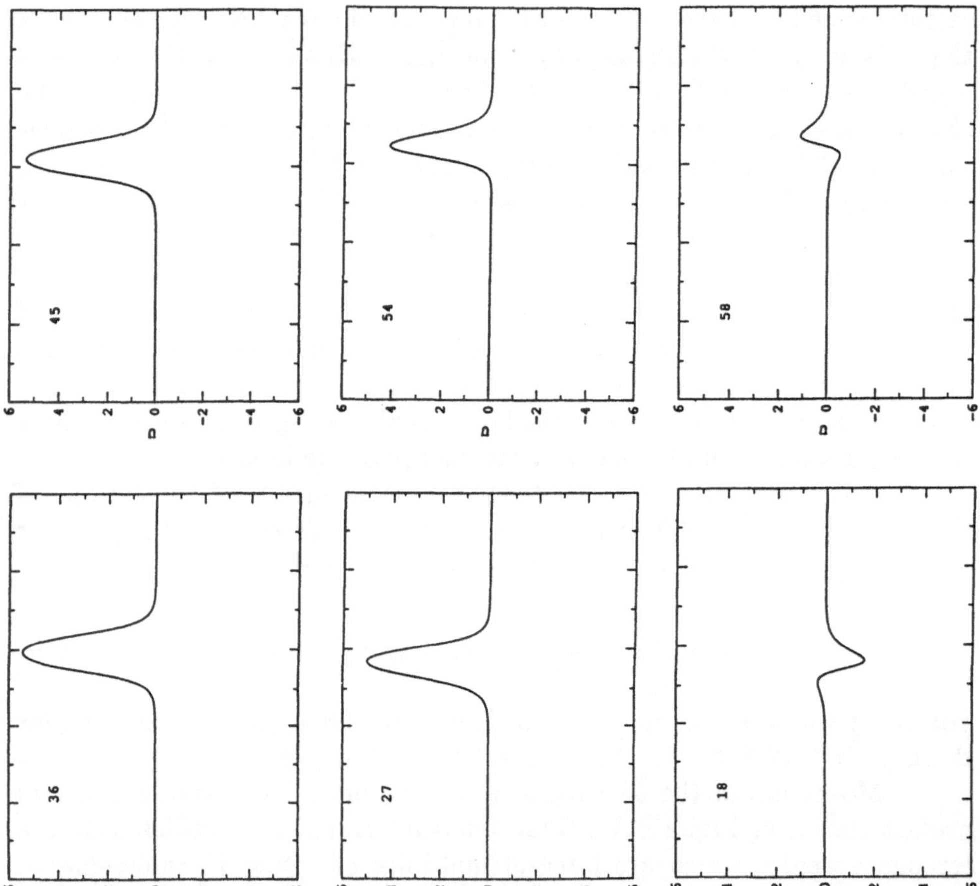

45
54
58
36
27
18
D
6
4
2
0
-2
-4
-6

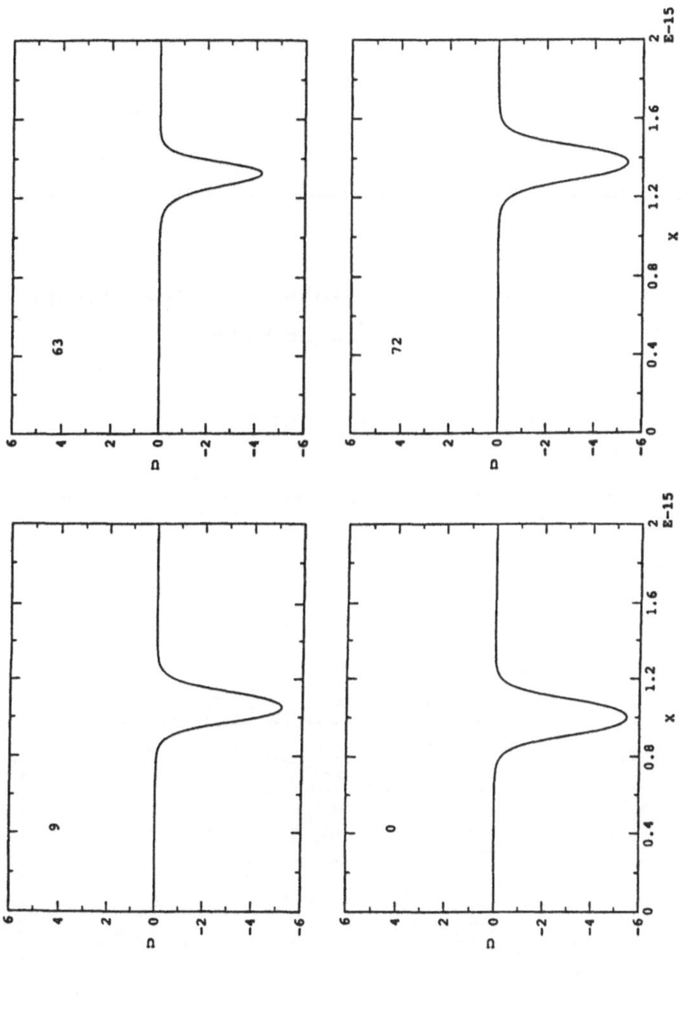

Fig. 8

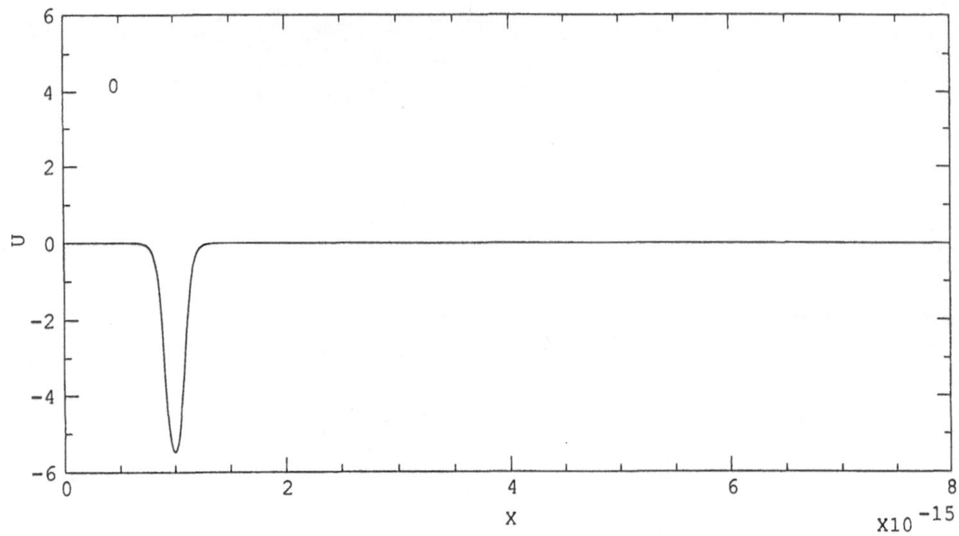

Fig. 9. Extended-range linear plot of initial profile ($I = 0$) of moving nucleon breather. No visible evidence of de Broglie waves.

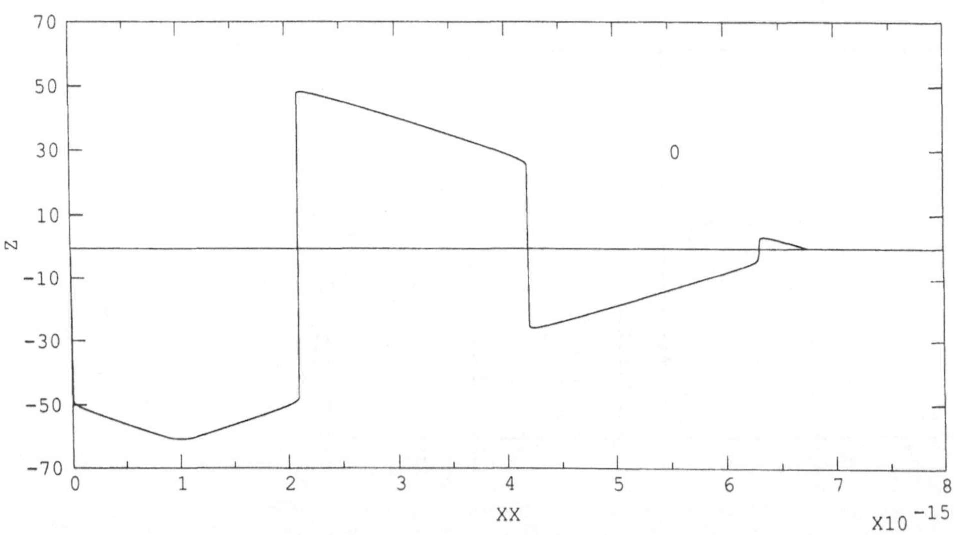

Fig. 10. Semilogarithmic plot of $\log u$ vs x for the same parameters as in Fig. 9. See text for a detailed discussion.

for the whole cycle portrayed in Fig. 11 we have $\Delta x = \lambda_b = 4.207\text{E}{-15}$ m and $\Delta T = 4.2068\text{E}{-24}$ s which give $v_p \doteq 1\text{E }9$ m/s $=(10/3)c = c/\beta$, thereby confirming the expected phase velocity of the de Broglie waves. The motion of the flat top apex in Fig. 11 toward the right corresponds to the motion of the central peak in Figs. 8 and 9 and of the apex in Fig. 10 and thus has the group velocity $v_g = \beta c$. Therefore we have the expected outcome that $v_g v_p = c^2$. It seems remarkable that Figs. 9 and 10 (and their extended counterparts in Figs. 8 and 11) which are apparently so different should represent the same basic set of values of $u(x)$.

Finally, it is instructive to make a semilogarithmic plot of a "whole" moving breather wave packet representing a one-dimensional moving particle. This is shown in the right side (b) of Fig. 12 where the axial length of the breather has been extended to ~ 4 de Broglie wavelengths symmetrical initially about $x_0 = xx_0 = 1\text{E}{-15}$ m.[4] A motion picture sequence of this breather wave packet shows clearly the movement of the apex toward the right at the group velocity v_g and of the de Broglie wave at the phase velocity v_p, both in phase with each other. Additionally, the wave packet has a characteristic envelope in the shape of a rhombus with one axis coincident with the x axis. This envelope is observed to move unchanged toward the right at the same group velocity as the apex. The congruent motion of the concentration of energy at the apex and the envelope of the wave packet has been confirmed by additional profiles (not shown) computed for times corresponding to 2, 3, $5\frac{1}{2}$ and 11 cycles ($I = 144, 216, 396, 792$), thereby strengthening the concept of breather-as-particle.

Enz's description of "the breather as a whole..." as quoted in the text (paragraph preceding Eq. (4)) is clearly demonstrated in Fig. 12. Here one can see the progression of the de Broglie phase wave ψ from left to right through the entire wave packet including the central "b" zone within which the major part of the mass-energy resides. One also has an overall view of the "...extended particle distributed continuously along the x axis..." This is in sharp contrast to the conventional quantum mechanics concept in which the point particle and its accompanying wave are considered to be separate but companion entities, the square of ψ not having any physical reality and serving only as a locator probability function.

Also shown in the left side (a) of Fig. 12 are conventional breather linear profiles corresponding to the semilog portrayal in the right side (b). Both sets of profiles in Fig. 12 (a) and (b) are generated from identical input parameters. The contrast in the amount and quality of the information portrayed in the two types of profile is striking. Thus in the semilogarithmic profiles we have a beautiful portrayal of both components of the double solution as envisioned by de Broglie and developed theoretically by Enz.

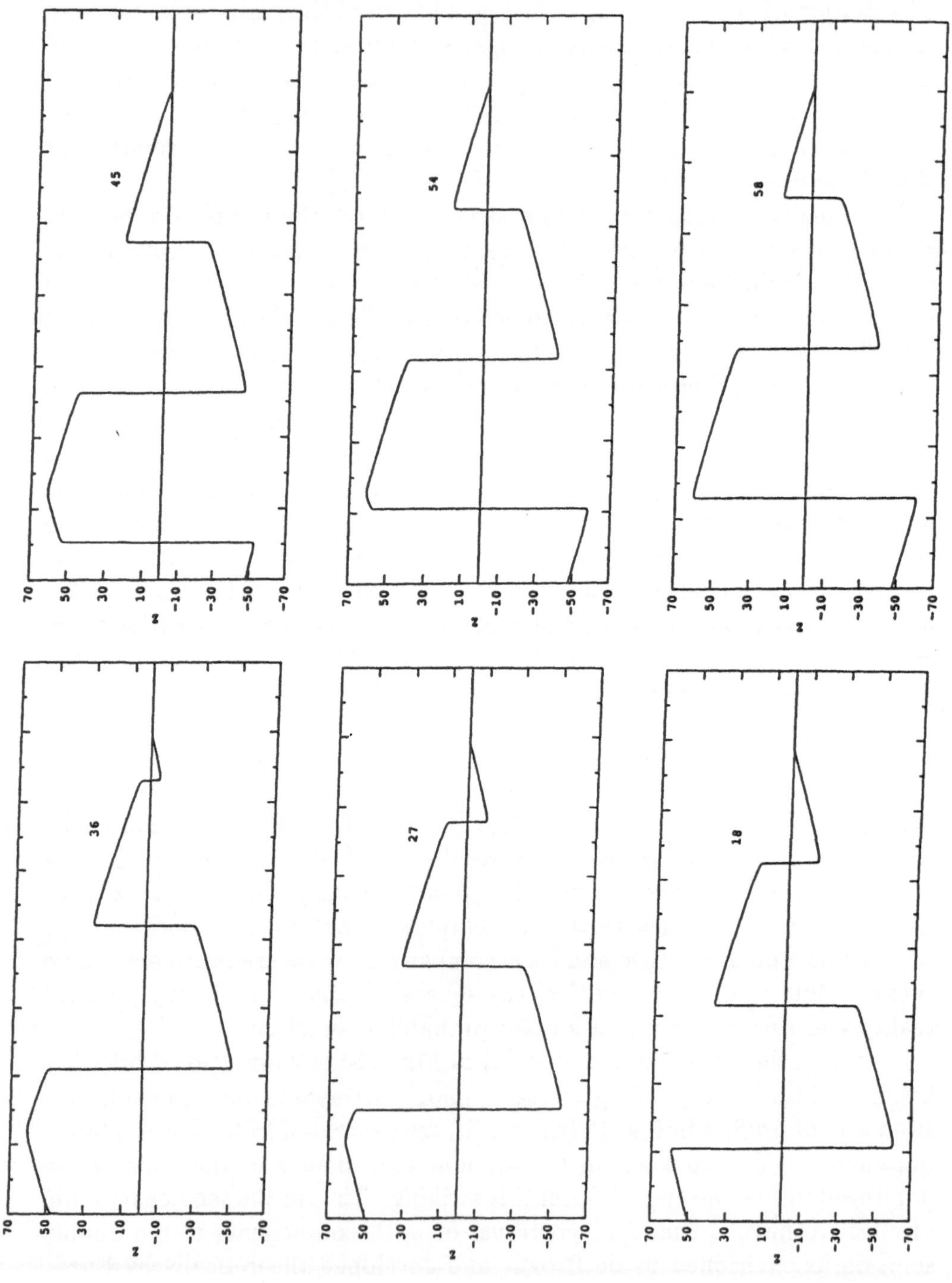
45
54
58
36
27
18
70 50 30 10 -10 -30 -50 -70

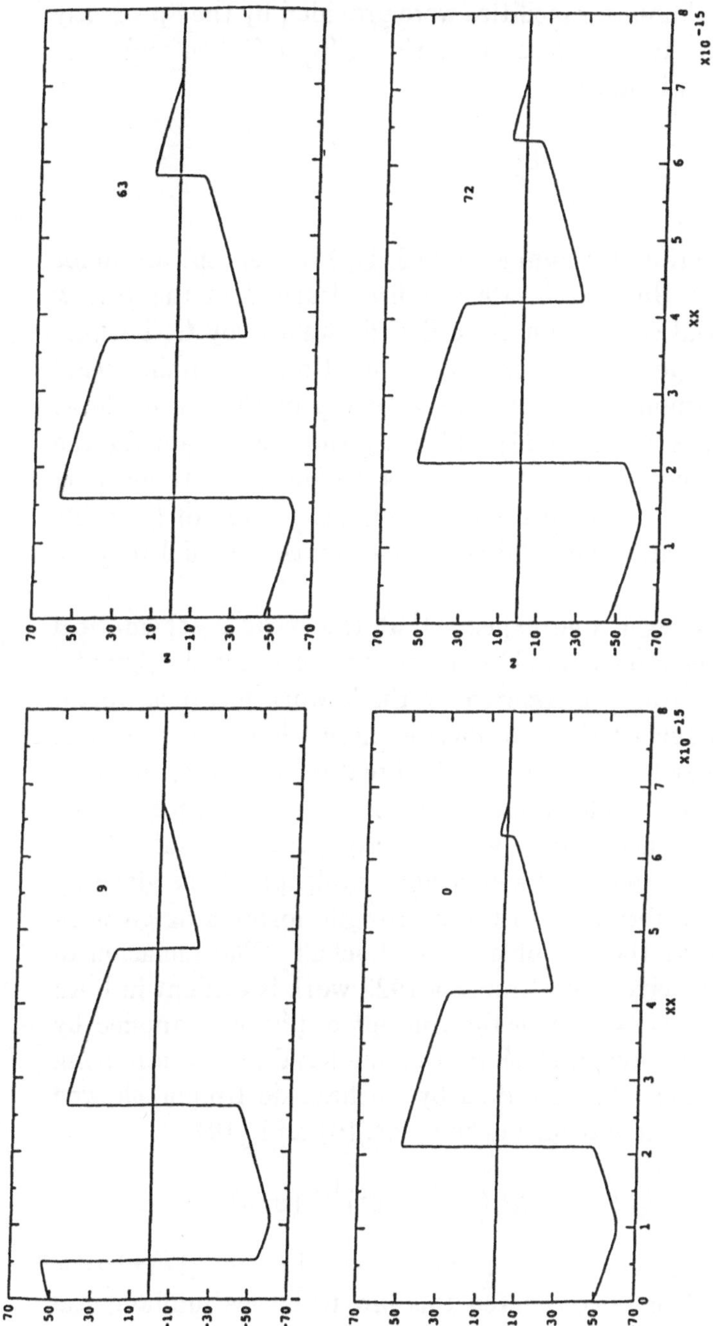

Fig. 11

ACKNOWLEDGMENTS

The author is grateful to U. Enz for helpful correspondence during the course of this work. Computer facilities were provided by the Speakeasy Computing Corporation. It is a pleasure to thank Stan Cohen and Tom Haux for programming assistance.

ADDENDUM

The above manuscript was completed in July, 1992, except for minor changes and corrections. In March, 1993, I first learned of the review paper "De Broglie's initial conception of de Broglie waves" by G. Lochak. [10] Section 2 of this paper is a summary of "de Broglie's main idea." This section provides—among other things—a survey of three pre-Thesis publications by de Broglie in 1923 [11] which lay the foundation for the 1924 theory of matter waves. The purpose of this Addendum is to use part of the mathematical and philosophical content of this section of Lochak's paper as a standard by which the validity of the breather model may be assessed.

At an elementary level, one expects that the basic mathematical properties of the breather model should conform to those found originally by de Broglie. Such conformity is assured by Enz's work but it is worthwhile for the reader to review these properties here. For example, it is evident by inspection that the formula for the breather period τ_b as given by Eqs. (4) and (5) in Sec. 2 above corresponds to the reciprocal of the de Broglie formula for frequency ν as given in Eq. (6) of Lochak. Likewise, by inspection the formula for the breather wavelength λ_b as given by Eqs. (6) and (7) above corresponds to the de Broglie formula as given in the footnote 1 equation at the top of p. 24 in Lochak. The inclusion of the special theory of relativity in de Broglie's 1923 work is evident in both of these examples. It is also evident in the concept of phase as applied by de Broglie to the case of a moving clock on the one hand and to a moving standing wave on the other. As described by Lochak, de Broglie showed that these two phases were identical, Lochak Eqs. (9) and (10),

$$\phi_{\text{clock}} = \phi_{\text{wave}} = \left(m_0 c^2/h\right)\left(1 - v^2/c^2\right)^{1/2}(x/v),$$

thereby establishing de Broglie's "law of accordance (or harmony) of phases." Lochak relates that de Broglie considered this law to be his fundamental achievement. For the breather the phase ϕ_b is evidently that of the sin[ ] function in Eq. (2) above,

$$\phi_b = \left[(r/d)\left(c\,(t - t_0) - \beta x\right)\left(1 - \beta^2\right)^{1/2}\right].$$

Making use of $r/d = 2\pi/\lambda_c, \lambda_c = h/m_0c, \beta = v/c, v = x/t$, and $t_0 = 0$, one gets

$$\phi_b = 2\pi\left[\left(m_0c^2/h\right)\left(1 - v^2/c^2\right)^{1/2}(x/v)\right].$$

The factor 2π appears in the breather phase because it is angular whereas in the Lochak summary it is cyclical. Thus it appears that the breather phase equals the de Broglie phase.

Other mathematical/physical properties include energy equivalence established by Enz setting the integrated energy for the stationary breather equal to the rest mass energy of the particle being modeled. The validity of energy in the breather model is also examined by Barut and Rusu. [12]

In addition to the linear mathematical aspects of the de Broglie concept and the breather which are in concordance, there remain the nonlinear properties of the breather which generates the localized hump shape of the u field. The hump is the unique characteristic of the breather which has long been identified as a manifestation of particle properties. It appears most prominently in linear plots of u as seen in several of the above figures, but in plots of the logarithm of u it appears only as the rounded-off apex of the rhombus-envelope profile. The tradeoff is that the de Broglie waves are lost in the linear plot but appear prominently in the logarithmic plot. Inspection of Eq. (2) shows that the hump results from the cosh[...] term in the denominator. Thus as the argument of the cosh function approaches 0, the function itself approaches 1 smoothly (no singularity) and hence the hump profile appears. The final profile is determined in part by the arctangent form of Eq (2). None of these nonlinear properties appear in the de Broglie theory. Their presence in the breather theory and the interpretation above of their significance provide a possible formulation and solution of the nonlinear problem envisioned by de Broglie. Other nonlinear equations may provide different mechanisms whereby a hump or other manifestation of a particle may be generated.

We turn now to the deeper philosophical aspects of de Broglie's concept of matter waves. As told by Lochak, "De Broglie's main idea is not dualism, but *coexistence* between waves and particles." Continuing (2+ pages later), Lochak recounts de Broglie's thinking thus: "De Broglie never considered that, in stating this law" (of phase harmony), "he had given any explanation of wave particle dualism: he only found an important formula which follows from the laws of relativity. But the question is *What* property is hidden behind this formula? What is the mysterious balance between corpuscle and wave (similar to the balance between a surf-rider and a sea wave), which is expressed in the formula?"

Coexistence (simultaneity) has already been described implicitly in the wave packet portrayal in Fig. 12 above and in the accompanying text, Sec 3.C. In light of the thinking of de Broglie as quoted above, one can now

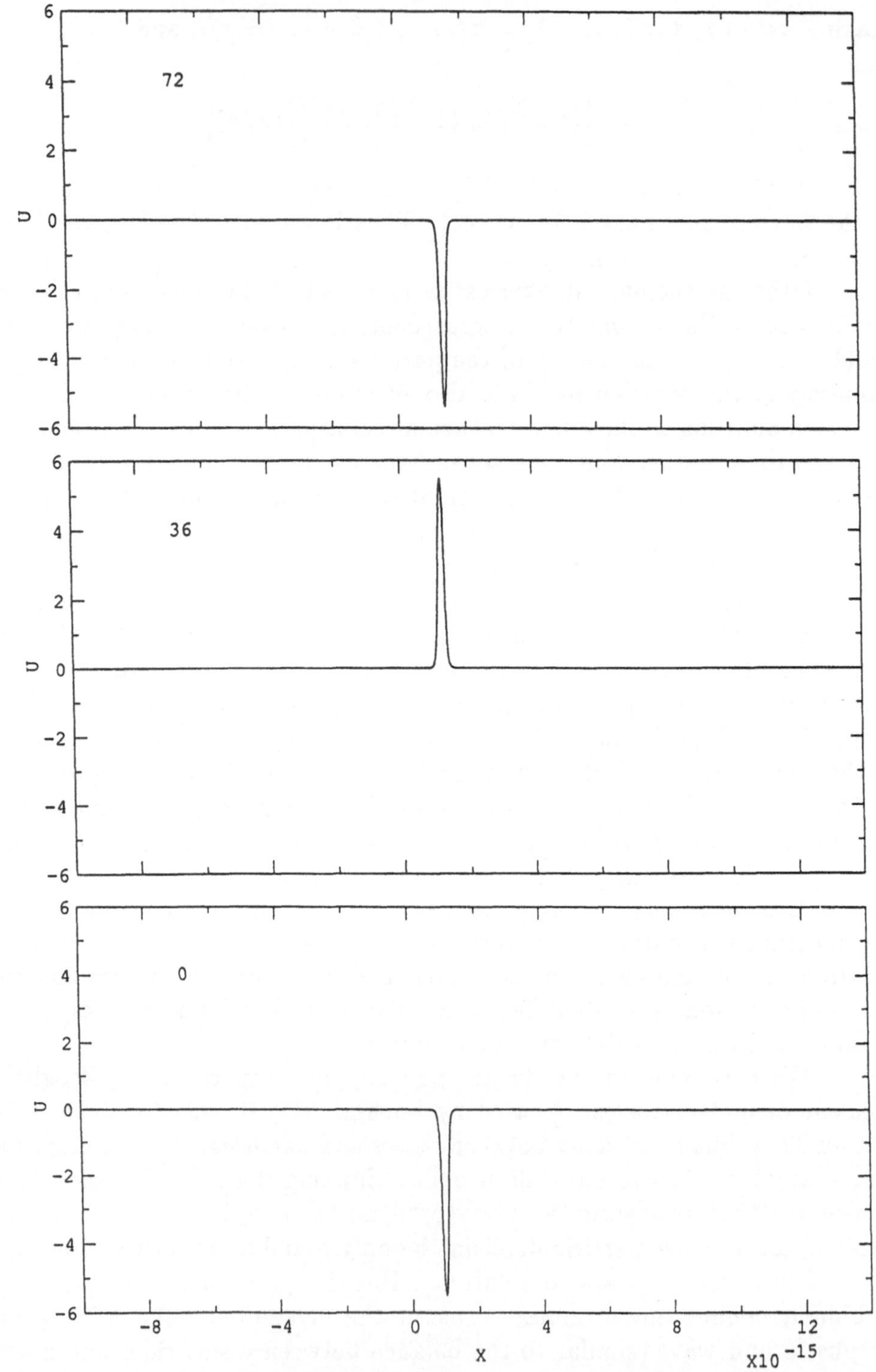

Fig. 12(a). Portrayal of a moving wave packet at times of 0, 1/2, and 1 cycle ($I = 0, 36, 72$) by two different methods: (a) Frames on left are linear plots of the breather wave function $u(x)$. Detailed description in text.

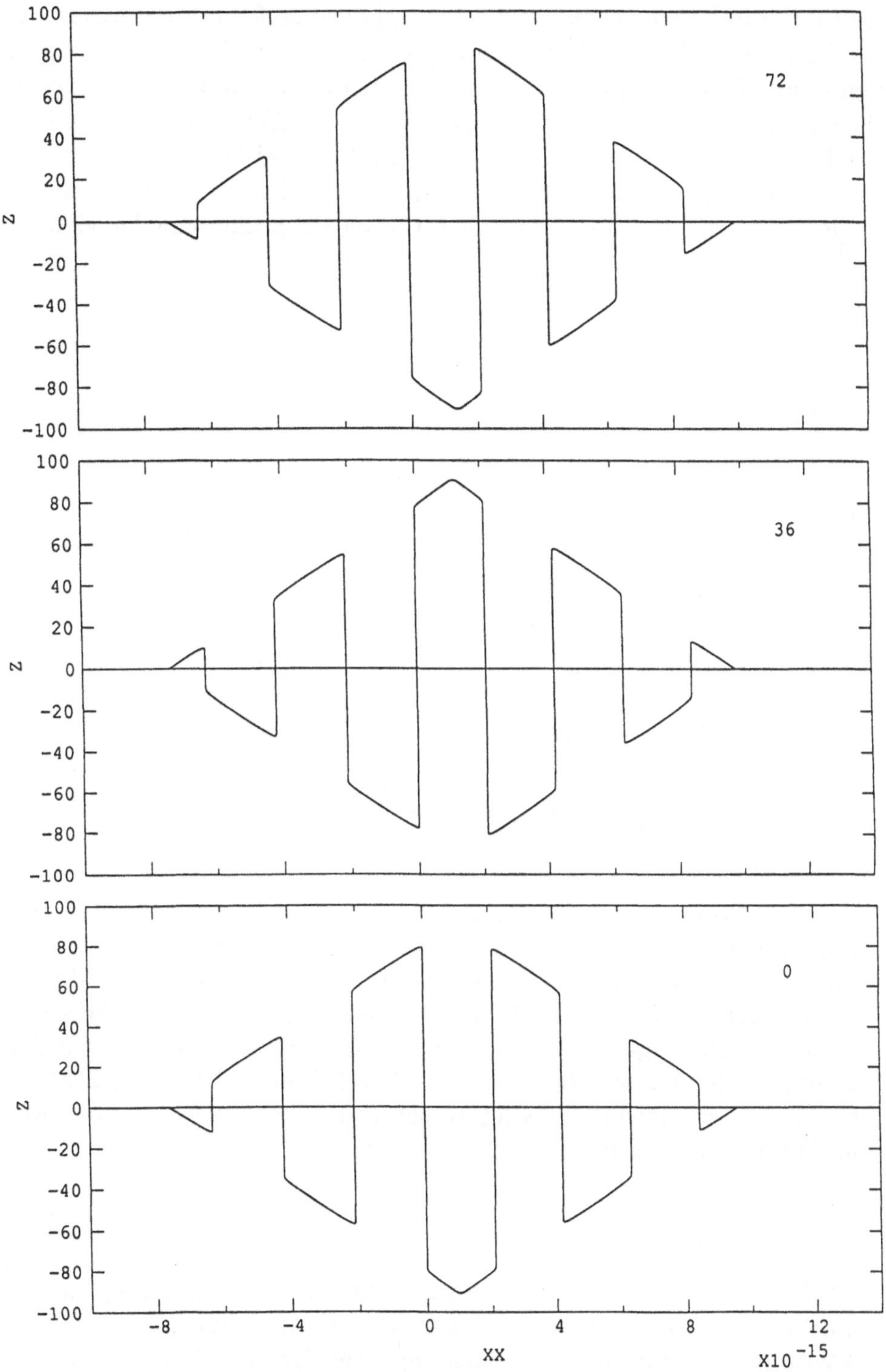

Fig. 12(b). Frames on right are semilogarithmic plots of $z = \log u(x)$ at the same times as in (a). Both sets of frames are generated from identical input parameters. Detailed description in text.

revisit the breather profiles and their significance for matter waves. This can be done with the help of a new profile shown in Fig. 13. This profile represents the same moving nucleon breather as shown in Fig. 12 except at the end of the second cycle ($I = 144$). Several labels have been added by hand to this profile, namely $\lambda_b, v_g, v_p, c, CM, b$ and the apex. The center of mass CM is at 1.76E−15 m in this case. In summary, coexistence as portrayed in Fig. 13 means that the entire profile can be viewed as

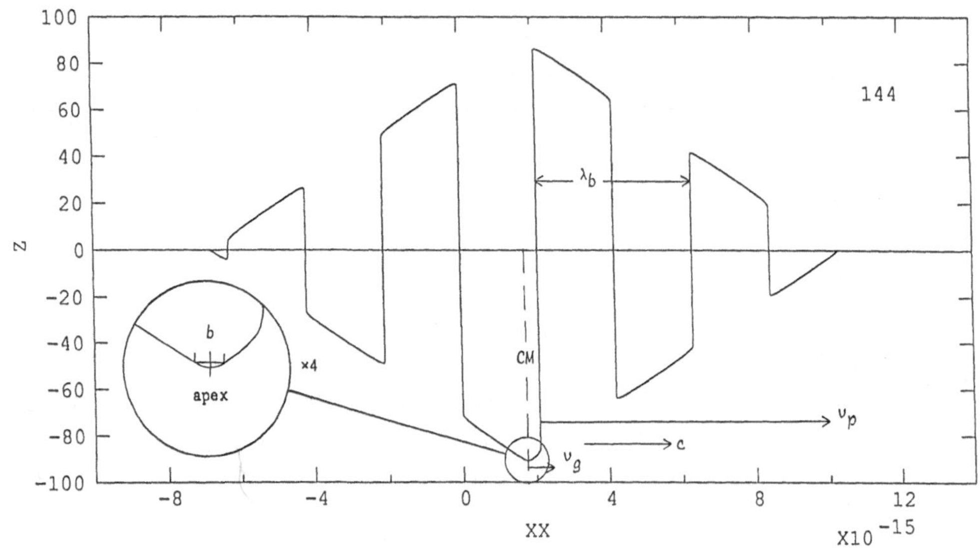

Fig. 13. Portrayal of continuation of same moving wave packet as in Fig. 12 except at the end of 2 complete cycles ($I = 144$). The whole wave packet including apex has moved toward the right an additional 1-cycle increment of 0.38E−15 m, placing the CM at $xx = 1.76\text{E}{-15}$ m. Other features highlighted in this figure include λ_b, v_g, v_p, c and b. The apex peak is at $z = -90.74$ and the half-height width $b = 0.213\text{E}{-15}$ m) is at $z = -90.44$. This figure shows the *coexistence* of both wave *and* particle as discussed in the text.

simultaneously (i) a moving standing wave having alternating half-wave segments of $+u$ and $-u$, and (ii) a moving particle having a continuously varying mass-energy density extending over four de Broglie wavelengths (theoretically from $-\infty$ to $+\infty$) and having an exceptionally high density in the apex zone centered at the CM. Thus, each half-wave segment of the profile represents a "piece" of both wave and particle simultaneously. Referring back to de Broglie's metaphorical "surf rider and sea wave,"

Fig. 13 shows that wave and particle are *not* separate entities as implied, but must be considered as melded into one single *coexisting* entity: wave *and* particle.

All the above applies to moving electrons as well as to moving nucleons, the only difference being that the horizontal scale for the electrons would be expanded by a factor of 1836 as shown in Fig. 5.

POSTSCRIPT

Up to this point this paper has been concerned only with a one-dimensional model, including the comparison with de Broglie's work. However, subsequent to the writing of the Addendum, it came to the author's attention that some recent studies have been made by Barut on three-dimensional classical models. Thus Barut uses linear wave equations from which he develops highly localized solutions which he calls wavelets. These have a size corresponding to the Compton wavelength rather than that of the larger conventional de Broglie wave packet and possess several important properties corresponding to those of particles.

In the earlier paper, [13] Barut constructs a "localized oscillating three-dimensional wave lump" (the wavelet) "representing a single quantum particle." The wavelet possesses appropriate particle attributes including total field energy, linear and angular momentum, non-dispersiveness and relativistic motion.

In the more recent paper, [14] Barut and Bracken take up an approach originally considered by de Broglie [15] involving the use of spherical standing waves as in a spherical resonant cavity. Barut develops this concept into a three-dimensional model of a moving particle. A significant feature of this classical approach is that the spherical standing waves are of two types called "electric" and "magnetic." These correspond, respectively, to the modes known to microwave engineers in the U.S.A. as TM (transverse magnetic) and TE (transverse electric.)

In view of the latter reference, [14] the author takes this opportunity to cite his own work on a standing-wave model of nuclear structure based on the standing electromagnetic waves as in a spherical cavity. [16,17,18] Being unaware of the early de Broglie work, [15] the author had no prior references to cite and his work was thus rejected by journal editors for various reasons such as having no legitimate basis and being too speculative. The author's model differs from that of Barut and Bracken in several respects, most of them beyond the scope of the present work. However, it is worth mentioning that an important similarity between the work of Barut and Bracken and the author is that the latter also makes use of the "electric" and "magnetic" modes and arbitrarily identifies them as corresponding, respectively, to protons and neutrons.

REFERENCES

1. L. L. de Broglie, *Non-Linear Wave Mechanics* (Elsevier, New York, 1960).
2. U. Enz, *Physica* **17D**, 116 (1985); *Physica* **21D**, 1 (1986).
3. N. J. Zabusky and M . D. Kruskal, *Phys. Rev. Lett.* **15**, 240 (1965).
4. G. L. Lamb, Jr., *Elements of Soliton Theory* (Wiley, New York, 1980).
5. G. Eilenberger, *Solitons: Mathematical Methods for Physicists* (Springer, Berlin, 1981).
6. The profiles in Ref. 5 are credited to Ch. Pöppe, *Diplomarbeit*, Universität Heidelberg, 1977. They are actually derivatives of the breather potential.
7. P. G. Drazin and R. S. Johnson, *Solitons: An Introduction* (Cambridge University Press, Cambridge, 1989).
8. J. Rubinstein, *J. Math. Phys.* **11**, 258 (1970).
9. M. J. Ablowitz et al., *Phys. Rev. Lett.* **30**, 1262 (1973).
10. G. Lochak, in *The Wave-Particle Dualism*, S. Diner et al., eds. (Reidel, Dordrecht, 1983).
11. L. de Broglie, *Compt. Rend. Acad. Sci.* **177**, 507, 548, 630 (1923).
12. A. O. Barut and P. Rusu, *Can. J. Phys.* **67**, 100 (1989).
13. A. O. Barut, *Found. Phys.* **20**, 1233 (1990).
14. A. O. Barut and A. J. Bracken, *Found. Phys.* **22**, 1267 (1992).
15. L. de Broglie, *Problèmes de Propagations Guidées des Ondes Électromagnétique* (Gauthiers-Villars, Paris, 1941).
16. A. J. Hatch, *Standing-Wave Model of Nuclear Structure* (1983, unpublished).
17. A. J. Hatch, *Binding Energy of Superposed Standing Waves* (1983, unpublished).
18. A. J. Hatch, *Calculation of Nuclear Radii* (1984, unpublished).

NOTES

1. Mailing address: 5460 South Cornell Avenue, Chicago, Illinois 60615, U.S.A.
2. The algorithm used here was developed by Stan Cohen. The computer null was set at 1E–60. For positive values of u, $z = \log u + 60$. For negative values of u, $z = \log(-u) - 60$.
3. The IBM laser printer used has a resolution of 300 dots per inch. The original figures had a length of 8 in. and a height of 4 in.
4. The value of 60 used in footnote 2 was changed to 90. These values impose arbitrary limits on the extent of the wave packet on the x axis. The theoretical limit is $x \to \pm\infty$.

PHENOMENOLOGY OF A SUBQUANTUM, REALISTIC, RELATIVISTIC THEORY

William M. Honig[1]

Curtin University
Perth, South Bentley, 6102, Western Australia

This paper outlines a unified realistic literal sub-quantum theory. Its operational relativistic invariance is based on the Lorentzian-Builder-Prokhovnik approach. The sub-quantum basis is a physical model for vacuum space consisting of the superposition of two oppositely charged and microscopically *continuous massless* fluids. Dual fluid models for the electron, photon, neutrino, proton, neutron, etc., consist of unique variations in the relative densities of the vacuum 2-fluids and their flow patterns. A point function global dual charge-flow 4-vector J_4 now characterises the space of fundamental particle models together with A_4, the associated 4-vector potential for each of the J_4 terms. The energy E necessary to assemble the fluids of a particular model gives the mass of each fundamental particle via $E = mc^2$. Mass is thus a *derived* parameter.

The metric of a model comes from the 3 or 4-space dimensions which are set up to follow the fluidic charge and flow patterns of a fluid model. The canonical curvature of GTR is then interpreted as the dual charge and flow curvature of a model. From the energy necessary to set a model the mass and the gravitational field of the model thus should be derived parameters. With the charge and mass thus unified, the dual fluid charge and flow distributions and their canonical space curvature of GTR is no longer a disembodied unexplained feature of matter. The gravitational mechanism should reside in the dual-charge and flow fluid distributions composing a fundamental particle model.

A number of new ideas and experimental predictions are presented, first among which is the electromagnetic "photex." This consists of the canonical half wavelength dipole field distributions considered as a "smoke-ring-like" discrete deformable entities which provide a physical model for and explanation of Planck's constant and is a physical model for the hidden variable of quantum mechanics.

Waves and Particles in Light and Matter, Edited by
A. van der Merwe and A. Garuccio, Plenum Press, New York, 1994

Key words: dual-charged fields, operational invariance, neo-Lorentzian relativity.

1. INTRODUCTION

It must first be noted that this short paper can be no more than a brief outline and some sample examples for the fluidic realistic sub-quantum theory presented, and the reader is referred to the book and papers which appear elsewhere [1–15]. In addition it must be pointed out that the prime requirement for such efforts lies in providing, inventing, and suggesting a sub-quantum phenomenology which must underlie any subsequent work. It is thus the dearth of such concepts which must be attended to first. The mathematics must of necessity come second, although much new mathematics have been uncovered in these efforts, primarily in number set theoretical non-Boolean logics and metrical transformation sequences, which are also listed in the sources above. Such new efforts are not and should not be mathematical reinterpretations or operations on details of the QM paradigm.

This summary of such a covariant sub-quantum theory shows how realistic, literal pictures of physical reality are possible. Of course, this writer is committed to these particular ideas, but only confirmed experimental predictions which are based on these ideas and which are not derivable from the STR/QM paradigms should provide the basis for their acceptance. Only by following such a procedure, it is strongly believed, will progress ever be made in the central mysteries of STR/QM. Although reinterpreting STR/QM appears to be a reasonable goal, it is a sterile one. That is, straightening out (epistemologocially or otherwise) the unsatisfactory, mysterious global nature of STR/QM does not appear to lead to a deeper (microscopic) knowledge. More radical approaches such as this or similar ones need to be examined. We believe that realistic microscopic ideas (such as those given here or others of a covariant realistic nature) have the fertility and heuristic power to again permit ongoing cognisance of microreality to suggest new experiments.

We present here some necessary and primary phenomenological features:

a. a two-fluid oppositely charged continuous ether;
b. fluidic realistic models for the fundamental particles and electromagnetic waves composed from the 2-fluid plenum;
c. a neo-Lorentzian or Builder version of STR which make the fluid models operationally covariant;
d. discrete electromagnetic field distributions which are the half wavelength dipole fields first pictured by Hertz, each of which is here called a photex;

e. relativity of the metric which makes space a relative concept, clarifies the difference between local and non-local phenomena, and gives a physical mechanism for the canonical space curvature [4];
f. constructing a realistic but non-Boolean logic to cover the present STR/QM paradigms [5].

Section 2 covers items a and c and discusses how covariance can be imputed to fluid models and how mass can be developed from fluids which in themselves do not possess this quality. Section 3 covers b and d and describes the fluidic model for the electron, the progresion to other particle models and the prototype discontinuous electromagnetic wave (dubbed the photex). It further shows how the shedding of toroidal vortices by the fluid electron in rectilinear acceleration and deceleration, which generates a sequence of half wavelength dipole fields is a good model for electromagnetic radiation generation since these discrete entities (photexi) can be considered either as particles or assembled into discrete finite coherent wave trains. They also provide the physical model for the hidden variable and for a stochastic medium made of myriads of such photexi. Items e and f are described in other papers [4,5].

This results in:

1. suggested fluidic models for the particles wherein the energy for the assembly of the model corresponds to the mass of the particle;
2. the complete calculation of the electron rest mass [1];
3. a realistic interpretation for h and for hidden variables [3,4,6,7];
4. calculation of h from Larmor's formula for the fluid electron model in rectilinear acceleration and deceleration [1];
5. a number of suggested experiments [1–3];
6. a correspondence between fluid flow curvature and the canonical curvature tensor [2,3].

2. A TWO-FLUID ETHER AND ITS COVARIANCE

It is proposed that empty vacuum space consists of two superposed quiescent massless positive (+) and negative (−) charged microscopically *continuous* fluids. Each of these quiescent fluids are set up with a charge density equal to that of the classical electron. Quiescent vacuum space thus becomes similar to a neutral plasma. These fluids can be used to characterise massive models of the fundamental particles including their fields and all their other parameters. The dual fluid basis for the composition of vacuum space is operationally covariant and permits the construction from these fluids of operationally covariant fluidic massive models of the fundamental particles and of electromagnetic radiation fields (which fields are discontinuous) and which exhibit quantum behaviour. With such a 2-fluid plenum it is also easy to account for the electric and magnetic fields of electromagnetic waves in regions remote from their point of generation. The progression to other fundamental particles has been given [1–3].

With this dual fluid vacuum space plenum, net positive and negative charge distributions and velocity flow fields are possible in all of space. The use of dual charged fields makes possible two useful symmetries. First, any particular complete fluid model (both for charged and uncharged particles and electromagnetic waves) will have a *net charge of zero*. Second, the reversal of all fluid densities and flows gives the anti-particle fluid model.

For model construction relative density variation and velocity flow fields can be set up for a particular fluidic fundamental particle model by rearranging the vacuum space quiescent dual fluids. This results in extended fluid models of the canonical particles and their fields. Thus the proposed electron spinning droplet model, for example, has a spinning negative internal fluid charge which is balanced by an equal amount of positive external fluid charge which is used for its external electric field characterisation. Using these ideas the complete mass of the electron and its anomalous magnetic moment can be easily accounted for [1].

These ideas bring clear physical meaning to a fluidic net charge 4-vector, J_4, which characterises every point in space. Accordingly, the associated fluidic 4-vector potential, A_4, also has a clear phyiscal meaning; it is the energy density necessary to set up at any point in space the fluidic net charge density and charge flow fields of each particle model. A correspondence principle can be set up for this method whereby the canonical J_4 and A_4 vectors which are correct for the external fields of particles and for waves can be extended into the postulated fluid internal charge and flow modes of a model. Since such realistic models appear to be restricted to the particular rest frame of each fluid model these ideas have been ruled out up till now by the canonical requirements of covariance. It is the introduction of an operational covariance based on the Builder-Lorentz version of STR which makes such fluidic models possible [1–4].

Electromagnetic (em) waves are composed of a prototype discontinuous em wave entity which is a fluidic half wavelength electromagnetic dipole field distribution. These separate em dipole field distributions are discontinuous ever expanding "smoke rings" (each called the photex). They are formed in the dual fluid plenum by the fluidic electron droplet model in accelerated (or decelerated) vortex-shedding motion. Contiguous assemblies of such field distributions are the models for em continuous wave trains. Arbitrary spatial assemblies of these field configurations are sufficiently general to express arbitrary em field distributions.

Even at this qualitatitve level this plenum can be seen to be capable of being populated by myriads of such photexi which can thus be a model for the ghost (*gespenster*) waves of Einstein or for a stochastic subquantum medium. Such a vacuum plenum can also be seen to support the idea of a chaotic medium consisting of many other kinds of disturbances in the neutrality and velocity fields of the quiescent 2-fluid naked vacuum space plenum.

Such a revision of the basis from which the fundamental particles can be constructed is a major change with respect to present ideas in this subject. A very detailed and careful discussion of this new phenomenology is, of course, necessary. Much of this has been given elsewhere [1–15] and is summarised here.

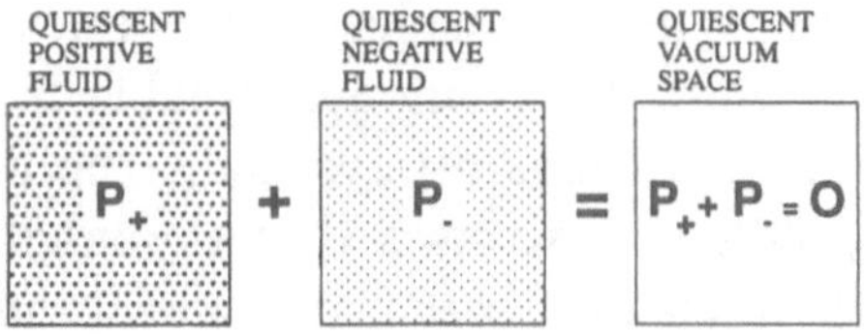

Fig. 1. The 2-fluid model for quiscent vacuum space.

The fluid plenum proposed here is not a materially testable ether and so its qualities differ radically from the canonical non-relativistic and relativistic ethers. Its presence cannot be directly experimentally confirmed. The fluidic fundamental particle models constructed from these fluids, however, do possess mass and are material objects which can cause effects (many so far unmeasured) which must be capable of experimental confirmation.

The dual fluid ether shown in Fig. 1. consists of separate uniform positive and negative charged (otherwise featureless) infinitely divisible fluids. Their quiescent densities are taken as equal to the charge density of the classical electron. These fluids are to fill all of space so that their superposition confers on space a neutrality.

This ether, a massless dual charge fluidic plenum, is covariant in the following way. Since the quiescent dual fluids are completely uniform then obviously each or both of them have no distinguishing variations evident or measurable from any rest frame moving at any velocity in any direction. The issue of electromagnetic waves, however, is another matter. These waves will be represented by fluidic model entities (the photexi) and it is the velocity of these electromagnetic entities with respect to the canonical particle fluid models of which any physical rest frame must be composed, which defines their velocity, c; the plenum 2-fluid is not directly involved. The waves consist, as will be shown, of dynamic fluidic entities, the photexi (evolving toroidal smoke rings), which are generated by the shedding of toroidal vortices of the fluidic spinning droplet electron fluidic models when they undergo acceleration or deceleration, as in collisions.

The velocity of the photexi with respect to the rest frame of the generating electron droplet is c in the radial direction. All rest frames would give the same answer for c if we adopt the convention that the absolute rest frame defines the absolute value of c and all other rest frames give the same answer because the Lorentz distortions are literal. The Lorentz space and time distortions will also be imputed to distortions of the fluidic models for the fundamental particles when moving at an arbitrary speed ($v < c$) with respect to the absolute or cosmological rest frame. This is, however, simpler to understand using Builder-Lorentz STR since there the velocity c is directly defined as c in the absolute rest frame and although it is $c \pm v$ in the other rest frames, this is unmeasureable in the other rest frames because the Lorentz distortions always make the final value come out as c. This version of STR also stresses that the inhabitants of an arbitrary rest frame may take their own rest frame and its models for fundamental particles as *operationally identical* to what they would be in the absolute rest frame.

3. FLUID MODELS

The model for the electron, see Figs. 2 and 3, is a spinning spherical droplet of the negative charge of the original plenum fluid which is surrounded outside the droplet by the equal amount of positive charge (which was scooped out of the original 2-fluid space) but falling off with distance in such a way that the external electric field of the electron will derive from this. The droplet internal velocity field gives the droplet a solenoidal rotation with an equatorial velocity which approaches c; thus it may be written as $c - \Delta$. The external (+) charge (from which its external electric field derives) has a non-solenoidal velocity field. The total net charge of the complete model is zero and the energy to assemble this model divided by c^2 is the mass of the electron. Mass thus becomes a derived quantity and energy is the active agent for the fluid model constructions. Charge and potential 4-vectors can then be set up for all parts of this model. With such a model it becomes a simple matter to account for the full self energy of the electron, for its Poincaré stress, and its anomalous magnetic moment [1]. A new correspondence principle aids in constructing these models and in reconciling the canonical properties of particles with the relevant parts of the fluid models. Thus the canonical external field of the electron is

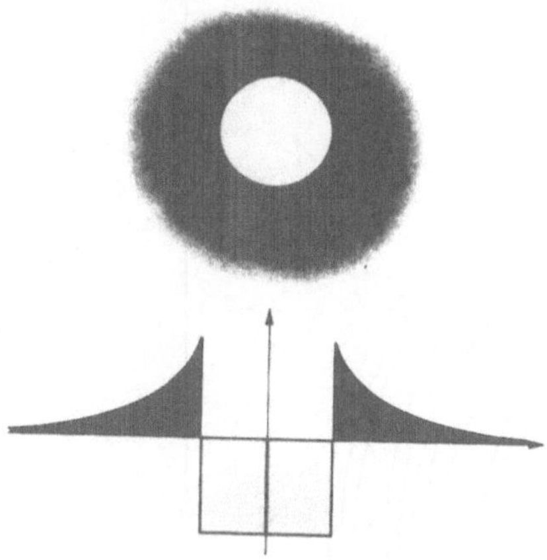

Fig. 2. Electron-droplet model:
Charge distribution.

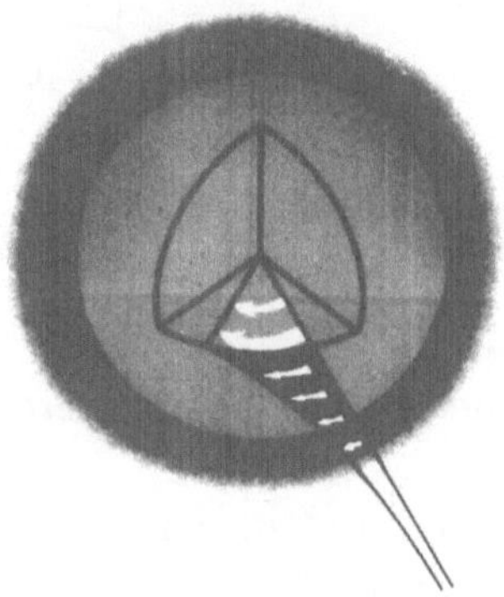

Fig. 3. Electron-droplet model:
Flow velocity distribution.

used to deduce the fluid characteristics external to the fluid model droplet using mainly the A_4 4-vector potential which has always had a fluidic interpretation. This is then extended inside the droplet in such a manner that the values for the Poincaré stress, the anomalous magnetic moment, and the rest mass of the electron were simultaneously correct. The final fluid model thus includes all the physical details of the fluid flow.

EM waves are developed from the simplest EM generating motion of the electron droplet: its rectilinear acceleration and deceleration. The field of hydrodynamics is used to suggest that the original Hertzian dipole wave pictures are caused in a similar way to the non-linear vortex shedding behaviour of spinning spheres immersed in water flow which are kicked back and forth along their polar axes. In hydrodynamics each such vortex is a

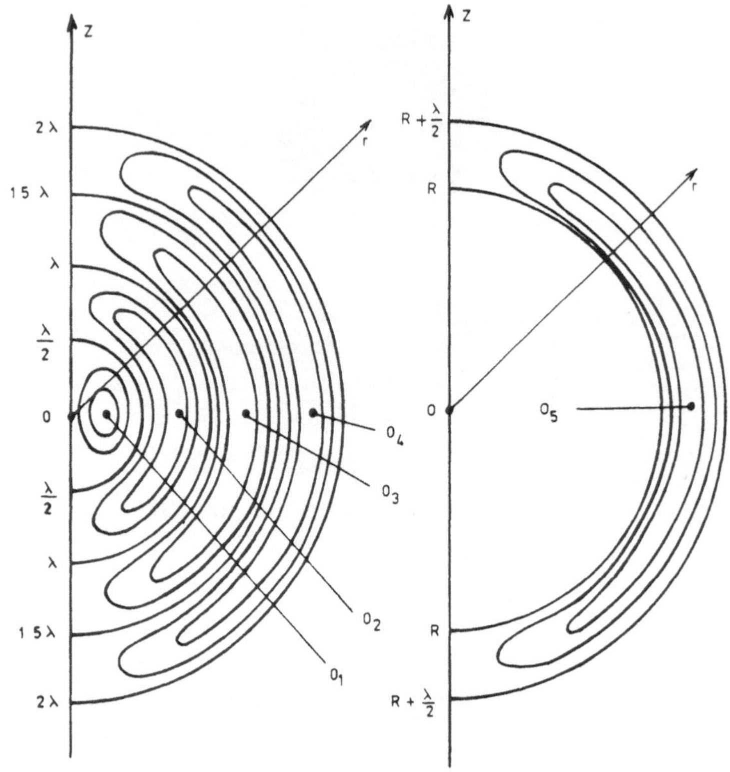

Fig. 4. Hertzian dipole and photex sketches - right hemisphere

discrete entity which is generated in a non-linear fashion and then takes on its independent existence; see Fig. 4.

What has been described is the generation mechanism. The absorption mechanism would be the time reversal of this; such a dipole half wavelength EM field distribution would be removed from say a large magnitude EM wave incident upon the electron droplet. Absorption appears to be an entropy increasing process.

The great importance of these half wavelength dipole field distributions must be emphasised. Each photex is physically complete, no need exists for plane waves or other non ideal representations. Since the minimum energy per cycle per second of EM waves derives from the value of h as about 4×10^{-15} electron volts; the energy for a half cycle and thus for each vortex (photex) is about 2×10^{-15} electron volts. It becomes possible now using the above photex shedding electron droplet model to derive h from the value of e, c, and the fine structure constant (which in fluidic terms is a winding number for the rolling up of the photex vortex) [1].

When one speaks of the minimum energy of this photex, this half cycle of EM energy, one is projecting oneself into the non-physical rest frame of this EM vortex. In a our physical rest frames where EM energy is always going by at the speed c, the correct designation is 2×10^{-15} ev per half cycle per second.

Since mass is a derived quantity from the energy necessary to set up the photex, the mass of the photex is about 10^{-47} gm. This appears to correspond quite well to the so called "rest mass of the photon" which has been discussed by many others (see discussion and references for this in Ref. 1.) A rest frame for the photex can be defined where the photex is a local canonical particle to which this mass can be imputed. This unifies the idea of rest mass for EM energy (of this photex) with the canonical concept of mass so that no separate spatial mechanism for this mass is necessary.

Consider such a moving droplet rebounding back and forth between 2 rigid walls in the manner of a tennis ball; see Fig. 5. At the wall the droplet will have to decelerate to zero velocity and it will then accelerate to the negative of its approach velocity. According to the fluid picture for the generation of the vortex given above, the electron droplet's deceleration and its acceleration should result in the generation of a photex pair for each rebound or collision. This should decrement the energy of the electron by about 4×10^{-15} ev for each such rebound. About 0.25×10^{15} such collisions would reduce the electron kinetic energy by one ev and generate an equally huge number of photexi pairs. Thus, electron droplet collisions, provide the means for filling vacuum space with myriads of such photexi pairs. This implies that all electron collisions are inelastic to the extent of a 4×10^{-15} ev loss per collision.

The essence of the photex corresponds closely to the essence of the photon in that both photon and photex are emitted and absorbed in a discrete manner. Whereas, however, the photex is spatially extended and wavelike, the photon is not. Photon experiments over the past 80 years have amply demonstrated that it is a localised, discrete, and ballistic entity. Photon energies via $E = hf$ lie in the range greater than about 0.1 ev. The photon, discussed elsewhere [1–4] arises in atomic excitations and emissions and is structurally similar to a localised stable spherical EM field distribution. The photon and photex can correspond to the de Broglie idea of a particle and carrier wave.

Strictly speaking, it is quite wrong to refer to photon particles as photons or photon particles. This is because we are inhabitants of physical rest frames, thus we cannot experience or observe photons or photon particles. They can only be observed and should only be referred to as photons per second or photon particles per second. This is not a grammatical or linguistic matter, it is the most important requirement of relativistic invariance. It appears to do no great harm to refer to them in the aforementioned manner (as photons or photon particles). This is because the canonical photon particle has since its introduction amply demonstrated in the laboratory via the many ballistic test of photon interactions that not

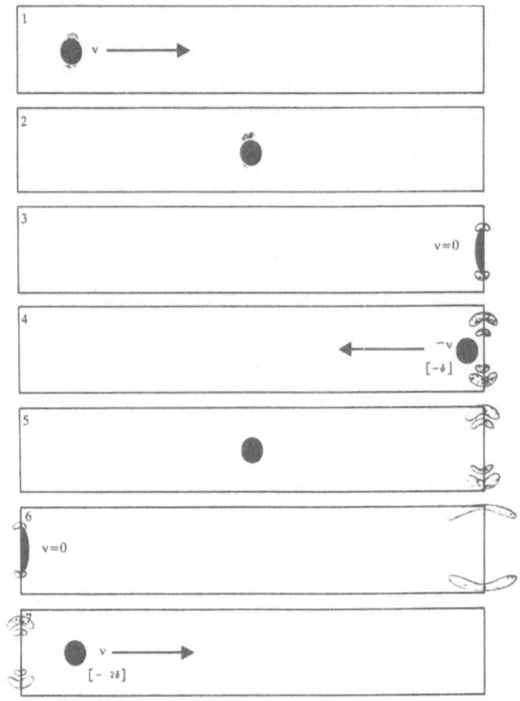

Fig. 5. Model for electron rebounds from collisions - photex pair generation.

only is it discrete but it can be treated like a point-like particle.

On the other hand, the photex which is also discrete is an extended space-like continuously expanding entity. We believe, however, that it is detrimental to refer to the photex in the above manner (as photexi or photex particles) and that they should be referred to as photexi per second or photex particles per second. This is because the first mode of expression confuses two kinds of rest frames.

In the physical rest frames which we inhabit we are used to treating objects such as electrons, protons, etc., as small objects and for most theoretical and experimental considerations they can be mathematically replaced by, say, a 3-dimensional delta function. The word photon came into existence to refer via $E = hf$ to discrete amounts of EM energy which can be absorbed or emitted. Since waves have been canonically assumed to possess only the quality of spatial continuity,with which it is not possible to associate discreteness, then discreteness was instead imputed to objects like, say, electrons. Such canoncial localised electron-like objects are spatially discrete and local. This means that if such an object is situated for example around the origin of a Cartesian coordinate set then a radial gauge shrinkage of the space and of the discrete localised object will result in the canonical conception of the point-like particle. Photexi, however, are both discrete and wavelike. The importance of this consideration is that if one refers to or observes an object as a photex (and not as a photex per second) *one must be in the rest frame of the photex and it should appear as a point-like particle only in its own rest frame.* Such a rest frame can be termed a non-physical or conceptual rest frame because we physical beings cannot enter (but may contemplate) such a rest frame. The consequence of this is the question: Can a continuously expanding non-local object (such as the photex) seen in our physical rest frame simultaneously appear to an approriate observer to be the canonical (local) particle in its own rest frame? We summarise here how the answer to this question can be "yes."

Consider the photex, the half wavelength dipole field distribution which continuously evolves in time into the larger and larger kidney shaped entities of Fig. 4 and which is an extended discrete object. This could be considered as a localised or point particle if the metric of space can be considered as a relative measure. The clue to such a relative metric lies in what one means by the concept of particle. In the case of the electron which is here taken as a spherical droplet, a radial gauge change applied to the droplet which shrinks it to a vanishingly small size permits that it be replaced for many considerations by a three dimensional delta function and thus to its canonical designation as a particle [1,4].

If one accepts the idea that electrons always shed these photexi electromagnetic vortices in collisions, rebounds, and other means for acceleration/deceleration (due to electric and magnetic fields) then this is an objective new phenomenon. This would have to be capable of explaining known quantum behaviour. Although photex energy (2×10^{-15} ev) is quite small, the energy of relatively small numbers of them is able to affect electron motion, as Selleri, Dewdney, and Hiley have discussed.

The thickness, or the half wavelength of the photex appears to be calculable from the de Broglie relation in the form

$$p = mv = (h/2)/(2/\lambda), \tag{1}$$

where p is the momentum of the electron, m is its mass, v is its velocity, and $2/\lambda$ the wavenumber of the EM dipole field distribution (the photex). This connects the velocity of the droplet electron with the half wavelength of the photex it will generate upon collisions, etc.

The existence of photex toroids could be tested experimentally so as to establish the experimental validity of these models. This could consist of setting up dilute electron beams or even solitary electrons with kinetic energies of say, 50 electron volts. The electrons could be put into a vacuum chamber, so as to ballistically rebound back and forth between two repelling walls a huge number of times. For each 0.25×10^{15} such rebounds the electron should lose on electron volt of kinetic energy. Thus, for this

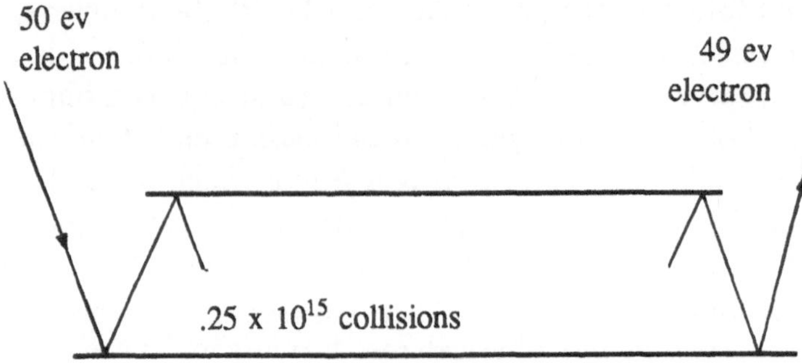

Fig. 6. Inelastic electron collision experiment.

example, the electron would come out of this test with an energy of 49 ev; see Fig. 6.

The tennis ball example shown in Fig. 5 is further examined. In its deceleration to zero velocity at the wall it would flatten to an extent depending on its approach velocity. The wall could be replaced by an identical electron droplet moving from beyond the wall in the opposite direction to the original droplet with its point of impact the same as the original droplet's point of impact with the wall. This would be a more realistic example. In either case this example provides a physical interpretation for the commutation relation

$$xp - p'x = h, \tag{2}$$

where i and 2π have been suppressed. The first xp term could express the situation at a small distance x before the droplet with momentum p hits the wall, and the second $p'x$ term is the expression after the rebound from the wall at the same position x when the momentum p has been reversed. Here the momentum p' should be only slightly less than the original incident momentum. The photex pair generated at the wall decrements the electron kinetic energy by only 4×10^{-15} ev. For electron kinetic energies of more than one ev, for example, the first and second p terms differ by an extremely small amount so that practically the prime symbol in Eq. (2) could be dropped.

If the photex is indeed the hidden objective variable of QM, the quantum potential should derive from the vortex shedding photex generation phenomenon of a droplet electron.

REFERENCES

1. W. M. Honig, *The Quantum and Beyond* (Philosophical Library, New York, 1986).

2. W. M. Honig, "An electromagnetic world picture, Part I: Massless dual charged fluids for modelling vacuum space, fundamental particles, and electromagnetic waves," *Physics Essays* **4** 583–590 (1991).
3. W. M. Honig, "An electromagnetic world picture, Part II: Planck's constant, and the discrete electromagnetic wave model - the photex; a physical model for the QM hidden variable," *Physics Essays* **5**, 254–261, (1992).
4. W. M. Honig, "An electromagnetic world picture, Part III: Subjectivity of space, relative metrics, and the locality-nonlocality conundrum," *Physics Essays* **5**, 514–525 (1992).
5. W. M. Honig, "Logical organisation of knowledge with inconsistent and undecidable algorithms, using imaginary and transfinite exponential number forms in a non-Boolean field: Part 1: Basic principles," *IEEE Trans. in Knowledge and Data Engineering* **5**, 190–203, (1993).
6. W. M. Honig, "Physical models for non-local particles, hidden variables, and all that," in *Problems in Quantum Physics* (World Scientific, Singapore, 1988).
7. W. M. Honig, "On the physical meaning of Planck's constant h from a realistic subquantum theory," *ibid.*
8. W. M. Honig, "The locally consistent and globally inconsistent axioms in STR and QM: Using exponential, imaginary, and transfinite number fields and the forms $e^{i\phi}$," in *Nature, Cognition, and System, Volume 3*, M. E. Carvallo, ed. (Kluwer, Dordrecht, 1993).
9. W. M. Honig, "The relativity of the metric and geometry," *ibid.*
10. W. M. Honig, "A minimum photon 'rest mass' - using Planck's constant and discontinuous electromagnetic waves," *Found. Phys.* **4**, 367–380 (1974).
11. W. M. Honig, "Gödel axiom mappings in special relativity and quantum-electromagnetic theory," *Found. Phys.* **6**, 37–57 (1976).
12. W. M. Honig, "Transfinite ordinals as axiom number symbols of quantum and electromagnetic wave functions," *Int. J. Theor. Phy.* **15**, 87–90 (1976).
13. W. M. Honig, "Photon rest frames and null geodesics," *Int. J. Theor. Phys.* **15**, 673–676 (1976).
14. W. M. Honig, "Relativity of the metric," *Found. Phys.* **7** 549–572 (1977).
15. W. M. Honig, "Quaternionic electromagnetic wave equation and a dual charge-filled space," *Nuovo Cimento* **19**, 137–140 (1977).

NOTE

1. Direct mail address: P.O. Box 361, S. Perth, 6151, Australia. E-mail address: RHONIGW@cc.curtin.edu.au

p=h/λ? W= hν? A RIDDLE PRIOR TO ANY ATTEMPT AT GRAND UNIFICATION

Mioara Mugur-Schächter[1]

Laboratoire de Mécanique Quantique
et Structures de l'Information
University of Reims
F51062 Reims Cedex, France

It is shown that: (a) De Broglie's quantum relation $W = mc^2 = h\nu$ does not have the same relativistic variance as Einstein's quantum relation $W = h\nu$ assumed for light. (b) If "matter waves" possess *physical* reality and permit a *local* representation of microphenomena then, instead of $p = h/\lambda$, another relation has to be admitted, which numerically differs significantly from $p = h/\lambda$ only for nearly null values of the relative velocity of observation u, but the difference becoming infinite for $u = 0$. Then quantum mechanics has to be reorganized. However, the splitting of the quantum relation subsists if the Einstein variance of macroscopic mass and mechanical energy holds rigorously also at the microscopic level of description. (c) If furthermore the same variance is valid for a quantum of light energy and a quantum of heavy energy, then the definition of a *micro*-mass must involve a factor that depends on the phase velocity of the corresponding corpuscular wave; notwithstanding the tiny numerical difference with Einstein's definition introduced by this factor, this modified definition of micromass, if true, requires reconstruction of also the relativistic representation, at the basic microscopic level of description. In *either* case—(b) alone, or (b) and (c)—a real and local nature assigned to the corpuscular field would entail a deeply unified representation of fields and heavy matter, at the microscopic, macroscopic, or cosmic level of description.

Key words: quantum relation, de Broglie wavelength, mass, unification.

1. INTRODUCTION

This work develops further consequences of results established before [1] and reproduced here.

Waves and Particles in Light and Matter, Edited by
A. van der Merwe and A. Garuccio, Plenum Press, New York, 1994

The whole structure of quantum mechanics, and even of modern physics as a whole, is indissolubly tied to the acceptance of the two correlated relations $p = h/\lambda$ and $W = mc^2 = h\nu$. These have been introduced by Louis de Broglie, in his thesis [2], on the basis of considerations of relativistic variance. Such a genesis should insure a fundamental harmony between quantum mechanics and relativity. But this is not the case.

In recent years the question whether the de Broglie "corpuscular" waves are *physical* waves has been much debated in connection with the locality problem [3–6], and experimental investigations have been devoted to it. But, as far as I know, a connection between this question and the structure of the basic formulas $p = h/\lambda$ and $W = mc^2 = h\nu$ has never been asserted. We shall show that there exists a most intimate connection and that it entails very fundamental consequences.

2. THE PROBLEM OF THE QUANTUM RELATION

In *macroscopic* relativistic electromagnetism, the transformation law admitted for the wave frequency $\nu_{\ell w}$ of a light wave (ℓ: light; w: wave) is (in one spatial dimension) $\nu_{\ell w} = \left(\nu_{o\ell w}/\sqrt{1-\beta^2}\right)(1+u/c)$ (o: index of the proper referential; u: the relative velocity of observation; c: the phase-velocity of light). To explain the photoelectric effect, Einstein has "heuristically" postulated [7] the existence of *microscopic* quanta of light, photons, of which the energy W_ℓ satisfies the "quantum relation" $W_\ell = h\nu_{\ell w}$.

According to de Broglie's treatment [2], the relativistic variance of the wave frequency ν_{cw} of a corpuscular wave (c *as an index*: corpuscular, (not to be confused with the velocity of light)) is $\nu_{cw} = \nu_{cwo}/\sqrt{1-\beta^2}$, which is the form of the transformation laws $m = m_o c^2/\sqrt{1-\beta^2}$ and $W_c = m_o c^2/\sqrt{1-\beta^2}$ first admitted in the theory of relativity for, respectively, the mass and the energy of a *macroscopic* heavy body. Furthermore, the "quantum relation", first asserted only for microscopic quanta of light-energy, is assumed to continue to hold also for microscopic and *heavy* quanta of energy: $W_c = mc^2 = m_o c^2/\sqrt{1-\beta^2} = h\nu_{cwo}/\sqrt{1-\beta^2} = h\nu_{cw}$. However:

The product $h\nu_{\ell w}$ does not transform like the product $h\nu_{cw}$.

What does this mean? Are we to accept that the concept of a quantum of light-energy is basically different from the concept of a quantum of heavy energy? Einstein's "equivalence" relation $W_c = mc^2$ suggests the contrary, since it has been derived [8] from considerations on a process of emission, and it is systematically applied to the processes of creation and annihilation. *Anyhow* the fact is this:

> *In present-day physics, the microscopic quantified electromagnetism and the microscopic (quantum) mechanics are basically disjoint with respect to the representation of the quanta of energy.*

In the vertical transition from the macroscopic level of description to the microscopic one, and in the horizontal transition from the representation of electromagnetic fields to that of heavy substance, the involved most fundamental notions of wave frequency, mass, and energy have incorporated obscurities. We shall try to put the conceptual situation under a magnifying lens. To do this we go back to the sources.

3. EXEGESIS

So Einstein has postulated the relation $W_\ell = h\nu_{\ell w}$ exclusively for photons and de Broglie extended this relation to heavy microsystems. Now, this extension is tied logically with the assumption that, in the rest frame, the extended periodic phenomenon associated by de Broglie with a heavy corpuscle takes on a *non*-progressive form. The structure of de Broglie's approach can be seen clearly in the following text [9] (pp. 1–5), where de Broglie himself presents the essence of his thesis (our translation from the French edition; de Broglie's notations and numbering of the formulas are left unchanged which entails an inversion of order between (11) and (9), (10)):

> Let us imagine a corpuscle that moves with uniform rectilinear motion along a certain direction, in the absence of any external field. We shall fix our attention exclusively on the state of motion of the corpuscle, disregarding its position in space. This motion will take place along some given direction, which we choose as the z axis, and it will be defined by two quantities, the energy and the momentum, of which the relativistic expressions as functions of the rest mass m_o of the corpuscle are given by the formulae
>
> $$W = m_o c^2/\sqrt{1-\beta^2}, \quad p = m_o v/\sqrt{1-\beta^2}, \quad \beta = v/c \quad (1)$$
>
> whence the formula
>
> $$p = |\mathbf{p}| = (W/c^2)|\mathbf{v}| = (W/c^2)v \quad (2)$$
>
> is derived.

In this way the state of motion is defined for a certain observer A tied to a Galilean frame of reference, an observer who makes use of a time t and of rectangular coordinates x, y, z.

Consider now another observer B having with respect to the first one the velocity v with direction Oz, in other words, an observer accompanying the corpuscle. We can assume that B has chosen an axis $O_o z_o$ that glides along Oz and axes Ox_o and Oy_o respectively parallel to Ox and Oy. This being assumed, the coordinates x_o, y_o, z_o, t_o of space and time of B are related to the coordinates x, y, z, t of A by the well-known simple Lorentz transformation

$$\begin{aligned} &x_o = x, \quad y_o = y, \quad z_o = (z - vt)/\sqrt{1-\beta^2}, \\ &t_o = \left[t - (\beta/c)z\right]/\sqrt{1-\beta^2} \end{aligned} \tag{3}$$

Now, for the observer B the velocity of the corpuscle is zero: so he takes the energy and the momentum to have the values

$$W_o = m_o c^2, \quad p_o = 0 \tag{4}$$

According to our basic idea, we must now try to introduce a periodic element, and we shall try to define it first in the rest frame of the corpuscle, that is, in the system of the observer B. Since in this system all is at rest, it is natural to define there the desired periodic element in the form of a stationary wave. We therefore define the periodic element by the quantity, supposed to be a scalar,

$$\psi_o = a_o \exp\left(2\pi i \nu_o t_o\right) \tag{5}$$

which has the form of the complex representation of a stationary wave. ψ_o oscillates as a function of the proper time with a frequency ν_0 characteristic of the nature of the envisaged corpuscle. We shall assume that a_o is a constant (in general complex), so that ψ_o will have at t_o the same value at any point of the rest frame of the observer B....

.... What value is it convenient to assign to the proper frequency ν_o? Evidently we must try to define it starting from a quantity that characterizes the corpuscle in the rest frame of B: but, in this frame, only one non null quantity is available, the energy $W_o = m_o c^2$. Given the role played by Plank's constant in all the quantum problems, it is natural to postulate

$$\nu_o = W_o/h = (m_o c^2)/h \qquad (6)$$

the analog of Einstein's relations for photons.

How will the periodic element defined above for the observer B manifest itself for the observer A? Supposing, which is natural here, that the element ψ is an invariant, it will suffice, in order to obtain its expression for A, to substitute in its expression for B the value of t_o yielded by the fourth Lorentz equation (3), which entails

$$\psi = a_o \exp 2\pi i\nu(t - z/V) \qquad (7)$$

if one sets

$$\nu = \frac{\nu_o}{\sqrt{1-\beta^2}} \quad \text{and} \quad V = c/\beta = c^2/v \qquad (8)$$

Thus, for the observer A who sees the corpuscle traveling with velocity $\boldsymbol{v}$ in the directon Oz, the phases of the periodic phenomenon ψ are distributed as those of a plane monochromatic wave for which the frequency ν and the phase velocity V would have the values [8]....

.... Comparison of the first relations (1) and (8) yields

$$W = h\nu \qquad (11)$$

a relation which evidently must be valid in any Galilean frame, since the observer A is any Galilean observer.

So, in de Broglie's approach, the "corpuscular" quantum relation (11) is a consequence deduced from: (a) the Einstein-Lorentz transformation of time; (b) the relativistic definitions of mechanical energy and momentum; (c) the direct assumption (6) $W_o = h\nu_o = m_o c^2$ of the quantum relation for a heavy quantum, for the particular observer B *tied* to the corpuscle; (d) *the postulation of the "stationary" form* (5) for the "periodic element ψ" associated with the proper mass m_o, as it is perceived by B; (e) the assumption that "the element ψ is an invariant," which here, as (7) shows, can only mean that the *definition* of ψ, as a function of the more basic wave descriptors "amplitude," "frequency," and "phase velocity," must be the same for any observer, namely that of a plane *progressive* wave.

Here enters in a crucial remark. In Einstein's relativity the "most real" aspects are the "intrinsic" ones, those which, like rest mass, rest energy, or existence of some spacetime position, characterize the system in its *rest* frame. The other characteristics, like velocity, the numerical value of the spacetime coordinates, magnetic field, etc., are regarded only as observational characters devoid of "intrinsic reality." But according to (5) the WAVE-like aspect of de Broglie's "periodic element" ψ ***vanishes*** in the rest frame, where it transmutes into a pulsation independent of the space coordinate x_o. So:

> *The degree of reality assigned by de Broglie to the WAVE-like aspect of the "periodic element ψ" associated with a piece of heavy energy of rest mass m_o, is of the same order as the degree of reality of a velocity or a magnetic field: a characteristic which entirely vanishes in the rest frame and thus is entirely generated by the relative state of observation.*

De Broglie conveys in a very striking way the peculiar view he holds concerning this question [2, pp. 3–4]:

> We can form a representation of the repartition of the values of ψ_o by imagining an infinity of small clocks placed at all points of the rest frame of the corpuscle, mutually synchronized and possessing a period $T_o = 1/\nu_o$. These small clocks somehow represent in each point the "phase" of the periodic phenomenon which is everywhere the same for the observer B at a same moment t_o of his proper time....
>
> .... In consequence of the relativistic phenomenon of the slowing down of moving clocks, each one of these clocks appears to the observer A as having a diminished frequency
>
> $$\nu_H = \nu_o\sqrt{1-\beta^2} \qquad (9)$$
>
> but the distribution of the ensemble of all the phases of all the clocks is given for A by the formula (7), that is, it coincides with the distribution of the phases of a plane monochromatic wave of which the frequency ν and the phase velocity V are given by (8).
>
> By comparing the expressions (8) and (9), one can note the essential difference between the apparent frequency ν_H of an individual moving clock, which is *diminished* by the influence of motion, and the frequency ν of the associated wave, which is increased by this influence. This difference between the relativistic variances of the frequency of a clock and the frequency of a wave is essential: It had strongly drawn my attention upon it and it is by meditating on it that I had been oriented in my researches.

What precedes can be summarized by saying that the corpuscle attached to one of the small clocks glides with respect to the phase of the wave with a velocity $V - v = c\left(1-\beta^2\right)/\beta$, so that it remains constantly in phase with the wave.

Let us reconsider this last idea in a more precise form. Among the infinity of small clocks imagined above, suppose that one plays a particular role. This will be the regulating clock which we shall identify with the corpuscle, while the other clocks represent the phase of the wave-like phenomenon of which the corpuscle is the center. In the rest frame all the clocks are immobile and have the same frequency ν_o. In the frame of the observer who sees all the clocks passing by with velocity v, the ensemble of the phases of these clocks is given by the factor $\nu(t - z/V)$, with the definitions (8). During a time dt, the regulating clock suffers a displacement $v\,dt$ in the direction of Oz, and its reading undergoes a change $\nu_o(1-\beta^2)^{1/2}\,dt$. And the phase of the wave at the point where this clock is located varies by the amount

$$\left(\frac{\nu_o}{\sqrt{1-\beta^2}}\right)(dt - v\,dt/V).$$

Since these two changes must be equal, one has

$$\sqrt{1-\beta^2} = \left(1/\sqrt{1-\beta^2}\right)(1 - v/V), \text{ where } \beta^2 = v/V \quad (10)$$

in agreement with the second relation (8).

So indeed de Broglie's famous "wave" was not initially conceived as an entity *invariantly*, "intrinsically" endowed with a wave-like aspect! But let us now finish the quotation.

Defining, as usual, the wavelength by the formula $\lambda = V/\nu$, one finds the value

$$\lambda = (c^2/v)(h/W) = h/p \quad (12)$$

We thus have found the two fundamental formulae (11) and (12) which define the frequency and the wavelength of the wave associated with the corpuscle, starting from its energy and momentum. For velocities that are small with respect to that of the light in vacuum, the formula (12) acquires the approximate form

$$\lambda = h/mv \quad (13)$$

For a particle with velocity c (or undistinguishable from c),

$$v = V = c, \quad W = h\nu, \quad p = h\nu/c. \tag{14}$$

So one finds indeed the fundamental formulae of the theory of quanta of light (Einstein, 1905) valid for photons.

4. CRITICAL REMARKS

4.1. Logical Independence Between Einstein's Quantum Relation and that of de Broglie

The last proposition quoted above is misleading. It suggests that de Broglie's formulae (11) and (12) somehow would *entail* the formulae (14) valid for photons. In fact, the radical $\sqrt{1-\beta^2}$ becomes 0 when v becomes c, and when this happens, in order to avoid divergence, one has to set also $m_o = 0$. Thus the expression $W = mc^2 = \left(m_o/\sqrt{1-\beta^2}\right)c^2$ becomes undeterminate. Therefore the second relation (14) cannot be *derived* from (11) written as $W = mc^2 = h\nu$. (*Mutatis mutandis*, the same remark holds concerning the connectibility of (12) with the third relation (14)). So the "quantum relation" (11) introduced by de Broglie is neither confirmed nor invalidated by the relations (14) concerning photons: De Broglie's relation (11) is a consequence solely of the assumptions (a), (b), (c), (d), and (e) we listed below it. Thus, contrary to what seems to be ofted believed:

> *Einstein's postulate $W = h\nu$ of a quantum relation for light quanta and de Broglie's postulate $W = mc^2 = h\nu$ of a quantum relation for a microscopic heavy body, are LOGICALLY INDEPENDENT.*

This allows us to understand how it is possible that these two quantum relations have different relativistic variances.

4.2. Inconceivable Physical Meaning for de Broglie's "Periodic Element" ϕ

De Broglie's corpuscular "wave," which in fact is a "periodic element" with only observational wavelike aspects, is an abstract concept. Notwithstanding de Broglie's considerations concerning an infinity of small clocks, it seems difficult to fashion an acceptable *physical* interpretation. What physical counterpart could there exist for this infinity of small clocks? The proper form (5) cannot be regarded as expressing a physically *local-*

ized, clocklike periodicity, that "exists" exclusively at the location (formally implicit in (5)) of the mass m_o. The physical support of the wave-like distribution "of phase" that appears formally via the substitution $t_o = [t - (\beta/c)z]/\sqrt{1-\beta^2}$ must exist already in the object of the representation (5), it cannot arise *ex nihilo*. So, as it follows also from de Broglie's requirement that "the element ψ shall be an invariant," *(5) must be regarded as a particular case of (7)*. But this is possible only if, to obtain (5), we insert in (7) an infinite proper value V_o of the phase velocity. Which yields an *extended* and *non*-progressive periodicity of proper frequency ν_o *animating as a sole block the entire space*, so that, at any given time t_o, the same phase will prevail everywhere, independent of the distance to the localized piece of energy $m_o c^2$, like a monolithic feeble pulsation. But the *genesis* and subsistence of such a process cannot be imagined, *the time of the process is eliminated*, the process just is there, permanently, huge, rigid. In an approach the aim of which is to introduce a *model* for what is called a microsystem, this idea is a weird one to accept.

4.3 A Physical Phase Velocity with Infinite Proper Value and General Expression $V = c^2/v$ is *Not* a Relativistic Concept

Real Propagation or Action at a Distance: Much confusion is raised by the currrent assertion that "phase velocities have a non-energetic character." A phase velocity *can* be a purely observational effect (as in de Broglie's small clocks analogy), and *then* indeed it has not an energetic character because there is no "real" propagation (correlatively, it vanishes in the rest frame). But it also *can* correspond to a "real" propagation—one that exists in the rest frame—and *then* it *does* possess an energetic character (think of the kinetic effects of ocean-waves, on a small boat). *This* is the case for the phase velocity of a real field. The notion that a physical field "really" propagates, and that it propagates with *finite* velocity, is one of the essential (though more or less implicit) ideas of relativity.

Now, a *physical*, infinitely extended and non-progressive *proper* pulsation of the form (5) $\psi_o = a_o \exp(2\pi i \nu_o t_o)$ *implies action at a distance*, "non-locality," thereby *directly contradicting* relativity, since, as has been already remarked: (1) the consistency between the standard form (7) $\psi(x,t) = a \exp 2\pi i\nu(t - x/V)$ of a progressive wave, and the proper form (5) $\psi_o = a_o \exp(2\pi i \nu_o t_o)$—which is the *sine qua non* condition for obtaining the relations (8) and thus also the fundamental relation (12) $\lambda = h/p$—can be insured *only* by setting in (7) $V_o = \infty$; (2) by definition, the proper value of a quantity is an "intrinsic" physical reality, not a purely observational effect.

So we must definitely dismiss all the superficial and inertial repetitions of the self-contradicting arguments that protect the acceptance of infinite phase velocities for corpuscular waves which, on the other hand, are conceived of as physical waves:

> *What is at stake here is the physical character of the corpuscular waves and the locality of the representation of microphenomena. If we accept the proper representation (5) we negate both this physical character and this locality.*

In de Broglie's approach, the two distinct concepts of an observational wave, or a physical wave, *got mixed.* The abstract and nonlocal character of quantum mechanics as it now stands stems mainly from de Broglie's proper representation (5).

Variance: No matter whether it is a phase velocity or a mechanical velocity, *a velocity is just a velocity*, i.e., by definition, a ratio between a covered distance and the time taken to cover it. So *any* velocity has to be assigned one same variance, namely the one that follows from the definition of the general concept of velocity and the basic laws accepted for the transformation of space and time coordinates. Whereas a phase velocity $V = c^2/v$ does *not* possess *specifically* the Lorentz-Einstein variance of a velocity: with de Broglie's hypothesis $V_o = \infty$ the general expression of the Lorentz-Einstein variance of a velocity, $v = (v_0 + u)/\left[1 + (v_0 u)/c^2\right]$, yields $V = \infty/\infty$, which is compatible with *anything*, thus also in particular with de Broglie's relation $V = c^2/u$, but not specifically with it. Between de Broglie's relation $V = c^2/v$ and the Lorentz-Einstein variance of a velocity there simply is no *necessary* connection.

> *The "scandal" raised by the Michelson-Morlay experiment resided precisely in the fact that the PHASE velocity of light (which in the case of light is also the velocity of a "signal" or "packet") did not obey the (Galilean) variance then accepted for ANY velocity.*

So, why should we at present accept for the phase velocity of a corpuscular wave a variance that is not specifically the Einstein-Lorentz variance nowadays accepted as characteristic of the concept of velocity?

4.4. Conclusion

One of the most remarkable features of de Broglie's Thesis is the way in which a *distinction* has been made there, on the *basis* of considerations of variance, between:

—clock-frequencies ν_{II} that can exist inside an arbitrarily small spatial

domain, possessing in this sense a purely temporal nature;
—wave-frequencies ν_w that possess on the contrary an essentially spatial character, since, being *defined* as $\nu_w = V/\lambda$, they are conceivable *only* in connection with the concepts of a wavelength λ and a phase velocity V that involve non-removably the notion of spatial distance.

However, though it is founded on such subtle relativistic considerations, the treatment from de Broglie's thesis is not fully relativistic. By the assumption of the proper form (5), de Broglie's wave either is a non-physical, only observational wave, or if it is regarded as a physical wave, it has a non-local character that violates relativity. In these conditions it does not suffice to imagine and to perform experiments in order to establish whether yes or not the corpusclar waves are physical and local. Correlatively, a conceptual-formal reconstruction is necessary. We shall now attempt it.

5. RECONSTRUCTION

5.1. Basic Assumptions

Our approach is strictly *INDIVIDUAL*. No sort of deviation from individuality is tolerated: no slipping into the notion of wave-packet, which involves *variation* of the value of the relative velocity of observations (Ref. 2, pp. 25–26) and *multiplicity* of wave-frequencies, for *one* mass; and no sort of probabilistic feature. The construction is founded on a methodological principle and three physical hypotheses that constitute a physical and local reformulation of de Broglie's model.

The Methodological Principle of "Definition by Variance." A principle of relativity, the Galilean one as much as that of Einstein, makes sense only if one admits also the following principle of "definition by variance" (PDV):

> (a) *The definition of a concept is an invariant.*
> (b) *The "nature" of a quantity, i.e., the denomination applicable to it, is decided by its transformation law, such as this follows from the definition of the general concept labeled by that denomination and the basic laws of transformation posited for the coordinates of space and time.* (PDV)

This can be regarded as a methodological principle, to be explicitly adjoined to any (physical) principle of relativity.

Physical Hypotheses: We admit the whole essence of de Broglie's model, namely that a "microsystem" involves a localized "corpuscle" with an inner clocklike periodicity with frequency ν_H, and a surrounnding "corpuscular wave," the localized energy of the corpuscle being tied to its inner clocklike periodicity ν_H, to its "mass" m, and to the wavelike frequency ν_{cw} of the surrounding corpuscular wave. But the model is restated as follows.

Like de Broglie, we admit the hypothesis H_1 that:

The mass m of a heavy microsystem involves a localized inner clocklike periodicity with frequency ν_H. $\qquad (\mathrm{H}_1)$

But the fact that, inside the rest frame, the mechanical displacement of the corpuscle vanishes, is *not* posited to entail that there "all is at rest," nor to support the assertion that "it is natural to define there the desired periodic element in the form of a stationary wave." Even inside the rest frame, de Broglie's clocklike process *interior* to the localized piece of energy *develops in time.* And *nothing* prevents us from posing furthermore that inside the rest frame the surrounding corpuscular wave also evolves: One can choose to imagine that it is generated progressively by the inner clocklike periodicity, hence that correlatively it spreads progressively through space with respect to the proper location of the corpuscle (energetic aspects will be considered later). So we admit the following hypothesis H_2:

The proper value V_o of the phase velocity V of a corpuscular wave is finite. $\qquad (\mathrm{H}_2)$

Finally, for consistency between the localized inner clocklike frequency ν_H and the wave frequency ν_{cw} of the surrounding corpuscular wave, and because furthermore inside the proper frame indeed "only one non null (mechanical) quantity is available, the energy $W_o = m_o c^2$," we admit also the following hypothesis H_3 (de Broglie's hypothesis (6)):

In the rest frame, the proper values ν_{wo} and ν_{Ho} of, respectively, the wave frequency ν_{cw} and the clock frequency ν_H of a microsystem, have the same numerical value $\nu_{cwo} = \nu_{Ho}$, the proper mass m_o being connected with this value accordingly to the equations $\nu_{cwo} = \nu_{Ho} = W_o/h = (m_o c^2)/h$. (H_3)

By H_3, as in de Broglie's treatment, the quantum relation $W_\ell = h\nu_{\ell w}$ assumed by Einstein for photons is extended to also the quantum of heavy energy $m_o c^2$, but *only* inside the rest frame. In other frames we must be prepared to find some modification, since de Broglie's general extension (14) is logically connected with the form (5) of ψ_o, which our hypothesis H_2 excludes.

5.2. The New Representation in Purely Undulatory Terms

Representation in the Rest Frame: From now on (for clear graphic distinction between frequencies ν and relative velocities of observation) a relative velocity of observation will be denoted by u instead of v. We work in only one spatial dimension, denoted by x.

Consider a corpuscle of proper mass m_o. In consequence of H_2, the individual corpuscular wave ψ_I (I: individual) associated with the piece of energy $m_o c^2$ can be represented in the rest frame by the wave function

$$\psi_{Io}(x_o, t_o) = (K/x_o) \exp 2\pi i \nu_{ocw}(t_o - x_o/V_o) \qquad (5')$$

with V_o finite. [For reasons of symmetry, we consider instead of a plane progressive wave, a spherical one (projected on the one spatial dimension labeled by x) of which the amplitude function $A_o = (K/x_o)$ (K a constant) is by definition continuous and finite at the location $x_o = 0$ of the mass m]. The form (5′) expresses that the wavelike aspect of an individual corpuscular wave is now asserted to be an "intrinsic," real character, of a physical wave.

Passage to Another frame: Consider now a frame that is fixed in the laboratory, thus has a non-null velocity of observation $u_x = u$ with respect to the corpuscle. For the observer attached to this frame, according to the Einstein-Lorentz transformations, $x_o = (x - ut)/\sqrt{1-\beta^2}, y_o = y, z_o = z, t_o = [t - (\beta/c)x]/\sqrt{1-\beta^2}$. So the projection (5′) becomes

$$\begin{aligned}\psi_I(x,t) = &\frac{K}{(x-ut)/\sqrt{1-\beta^2}} \times \\ &\exp 2\pi i \nu_{ocw}\left[t - (\beta/c)x/\sqrt{1-\beta^2} - (x-ut)/V_o(\sqrt{1-\beta^2})\right]\end{aligned} \qquad (15)$$

Let us now concentrate for a while exclusively on the phase function.

The coefficients, in the phase of $\psi_I(x,t)$, of the time coordinate t and the space coordinate x, are, respectively

$$\left(2\pi i \nu_{ocw}/\sqrt{1-\beta^2}\right)(1 + u/V_o) = \left(2\pi i \nu_{ocw}/\sqrt{1-\beta^2}\right)((V_o + u)V_o),$$
$$\left(2\pi i \nu_{ocw}/\sqrt{1-\beta^2}\right)\left[(uV_o + c^2)/c^2 V_o\right]$$

In the standard form (7) $\psi_I(x,t) = a \exp 2\pi i \nu(t - x/V)$ the phase velocity V is the inverse of the coefficient of x inside a bracket which admits the coefficient of t as a common factor. Thus it is convenient to rewrite the coefficient of x in the form

$$\left(2\pi i\nu_{ocw}/V_o\sqrt{1-\beta^2}\right)(V_o+u)\left((uV_o+c^2)/c^2(V_o+u)\right) =$$
$$\left(2\pi i\nu_{ocw}/\sqrt{1-\beta^2}\right)(1+u/V_o)\left[(1+uV_o+c^2)/(V_o+u)\right]$$

The expression (15) acquires in the new frame also the standard form $\psi_I(x,t) = a\exp 2\pi i\nu(t-x/V)$ of a progressive wave—required to be an invariant—only if one imposes convenient transformation laws for ν_{cw} and for V, as follows.

The Transformation Law for the Wave Frequency: For ν_{cw} we must postulate

$$\nu_{cw} = \left(\nu_{ocw}/(\sqrt{1-\beta^2})\right)\left[(V_o+u)/V_o\right] = \left(\nu_{ocw}/\sqrt{1-\beta^2}\right)(1+u/V_o) \tag{16}$$

where the ratio u/V_o can be positive or negative, because the observer always perceives the front of the wave as approaching him whereas the source is perceived either as approaching or receding. This is different from de Broglie's transformation law (8), $\nu = \nu_o/\sqrt{1-\beta^2}$. Namely—*in agreement with PDV*—it is now the *same* transformation law as the one admitted in electromagnetism for the frequency $\nu_{\ell w} = \left(\nu_{o\ell w}/\sqrt{1-\beta^2}\right)$ $(1+u/c)$ of a light wave involving photons. Of course, one might remark that (8) also has, in agreement with PDV, the form (16), namely for the particular proper value $V_o = \infty$ of V_o. This, however, would be a fallacious argument: *a finite value V_o entails a Doppler effect of the corpuscular waves*, which is far from trivial and does not happen with $V_o = \infty$, which can be regarded as tied with the fact that $V_0 = \infty$ involves the concept of an only observationally wavelike phenomenon, not the concept of a physical wave.

In our framework the Lorentz-Einstein transformation laws for the space and time coordinates entail explicitly a relativistic Doppler effect for the GENERAL *concept of wave-frequency of a* PHYSICAL *wave, whether photonic or "corpuscular."*

The Transformation Law for the Phase Velocity: For V, in order to obtain from (15) the standard form (7), we must require

$$V = \left[(V_o+u)c^2\right]/uV_o+c^2) = (V_o+u)\left[1+(V_ou)/c^2\right] \tag{17}$$

Once more, *in agreement with PDV*, the expression (17) has the Einstein form $v = (v_o+u)/\left[1+(v_ou)/c^2\right]$ characteristic of the general concept of a velocity. For $u = c$ we get $V(u = c, V_o) = (V_o + c)c^2/cV_o + c^2) = c$, whatever the value V_o might be. According to (17), if $V_o > c, V > c$ and if $V_o < c, V < c$. So, in contradistinction to de Broglie's law $uV = c^2$, *in abstracto* the law (17) does not exclude for the corpuscular waves phase velocities smaller than c. However, as we show later, experience seems to *do* that.

Notice that Eq. (17) admits a solution $V = \infty$, namely for $u = -c^2/V_o$. But the significance of this solution is *directly opposed* to that from de Broglie's treatment: Since, by definition, a proper value is "intrinsic," "most real," in de Broglie's treatment V_o is "intrinsically" infinite but can acquire by observational effects a finite appearance tied to a wavelike semblance; whereas here V_o is posed to be "intrinsically" finite, but (in just one frame) it can acquire by observational effects an inifinite *appearance* connected with the effacement of the wavelike character: another manifestation of the distinct natures of an observationally wavelike phenomenon and a physical wave.

The Transformation Law for the Corpuscular Wavelength: We start from the standard definition $\lambda = V/\nu$. This yields

$$\begin{aligned}\lambda_c = V/\nu_{cw} &= \left[\left[(V_o + u)c^2\right]/(uV_o + c^2)\right]\left[\left[V_o\sqrt{1-\beta^2}\right]/\left[\nu_{ocw}(V_o + u)\right]\right] \\ &= \lambda_{co}\left[(\sqrt{1-\beta^2})/\left[(1 + uV_o/c^2)\right]\right] \qquad (18)\end{aligned}$$

which—again in agreement with PDV—has the same form as for light waves and yields another expression of a corpuscular Doppler effect.

The Harmony of Phases: We arrive now at the central consequence of the hypothesis H_2: In de Broglie's treatment, the equality $uV = c^2$ is the condition to be accepted in order to insure a lasting harmony between the phase of the inner, localized, clocklike perodicity of the piece of energy mc^2, with clock frequency ν_H, and the phase of the extended wavelike periodicity of the corpuscular wave, with wave frequency ν_{cw}. In our treatment, the product uV has to be written in agreement with the transformation law (17), so the condition $uV = c^2$ ceases to hold in general. Is then the harmony of phases still preserved? De Broglie assigned the utmost importance to this question.

For the variations during dt, of the phase Φ_H of the inner clock periodicity and of the phase Φ_w of the wavelike periodicity of the corpuscular wave, we have respectively [with $(dx/dt)dt = udt$, and making use of (16) and (17):

$$\begin{aligned} d\Phi_H &= 2\pi\nu_H dt = 2\pi\nu_{Ho}\sqrt{1-\beta^2}dt \\ d\Phi_w &= 2\pi\nu_{cw}\left[dt - ((dx/dt)dt/V)\right] \\ &= 2\pi\left[\nu_{ocw}/\sqrt{1-\beta^2}\right] dt \left[(1+u/V_o) - u(u/c^2 + 1/V_o)\right]. \end{aligned}$$

According to H_3 we have $\nu_{Ho} = \nu_{ocw}$. So the condition of harmony of phases, $d\Phi_H(t) = d\Phi_w(t)$ for any t, is

$$2\pi\sqrt{1-\beta^2}dt = \left[2\pi/\sqrt{1-\beta^2}\right] dt \left[(1+u/V_o) - u(u/c^2 + 1/V_o)\right]$$

which reduces to an *identity*.

> *While in de Broglie's treatment the harmony of phases required the INDEPENDENT acceptance of the CONDITION $uV = c^2$, with our assumptions the harmony of phases is insured UNCONDITIONALLY: It appears as a spontaneous consequence of the Lorentz-Einstein transformations for the space and time coordinates in the case of a progressive wave emitted by a localized source.*

This is a very striking result. It finally *proves* that the assumption H_2, entailing for ψ_{Io} the proper form (5′) characteristic of a physical wave, is indeed in far deeper agreement with Einstein relativity than de Broglie's assumption (5) of a merely observational wavelike phenomenon.

Conclusion. The new representation in purely undulatory terms shows—by the results (16), (17), (18) and the "automatic" harmony of phases—that the radical distinction between physical waves and only observational waves is as important as de Broglie's distinction between clock-frequencies and wave-frequencies.

5.3. How Connect with the *Mechanical* Qualifications?

We should now connect, like de Broglie, the purely undulatory representation obtained above, with mechanical qualifications (heavy energy and momentum, mass). So now we stumble on the problem of the quantum relation.

In de Broglie's treatment, the transformation laws $W_c = m_o c^2/\sqrt{1-\beta^2}$, $m = m_o c^2/\sqrt{1-\beta^2}$, first admitted in the theory of relativity only for *macroscopic* heavy bodies, have been assumed to continue to hold *unchanged* for microscopic heavy bodies also. This assumption, associated with the non-progressive proper form (5) $\psi_o = a_o \exp(2\pi i \nu_o t_o)$ assigned in the rest frame to a corpuscular wave, entailed for the corpuscular wave frequency the transformation law $\nu_{cw} = \nu_{cwo}/\sqrt{1-\beta^2}$, which is of the *same* form as the transformation law admitted for mass and for heavy energy. It seems now clear that *precisely this, no doubt, was the very* AIM *for which de Broglie posited the proper form* $\psi_o = a_o \exp(2\pi i \nu_o t_o)$, *in spite of the fact that this form compelled him to elaborate a sophisticated and hybrid interpretation for the "periodic element"* ψ *and for the harmony of phases*: It was the unique way of insuring for heavy microsystems a covariant quantum relation *without* questioning the Einstein definition of mass. Indeed, in de Broglie's approach, if one assumes in the rest frame the quantum relation $W_{co} = m_o c^2 = h\nu_{cwo}$, then this same quantum relation, $W_c = mc^2 = h\nu_{cw}$, holds in any frame, and covariantly with mass, which permits one to first translate ν_{cw} in mechanical terms by writing $\nu_{cw} = mc^2/h$ and then to continue the translation and to find $\lambda_c = h/p$.

But, as we remarked from the start, *de Broglie's quantum relation does not hold also for light waves.*

Within the framework [PDV, H_1, H_2, H_3], in order to be able to continue our reconstruction, we are now in our turn obliged to deal with this problem.

5.4. What choice?

Inside [PDV, H_1, H_2, H_3] the corpuscular wave frequency ν_{cw} has the same variance (16) as the wave frequency ν_{cw} of a light wave $\nu_{\ell w}$. In these conditions there exist the following possibilities concerning the quantum relation:

A. If, like de Broglie, we assume that a microscopic mass has the same variance $m = m_o/\sqrt{1-\beta^2}$ as a macroscopic mass, then, instead of de Broglie's quantum relation $W_c = h\nu_{cw}$, for a heavy microsystem we must write the *modified* quantum relation

$$W_c = mc^2 = m_o c^2/\sqrt{1-\beta^2} = h\nu_{cwo}/\sqrt{1-\beta^2} = h\nu_{cw}/(1+u/V_o) \quad (19)$$

It is most important to realize that, as it is stressed by the intermediary writing from (19), *numerically* the relation (19) asserts the *same* thing as de Broglie's quantum relation, namely that the equality

$$\left[W_c = m_o c^2/\sqrt{1-\beta^2}\right] = h\nu_{cwo}/\sqrt{1-\beta^2}$$

does hold. Only the *connection* between the values of the involved heavy energy and the corresponding corpuscular wave frequency changes, in consequence of the fact that, according to (16), $\nu_{cwo}/\sqrt{1-\beta^2} \neq \nu_{cw}$, so that one has

[a quantum W_c of microscopic heavy energy] $\neq$ [$h\times$ the corresponding wave frequency ν_{cw}].

If we adopt this position, then—according to PDV—quantified microscopic electromagnetism and microscopic (quantum) mechanics remain basically disjoint with respect to the representation of what is called, too uniformly in this case, "a quantum of energy."

Is this an acceptable situation? De Broglie's distinction between clocklike frequencies and wavelike frequencies has played a fundamental role in his initiation of quantum mechanics. Are there perhaps reasons why we *should* indeed distinguish, in the sense of PDV, between a basic concept of a piece of heavy energy, able to conserve a constant location inside any given frame of reference, and *another* basic concept, of a piece of field energy, which in any frame keeps changing its location with the speed c? The possibility, as soon as it has been conceived, captures the attention: In contradistinction to the concepts of velocity and of wave frequency, which are "simple," in the sense that they are related in a very clear way to the most basic notions of distance and duration, the concept of energy—at a fundamental level—resists analysis. It raises the question of conservation and it invokes the concept of "action," which circularly brings us back to "energy" as a *capacity* for "action," which finally seems to be just a sort of potential of "materialized time," i.e., of "change." But the class of changes is infinite, etc. In short, in the presence of such complexity and at the extreme limit where we try to represent how heavy matter merges with non heavy substance, it would be naive to simply dismiss *a priori* the possible relevance of two distinct quantum relations.

B. If however we *do* require uniformly the same quantum relation for quanta of light $h\nu_\ell$ and quanta $h\nu_{cw} = mc^2$ of heavy energy, then *ipso facto* we are obliged to assume that a microscopic mass—let us introduce the specific notation m_μ—has the variance

$$m_\mu = \frac{m_o}{\sqrt{1-\beta^2}}(1 + u/V_o), \tag{20}$$

i.e., a variance of the same form as that of the wave frequencies, ν_{cw} and $\nu_{\ell w}$, instead of that, $mc^2 = m_o c^2/\sqrt{1-\beta^2}$, of the mass of a macroscopic heavy body.

But are there conceptual and physical reasons that permit us to contemplate different variances for macroscopic and miscroscopic masses? The answer is *yes*:

(1) At the time when Einstein constructed his theory of relativity, quantum mechanics did not exist. So Einstein did not refer his representations to parameters qualifying a corpuscular wave associated with any

microscopic mass m_μ. But nowadays we know that such a wave exists, and nothing excludes that it contributes to the resistance to modification of the state of movement. So nowadays:

> *In a relativistic description which is rigorous in principle, it simply is a necessity to a priori admit the possibility of some dependence of a micro-mass m_μ on the corresponding corpuscular wave also. The contrary would amount to an exclusion both arbitrary and improbable.*

(2) Then the next question is whether or not such a dependence can be conceived of consistently with Einstein's macroscopic relativistic mechanics. To begin with, we remark that a derivation of the Einstein law $m = m_o c^2/\sqrt{1-\beta^2}$ that eliminates the possibility of any other form, simply *cannot exist.* This is so because the Lorentz-Einstein transformation law for the space and time coordinates plus the principle of relativity (the condition of form-invariance for the dynamical equations) entail precisely the law $m = m_o c^2/\sqrt{1-\beta^2}$ *only in so far* that one assumes implicitly or explicitly that "all" the relevant parameters have been taken into account. Which, of course, is impossible. Consider, for instance, Bohm's derivation in his book *The Special Theory of Relativity* [10, pp. 81–90)]. Assuming the principle of relativity and the Lorentz-Einstein transformations for space-time coordinates, and considering a system of two particles as seen from two distinct frames, Bohm derives first the logarithmic equation

$$\ln(m_1'/m_1) - \ln R(v_1, v_1') = \text{const.} \qquad (L)$$

[p. 86, Eq. (18–25)] where m_1, m_1' and v_1, v_1' are, respectively, the masses and the velocities of *one* of the two particles, as seen in the two considered frames, and the constant is taken to be 1. Einstein's law $m = m_o c^2/\sqrt{1-\beta^2}$ is then derived from Eq. (L). But one readily finds out that in fact Bohm's derivation of the intermediary equation (L) holds as well for a whole family of expressions obtained by adding in the second member of (L) *any* function of the parameters on which the first member does not depend. Notice now that the first member of Bohm's equation does not depend either on the relative velocity of observation (in Bohm's notation this is V', which, in the case of only one microsystem and an observer, becomes just our relative velocity of observation u) or on our finite proper value V_o of the phase velocity of the corpuscular wave connected with m_1 (a parameter which has not even been envisaged by Bohm in this context). Thus one is free to add to the second member of (L) any function $F(u, V_o, \alpha)$, where α designates any other parameter characterizing the corpuscular waves connected with the considered masses. To each such function will correspond a definition of the variance of a miscroscopic mass that is consistent with Einstein's variance $m = m_o c^2/\sqrt{1-\beta^2}$ and nevertheless introduces a dependence on the corresponding corpuscular wave: an infinity of possible definitions.

(**3**) So the next question is how to choose $F(u, V_o, \alpha)$. In this respect there is no choice within the framework [PDV, H_1, H_2, H_3] and given the presupposed aim of insuring a unified quantum relation: only the variance $m_\mu = m_o/\sqrt{1-\beta^2}(1 + u/V_o)$ assumed for a microscopic mass m_μ leads to a unified quantum relation. So the convenient term to be added to the second member of (L) is $F(u, V_o, \alpha) = F(u, V_o) = \ln(1 + u/V_o)$, which entails (20).

The definition (20) involves a relativistic Doppler effect also for mass.

If this line of thought is adopted, then PDV suggests, in order to avoid a non-necessary increase of the number of fundamental concepts, to introduce the following new postulate for mass variance, PMV:

The RIGOROUS definition of the GENERAL concept of "mass" is expressed by the transformation law (20) $m = (m_o/\sqrt{1-\beta^2})(1{+}u/V_o)$ (PMV)

This amounts to regarding Einstein's variance for a macroscopic mass as some approximation stemming from mean effects. Indeed, for macroscopic heavy bodies, where the sign and norm of the various involved velocities of observation u always fluctuate according to some distribution law, the mean effects can be assumed to achieve—in general at least—a quasi perfect overall annihilation of the contributions of the various involved factors $(1 + u/V_o)$, thus leading to Einstein's law. Completed in this way, our framework would become [PDV, H_1, H_2, H_3, PMV].

Numerically, a variance (20) appears as only a "correction" of Einstein's definition. And an extremely tiny one no doubt, since Einstein's definition resisted up to now so many tests: We have to admit *a priori* that, for any microsystem on which relativistic measurements of the mass have been carried out, the value of V_o, though finite, was bigger than c, and *much* bigger. Nevertheless, if this numerically so tiny modification (20) were true, its conceptual consequences would be quite fundamental, as will become clearer below.

Each one of the hypotheses A and B, when, within $[PDV, H_1, H_2, H_3]$, it is utilized for connecting the wave descriptors ν_{cw} and λ_c, with corpuscular descriptors, entails specific consequences. These are as follows.

5.4. Summary of A and B and Corresponding Translations in Mechanical Terms of the New Representation

HYPOTHESIS A

Einstein Variance for Miscroscopic Mass: We assume that "mass" is uniformly characterized by the variance $mc^2 = m_o c^2/\sqrt{1-\beta^2}$ at both the macroscopic and the microscopic level of description. So we have $W_c = mc^2 = m_o c^2/\sqrt{1-\beta^2}, p_c = mu = (m_o c^2)/(\sqrt{1-\beta^2})u = W_c u/c^2$, and the macroscopic relativistic dynamics remains rigorously valid at the microscopic level of description also.

Two Different Quantum Relations: For a light quantum we have

$$W_\ell = h\nu_{\ell w} = h\left(\nu_{o\ell w}/\sqrt{1-\beta^2}\right)(1+u/c)$$

For a heavy quantum we have (19)

$$W_c = mc^2 = m_o c^2/\sqrt{1-\beta^2} = (h\nu_{cwo})/\sqrt{1-\beta^2} = (h\nu_{cw})/(1+u/V_o)$$

This duality would have to be taken into account in the descriptions of interactions between light and heavy matter.

The Corpuscular Wave Frequency in Mechanical Terms: From (19) we draw (A: hypothesis A)

$$(\nu_{cw})_A = [W_c(1+u/V_o)]/h = \left[(m_o/\sqrt{1-\beta^2})(c^2/h)\right](1+u/V_o) \quad (21)$$

The Corpuscular Wavelength in Mechanical Terms: The general definition $V = \lambda\nu$ and Eq. (21) entail for λ_c, instead of de Broglie's relation (12), the modified relation

$$(\lambda_c)_A = h/(p_c + W_c/V_o) \quad (22)$$

For $u = 0$, (22) yields

$$(\lambda_c)_A(u=0) = \lambda_{co} = h(V_o/W_{co}) = \text{finite} \quad (22')$$

In contradistinction to de Broglie's formula, which yields $\lambda(u = 0) = \lambda_{co} = (h/m.0) = \infty$, the modified expression (22) entails a FINITE value λ_{co}.

Though it is not conceivable to—directly—*observe* interferences inside the proper frame of reference, this result can be regarded as a limiting assertion concerning a series of results obtained with a series of relative velocities of observation which decrease toward 0.

The Individual Wave Function in Mechanical Terms: The transform of the form (5′), rewritten in terms of the mechanical energy W_c and the corresponding momentum p_c by using (21) and (22), is

$$(\psi_I)_A(x,t) = \frac{K}{(x-ut)/\sqrt{1-\beta^2}} \exp(2\pi i/h)\,[W_c(1+u/V_o)t - (p_c + W_c/V_0)x] \tag{23}$$

Since with $W_c = mc^2 = m_oc^2/\sqrt{1-\beta^2}$ and $p_c = mu = (m_oc^2/\sqrt{1-\beta^2})u = W_cu/c^2$ Einstein's momentum-energy 4-vector remains the same, it might be that with the hypothesis A the Klein-Gordon equation can be maintained. But certainly, in consequence of the modified connections (21) and (22) between corpuscular wave descriptors and mechanical descriptors and of the non-probabilistic, the strictly *individual* significance of $(\psi_I)_A(x,t)$, quantum mechanics would have to be entirely reorganized *even* if the Klein-Gordon equation were found still pertinent.

HYPOTHESIS B

A New Variance – Basic – for Microscopic Mass: For a microscopic mass $m = m_\mu$ the assumed variance is (20)

$$m_\mu = m_o(1/\sqrt{1-\beta^2})(1+u/V_o)$$

which, according to PMV, is regarded as the rigorous variance defining the general concept of mass.

A Universal Quantum Relation: For the mechanical energy $W_{c\mu}$ and the momentum p_μ associated with a microscopic mass m_μ, the transformation law (20) entails, respectively, the new definitions, depending explicitly on the proper value V_o of the phase velocity of the corresponding corpuscular wave

$$\begin{aligned} W_{c\mu} &= m_\mu c^2 = \frac{m_oc^2}{\sqrt{1-\beta^2}}(1+u/V_o), \\ p_\mu &= W_cu/c^2 = \frac{m_o}{\sqrt{1-\beta^2}}(1+u/V_o)u = m_\mu u \end{aligned} \tag{24}$$

With these definitions the pair $(p_\mu, W_{c\mu})$ can no more be regarded as defining a 4-vector: There is no more invariant norm. So acceptance of [(20),

(24)] would entail a radical reconstruction of the relativistic representation at the basic, the microscopic level of description. "Force," "inertia," "momentum," "energy," would have to be reconceived in explicit connection with the corpuscular waves. On the other hand, the first definition (24) entails now, as required, a *universal* quantum relation

$$W = h\nu = (W_o/\sqrt{1-\beta^2})(1+u/V_o) = h\nu_o/(\sqrt{1-\beta^2}))(1+u/V_o), \quad (25)$$

valid *no matter* whether the pair of values (W, ν) refers to a light wave or to a corpuscular wave. This, beneath the description of fields and that of heavy systems—whether microscopic, macroscopic, or cosmic—displays a basic unity of principle concerning the fundamental concept of energy. And:

> *The corpuscular Doppler effect involved in (16), (18), and(19), and now also in (20) and (24)—if it really exists—might play a crucial role in astrophysics and cosmology.*

The Corpuscular Wave Frequency in Mechanical Terms: The universal quantum relation (25) allows us now to write for ν_{wc} (B: hypothesis B)

$$(\nu_{cw})_B = W_{c\mu}/h = m_\mu c^2/h = \left(\frac{m_o}{\sqrt{1-\beta^2}}\right)\left(\frac{c^2}{h}\right)(1+u/V_o) = (\nu_{cw})_A \quad (26)$$

which *numerically*, as indicated, identifies of course with $(\nu_{cw})_A$, since in both cases the basic expression is (16).

The Corpuscular Wavelength in Mechanical Terms: From the basic definition $\lambda = V/\nu_w$ and on making use of (26) and the second relation (25), we get:

$$(\lambda_c)_B = V/\nu_{cw} = h(V/W_{c\mu}) = (h/p_\mu)(Vu/c^2) \quad (27)$$

For $u = 0$ this yields the *same* value of as (22′):

$$\begin{aligned}(\lambda_c)_B(u=0) &= \lambda_{co} = V_o/v_{cwo} = h\,(V_o/W_{c\mu o}) \\ &= h(V_o/m_{c\mu}c^2) = (\lambda_c)_A(u=0) = \textit{FINITE}\end{aligned} \quad (27')$$

The Individual Wave Function in Mechanical Terms. From (26) and (27), the transform of the form (5′), expressed in mechanical terms, becomes

$$(\psi_I)_B(x,t) = \frac{K}{(x-ut)/\sqrt{1-\beta^2}} \exp 2\pi i(W_{c\mu}/h)(t - x/V) \quad (28)$$

The phase function, considered *separately*, satisfies the very simple covariant equation $\partial\psi/\partial t = -V\,\partial\psi/\partial x$. But because with (20) and (24) the difference $(W_{c\mu})^2 - c^2(p_\mu)^2$ ceases to be invariant, with the hypothesis B the Klein-Gordon equation certainly cannot be considered any more. Quantum mechanics *and* microscopic relativity would have to be reorganized connectedly and so as to yield Einstein's relativity as a mean valid only on the macroscopic level of description.

6. EXPERIMENTAL TESTS

6.1. Preliminary Considerations

The results obtained can be investigated experimentally by:
— measurements of corpuscular wave-lengths,
— search of corpuscular Doppler effects,
— measurements of microscopic mass or microscopic corpuscular energy.

The question that arises preliminarily is why, over nearly 70 years, there has never been a hint about the relevance of one or the other of all the modified definitions introduced by the hypotheses A or B. The common explanation would be that V_o, if it is indeed finite, must be very big with respect to c. Now, this can be checked preliminarily by approximate estimates based on already available data.

Consider first the corpuscular wavelengths.

To examine the situation in principle, we introduce the following simplified notations. λ_B: de Broglie's wave length; λ_M: the modified corpuscular wavelength; $\lambda_{B-M}(u)$: the difference $(\lambda_B - \lambda_M)$ as a function of u, for any fixed type of microsystem. With (17) and (18) one can eliminate V from $\lambda_{B-M}(u)$. We thus get

$$\lambda_{B-M}(u) = (h/m)\left[\frac{1}{u} - \frac{V_o + u}{c^2(1 + u/V_o)(1 + uV_o/c^2)}\right] \tag{29}$$

We lack any basis of principle for deriving the value of V_o or for establishing if, or how, V_o is connected with the involved proper mass m_o or with other characteristics of the studied microsystem (charge, spin, etc.). So we cannot estimate theoretically the difference $\lambda_B - \lambda_M(u)$. But we know that, *whatever* this difference is, it increases toward ∞ when u falls toward 0, since λ_M tends toward the finite value [(22′), (27′)] $\lambda_M(u = 0) = \lambda_{Mo}$ while λ_B tends toward ∞. The *very* low relative velocities of observation u play the crucial role. So let us stick to exclusively the following main aim:

Pin down the limiting value λ_{Mo} for various types of particles

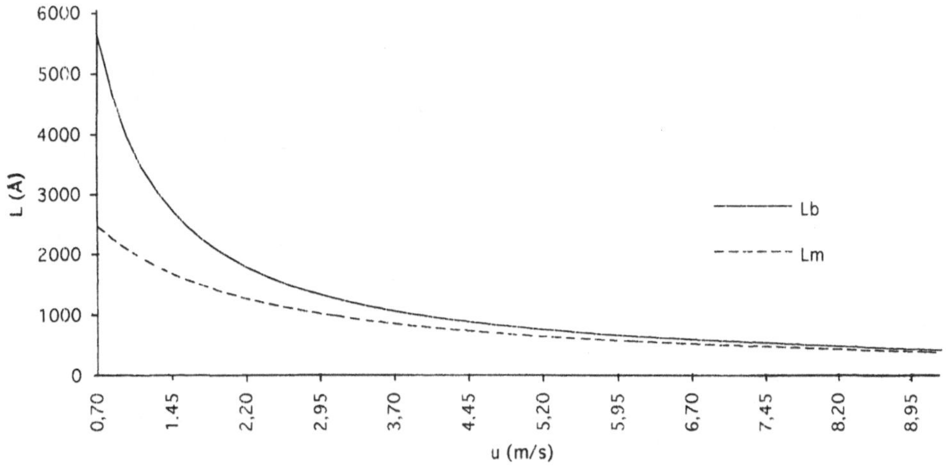

Fig. 1. Qualitative aspects of $(\lambda_B - \lambda_M)(u)$.

(at least for neutrons and electrons), in order to see whether or not it is FINITE.

A more detailed aim is to establish the form of the function $(\lambda_B - \lambda_M)(u)$ for various types of microsystems, to extract the limiting value V_o of V when u tends toward 0, and to establish the dependences of V_o on m_o, charge, etc.; or possibly, the *independence* of V_o from these, i.e., a character of universal constant. These aims could be realized by using corpuscular interferometry.

The qualitative aspects of the situation are plotted in Fig. 1.

To obtain preliminary indications on the order of magnitude of V_o, one can proceed along the following lines. For neutrons, Klein and Werner [11] report the value u=3955ms^{-1} for $\lambda_c = 1A$. Let us take this value as a reference. *If* we suppose that, *factually*, the reported value $1A$ of λ_c is *rigorously* true for neutrons at 3955ms^{-1} and that this does not contradict de Broglie's formula (if it did, the authors would have noticed), then inside *our* framework we have to *require* that $\lambda_{B-M}(3955\text{ms}^{-1}) = 0$ within the declared precision ($\Delta\lambda = x$Angström) concerning the estimation of λ. On the basis of this requirement, one can research by trial the *smallest* value of V_o such that

$$\lambda_M(V_o) \geq (\lambda_B - \Delta\lambda)$$

which sets for the proper value V_o of the phase velocity of a neutron corpuscular wave a *lower bound* $V_{o\min}$, founded on the assumptions employed. Assuming a precision of $\Delta\lambda = 10^{-3}A$, one finds (for $m_o = 1.67492 \times 10^{-27}$kg, $u = 3955\text{ms}^{-1}, \lambda = 1A, \Delta\lambda = 10^{-3}A$)

$$V_{o\min} \cong 2.26 \times 10^{15}\text{ms}^{-1} \tag{30}$$

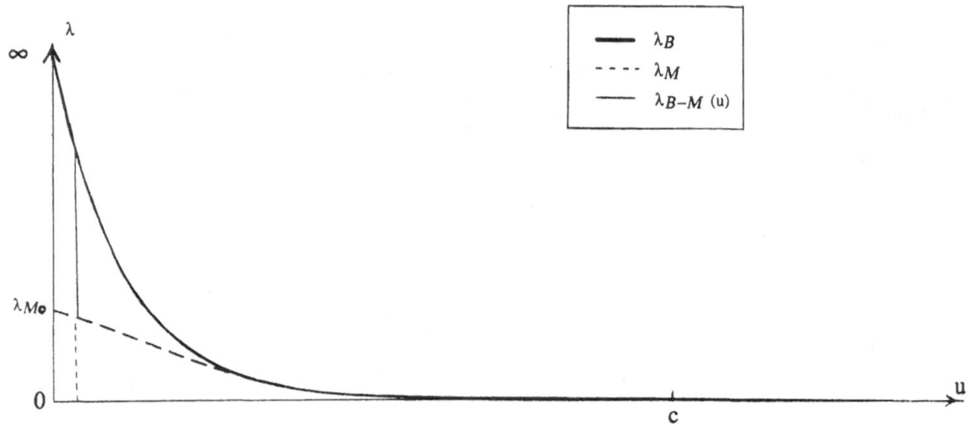

Fig. 2. $(\lambda_B - \lambda_M)(u)$ correspondingly to (30) and very near $u = 0$.

which is nearly the *square* of the value of the velocity of light: This—insofar that it is true—can explain why the modified expressions for the corpuscular wavelength, as well as those of a modified mass and energy, *can* indeed have remained unknown. But the estimation (30) is uncertain because it varies strongly with the precision assigned to the estimation of λ_B.

As for the value V_o itself, V_{omin} can be *very* far from it: It can be seen from (29) that if V_{omin} fits with the available data, then also any bigger value $V_o > V_{omin}$ will fit, and *better*. And the difference $V_o - V_{omin}$ can be very big. Indeed, inside the interesting zone of u values, namely in the close vicinity of $u = 0, \lambda_{B-M}(u)$ and V_o vary very rapidly with u (study (29) and Fig. 2), hence precisely there a tiny imprecision in u can introduce enormous errors. But uncontrolled imprecisions in u—and quite especially in very small u—are currently involved in the available data, because u is usually estimated from *thermodynamic* considerations, so that usually it has only the significance of a *mean*. It follows that:

Only measurements specifically dedicated to the verification of (27′) can produce a serious decision concerning the finiteness of λ_{co} and thus of V_o.

Considering now the definition (20) of a microscopic mass, with V_o presumably of the order $2.25 \times 10^{15}\text{ms}^{-1}$ at least, the irrelevance of the available data seems clear. So, again, only specifically appropriated experiments could be conclusive.

6.2. Procedures

Measurements of Corpusuclar Wavelengths: Since $\lambda_{B-M}(u)$ is non-negligible only at very low values of u, the difficulties are at least twofold: (a) It is very difficult to realize monochromatic sources of very low kinetic energy. (b) At a relative velocity of observation $u = 0$, a device for corpuscular interferometry ceases to register. However, with a *sliding* source or/and interferometric device, increasingly low values of the *relative* velocity of observation u might be realizable, *without* having to face the impossibility of going down toward vanishing kinetic energies of the studied microsystems. And, for a fixed type of microsystem, the value of the wavelength for a zero relative velocity of observation can be approached as only a limit of a series of registrations corresponding to non-zero u's. In this way one can hope to overcome the main difficulties.

Corpuscular Doppler Effects: The difference between the case $+u$ and the case $-u$ in the expressions (20), (24) and (26), if it exists, might be difficult to be brought into evidence. However some strategy able to discriminate might perhaps be imagined.

Modified Micromass and Energy: Quite *independently* of any consideration on corpuscular wavelength or wave frequency, the conjecture B on a modified micromass (20) $m_\mu = (m_o/\sqrt{1-\beta^2})(1+u/V_o)$ and the corresponding modified corpuscular energy (24) $W_{c\mu} = ((m_o c^2)/\sqrt{1-\beta^2})(1+u/V_o)$, might come out to be observable by high energy measurements, with the help of high precision techniques [12].

The following summary brings into evidence that—*whatever* their results the experiments proposed here are crucial.

7. WHAT IS AT STAKE?

We have made a conceptual-formal investigation on the physical and local character of what is called "corpuscular waves." The conclusions are as follows. If the proper value of the phase velocity of a corpuscular wave is infinite, then the relativistic variance assigned to the frequency of a corpuscular wave is of the same form as the Einstein variance for mass and for mechanical heavy energy. In this case, the corpuscular waves must be regarded as abstract, the representation of microphenomena is nonlocal, and two distinct variances must be admitted for the quanta of energy, thus two distinct quantum relations, one for the quanta of light energy and the other for the quanta of heavy energy. Such is the description offered to us by quantum mechanics as it now stands.

If the proper value of the phase velocity of a corpuscular wave is finite, then the relativistic variance assigned to the frequency of a corpuscular wave is of the same form as that of the frequency of a light wave. In this case the corpuscular wave of a microsystem can be regarded as a physical perturbation that propagates "really" (also inside the rest frame of that microsystem). Then quantum mechanics has to be reorganized. And, with V_o finite, the reconstructed representation of microphenomena would be *local* no matter how big V_0 is. As for the experiments performed so far concerning locality, they are all unconclusive for $V_0 \gg c$.

But the duality of the quantum relation *subsists* if the Einstein variance for mass and mechanical heavy energy is assumed to hold rigorously for also a microscopic mass and for the corresponding microscopic mechanical energy.

If, on the contrary, microscopic mass and heavy energy have the same variance as the frequency of any wave, of light as well as corpuscular, then one universal quantum relation is valid for any quantum of energy, ponderable or not. But in this case, at the microscopic level of description, the relativistic representation of reality has to be reconstructed consistently with the reconstruction of quantum mechanics, and so as to obtain Einstein relativity as a mean valid on the macroscopic level of description.

In *any* case, within a new-representation of microphenomena founded on a physical and local character of the corpuscular field, the general concept of force would have to obtain a *basic* definition, *explicitly connected, via mass, with the wave frequency* and the proper phase velocity V_o of the corpuscular wave, and with the universal constants *c* and *h*. And since gravitation is tied to mass, and mass is connected with the frequency of the corpuscular wave, the corpuscular field must be regarded as the *THE UNIVERSAL SUBSTRATUM OF THE GRAVITATIONAL FIELD.*

REFERENCES

1. M. Mugur-Schäcter, "Quantum mechanics and relativity: Attempt at a new start," *Found. Phys. Lett.* **2**, 17 (1989).
2. L. de Broglie, *Thèse*, Paris, 1924.
3. F. Selleri, *The Wave-Particle Duality*, F. Selleri, ed. (Plenum, New York, 1992).
4. J. R. Croca, A. Garruccio, V. L. Lepore, and R. N. Moreira, *Found. Phys. Lett.* **66**, 1111 (1991).
5. R. Gosh and L. Mandel, *Phys. Rev. Lett.* **59**, 1903 (1987); Z. Y. Ou and L. Mandel, *Phys. Rev. Lett.* **61**, 50 (1988); Z. Y. Ou and L. Mandel, *Phys. Rev. Lett.* **62**, 2941 (1989); J. G. Rarity, P. R. Tapster, E. Jakeman, T. Larchuck, R. A. Campos, M. C. Teich, and B. E. A. Saleh,

Phys. Rev. Lett. **65**, 1348 (1990); L. J. Wang, Z. Y. Ou and L. Mandel, *Phys. Rev. Lett.* **66**, 1111 (1991).
6. F. Selleri, *Quantum Paradoxes and Physical Reality*, A. van der Merwe, ed. (Kluwer Academic, Dordrecht, 1990).
7. E. Einstein, "Über einem die Erzeugung und Umwandlung des Lichtes betreffenden heuristischen Gesichtspunk," *Ann. Phys. (Leipzig)* **17**, 132 (1905).
8. E. Einstein, "Zur Elektrodynamik bewegter Köper", *Ann. Phys. (Leipzig)* **18**, 639 (1905).
9. L. de Broglie, "Tentative d'interprétation causale et nonlinéaire de la mécanique ondulatoire" (Gauthiers-Villars, Paris, 1956).
10. D. Bohm, *The Special Theory of Relativity* (Benjamin, New York, 1965).
11. A. G. Klein and S. A. Werner, "Neutron optics," *Rep. Prog. Phys.* **46** (1983).
12. R. Folman, "An informal sketch of a search for possible deviations from known relativistic mass," private communication, 1993.

INFORMATIONAL EXPERIMENTS WITH MICROPARTICLES AND ATOMS[1]

Raoul Nakhmanson[2]

Institute for Didactics of Physics
Justus-Liebig University
6300 Giessen, Germany

Accepting information as a physical category and ascribing to inanimate matter some spirit (consciousness, intelligence) allows to explain quantum-mechanical phenomena, including delayed-choice and EPR-Bohm-Bell experiments, as well as irreversibility of time, remaining on the basis of local realism, and suggest essentially new experiments with microparticles and atoms in which information plays the principal role.

Key words: microparticles, quantum mechanics, information, intelligence.

Historically formed physics is only a part of human knowledge. It searches laws of behaviour of so-called "inanimate" matter, and does it with a success. The behaviour of "animate" matter is also a subject of physics, but only if it does not include spiritual processes. The consequent departments of physics are biomechanics and biophysics. The spiritual phenomena lay beyond the limits of the traditional physics. In this respect many prominent physicists, including Einstein and Bohr, had similar opinions.

The division of "*being*" in spirit and matter is ascribed to Descartes. The sages of ancient times (and of the modern East) did not make such a division and thought all matter was animated. For example, Aristoteles thought that any matter had not only "*potentia*"—ability to move, but also "*entelechia*"—a wish and purpose to move. I think this is a correct point of view. Spirit can not be separated from matter, a boundary line between them cannot be drawn. The Cartesian dualism is only a very useful method of analysis, no more.

Omitting the middle age and the golden age of classical physics, I jump to the transition period and cite Emil Borel, an acquaintance of the de Broglie family. In the book *Le Hasard*, isued in 1913, the year of Bohr's atom model, Borel remarked that humans cause entropy to decrease in small volumes by means of processes accompanied by increasing of entropy in big volumes, and then goes on as follows;

> In other words, the structure of the universe is becoming more and more subtle..., it is probable that similar phenomena are going on on other scales as well, too large or too small to be accessible to us. Thus, the evolution of the universe may be represented as a gradual complication of its structure, accessible to understanding and use of beings of lesser and lesser size. As there is no absolute standard of length, we may not be afraid of such a lessening of scales; it seems to us presently that beings of molecular sizes and, all the more, beings so small in respect to molecules as we are to sun, are objects scarcely deserving our attention; but it is quite possible that the progressing complexity of the universe will create or has already created some beings with an organisation much more complex than ours.

Ascribing to "inanimate" matter some spirit (other terms: consciousness, intelligence) is not alien to modern science. Similar meanings were expressed by well known physicists, such as Charles Galton Darwin, Eddington, Heisenberg, Schrödinger, Jordan, Margenau, Wigner, Cochran, and others. [1]

Quantum-mechanical indeterminism looks like free will, i.e., it allows a micro-object to choose freely among several possible alternatives. But quantum mechanics does not search this way, it only postulates indeterminism, and *a priori* declares it as absolutely random. Thus, quantum mechanics closed its eyes on possible purposefullness of any object's behaviour, *a priori* replacing it by some chaos.

The notion "information" is a good connecting link between physics and the humanities. Firstly, it is used in communication technique and is formalized in the "theory of information." Secondly, it calls forth some suppositions about sources and receivers of information which are intelligent or supported by some intelligence. Thirdly, interchange and treatment of information are the essence of spiritual life of individuals and society as a whole.

Is information of physical category? Or, in other words, does information play a role in a world of "inanimate" objects? This question, in its latent form, has existed in physics for more than hundred years and is connected with discussions around Maxwell's demon of thermodynamics, the passive observer of the theory of relativity, the active experimenter of quantum mechanics, and the coincidence of Boltzmann's and Shannon's formulae for entropy. Accepting information as an inherent attribute of matter, we ascribe to "inanimate" matter some intelligence. This question has a direct relation to irreversibility of time and to the crisis of notions of reality and locality in modern physics.

The clues supporting the opinion that information like mass and energy is an inherent attribute of matter (which, in it's turn, is intelligent) are the following:

(1) the Heaviside-Einstein formula $E = mc^2$, because c is the maximal velocity for the spreading of information;
(2) the informational character of the wave function ψ, describing probability, in contrast to classical potentials describing "tangible" fields transporting energy and impulse;
(3) "teleological" movement of matter seemingly in the principle of the least action; and
(4) quantum-mechanical stochastics which can be seen as optimal tactics of behaviour of disconnected members of a quantum-mechanical ensemble searching all possible alternatives.

The irreversibility of time and the second law of thermodynamics can be explained if the state with maximal entropy has some purposeful advantages. For example, molecules of gas "inspect" the space better when they are distributed in a vessel homogeneously. Accordingly, they keep such a distribution (using control of collision parameters) and do not gather.

The acceptance of spirit or even intelligence of microparticles leads us to the acceptance of their complex structure. Do they have enough capacitance? Theoretically it was shown (firstly perhaps by Markov [2]) that one microparticle can contain a whole universum ("fridmon," "baby-universe"). If one goes not farther than Planck length ($\approx 10^{-35}$ m), one finds in a micro-object ($\emptyset \approx 10^{-15}$ to 10^{-20} m) of the order of 10^{45} – 10^{60} "Planck cells," that is much more not only than the number of neutrons in the human brain but than the number of atoms contained in all living beings, too.

A consistent development of this idea permits explaining quantum-mechanical phenomena remaining on the basis of local realism, and suggests essentially new experiments with microparticles and atoms in which information plays the principal role.

The ψ function in this conception is a strategy function. It reflects an optimal behavior of particles, both individual and social (ensemble, universum). Where is this function? Of course, it is not in the real 3-dimensional space; it is in imaginary configurational space, which, in its turn, is in the imagination (consciousness) of the particle. When the particle receives new information (it takes place by any interaction with micro- or macroobjects),

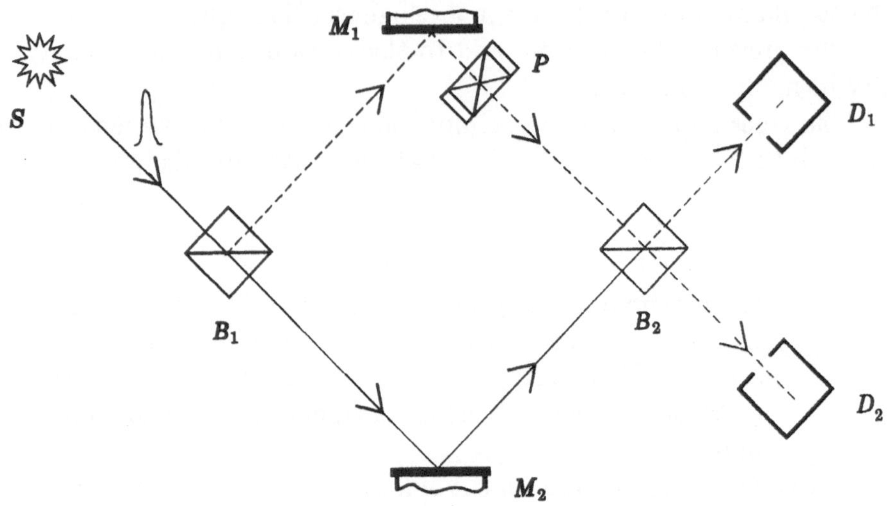

Fig. 1. "Delayed-choice" experiment with Mach-Zender interferometer. S is the source of photons; B_1, B_2 are the beam-splitters; M_1, M_2 are the mirrors; D_1, D_2 are the detectors; P is the Pockels cell.

it changes its strategy. Thus occurs the so-called "collapse" of the ψ function. It occurs in the consciousness of the particle (but not in the observer's consciousness, as was suggested by von Neumann and Wigner), i.e., locally and realistically. In a human, consciousness collapses only its knowledge about the ψ function.

The flows of particles carry not only mass and energy, but also information. It must be emphasized that the Shannon formula for information gives only a "static" approximation. In accordance with this formula, today's and yesterday's newspapers contain approximately the same information. The complex object having the same mass, energy, entropy, and Shannon information can contain different "actual" information. The different states of such an object as a functional series of time are an analogue of the series of newspapers. Therefore it is possible during the interaction to transfer information without a change of the mass, energy, and entropy of interacting objects. These flows of "actual" information mean the "spirit" life of matter.

Properly speaking, the same is true for traditional live ojbects, e.g., human beings. The flows of mass, energy and entropy have essential convergences only for growing and learning children. For an adult and educated person these flows pass mainly through him and change only the "actual" information.

Let me explain this by the example of the "delayed choice" experiment with the Mach-Zender interferometer, Fig. 1, which was discussed by

Wheeler [3] and really performed in Maryland, Munich, and Frankfurt [4,5]. Here S is the source of particles (photons), B_1 and B_2 are the beamsplitters, M_1 and M_2 are the mirrors, D_1 and D_2 are detectors, and P is the Pockels cell. The flow of photons is so weak that no more than one photon is in the interferometer as a rule.

The Pockels cell is at all times open (closed), but anytime for a short moment (several nanoseconds) when a photon can pass it, it is closed (opened). As a result of experiments having been performed, the interference after D_2 appears only if the Pockels cell was open at the appropriate moment, independent of its state at other times.

This result can not be explained by remaining on the level of classical realism and thinking that, after B_1, a photon really chooses one path (upper or lower in Fig. 1) in the interferometer, because if the photon really traveled the lower path (this "must be" in 50% of the cases), it is too far (several meters) from the Pockels cell, and at the above mentioned "appropriate moment" is under no influence from the cell (superluminal action is supposed to be impossible). The photon in the interferometer looks like a "Great Smoky Dragon" (Fig. 2, drawing by Field Gilbert).

From a new point of view all this looks different. When a photon meets the first beamsplitter B_1, it has a choice. Simultaneously it receives from B_1 the fresh information about the *past* of the world, particularly about the interferometer and Pockels cell. The physical conditions of B_1 induce a 50% choice. But the decision must be at random: It is the optimal tactics for an ensemble of disconnected photons. The consciousness of the photon works only to find an optimal strategy, i.e., ψ function (in this case a 50% choice of path after B_1, and interference after B_2, if Pockels cell appears open). To find it, the photon solves a variation problem (e.g., the wave equation).

Let us suggest that after B_1 the photon takes the lower path. It meets the mirror M_2 and becomes some new information about the *past* of the world. But in respect to interferometer, it is the same information, so this part of the wave function stays the same. Finally our photon meets the beamsplitter B_2 and must again choose its path. Simultaneously it receives from B_2 the fresher information about the *past* of the world, including an actual history of the upper arm of the interferometer with the state of Pockels cell at the "appropriate moment." This information is brought to B_2 with luminar velocity, e.g., by virtual photons in solids and by thermal photons in vacuum/air: At room temperature the characteristic time of several nanoseconds is enough for B_2 to receive from P several thousands of thermal photons. Actually, the number of thermal photons emitted by one and arriving at other macroobject is

$$N \approx 5 \cdot 10^{10} S_1 S_2 T^3 \tau / L^2,$$

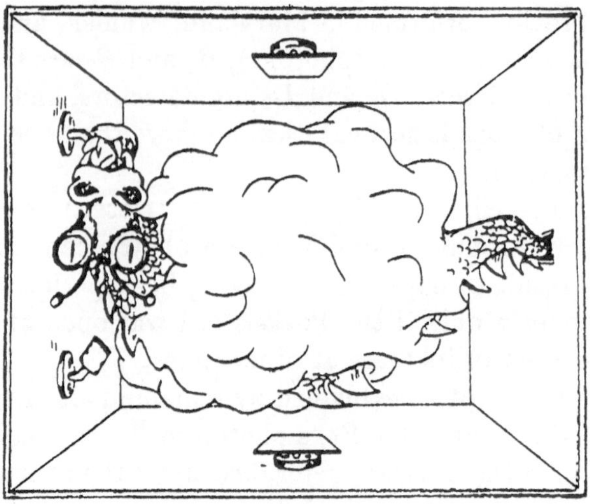

Fig. 2. "Great Smoky Dragon" (from Ref. 4; drawing by Field Gilbert). Within the standard (Copenhagen) interpretation, in spite of our knowing of input and output of a photon (the tail and the mouth of the dragon, respectively), we neither know nor have the right to say how the photon came. As it is explained in the text, this dragon exists only in the photon's imagination (and in imagination of physicist), not in the interferometer.

where S_1 and S_2 are the effective areas of the first and the second macro-objects, respectively, T is the absolute temperature of S_1, τ is the characteristic time, and L is the distance between S_1 and S_2. Substituting the typical values $S_1 = S_2 = 2\ \text{cm}^2, T = 293°\text{K}$, $\tau = 10^{-8}$ sec, and $L = 10^3$ cm, one finds $N \approx 5000$.

Now our photon has all the necessary information to make a decision, namely, to prefer a direction of constructive interference (if the cell at the "appropriate moment" was opened) or to make a 50% random choice between two possible directions (if the cell at the "appropriate moment" was closed). If the state of the Pockels cell in the "appropriate moment" was different than before, a reduction of the strategy (i.e., a reduction of wave function ψ relating to the experiment) takes place in the consciousness of the photon after interaction with the beamsplitter B_2. The "Great Smoky Dragon" does not exist in the real interferometer. One can speak only about the Dragon's images in the consciousness of the particle (primarily) and that of the physicist (secondarily).

To have a success in the "delayed choice" experiments, one can try to cut off the informational contact between the Pockels cell and the beamsplitter B_2, e.g., by introducing a deep-cooled filter. If the filter is cooled by liquid helium, no photons come from it to B_2 during the characteristic time (several nanoseconds). Besides, to prevent a prediction of a state of Pockels cell by the B_2, it is better to control the cell using not a regular but "good" random generator.

The situation discussed by Einstein, Podolsky, and Rosen, and in modern form by Bohm and Bell, is more complex. Here two particles flying in the opposite directions have a common correlated wave function = common correlated strategy. As a consequence, the result of interaction of one particle with some measurement apparatus must be correlated with the result of the interaction of another particle with another measurement apparatus, in spite of a large space separating the two apparatuses. Such correlation is possible only if each apparatus knows the state of the other one; more correctly, if each apparatus (and/or the particle interacting with it) can *predict* the state of the other apparatus at the "appropriate moment" of measuring with a good probability. In the world of communicating matter, like in human and other biological societies, such prediction is a natural attribute of existence, as well as the common wave function reflecting this intercommunication of whole matter of the universe. In all experiments made up to now, such *predictions* could take place, without supposing a superluminal velocity of communication, because the states of the measurement apparatuses have been changed very slowly [6] or periodically [7]. Again one can try to perform such experiments with a "good" random control of measurement apparatuses.

It is interesting to note that the authors of experimental researches felt the advantage of random control (as it seems more intuitive, because they do not discuss it) and sometimes used it [4]. The peculiarity of random signal series is non-predicting of its next term. Therefore, these authors felt a possibility of "inanimated" matter to predict the future, and have tried to restrict it.

In the "delayed choice" and EPR experiments one tries to eliminate the informational contacts already existing in nature. It is possible to go further and try to "speak" with a quantum object itself. The first suggestions of such experiments have been made in [8]. In this case a content and code of a "message" are defined by the experimenter, making his task more difficult but more interesting. One can try, e.g., to use some universal language, like mathematics or music, and hope for an "attractivity" of information, even if it can not be decoded. Using feedback, the sending of a new message can be made dependent on the "free-choice" of a microparticle when it passes through the beamsplitter or jumps from one energy level to others.

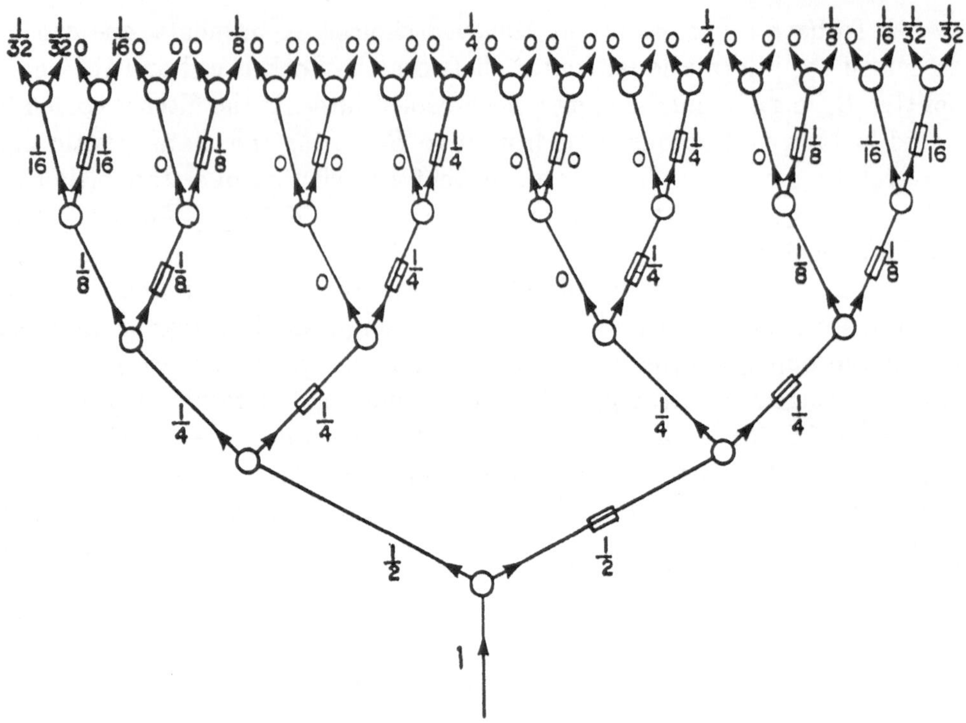

Fig. 3. Binary-tree experiment. Circles stand for beamsplitters, rectangles denote informational cells. Ciphers show the probability of detecting a particle in the case of the most rapid formation of a rigid, conservative condititional reflex.

Figure 3 shows the scheme of "binary-tree" experiment. The initial beam of micro-objects (particles, atoms) enters into a system of beamsplitters (shown by circles). They can be semi-transparent/semi-reflecting mirrors for photons, crystals for electrons or neutrons, Stern-Gerlach apparatuses for atoms, etc. Figure 3 shows only five rows of beamsplitters, but, in principle, there can be as many as experimentally feasible. According to present-day theoretical ideas and practical experience, each of the output beams has the same intensity, namely, 1/32 of intensity of the initial beam (real beamsplitters may have, of course, some absorption, but here it is not a matter of principle).

Now let us introduce into each of the right channels an "information cell" (shown by rectangles), which is a device leaving unchanged the intensity of the beam passing it, but offering some information to particles. For polarized photons such a cell may be a set of transparent plates fixed at the Brewster angle, and the information can be coded, say, by differences in the materials of plates, their thickness, or distance between them. For example, the information cell may include 10 glasses of thickness A and 10 glasses of thickness B arranged in the following way:

$$BBABBBABBA\ AABAAABAAB.$$

In binary code (direct and inverse) they reprsent the number 0010001001 = 137, which is the reciprocal of the fine-structure constant. Absorption brought in by the information cells may be compensated by introducing into each of the left channels a compensating cell bearing no information, more precisely, bearing less, or less significant—in our judgment—information (e.g., the same glasses staying random or periodically). The information in each next row is a sequel of that in the preciding row.

The commonly accepted point of view is that the introduction of information cells, together with compensators, will not change the uniform probability distribution of particles in the output beams. But if particles are intelligent, and are able to notice the information offered to them, they may become interested in it. After a number of rows, the particles should notice that the information is offered only in right channels, and should prefer the choice of right channels in passing through the following beamsplitters. In other words, particles could develop a "conditional reflex," of essentially the same kind as in behaviour experiments on living beings. Such an intelligent behaviour of particles should lead to a change of their distribution in the output beams. For example, if the conditional reflex appears immediately and the particles are "conservative," i.e., they are no more of interest to the left channels, the distribution of probability in branches of the binary tree is like the one shown in Fig. 3.

Deviation from the uniform distribution of particles in the output beams will mean that the particles at least recognize the offered information and have an interest in it. This, however, still does not mean that the particles understand the information offered: People of the modern era were interested in ancient hieroglyphic symbols long before they learned how to interpret them. To establish a deciphering stage, one can, starting from some row of the binary tree, introduce some specific "requests" into the information cells. For example, one can "ask" particles to choose a left channel after the next beamsplitter rather than a right channel. The honouring of such kind of requests can easily be detected by an experimenter. However, the possibilities of an experiment typified in Fig. 3 are not exhausted by this passive level of communication. Purposefully choosing direction at each subsequent beamsplitter, the particle, in its turn, can send information to the experimenter using "right" and "left" as a binary code.

The experiment Fig. 3 can be called "coordinate-impulse" one to distinguish it from the "energy-time" experiment which scheme is shown in Fig. 4. Here a four-level quantum system, e.g., an atom, with one low (1), one high (2), and two intermediate (3,4) energy levels is pumped by intensive radiation inducing the 1 → 2 transition, so that the atom stays not in the state 1 but immediately goes into the state 2. From it, the atom transits (spontaneously or light induced) to the state 3 or 4, and later transits to the state 1 completing the cycle. The radiations corresponding to some of transitions $2 \rightarrow 3, 2 \rightarrow 4, 3 \rightarrow 1$ and $4 \rightarrow 1$ are detected (in Fig. 4 two detectors are shown). Besides, there is an informational action on the atom, e.g., by modulation of light coming from the source S. The modulator M is controlled by the source of information SI, which, in turn, is connected with one or more detectors to close the feedback loop.

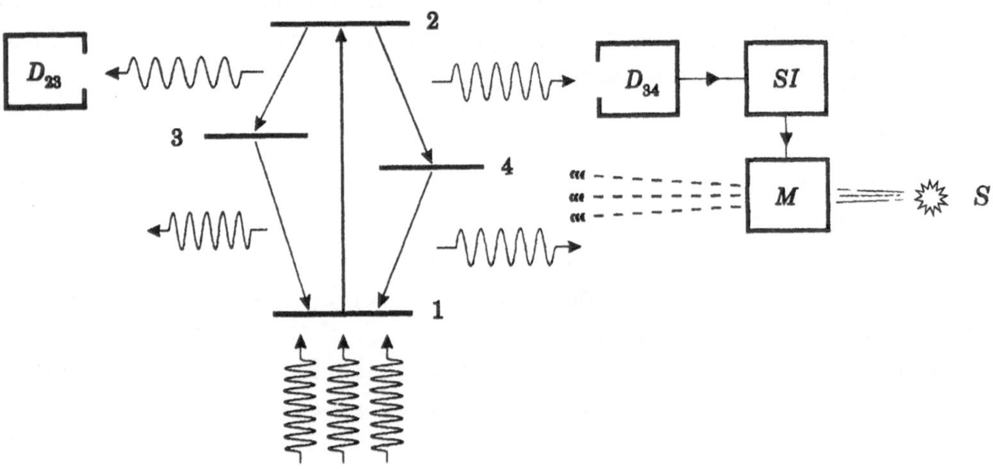

Fig. 4. Informational experiment with a single atom. 1, 2, 3, 4 are the energy levels; D_{23}, D_{24} are detectors; S is the source of light; M is the modulator; SI is the source of information.

The feedback works in such a way as to stimulate a rate and channel of transitions, in the case of Fig. 4, the 2 → 4 → 1 transitions. The source SI sends repeatedly the same message, for example, one line of a page or one circle of gramophone record/compact disc. Only when SI receives a signal from detector D_{24}, it changes the message to a new one, namely, the next line or the next circle. If the atom has an intelligence and is interested in new information, it develops a conditional reflex and will prefer the 2 → 4 transition to the 2 → 3 one. Besides, the rates of both 2 → 4 and 4 → 1 transitions must increase. All this can be registered by the experimenter. To be sure that the effect is connected with information, one can make a control experiment replacing information by "indefinite" noise, etc.

If the detector D_{24} has a small aperture, the space orientation of emitted photons toward to D_{24} can also develop itself as a part of conditional reflex. In spite of this interesting possibility, one must prefer to use effective detecting of emitted photons to facilitate the developing of conditional reflex. Perhaps the combination of an ion trap (e.g., Penning or Paul trap) and resonator ("single atom MASER" [9]) gives a good condition for informational experiments with single atoms.

It seems that a progress in semiconductor device technology can also be used for the same experiments. In a small metal-oxide-silicon field-effect transistor (MOSFET) one can observe random telegraph signals corresponding to charge and discharge of one electron trap locating at the Si-SiO_2 interface [10]. As in the previous case, the new information may be sent to MOSFET (i.e., to the trap), by modulation of light or RF, only after the next capture or/and emission of an electron by the trap. It closes the feedback loop, and the experimenter may find increasing of the capture or/and emission rate.

Like the scheme of Fig. 3, in two last cases one may hope to observe not only an interest of a quantum object (atom, "surface state") to receive a new information, but decoding of it also, as well as sending of messages from the objects to the experimenter.

The noted physician Erasmus Darwin (grandfather of the great naturalist Charles Robert Darwin) played the flute for his flowers. He did not see any reaction, but after two hundred years his idea was vindicated experimentally. As it seems, this example encourages us to respect of intuition of physicist Charles Galton Darwin (Ch. R. Darwin's grandson), who wrote in 1919:

> The great positive successes of the quantum theory have accentuated all along, not merely its value, but also the essential contradictions over which it rests.... It may be that it will prove necessary to make fundamental changes in our ideas ...or even in the last resort to endow electrons with free will.

Namely we should play music for electrons. Why not?

REFERENCES

1. M. Jammer, *The Conceptual Development of Quantum Mechanics* (McGraw-Hill, New York 1966); F. Selleri, *Die Debatte um die Quantentheorie* (Vieweg, Braunschweig, 1990).
2. M. A. Markov, *On the Nature of Matter* (Moscow, 1963) (in Russian); *Ann. Phys.* **59**, 109 (1970).

3. J. A. Wheeler, in *Mathematical Foundations of Quantum Theory*, A. R. Marlow, ed. (Academic, New York, 1978).
4. C. O. Alley, O. G. Jakubowicz, and W. C. Wickers, in *Proceedings, 2nd International Symposium on Foundations of Quantum Mechanics*, M. Namiki et al., eds. (Physical Society of Japan, Tokyo, 1987), p. 36.
5. T. Hellmuth, H. Walther, A. Zajonc, and W. Schleich, *Phys. Rev.* **35A**, 2532 (1987). J. Baldzuhn, E. Mohler, and W. Martienssen, *Z. Phys. Cond. Matt.* **77B**, 347 (1989).
6. S. J. Freedman and J. F. Clauser, *Phys. Rev. Lett.* **28**, 938 (1972). J. F. Clauser, *ibid.* **37**, 1223, 1976; *Nuovo Cimento* **B33**, 740 (1976). E. S. Fry and R. S. Thompson, *Phys. Rev. Lett.* **37**, 465 (1976). A. Aspect, P. Grangier, and G. Roger, *ibid.* **49**, 91 (1982). W. Perrie, A. J. Duncan, H. J. Beyer, and H. Kleinpoppen, *ibid.* **54**, 1790 (1985).
7. A. Aspect, J. Dalibard, and G. Roger, *Phys. Rev. Lett.* **49**, 1804 (1982).
8. R. S. Nakhmanson, Preprint 38–79, Institute of Semiconductor Physics, Novosibirsk, 1980; see also A. A. Berezin and R. S. Nakhmanson, *Physics Essays* **3**, 331 (1990).
9. G. Rempe, W. Schleich, M. O. Scully, and H. Walther, in *Proceedings, 3rd International Symposium on Foundations of Quantum Mechanics* (Physical Society of Japan, Tokyo, 1989).
10. K. S. Ralls, W. J. Skocpol, L. D. Jackel, R. E. Howard, L. A. Fetter, R. W. Epworth, and D. M. Tennant, *Phys. Rev. Lett.* **52**, 228 (1984).

NOTES

1. Shortened version of a report which was read on September 29, 1992 in Trani Workshop (Italy).
2. Present address: Waldschmidtstrasse 131, 60314 Frankfurt, Germany.

THE PRINCIPLES OF CLASSICAL MECHANICS AND THEIR ACTUALITY IN CONTEMPORARY MICROPHYSICS

Jean Reignier

Vrije Universiteit Brussel
Theoretische Natuurkunde
Pleinlaan 2
B-1050 Brussels, Belgium

The principles of classical and quantum mechanics are presented in a new way which emphasizes the different conceptual approaches to localization in space for both theories. It is shown that the idea that matter is permanently localized in space is fundamental in classical mechanics but not in quantum mechanics. We discuss an experimental situation where spin-1/2 particles exhibit a non-local behaviour.

Key words: Classical mechanics, quantum mechanics, locality.

1. INTRODUCTION

In this essay, I would like to discuss the role of space in the basic formulations of classical and quantum mechanics. I want to emphasize that the "truism" of a permanent localization of material objects should be considered as an axiom, and, even more, as one of the primary axioms of classical mechanics. Of course, all evidence coming from the external world and directly perceived by our senses is in accordance with this principle. As a consequence, it is so deeply rooted in the human brain that we simply can't imagine how matter could behave differently. However, it turns out that quantum mechanics is formulated in a way that doesn't refer to this principle of a permanent localization. And it turns out that quantum mechanics eventually contradicts this principle. This is the reason why quantum mechanics is perceived as a rather strange theory, being, on the one hand, extremely efficient in its description of physical phenomena, and going, on the other hand, against common sense. Confronted with this situation, many physicists adopt a rather schizophrenic attitude: On the one hand, they use the quantum formalism as it is, and they take great pride

Waves and Particles in Light and Matter, Edited by
A. van der Merwe and A. Garuccio, Plenum Press, New York, 1994

in its achievements. On the other hand, they keep in mind the classical image that particles are small objects moving continuously in space. They generally know about the contradiction but they evade the question by a clever, light-hearted answer: "...after all, nobody understands quantum mechanics."

Would it be possible to remedy this unsatisfactory situation? I don't know the answer, but my guess is that there will never be a complete cure because the idea of a permanent localization of material objects is so deeply rooted in our brain that it will always reappear, if only in our language. Nevertheless, I believe that a clearer comprehension of the principles of both theories, the classical and the quantum mechanics, can help. We can at least try to understand better the origin of the difficulty. In particular, it will become clear that the principle of a permanent localization is not only irrelevant for the quantum description but that, furthermore, it may turn out to be wrong in specific cases. This should not be felt as a major difficulty as long as it does not affect the use of the principle at the macroscopic level and in particular in everyday life. After all, the world is as it is, and the best we can do is to give a reasonable description of this world with the tools at our disposal. The extrapolation to the microscopic scale of rules obtained from our direct perception is certainly such a tool. But mathematics provides us with many other tools. If it happens that these different tools lead us to contradictory descriptions, one has to investigate which tools should be kept as the best ones. I believe that this is the normal way to improve our description of the world, i.e. the normal approach to a true realism.

2. THE PRINCIPLES OF CLASSICAL MECHANICS

What do we mean by "the principles of classical mechanics"? This problem was discussed at length at the end of the 19th century. Heinrich Hertz [1] distinguishes three different approaches:

(i) the classical or Newtonian scheme based on the four basic concepts of space, time, mass, and force;

(ii) the energetic or Hamiltonian scheme based on the four basic concepts of space, time, mass, and energy;

(iii) the Hertzian scheme where force and energy disappear as basic concepts and are replaced by geometrical or kinematical constraints between the masses; the scheme eventually uses only the three basic concepts of space, time, and mass.

The analysis made by Hertz clearly shows that force and energy are somewhat disturbing concepts, because they can't be reduced completely to space and time concepts. One has to consider them as new concepts about which one states properties which are nothing else than definitions, and which are only roughly established by direct experiment. Furthermore, force and energy introduce some difficulty with respect to the action at a distance, the spatial localization of the energy, etc. Henri Poincaré [2,3] completed this analysis by noticing that the Hertzian scheme itself had to pay a rather high price in order to solve these problems: It had to introduce unobservable masses with unobservable motions.

I shall not come back on this discussion. I shall essentially adopt the Newtonian scheme simply because it is the most general one. It contains as a particular case the scheme of d'Alembert with its ideal constraints, which in turn contains the Hamiltonian scheme (or equivalently the Lagrangian scheme)[1]. I shall however postpone slightly the introduction of the concept of force, in order to go as far as possible with the concepts of space, time, and matter only. In this way, mechanics appears primarily as the detailed description of the evolution in spacetime of a small fraction of the matter of the universe, which we agree to study as a separate entity. This statement contains already several principles that I shall discuss in more detail in the next sections.

In the presentation of the principles of classical mechanics, I shall distinguish between primary principles (called A, B, C) and secondary principles (called 1, 2, 3, 4). The primary principles are generally not mentioned in textbooks because they seem too evident. However, this is a very serious omission if one aims at proceeding with a comparison to quantum mechanics, where these principles are interpreted in a quite different way. The secondary principles are closely related to the laws about the forces formulated by Stevin, Galileo, and Newton. Regarding the problem of locality, they lead to the same kind of difficulty in classical and in quantum mechanics.

3. SPACE, TIME, AND MATTER

The first main principle (*principle A*) states the goal of mechanics: Mechanics aims to describe the evolution of permanent material objects in a spacetime framework. In this description, the permanent objects that are supposed to modelize the objects of every day life are in fact geometrical objects properly defined in space. We call them point particles or solid bodies, and they can be bound together by constraints in order to form more complex objects. This rather simple representation of matter reproduces faithfully the essential property that matter is always localized, i.e., each object is at each instant of time somewhere in space.

At the fundamental level where one introduces this principle A, the mathematical structure of spacetime is largely irrelevant. Space and time can be taken in the sense of the definition given by Kant [4] in his transcendental aesthetic:

"Space is a necessary a priori representation, which underlies all outer intuitions. We can never represent ourselves the absence of space,

though we can quite well think of it as empty of objects. It is to be considered as the condition for the possibility of the phenomena and not as a determination that depends on them, and it is an a priori representation which gives a foundation to external phenomena."

"Time is a necessary representation which underlies all intuitions. We cannot remove time itself with respect to the phenomena, though we can quite well think of time as void of appearances. Time is given a priori. Itself alone contains the reality of phenomena. These can well all disappear but time itself (as a general condition for their possibility) cannot be suppressed."

This "a priori" about a general spacetime framework is permanent, from the early birth of science to our days. Only the structure of that spacetime, i.e., the mathematical description we give of it, has considerably changed. This evolution should be considered as the result of our experimental study of spacetime (in the spirit of the exploration of space described by H. Poincaré [3,5] in relation with the very existence of solid bodies). One can summarize this evolution of our knowledge and description of the space-time structure through the following steps[2]:

- From Newton to the end of the 19th century, space is a three dimensional euclidian space ($\mathbb{R}^3$), completely independent of time; mathematically, spacetime is $\mathbb{R}^3 \otimes \mathbb{R}$, fully independent of the matter it may contain.
- For Minkowski (1908), the objective description of physical phenomena requires that spacetime be described as a four-dimensional continuum with the invariant quadratic form

$$ds^2 = c^2 dt^2 - dx^2 - dy^2 - dz^2 = g_{\mu\nu}\, dx^\mu\, dx^\nu, \quad \mu, \nu = 0, 1, 2, 3,$$

  where c is the speed of light in vacuum; again, spacetime is independent of the matter it may contain.
- For Einstein (1916), the structure is closely related to the presence of matter; the $g_{\mu\nu}$ of Minkowski become ten functions $g_{\mu\nu}(x)$ called "gravific potentials".
- This exploration of the structure of spacetime is going on today, and some authors suggest (see, for example, [6]) that the structure is a fluctuating one, due to the combined action of gravity and quantum effects.

For a critical discussion of the first principle one must state from the very beginning which structure of spacetime is adopted. In this paper, devoted to a discussion of non-relativistic classical and quantum mechanics, I shall adopt the Newtonian representation of spacetime ($\mathbb{R}^3 \otimes \mathbb{R}$). The analytical description of the material objects in this space requires the choice of a frame of reference, i.e., a spatial frame and a clock (i.e., a specific motion). Although this frame of reference can be arbitrarily chosen, it turns out that some of the secondary principles are only true if the frame is a so-called inertial frame (see Sec.5). Once this choice is made, the position of any geometric point of a material object with respect to this frame of reference becomes an objective property of this object. If one assumes furthermore, in full agreement with our daily experience, that such a position evolves smoothly in time, one can deduce from principle A and from ordinary mathematics, new functions of time which are called velocity $\vec{v}(t) = \dot{\vec{r}}(t)$, acceleration $\vec{a}(t) = \ddot{\vec{r}}(t)$, superacceleration $\dot{\vec{a}}(t) = \dddot{\vec{r}}(t)$, etc. These functions are to be considered as new objective properties of the material object (again, with respect to the chosen frame of reference). Of course, one has to make a difference between the very existence of all these functions, which is objective, and the knowledge one can have of them which depends on observation. In order to determine the position of a particle at some instant of time t, the observer acts during a small time interval τ around t and he succeeds to localize the particle in a small volume around an average position $\vec{r}$; the result is thus a knowledge of $\vec{r}(t)$ affected by an error τ in time and an error $\Delta\vec{r}$ in space. Furthermore, it is assumed that this act of measurement does not affect the kinematical properties of the observed particle. In order to measure the velocity, the observer has to repeat this operation at a later time $t + \delta t$ and to perform an appropriate "physical limit" of the ratio $(\vec{r}(t+\delta t) - \vec{r}(t)/\delta t$ when δt goes "physically" to zero, in the sense that δt remains large in comparison with the small interval of time τ required for a position measurement. All these operations are quite conceivable and possible with a very high degree of precision for the ordinary material objects of mechanics. But, at the turn of the century, the discovery of Planck's quantum of action $\hbar \sim 10^{-27}$ erg·sec has changed this aspect of the problem. If one admits that any observation of a physical object requires some interaction between the object and the observation apparatus, and if an irreducible unit of action $\hbar$ does exist, then these rules of measurement lead to a limitation of our knowledge of the position and the velocity at one instant of time. Indeed, the detailed analysis of the measurement process of $\vec{r}$ and $\vec{v}$ at one instant of time t, taking into account the uncontrollable exchange of at least one quantum of action at each interaction, leads to the Heisenberg relation

$$\Delta x \cdot \Delta v \gtrsim \frac{\hbar}{m}$$

where m is the mass of the material object one observes[3]. This physical constraint is completely negligable for all "macroscopic" objects, even for the smallest ones which are directly accessible to our senses (e.g., a drop of water of 1 micron has a mass of 0.5×10^{-12}gm). It is therefore safe to consider the empirical kinematics based on our measurement as a true representation of the mathematical kinematics of principle A. Conversely, this legitimizes principle A, because we can really construct such kinematics, and principle A can therefore be considered as an absolute truth for macroscopic objects. This is not valid any more for these extrapolations of the idea of matter that we call electron, proton, or atom, the mass of which ranges from 10^{-27}gm to (say) 10^{-22}gm. The Heisenberg relation gives then $\Delta x \cdot \Delta v$ ranging from 1 cm^2 sec^{-1} to 10^{-5} cm^2sec^{-1} (respectively) for objects which are presumably not larger than 10^{-8}cm. Such a large correlation of the errors excludes practically the construction of an empirical kinematics. Therefore, there exists no empirical justification for the fundamental assumption of the existence of a permanent coordinate $\vec{r}(t)$ for these objects. It is too early to go further and make a statement concerning the existence of $\vec{r}(t)$, and I shall come back to this point later on. But one can certainly state that it seems rather inconvenient to base a theory on a mathematical kinematics which cannot possibly be approximated by an empirical one.

What about principle A in quantum mechanics? Just as classical mechanics, quantum mechanics admits the permanent existence of material objects in spacetime[4]. But confronted with the major difficulty just discussed, quantum mechanics interprets principle A in a different way. For quantum mechanics, the objective reality is that any observer, at any time, can indeed define a position of the particle if he makes a measurement, i.e., if he makes the particle interact (directly of indirectly) with a macroscopic apparatus which eventually **defines** the position. Such a measurement is always affected by some errors which depend only on the apparatus, so that, just as in classical mechanics one gets an actual position $\vec{r}(t)$ with errors $\Delta\vec{r}$ and τ, and these characteristics are **defined** by the macroscopic apparatus one uses. Such a measurement can (in principle) be repeated as often as it pleases and, therefore, the objective reality is very close to the objective reality of classical mechanics: For a set of time instants t_i (with error τ), one has localized the particle inside volumes Ω_i, i.e. one has found positions $\vec{r}(t_i)$ with errors $\Delta\vec{r}_i$. Of course, one can always imagine a polygonal line which joins these successive points and define in this way a trajectory. This is the interpolation made by classical mechanics. But for an object strongly sensitive to the Heisenberg relation, such an interpolation becomes meaningless because it does not correctly approximate the assumed continuous property $\vec{r}(t)$. It is a purely mathematical operation which does not reflect a physical reality. Therefore, quantum mechanics rejects the idea of a continuous trajectory as a basic element for the description of the phenomena.

At the Solvay Conference of 1927, where these ideas were debated for the first time, H.A. Lorentz reacted strongly against this new point of view ([8], p.248):

> "Pour moi, un électron est un corpuscule qui, à un instant donné, se trouve en un point déterminé de l'espace, et si j'ai eu l'idée que à un moment suivant ce corpuscule se trouve ailleurs, je dois songer à sa trajectoire, qui est une ligne d'espace. Et si cet électron rencontre un atome et y pénètre, et qu'après plusieurs aventures il quitte cet atome, je me forge une théorie dans laquelle cet électron conserve son individualité; c'est-à-dire que j'imagine une ligne suivant laquelle cet électron passe à travers cet atome."

This is a very clear statement which renders perfectly the normal reaction of each of us. However, it should be remarked that the strongest words – "I have to think of a trajectory," "I have to build a theory," "I imagine a line" – describe human acts which add something to the experimental fact. Again, the fact that it is quite legitimate to proceed that way at the macroscopic level is not a sufficient condition to assess its validity at the microscopical scale. One has to look for some experimental evidence and up to now no such evidence exists. On the contrary, I shall discuss in Sec.6 an experimental situation which can be considered as evidence against the idea that such a continuous position $\vec{r}(t)$ exists physically.

But then, what about the common-sense statement of a continuous position $\vec{r}(t)$ between observations? For quantum mechanics, this is to be replaced by a probabilistic statement: To each volume of space and at any instant of time, one can assign a definite probability to find the particle in that volume at this instant of time. This probability depends on the last previous observation and on the dynamics. The global certitude of the permanent existence of the particle is asserted by the normalization to one of the integral on the full space of the probability distribution. But no function $\vec{r}(t)$ representing the true presence of the particle at point $\vec{r}$ on time t exists in the basic formalism. At the very most, one can consider a position $\vec{r}$ as a stochastic variable with no particular value, except when a detection occurs.

4. ISOLATED SYSTEMS

The second main principle (*principle B*) can be called "principle of isolation of a system." It amounts to suppose that one can divide the world into two parts: (i) a rather minute part, called "the system," which will be described with as many details as the theory demands; (ii) a second part, called "the rest of the world," which is not described except for its influence on "the system" and which is either neglected or described by a very simple parametrisation.

This principle B appears to be an absolute necessity for the scientific activity in so far as it seems simply impossible to describe all the world with the details required by a scientifically meaningful theory. One can also say that it puts severe restrictions on the field of applications of definite theories. It is so commonly applied in physics that one generally forgets to mention it. However, strictly speaking, this principle is neither true nor false, this appreciation being only a matter of precision. Consider, for instance, the problem of a man who jumps from the top of a building. In a very simple approximation we can consider the man alone, and his centre of gravity will fall like a particle in the field of gravity. A better approximation will certainly be to take some account of the "rest of the world," represented by the air and its parametrized influence on the particle or rather on the solid body "man." Now, if the man gesticulates, or if he opens a parachute, or if he carries some mechanical device to move in the air, the problem of the parametrization of the "rest of the world" changes completely. The system is not any more the man alone; it is the man and the surrounding air, and it requires therefore a completely different description. One can consider as many examples as one wishes (think for example of the earth-sun problem, with as many planets as you like included in the rest of the world, with the stones and dust and radiations which permanently change the mass of both bodies, etc.). One can easily convince oneself that principle B represents a compromise which depends on the precision one wishes to reach and our capability of treating the problem. This compromise is subject to the following general rules: On the one hand, the isolation of a system is an absolute necessity and, on the other hand, the better one describes the system and the parametrization of the influence of the rest of the world, the higher the complexity of the problem to be solved.

Quantum mechanics also has to consider principle B; and quantum mechanics admits the very possibility to define a "system" and to describe this isolated system with a dynamics which I shall consider in the next paragraph. But it runs immediately into a big difficulty at the level of the measurement, i.e., when one leaves the probabilistic description to meet the objective realistic properties. A measurement requires the effective presence of a macroscopic apparatus which interacts with the system, and therefore the system can no longer be considered as isolated. Therefore, the description of the measurement process is necessarily different from the description of the original isolated system. There are two possible alternative solutions: Either one keeps the general rules governing the evolution of isolated systems, considering however that the system is now composed of the original system and the apparatus; or one defines new rules in order to describe the original system in this specific situation of non-isolation. The first alternative implies that one tries to describe in a detailed way, with the usual rules of quantum mechanics, a very complex system containing a very large number of particles. This is a tremendous problem, of the same order of difficulty as the detailed classical description of the motion of the molecules in one mole of gas. Such a problem cannot possibly be handled without drastic simplifications. In spite of many efforts, no real satisfactory result has been obtained up to now. The second alternative is perhaps not too pleasant because it is perceived by many physicists as a renouncement to the idea of the unity of physics. They consider it at best as a "temporary" empirism which will be improved in the future. My point of view is rather different. We have to distinguish between two radically different situations. Before and (occasionally) after the measurement, we have to give a description of an isolated system. While during the measurement we have to give a description of what happens to a system that has been and will (possibly) be isolated but which is not presently in that situation. Therefore, a change of the rules of description is not a renouncement but rather an extension of the theory. This extension must in any case fulfill two conditions: Firstly, it must preserve the general framework of the description of the original isolated system in such a way that the measurement episode remains compatible with a "before" and an "after" described by the rules of this isolated system; secondly, it must lead to results which are empirically correct. It is well-known that quantum mechanics succeeds to fulfill the first condition within the mathematical framework of the Hilbert-space description of an isolated system. Quantum mechanics represents the normal evolution of the isolated system by a unitary transformation, and it represents the episode of the measurement by an orthogonal projection. Notice that this second type of evolution is properly "timeless" and not "instantaneous" (as is often said): Time is abolished for the old isolated system (not for us!) because the measurement is a link that connects two situations where time is meaningful (time is the parameter of a unitary transformation in the Hilbert-space associated with the isolated system), while "during the measurement" (in our sense), time cannot possibly be assigned to some (not too well-defined) part of the global system. It is also well-known that this solution proposed by quantum mechanics proves to be empirically correct. Therefore, I think, that for the time being, we can accept this solution as a true representation of the world, at least in the sense enounced by H. von Helmholtz:

"Our representations of the external world are true if they give us sufficient indications on the consequences of our acts with respect to this external world and if they allow us to derive exact conclusions about the modifications which we can expect from these acts." (Quoted from [10]).

In classical mechanics, the distance between material objects is an essential element of the isolation. If all parts of a system are far enough from all the other objects of the rest of the world, the system becomes isolated. If some part of an isolated system is far enough from the rest of the system, it becomes an isolated system on its own. This particular realization of isolated system was checked with a high degree of precision on the macroscopic objects, and as such it became an essential tool for our understanding of the world. What about this problem in quantum mechanics where the concept of position has a completely different status as in classical mechanics? Can we keep this idea that parts of a system that are "far from each other" are to be considered as isolated systems? It depends of course of what we are calling distance. If it is the physically well defined distance after that these parts have been submitted to a detection, then it is nothing else than the distance between the corresponding macroscopic apparatus. And there is no doubt that, after that detection occurs, the parts of the system are isolated. Otherwise, our understanding of the macroscopic world itself would get into trouble. If it is the distance computed from the stochastic variables "position" associated to "parts" of the system, then there is no objective reason why these parts would be isolated. In particular, strong correlations can exist between other stochastic variables associated with these parts. And after a detection which will state that the parts are now far from each other and isolated, the statistical ensembles found for the latter stochastic variables can be correlated in a rather surprising way, simply because they reflect the correlation of probabilities existing before the detection. This point was nicely illustrated in the EPR experiments, where the polarisations of particles detected far from each other show correlations which are not understandable on basis of the idea that these particles were isolated systems before detection occurs.

5. THE STATE OF AN ISOLATED SYSTEM. THE SECONDARY PRINCIPLES

Classical mechanics has given to science the concept of "state of an isolated system at one definite instant of time." This concept proved to be one of the most useful concepts for the development of science, although the justification of its practical use is by far not evident and sometimes controversial. By definition, the state is a limited set of data concerning the system considered at one instant of time, which is necessary and sufficient in order to recover all the physical properties of the system at this same instant of time. The concept of state proceeds as evidence from a simplification of the reality. Once again, physics is rather simple and works very well because the simplified description is sufficient in order to grasp a good idea of what the system really is. But it is by no means evident that such a simplified view can be extended to other domains.

Since the time of Galileo, one admits that the "mechanical state" is the set of positions and velocities of the constituents of the system (*principle C*). For a system of N particles, the state at time t is the set $\{\vec{r}_\alpha(t), \vec{v}_\alpha(t)\ ;\ \alpha = 1, 2, \ldots N\}$. If the system contains also solid bodies, the data $\vec{r}_\alpha(t)$ and $\vec{v}_\alpha(t)$ concern one point of each solid body, and one has to add to the state three coordinates of orientation (Euler angles) and three coordinates of instantaneous rotation of the solid body around this point. The simplification is already evident: Physical quantities like the calorific contents or the colour of the rigid body do not proceed from this concept of state. Furthermore, the usefulness of this concept of mechanical state is limited to systems with a rather small number of particles or rigid bodies, for otherwise the state becomes inaccessible to experiment or cannot be used in practical calculations. For a tiny (macroscopic) fraction of a gas, this concept of state becomes totally useless; it is then replaced by the "thermodynamical state" or by the "statistical state" with even stronger limitations of our detailed knowledge of the system. For more complex systems (like living systems, societies, etc.) the concept of state ceases practically to exist.

Let us now see why this concept of state is so important in classical mechanics. According to principle A, the acceleration of a particle is an objective mechanical property of this particle. According to principle C, it can be expressed as a function of the state. This means, for a system of N particles

$$\forall \alpha = 1, 2, \ldots, N : \quad \vec{a_\alpha}(t) = \vec{\phi_\alpha}\left(\left\{\vec{r_\beta}(t), \vec{v_\beta}(t)\ ;\ \beta = 1, 2, \ldots, N\right\}, t\right),$$

where $\vec{\phi_\alpha}$ represents the acceleration function of particle α, i.e., the explicit dependence at time t of this property "acceleration" on the state. If these acceleration functions are known, one gets in that way a system of $6N$ first-order differential equations for the state which, together with the knowledge of the state at some initial time t_0, will, in principle, define the state at all other times:

$$\forall \alpha : \quad \frac{d\vec{r_\alpha}}{dt} = \vec{v_\alpha} \quad , \quad \frac{d\vec{v_\alpha}}{dt} = \vec{\phi_\alpha},$$

$$t = t_0 \ :\ \left\{\vec{r}_\beta, \vec{v}_\beta\ ;\ \beta = 1, 2, \ldots, N\right\}_0 .$$

This is the most explicit formulation of the principle of causality, the one which has guided the deterministic dream of physicists since the time of Newton. Of course, we know today that this happens to be only a dream, because the integration of the differential system on significant time intervals $t - t_0$ can be very difficult or even impossible, and because a high sensitivity of the solution to the initial conditions can even lead to chaos. Nevertheless, simple integrable systems also exist and they did play an important role in the scientific development (ordinary mechanics, astronomy, etc.). For these systems, the concept of mechanical state is essential. However, it must be stressed that it introduces from the very beginning of the dynamics a major difficulty with respect to locality. If the acceleration of a particle α at some time t depends in one way or another on the position and the velocity of distant particles at the same time, then the theory implies action at a distance. Strictly speaking, one cannot avoid the difficulty. One can only temper it by considering that the acceleration function of a particle α is not too sensitive to the characteristics of the more distant particles, i.e., by using only models of the acceleration functions in agreement with the idea that an increasing distance gives isolation.

Indeed, the explicit knowledge of the acceleration functions $\overrightarrow{\phi_\alpha}$ proceeds from a modeling of the system under consideration, and it is a problem by itself. Since Newton, one admits general rules for this modelisation, and these general rules are what I call the secondary principles of classical mechanics (principles 1-4).

Principle 1. The rule of the vectorial addition of the acceleration functions corresponding to one and the same particle α under various "circumstances" whenever these circumstances happen together (for the same state).[5]

Principle 2. The rule of decomposition of the acceleration functions $\overrightarrow{\phi_\alpha}$ into two parts: An internal part where the interaction between the particles of the system is represented as a sum of two-body interactions[6]; and an external part where one describes the influence of the rest of the world on each particle as if it were alone:

$$\overrightarrow{\phi_\alpha}\left(\left\{\overrightarrow{r_\beta}, \overrightarrow{v_\beta}\right\};t\right) = \sum_{\beta\neq\alpha} \overrightarrow{\phi}_{\beta/\alpha}\left(\overrightarrow{r_\alpha}, \overrightarrow{v_\alpha}; \overrightarrow{r_\beta}, \overrightarrow{v_\beta}\right) + \overrightarrow{\phi}_{\text{ext.}/\alpha}\left(\overrightarrow{r_\alpha}, \overrightarrow{v_\alpha}; t\right).$$

Principle 3. The action-reaction principle of Newton. The acceleration that α communicates to β (i.e. $\overrightarrow{\phi}_{\alpha/\beta}$) and the acceleration that β communicates to α (i.e., $\overrightarrow{\phi}_{\beta/\alpha}$) fall along the same line ($\overrightarrow{r_\alpha} - \overrightarrow{r_\beta}$), in opposite senses, and their strengths are in any circumstance in a constant ratio that depends only on α and β. This ratio was empirically found to be equal to the ratio of the weights (local weights) of the two particles. It can be used to define properly the mass of the particles after some convenient choice of units. One then writes successively:

$$\frac{|\phi_{\beta/\alpha}|}{|\phi_{\alpha/\beta}|} = c^{t} = \frac{\text{weight } \beta}{\text{weight } \alpha} =: \frac{m_\beta}{m_\alpha}.$$

Once the mass has been introduced[7], one recovers the Newtonian concept of force by multiplying the acceleration function by the mass:

$$\overrightarrow{F_\alpha} = m_\alpha \overrightarrow{\phi_\alpha}.$$

Principle 4. This principle assumes the existence of at least one frame of reference (i.e., a coordinate frame and a clock) where the previous principles are valid. This is the inertial frame, i.e., the aether and the absolute time rate of Newton. Furthermore, it assumes that the acceleration functions (the forces) are the same in all frames in uniform translation with respect to the aether and using the same time rate, so that all these frames of reference become equivalent for the mechanical description (Galilean relativity).

These principles have given such wonderful results in many applications of mechanics (ranging from astronomy to engineering) that they were considered as absolutely true during two centuries. However, it was found that their use in electrodynamics was not as successful. The action of a charged particle on another does also depend on its acceleration, the third principle seems to become invalid, one has to introduce a dynamical evolution of the electromagnetic field, etc. In other words, the secondary principles of classical mechanics, which are largely built on our immediate perception and intuition, do not survive when one tries to use them away from the circumstances of their creation.

Let us now come back to the concept of state of an isolated system and let us see how the principle of existence of such a state is represented in quantum mechanics. Of course, the quantum state must be essentially different from the mechanical state since no empirical kinematics exists which legitimizes such a state. Furthermore, the finality of the quantum state is to define probabilities and not objective properties. The concept of quantum state evolved rapidly during the years 1924-1927. The original works of L. de Broglie and E. Schrödinger led to the idea that a particle is essentially represented by a wave in ordinary space $\psi_t(\overrightarrow{r})$. One remembers how this clear and objective representation (the wave as a vibration of the aether) had to be replaced for systems of particles by a purely mathematical wave evolving in configuration space, $\psi_t(\overrightarrow{r_1}, \overrightarrow{r_2}, ...\overrightarrow{r_N})$. In both cases, the probabilistic interpretation requires that ψ be at any time a square integrable function of its spatial variables, i.e., that ψ be an element of the functional space $\mathcal{L}_2$ ($\mathcal{L}_2(\mathbb{R}^3)$ or $\mathcal{L}_2(\mathbb{R}^{3N})$) which is a particular representation of the abstract Hilbert space $\mathcal{H}$. The next step in the evolution of the concept of quantum state is based on the mathematical equivalence of all representations of Hilbert space. It reduces the concept of quantum state to a pure mathematical object: The quantum state ψ_t at some instant of time t is some "ray"[8] in some representation of Hilbert space. Which representation of Hilbert space? This is only a question of convenience! Because of the importance of ordinary space in the classical formulation of the problems we consider, the chosen representation is often $\mathcal{L}_2(\mathbf{R}^3)$ or other representations derived from this one, where one implements other properties of the particles (for example, $\prod_{i=1}^{N} \otimes \left(\mathcal{L}_2(\mathbf{R}^3) \otimes \mathbf{C}^2\right)_i$ for a system of N spin-1/2 particles). Remember however that the coordinate $\vec{r}_i$ present in the explicit representation of the state is by no means the actual position of the ith particle of the classical case.

An important property of the quantum state is the superposition principle. Although it can be seen as a legacy of the original Schrödinger linear evolution equation, it is certainly much more profound, as was emphasized by Dirac [15]. It must be considered as an essential secondary principle peculiar to quantum mechanics. This principle asserts that any linear combination of vectors corresponding to physical states[9] defines a new ray which again represents a physical state, and that such a linear combination keeps its meaning through the evolution of the system under consideration, whether it is isolated or subjected to a measurement.

The dynamical equation which determines the evolution of the quantum state follows then from the causality principle: for an isolated system, the state must evolve linearly and continuously from the initial state. In Hilbert space, this corresponds to a continuous unitary transformation. Such a transformation can always be described by a differential equation

$$i \frac{\partial \psi_t}{\partial t} = A \, \psi_t,$$

where A is some self-adjoint operator. Many good reasons, ranging from first principles to empirical results, lead physicists to the identification of $(\hbar \cdot A)$ with the "Hamiltonian operator" of the isolated quantum system. One obtains in that way the Schrödinger equation

$$i \, \hbar \frac{\partial \psi_t}{\partial t} = H \, \psi_t.$$

Among the many good reasons, I quote:

- the identification of this equation with the Schrœdinger-de Broglie wave equation for one or several particles (H is here the classical Hamiltonian transformed into a self-adjoint operator by the formal substitution of the momenta: $\vec{p_\alpha} \rightarrow \frac{\hbar}{i}\vec{\nabla_\alpha}$);
- the theorem of Ehrenfest, which shows that this rule makes that the quantum expectation values evolve as classical quantities;
- the principle of Noether that the Hamiltonian is the generator of time translations;
- the Bohr correspondence principle that the limit $\hbar \rightarrow 0$ means a limit to classical mechanics;
- and above all, the innumerable successes met in applications.

When writing this equation, the reference to a classical Hamiltonian is not a necessity, and one can easily construct models with no classical analogue. But it is true that a classical image helps to formulate the models of quantum mechanics. And because these classical images are nearly always formulated in space, one gives a special attention to the Hilbert space $\mathcal{L}^2(\mathbf{R}^3)$, which is therefore strongly privileged. Unfortunately, this very natural reason to introduce a mathematical coordinate $\vec{r}$ for each particle, maintains also the confusion that this coordinate could be related to the actual position of the particle. Let us finally remark that whenever we use a classical Hamiltonian, which is derived from the Newtonian scheme with its action at a distance, we implicitly also carry this difficulty in the quantum scheme.

I conclude this paragraph with a brief remark on the practical determination of an initial quantum state. It is generally defined through a "preparation experiment" where one lets the system interact with a convenient apparatus. According to the projection rule discussed previously, one acquires in that way the certainty that the system gets this or that property at the time of preparation: E.g., it is localized in this volume, or its momentum has that value (with some error), or its angular momentum is such, etc. These values are selected by the apparatus (with an error defined by its aperture) among the set of possible values defined by the spectrum of the self-adjoint operator which represents the measured physical quantity in the Hilbert space of the isolated system. For discrete spectra, one can in that way select one definite vector. For continuous spectra, one also selects some vector in each preparation experiment; but, because its characteristics are not completely known, repeated experiments do not give the same state. For the computation of statistical ensembles, one has then to use a somewhat more complex state, called "density matrix," which reflects this undeterminacy. The analogy with the classical case is complete.

6. NON-LOCALITY

Let us now come back to the problem of the localization of the microscopic particles. I described in Sec.2 the contrast between the two points of view: the classical one which emphasizes that common sense requires the existence of a permanent position of matter in space, even if this position is unknown to us; and the orthodox quantal one which assumes that such a position only exists when one proceeds to a detection, i.e., when one gets the particle to interact with a macroscopic apparatus which by itself defines the position of the particle. This problem has been central in the work of Louis de Broglie. His opinion has largely varied with the time, and this evolution as well as his deep motivations are nicely presented in [7] and [9]. De Broglie strongly insisted on the fact that wave packets are always limited and that one observes or measures the microscopic reality only through the intermediary observation of macroscopic phenomena caused by the local action of the particle. He also insisted on the point that only the assertion of a permanent localization can give us a clear idea of a particle. Both statements are certainly not questionable. But they do not proceed from the same level of objectivity. The first one is the objective formulation of a physical fact. The second one is rather a clear formulation of the limitation of the human capability of imagining things. In my opinion, only a precise formulation of what we call an objective localization, in conformity with the description that L. de Broglie gives of such an observation, can settle the answer to the question of a permanent localization of particles.

In classical physics, one considers that an entity is located in some volume Ω if it is possible to detect there the **totality** of one or another of the physical characteristics of this entity, e.g., its mass, or its electric charge, or its intrinsic magnetic moment, or its full energy, etc. According to this definition, not only a particle but also a system of particles and solid bodies, or even a classical wave packet, are localized entities. Whenever such an entity is located in a volume Ω, then it is not only impossible to find it elsewhere at the same instant of time, but it is even impossible to act on the future of this entity by means of local actions exerted at this instant in other volumes Ω', Ω'', ..., totally disjoint of Ω. This is the principle that no action at a distance really exists. It is important to make clear that the concept of a local action refers to any action which according to the laws of classical physics is for sure only efficient in the volume under consideration. As we have seen, action at a distance can indeed be present in classical mechanics because the forces depend on the state as a whole. However, forces with a limited range of action (local forces) give rise to local theories in the sense here above. Therefore, an experimental observation which is not in agreement with the equivalence here above should be interpreted as an evidence that the entity under observation is non-local.

This is the basic idea which led D. Aerts and myself to propose the following characterization of non-locality in the framework of non-relativistic physics [12]:

> "An entity is non-local if it can be prepared in a state which introduces several macroscopically separated regions of space where the following operations can alternatively be performed at some instant of time:
>
> – Either one performs a detection experiment; in that case, the whole entity will be found in one and only one of these regions (this one being possibly unknown before the detection occurs).
>
> – Or one tries to influence the entity by means of a local apparatus whose range of action, in its classical description, is limited to one region only ; in that case, the entity can be influenced at the same time from at least two of these regions."

In the same paper [12], we discuss a possible experiment of spin rotation which shows that spin-1/2 particles are to be considered as non-local entities[10]. The reasoning goes along the following lines: A single particle, represented by a small wave packet $\psi_A(x)\otimes\binom{\alpha}{\beta}$ enters a device which separates the wave packet in a coherent way into two small and macroscopicaly separated wave packets, each of them in a definite spin state[11]:

$$\psi = \alpha\, \psi_B(x) \otimes \begin{pmatrix} 1 \\ 0 \end{pmatrix} + \beta\, \psi_C(x) \otimes \begin{pmatrix} 0 \\ 1 \end{pmatrix}.$$

Let us clarify the notations: $A, B, C, \ldots$ are small, macroscopic, and macroscopically separated regions where one can install apparatuses whose range of action does not go beyond the region under consideration; $\psi_A(x)$ is a normalized very small wave packet, evolving in the region A with a certain mean velocity in such a way that crossing this region requires some finite time interval. It is clear that if one then proceeds to a detection experiment with ad hoc local apparatuses installed in B and C, the particle will be localized in B or in C, the two possibilities occurring with relative probabilities $|\alpha|^2 : |\beta|^2$. On the other hand, if one creates in B a local homogeneous and constant magnetic field $\overrightarrow{H_B}$, arbitrarily oriented but whose intensity is determined in such a way that the Larmor precession would cause a 2π rotation of the spin of the particle during the time required for the crossing of the B region by the wave packet ψ_B, then the corresponding spinor undergoes a change of sign in that time. Playing with this possibility in B and C, one creates later on the following alternative states:

$$\text{No field}: \quad \psi_0 = \alpha\psi_{B'}(x) \otimes \begin{pmatrix}1\\0\end{pmatrix} + \beta\psi_{C'}(x) \otimes \begin{pmatrix}0\\1\end{pmatrix},$$

$$\overrightarrow{H_B}\ \text{alone}: \quad \psi_1 = -\alpha\psi_{B'}(x) \otimes \begin{pmatrix}1\\0\end{pmatrix} + \beta\psi_{C'}(x) \otimes \begin{pmatrix}0\\1\end{pmatrix},$$

$$\overrightarrow{H_C}\ \text{alone}: \quad \psi_2 = \alpha\psi_{B'}(x) \otimes \begin{pmatrix}1\\0\end{pmatrix} - \beta\psi_{C'}(x) \otimes \begin{pmatrix}0\\1\end{pmatrix} = -\psi_1,$$

$$\overrightarrow{H_B}\ \text{and}\ \overrightarrow{H_C}: \quad \psi_3 = -\alpha\psi_{B'}(x) \otimes \begin{pmatrix}1\\0\end{pmatrix} - \beta\psi_{C'}(x) \otimes \begin{pmatrix}0\\1\end{pmatrix} = -\psi_0,$$

where B' and C' are neighbouring regions which follow B and C. Of course, ψ_0 and ψ_3 represent the same state, and alternatively ψ_1 and ψ_2 represent the same state; but these two states are different. A subsequent coherent reconstruction of a unique wave packet by an inverse device would give, alternatively,

$$\psi_0 = \psi_E(x) \otimes \begin{pmatrix}\alpha\\\beta\end{pmatrix} \quad \text{or} \quad \psi_1 = \psi_E(x) \otimes \begin{pmatrix}-\alpha\\\beta\end{pmatrix},$$

with a clearly different physical content. Now, ψ_1 is created by acting indifferently in B or in C; and ψ_0 is created by not acting, or by acting simultaneously in B and C, the action in B being counterbalanced by the action in C. Together with the possibility of a detection experiment which would indicate that the particle is fully and exclusively in B or in C, this leads to the conclusion that according to our definition the entity "spin-1/2 particle" is a non-local entity. This *gedanken experiment* is very close to the real experiments which were performed some years ago by Rauch et al. [13]. On the other hand, all the operations considered comply with the norms enounced by L. de Broglie concerning experiments with wave packets of finite extension. But the results contradict the idea of a permanent localization of the quantum entity called "spin-1/2 particle."

Once again, common sense will reject this conclusion on basis of the fact that macroscopic objects which are nothing else than aggregates of such microphysical particles are indeed permanently localized. Alternatively, I think that one should accept the conclusion of a non localization of the microscopic particles and understand the macroscopic localization in another way. The macroscopic localization would then result from the breaking of the natural and fundamental quantum correlations because of the higher complexity of such aggregates. From this point of view, we are grossly mistaken by our senses which lead us to consider the localization of objects as a simple and evident property of matter, while contrariwise it corresponds to some complex quantum chaos. This creates a psychological conditioning of the human brain which makes it difficult to us to apprehend serenely the microscopic reality.

ACKNOWLEDGEMENT

I would like to thank Dr. D. Aerts for the many discussions we had on this and many other subjects during the last years.

REFERENCES

[1] Hertz, H., *The principles of mechanics presented in a new form* (Dover, New York,1956); first German edition, 1894.

[2] Poincaré, H., "Les idées de Hertz sur la mécanique," *Revue Générale des Sciences* **8**, 734-743 (1897); in *Oeuvres de H. Poincaré* **VII**, 231-250 (Paris, 1952).

[3] Poincaré, H., *La science et l'hypothèse* (Flammarion, Paris, 1968); first edition: Flammarion, 1902.

[4] Kant, E., *Critique de la raison pure* (Quadrige/Presses Universitaires de France, 1990); first German edition, 1781.

[5] Poincaré, H., *La valeur de la science* (Du Cheval Ailé, Genève, 1946); first edition: Flammarion, 1905.

[6] Wheeler, J.A., "Information, physics, quantum: The search for links," in *Proceedings Third International Symposium on Foundations of Quantum Mechanics* (Japanese Physical Society, Tokyo, 1989, pp.354-358; Frenkel, A., "Spontaneous localizations of the wave function and classical behaviour," *Found. Phys.* **20**, 159-188 (1990).

[7] de Broglie, L., *Les incertitudes d'Heisenberg et l'interprétation probabiliste de la mécanique ondulatoire* (Gauthier-Villars, Paris,1982); English translation, van der Merwe, A., *Heisenberg's uncertainties and the probabilistic interpretation of wave mechanics* (Kluwer Academic, Dordrecht, 1990).

[8] *Electrons et photons* (Rapports et discussions du 5ième Conseil de Physique tenu à Bruxelles du 24 au 19 octobre 1927, Institut International de Physique Solvay)(Gauthier-Villars, Paris, 1928).

[9] de Broglie, L., *La théorie de la mesure en mécanique ondulatoire* (Gauthier-Villars, Paris, 1957).

[10] Cornu, L., *La mécanique. Les idées et les faits* (Flammarion, Paris, 1918), Chap.1, Considérations générales.

[11] Mach, E., *La mécanique. Exposé historique et critique de son développement* (Jacques Gabay, Paris, 1987); first German edition, 1883.
[12] Aerts, D., and Reignier, J., "On the problem of non-locality in quantum mechanics," *Helv. Phys. Acta* **64**, 527-547 (1991).
[13] Rauch, H., et al., "Verification of coherent spinor rotation of fermions," *Phys. Letters* **54A**, 425 (1975).
Rauch, H., "Neutron interferometric tests of quantum mechanics," *Helv. Phys. Acta* **61**, 589 (1988).
[14] Jammer, M., *Concepts of spaces. The history of theories of space in physics*, 2nd edition (Harvard University Press, Cambridge, Massachusetts, 1969).
[15] Dirac, P.A.M., *The principles of quantum mechanics* (Oxford University Press, Oxford, 1930), Chap.1, The principle of superposition.

NOTES

1. Alternatively, people could argue that the Lagrange-Hamilton scheme is more general because it allows to recover the Newtonian friction by taking appropriate limits for rather complex phenomena. This reductionist approach sounds like an interesting "program" but can hardly be carried out to produce a realistic value of the friction. Its application to real phenomena is by far less simple and successful than the phenomenological one.
2. This evolution was discussed in several books, e.g., [14].
3. Mass will be introduced later (see Sec.5, principle 3) and appears here as an anticipation. The reasoning is that an interaction that requires a time interval τ and the exchange of at least one quantum of action will cause an uncontrollable change of the kinetic energy of the particle of the order of $\hbar/\tau$.
4. I recall that QM is here considered in its non-relativistic approach. Relativistic quantum theory must also consider the problem of the creation and the annihilation of matter in the form of bosons or fermion-antifermion pairs. This can only be done in the framework of field theories, i.e., systems with infinite degrees of freedom.
5. It should be remarked that if this principle of vectorial addition of the acceleration is mathematically clear, its physical realization is not as clear. It represents a principle of independance and additivity of motions and becomes a principle of addition of forces after the introduction of the masses. But the classical rule of addition of forces can only be checked directly in statics (S. Stevin; cf.[11]).
6. These two-body acceleration functions are generally time independent and also often velocity independent.
7. One notices that in our approach, the famous equality of the so-called inertial and gravitational masses is not a mysterious and profound property of matter and gravitation but simply a definition: the inertial mass **is** the gravitational mass.
8. A ray can be represented by a vector normalized to 1 and defined up to an arbitrary phase factor (if α is a real number, ψ and $e^{i\alpha\psi}$ define the same ray and therefore the same state).

9. With the possible exception of states separated by a "superselection rule" which I disregard here.
10. Of course, the non-local character of quantum objects is not limited to the case of spin-1/2 particles. Other similar experiments can be discussed which do not explicitly involve the spin. The example of spin rotation was only chosen because of its simplicity and clarity.
11. Such a device is traditionally called a Stern-Gerlach apparatus. Today, laser techniques seem very promising for a much better realization of this situation.

ASCRIBED TO AND DESCRIBED BY: WHICH IS MORE IMPORTANT?[1]

Alberto Rimini

Università di Pavia
Dipartimento di Fisica Nucleare e Teorica
Via Bassi 6, 27100 Pavia, Italy

The theory of quantum measurement in the framework of the de Broglie-Bohm formulation of quantum mechanics is presented and discussed. It is shown that the de Broglie-Bohm theory allows a fully satisfactory description of quantum measurement. The importance of this feature of the theory is stressed.

Key words: quantum mechanics, plot-wave theory, measurement problem.

1. INTRODUCTION

In the standard formulation of quantum mechanics it is admitted that we can identify situations, which we call measurements, in which the general principle of evolution (the Schrödinger equation) for the state of the system (the wave function) is suspended and replaced by special rules. These rules are such that measurements have many possible outcomes and that in general, known the state of the system, one cannot foresee the actual outcome; one can only calculate a probability distribution for the various possible outcomes. Furthermore, the state of the system after the measurement is determined essentially (according to the reduction principle) by the actual outcome. The Copenhagen interpretation, which is the conceptual basis of the standard formulation, accepts plainly the aforementioned dualism of situations, so that the related dualism of evolution principles can also be accepted. Note that the rules for the measurement situation are such that there is no strict correspondence between the state of the system before the measurement and the outcome (the theory is not deterministic), while a strict correspondence is there between the outcome and the state after (a trace of the outcome is kept by the system).

The theory of quantum measurement tries to get rid of the dualism assuming the Schrödinger equation as the sole principle of evolution. The contextual inclusion of the measuring apparatus into the physical system should allow to reconcile the description with the standard postulates. As it is well known and as we shall review below, this program meets hard difficulties.

The de Broglie-Bohm pilot-wave theory is a version of quantum mechanics in which the state of the system consists in the wave function plus an additional variable, the configuration. For the state in the new sense an evolution principle is assumed which coincides with the Schrödinger equation for the wave-function part of the state. The specific value of the configuration in an individual measurement determines the outcome of the measurement. Furthermore, the configuration after the measurement is also in correspondence with the outcome, so that a trace of the outcome is kept by the system independently of the assumption of the reduction principle for the wave function. In other words, the outcome of the measurement is *ascribed to* and *described by* the configuration. "Ascribed to" transforms quantum mechanics into a theory deterministic in principle; "described by" allows one to set up a fully consistent and satisfactory theory of quantum measurement. Often the accent is put mainly on the first result; I want to draw the attention to the second achievement and its importance.

2. THEORY OF MEASUREMENT IN THE STANDARD FORMULATION

Returning to the standard formulation, let us include the measuring apparatus **A** together with the measured system **S** into the description. Different outcomes of the measurement correspond to different final pointer positions and consequently to different, macroscopically distinguishable states of **A** and **S + A**. This correspondence is enforced by the reduction principle. Let Ψ^{i} and Ψ^{f} be the states of **S + A** before and after the measurement, respectively. Making reference to the case of an "ideal measurement," to provide a concrete exemplification, the evolution from Ψ^{i} to Ψ^{f} is of the form

$$\Psi^{\mathrm{i}} = \sum\nolimits_m c_m \varphi_m \alpha^{\mathrm{i}} \quad \longrightarrow \quad \Psi^{\mathrm{f}} = \begin{cases} \varphi_1 \alpha_1^{\mathrm{f}}, & \text{with probability } |c_1|^2, \\ \varphi_2 \alpha_2^{\mathrm{f}}, & \text{with probability } |c_2|^2, \\ \dots\dots\dots\dots\dots\dots\dots, \end{cases} \tag{1}$$

with obvious meaning of the symbols. The reference to the ideal measurement is in no way essential for the arguments that follow. What really matters is that a measurement has several possible outcomes, that a definite outcome is there when the measurement is complete and that each outcome corresponds to a definite state of $\mathbf{S}+\mathbf{A}$.

Then one remembers that our program is to apply the Schrödinger equation to the system $\mathbf{S}+\mathbf{A}$. Let us consider the evolution

$$\Psi^{\mathrm{i}}=\sum_{m} c_m \varphi_m \alpha^{\mathrm{i}} \xrightarrow{\text{S. E.}} \Psi^{\mathrm{f}} \tag{2}$$

and ask whether this Ψ^{f} can be the state corresponding to one among the possible outcomes. The answer is obviously NO, because, if it were so, one would be able to calculate the outcome from the knowledge of Ψ^{i} and something would be wrong in the standard formulation. Ψ^{f} appearing in Eq. 2 is necessarily a state in which, in some sense, all outcomes coexist. Let us call it the *inconceivable state*. This conclusion is quite independent of any idealization of the dynamics of the measurement process. If, to exemplify, we refer again to an ideal measurement, then

$$\Psi^{\mathrm{i}}=\sum_{m} c_m \varphi_m \alpha^{\mathrm{i}} \xrightarrow{\text{S. E.}} \Psi^{\mathrm{f}}=\sum_{m} c_m \varphi_m \alpha_m^{\mathrm{f}}. \tag{3}$$

Let me designate by common sense requirement (CSR) the following statement: There is a correspondence between the outcome of a measurement and the description (state) of the system $\mathbf{S}+\mathbf{A}$ after the measurement. According to the discussion above, the standard formulation, including the reduction principle, satisfies CSR. On the other hand, if we pretend that the Schrödinger equation be the sole principle of evolution, we get the inconceivable state, i. e., a description of $\mathbf{S}+\mathbf{A}$ after the measurement in which there is *no counterpart* of the outcome, so that CSR is not satisfied. There are two ways[2] to recover CSR maintaining a single principle of evolution:

(1) leaving the evolution of Ψ, as given by Eq. (2), unaltered and adding to Ψ a new variable evolving in such a way that its value after the measurement is in correspondence with the outcome;
(2) changing the evolution of Ψ, i. e., assuming a new single principle of evolution essentially equivalent to the Schrödinger equation in ordinary situations and incorporating reduction in the measurement situations.

Line 1 is followed by the de Broglie-Bohm theory, line 2 by reduction theories based on spontaneous localization. In both cases one gets CSR satisfied. The difference is that in the de Broglie-Bohm theory there is also a strict correspondence between the state before the measurement and the future outcome, while no such a correspondence is there in reduction theories. Though the technical details may be complicated, the theory of quantum measurement (i. e., the description of the system $\mathbf{S}+\mathbf{A}$

through a single principle of evolution) is conceptually very simple in reduction theories: the state vector collapses stochastically in accordance with quantum-mechanical probabilities. The theory of quantum measurement in the pilot-wave theory, which is due to Bell [1,2] and Bohm and Hiley [3] is somewhat more involved. I will discuss it through a simple but significant example in the next section.

3. THEORY OF MEASUREMENT IN THE PILOT-WAVE FORMULATION

In the pilot–wave theory the state of an N–particle system is the pair $(\Psi, \boldsymbol{X})$, where $\boldsymbol{X} = x_1, \ldots, x_N$ is the point in configuration space and $\Psi(\boldsymbol{X})$ is a field in the same space.[3] If the configuration of the system or of a part of it is measured at a certain time, the system or its part are found as specified by the value of $\boldsymbol{X}$ at that time. The evolution of the pair is given by the equations

$$i\hbar \frac{d}{dt}\Psi = H\Psi, \tag{4}$$

$$\frac{d}{dt}\boldsymbol{X} = \boldsymbol{J}(\boldsymbol{X},t)/\rho(\boldsymbol{X},t), \tag{5}$$

where

$$\rho = |\Psi|^2, \tag{6}$$

$$\boldsymbol{J} = \hbar\,\mathrm{Im}\left(\Psi^* \frac{\boldsymbol{\nabla}}{M}\Psi\right), \qquad \frac{\boldsymbol{\nabla}}{M} := \left(\frac{\nabla_1}{m_1}, \ldots, \frac{\nabla_N}{m_N}\right). \tag{7}$$

Equation 5 is often called the guidance equation. Let us assume temporarily that, when an ensemble $\mathcal{E}$ of systems described by Ψ is prepared, the distribution σ of $\boldsymbol{X}$ be given by ρ. This is a consistent assumption, in the sense that the condition remains true while time is going on, because of the equation

$$\frac{\partial}{\partial t}\rho + \boldsymbol{\nabla}\cdot\boldsymbol{J} = 0, \tag{8}$$

which is a consequence of the Schrödinger equation, under the usual hypotheses on H. It may happen that the wave function splits into two parts, $\Psi = \Psi_1 + \Psi_2$, in such a way that the Schrödinger evolution keeps the two terms separated in configuration space during a certain time; then

$$\rho = \rho_1 + \rho_2, \tag{9a}$$

$$\boldsymbol{J} = \boldsymbol{J}_1 + \boldsymbol{J}_2, \tag{9b}$$

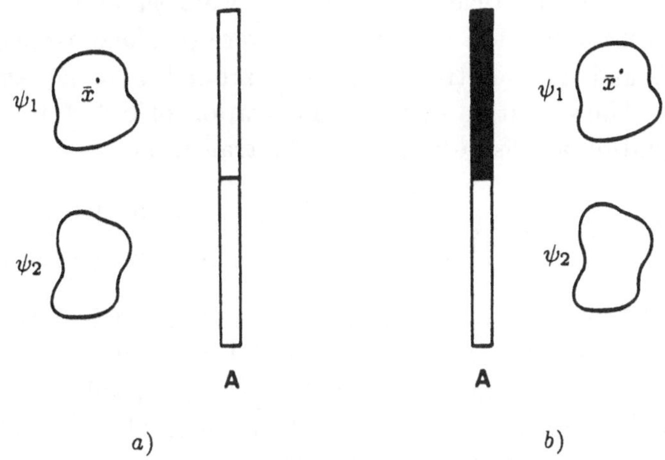

Fig. 1. Coarse position measurement in the pilot–wave theory: (*a*) before the measurement, (*b*) after the measurement.

where ρ_1, J_1 and ρ_2, J_2 are nonzero, respectively, in two domains in configuration space D_1 and D_2 disjoint during the considered time. According to Eq. 5, for the purpose of the evolution of a system whose X lays, say, in D_1 the term Ψ_2 can consistently be dropped.

Clearly, position has a special role in the pilot-wave theory. As a rule, any quantum measurement is, after all, a position measurement; if a variable does not commute with position, its measurement is achieved by a time-delayed position measurement. In the pilot-wave theory, it is so in principle. I shall discuss a simple example [4], in which a coarse measurement of the position of the particle **S** is performed by a pair **A** of detectors (see Fig. 1). The state variables of the system $\mathbf{S}+\mathbf{A}$ are the configuration x, Y and the wave function $\psi(x)\alpha(Y)$ or $\Psi(x,Y)$, where x is the position of the particle, Y is the configuration of the particles in the detectors, $\psi(x)$, $\alpha(Y)$ and $\Psi(x,Y)$ are the wave functions of the particle, of the detectors and, when necessary, of the entire system, respectively. A subset Z among the variables Y plays the role of pointer variable of **A**. The value, say, $Z = Z_1$ $(= Z_2)$ corresponds to the first (second) detector having revealed the particle. Let the particle wave function before the measurement be the superposition of two packets ψ_1 and ψ_2, each hitting one of the detectors. The evolution of the wave function of the system $\mathbf{S}+\mathbf{A}$ during the measurement is described by the transition

$$\Psi^{\mathrm{i}} = \big(\psi_1(x) + \psi_2(x)\big)\alpha^{\mathrm{i}}(Y) \quad \xrightarrow{\text{S. E.}} \quad \Psi^{\mathrm{f}} = \psi_1(x)\alpha_1^{\mathrm{f}}(Y) + \psi_2(x)\alpha_2^{\mathrm{f}}(Y), \tag{10}$$

where we have omitted to indicate the trivial time evolution bringing the particle through the detectors. The symbols α^{i}, α_1^{f}, and α_2^{f} denote the wave functions of the pair of detectors before the measurement, after the measurement if the particle wave function were ψ_1 alone, and after the measurement if the particle wave function were ψ_2 alone, respectively. The functions $\alpha_1^{\rm f}$ and $\alpha_2^{\rm f}$ are strongly peaked around $Z = Z_1$ and $Z = Z_2$, respectively. The evolution of the configuration of $\mathsf{S} + \mathsf{A}$ is ruled by the guidance equation and is described by the transition

$$\bar{x}, Y^{\rm i} \xrightarrow{\text{G. E.}} \begin{cases} \bar{x}, \bar{Z}^{\rm f} = Z_1, \ldots, & \text{if } \bar{x} \text{ lies within } \psi_1, \\ \bar{x}, \bar{Z}^{\rm f} = Z_2, \ldots, & \text{if } \bar{x} \text{ lies within } \psi_2, \end{cases} \tag{11}$$

where $\bar{x}$ is the position of the particle S for the particularly considered member of the ensemble $\mathcal{E}$, $Y^{\rm i}$ is the initial configuration of the pair of detectors, and $\bar{Z}^{\rm f}$ is the final value of the pointer variable of the detectors. After the measurement, each member of the ensemble $\mathcal{E}$ of $\mathsf{S} + \mathsf{A}$ systems is described by the wave function $\Psi^{\rm f}$ (the same for all members) and by the coordinates $\bar{x}, \bar{Z}^{\rm f}, \ldots$ (particular values for a particular member). The wave function $\Psi^{\rm f}$ is no more inconceivable because it is only a part of the complete state and CSR is satisfied.

One further problem, however, remains to be discussed. The condition $\sigma = |\Psi|^2$ is obviously satisfied by the ensemble $\mathcal{E}$. But, after the measurement, $\mathcal{E}$ can be split into two parts, $\mathcal{E}_1$ and $\mathcal{E}_2$, depending on whether the particle have been found in one detector or the other. Considering, e.g., $\mathcal{E}_1$, the systems contained in it are described, according to the above discussion, by the wave function $\Psi = \Psi_1 + \Psi_2 = \psi_1\alpha_1^{\rm f} + \psi_2\alpha_2^{\rm f}$, but the distribution $\sigma(x, Y)$ is that corresponding to Ψ_1 alone. Therefore, the ensemble $\mathcal{E}_1$, whose consideration is perfectly legitimate, violates the condition $\sigma = |\Psi|^2$. Something seems to be wrong, but it is not so. In fact the two terms in Ψ enjoy an important property: *effective incoherence*. Immediately after the measurement, Ψ_1 and Ψ_2 are obviously separated. Bohm and Hiley [3] correctly argue, using the irreversibility of the amplification which accompanies any quantum measurement, that this condition will remain true for all subsequent times with overwhelming probability, in spite of any possible attempt to rejoin the two terms. Then, as noted above, Ψ_2 can definitively be dropped from the wave function for the systems belonging to the subensemble $\mathcal{E}_1$, if one wishes. This circumstance does not cancel the fact that in the de Broglie-Bohm theory each individual physical system is described by a wave function which never suffered any reduction *and* and by the values of the coordinates of all particles in the system. Assumptions on the distribution σ of configurations in a given ensemble play no fundamental role and only serve the practical purpose of making statistical predictions about the ensemble. As is shown by the example of $\mathcal{E}_1$, the procedure used to prepare the ensemble indicates clearly the appropriate σ.

4. CONCLUSION

The de Broglie–Bohm theory is a version of quantum mechanics which is deterministic in principle and allows a fully consistent theory of quantum measurement. Since, in my opinion, determinism is a matter of taste while consistency is a prerequisite for a theory to be respectable, the latter feature is of prominent importance.

REFERENCES

1. J. S. Bell, CERN Preprint TH 1424 (1971); *Epist. Lett.*, July 1978, p. 1.
2. J. S. Bell, in *Quantum Gravity 2*, C. Isham, R. Penrose, and D. Sciama, eds. (Clarendon Press, Oxford, 1981), p. 611.
3. D. Bohm and B. J. Hiley, *Found. Phys.* **14**, 255 (1984).
4. O. Nicrosini and A. Rimini, in *Symposium on the Foundations of Modern Physics 1990*, P. Lahti and P. Mittelstaedt, eds. (World Scientific, Singapore, 1991), p. 280.

NOTES

1. Supported in part by Ministero dell'Università e della Ricerca Scientifica e Tecnologica and by Istituto Nazionale di Fisica Nucleare.
2. From a strictly logical point of view there is a third way, viz., to include the observer into the description. Then it is the content of the consciousness of the observer which discriminates among the various outcomes. This attitude gives rise to the relative-state or many-world or many-mind formulation. I forego considering this possibility because it upsets my personal mind.
3. Some authors, in order to avoid confusion between X as a state variable and X as the running argument of the other state variable Ψ, use different symbols in the two cases. This convention is not followed here.

QUANTIZATION OF GENERALIZED LAGRANGIANS: A NEW DERIVATION OF DIRAC'S EQUATION

Martín Rivas

Departamento de Física Teórica
Universidad del País Vasco, Euskal Herriko Unibertsitatea
Apdo. 644, 48080 Bilbao, Spain

Quantization of generalized Lagrangian systems suggests that elementary particle wave functions must be defined on the kinematical space. For spinning particles the center of mass and center of charge are different points. Spin-1/2 particles arise if the classical model rotates but no half-integer spins are related for systems with spin of orbital nature. Dirac's equation is obtained when quantizing the homogeneous space of the Poincaré group describing particles circling around their center of mass at the speed c.

Key words: Dirac's equation, classical spinning particles.

New Definition of Classical Particle

In previous works [1] we have found a Lagrangian formulation of classical spinning particles, where spin is produced by the *zitterbewegung* and rotational motion of the particle around its center of mass. The novelty with respect to other approaches is the definition of classical particles.

The usual canonical formulation defines a classical particle as a system whose phase space is a homogeneous space of Poincaré group. In our approach is the *kinematical space* of the system that has to be a homogeneous space of the kinematical group of space-time transformations.

Kinematical Space

For a generalized Lagrangian system $L(t, q_i, \ldots, q_i^{(k)})$, where the variables $q_i^{(k)} = d^k q_i/dt^k$, the variational problem is stated for the class of paths with fixed end points at initial time t_1 and final time t_2. The *kinematical space* of a system is the manifold spanned by time and initial (or final) boundary variables of variational problem. Thus kinematical space is spanned by the variables $x := (t, q_i, \ldots, q_i^{(k-1)})$.

Kinematical Groups

All kinematical groups considered, in particular the Galilei and Poincaré groups are ten-parameter groups. They are parameterized in an equivalent way in terms of the following parameters $(b, \boldsymbol{a}, \boldsymbol{v}, \boldsymbol{\alpha})$, where b and $\boldsymbol{a}$ with dimensions of time and length, represent the time and space translation respectively, $\boldsymbol{v}$ is the relative velocity among observers, and the non-dimensional magnitude $\boldsymbol{\alpha}$ represents their relative orientation, expressed in terms of a suitable parameterization of the rotation group.

Classical Spinning Particle

The kinematical space of a particle with highest structure is a ten-dimensional manifold whose variables share the same dimensionality as the above group parameterization. This manifold represents a spinning particle whose kinematical variables are identified with time t, position $\boldsymbol{r}$, velocity $\boldsymbol{u} = d\boldsymbol{r}/dt$ and orientation $\boldsymbol{\alpha}$. The Lagrangian depends also on the acceleration and angular velocity of the particle, and thus a generalized Lagrangian formalism arises. One of the salient features for a spinning particle is the existence of two position vectors. One defines the center of mass and the other is interpreted as the charge position.

Two Position Vectors

The kind of particles this formalism describes can be pictured as follows. Let us assume a uniform rigid conducting sphere of mass m and radius R. If charged, with total charge e uniformly distributed on its surface, the center of mass and center of charge are coincident at the center of sphere. Rotation can be described by the evolution of the corresponding orthogonal frame linked to the body. If while keeping the surface spherically symmetric the mass distribution inside the sphere manifests a tiny inhomogeneity, the center of mass becomes shifted from the center of sphere. The free motion of this model is a straight line trajectory for the center of mass at constant speed, the center of charge and thus the center of this apparent sphere spiraling around it and finally a possible rotation. It is important to realize that this kind of description is independent of the shape and size of the object and in the limit what we have are just two position vectors linked respectively to m and e and three orthogonal directions to describe orientation.

For relativistic particles, the center of mass $\boldsymbol{q}$ can never reach the speed of light, but for the charge position $\boldsymbol{r}$ we have found [2] three separate classes of particles for which velocity $d\boldsymbol{r}/dt$ can be less, equal, or greater than c. In fact, it is the quantization of a particle whose charge is circling around the center of mass with velocity c that leads to Dirac's equation.

Quantization

Feynman's quantization of the above generalized Lagrangian systems imply that Feynman's kernel is a function (or a distribution) defined on $X \times X$, i.e., $K(x_1, x_2)$ and thus wave functions must be squared-integrable functions defined on the kinematical space X. A general spinning particle wave function will be a function of ten variables $\Psi(t, \boldsymbol{r}, \boldsymbol{u}, \boldsymbol{\alpha})$.

Quantization of Particles with $u = c$

The kinematical group is Poincaré group. We have a homogeneous space spanned by variables $(t, \boldsymbol{r}, \boldsymbol{u}, \boldsymbol{\alpha})$ with domains $t \in \mathbb{R}$, $\boldsymbol{r} \in \mathbb{R}^3$, $\boldsymbol{u} \in \mathbb{R}^3$, but $u = c$ and $\boldsymbol{\alpha} = \alpha\boldsymbol{a}$ is the canonical parameterization of rotation group in terms of the rotated angle α and the unit vector $\boldsymbol{a}$ along rotation axis or in terms of $\boldsymbol{\rho} = \tan(\alpha/2)\boldsymbol{a}$.

Since $\boldsymbol{u} \cdot \dot{\boldsymbol{u}} = 0$, the particle describes a circle of radius $R_0 = S/mc$ with velocity c for the center of mass observer in a plane orthogonal to $\boldsymbol{S}$ which is constant in this frame.

The wave function of the system is a function $\Psi(t, \boldsymbol{r}, \boldsymbol{u}, \boldsymbol{\rho})$. The Poincaré group representation on these Hilbert space implies that the ten generators are given by:

$$H = i\hbar\frac{\partial}{\partial t}, \quad \boldsymbol{P} = \frac{\hbar}{i}\nabla, \quad \boldsymbol{J} = \frac{\hbar}{i}\boldsymbol{r} \times \nabla + \boldsymbol{S}, \quad \boldsymbol{K} = \frac{\hbar}{i}\frac{\boldsymbol{r}}{c}\frac{\partial}{\partial t} + \frac{\hbar}{i}ct\nabla - \frac{1}{c}\boldsymbol{u} \times \boldsymbol{S},$$

where the spin is the differential operator

$$\boldsymbol{S} = \frac{\hbar}{i}\boldsymbol{u} \times \nabla_u + \frac{\hbar}{2i}(\nabla_\rho + \boldsymbol{\rho} \times \nabla_\rho + \boldsymbol{\rho}(\boldsymbol{\rho} \cdot \nabla_\rho)) = \boldsymbol{S}_u + \boldsymbol{S}_\rho$$

∇_u and ∇_ρ being gradient operators with respect to $\boldsymbol{u}$ and $\boldsymbol{\rho}$ variables, respectively.

There are two Poincaré invariant operators, viz.,

$$KG := H^2 - c^2\boldsymbol{P}^2 - m^2c^4 = 0, \quad \text{(Klein-Gordon)},$$

that only differentiates the wave function with respect to position $\boldsymbol{r}$ and time t, and

$$D := H - \boldsymbol{P}\cdot\boldsymbol{u} - \frac{1}{c^2}(\frac{d\boldsymbol{u}}{dt}\times\boldsymbol{u})\cdot\boldsymbol{S} = 0, \quad \text{(Dirac)}.$$

Spin operator acts on the velocity and orientation variables, commutes with the Klein-Gordon operator, and thus we can find simultaneous eigenfunctions of KG, S^2 and S_3. The wave function can be separated:

$$\Psi(t,\boldsymbol{r},\boldsymbol{u},\boldsymbol{\rho}) = \sum_i \psi_i(t,\boldsymbol{r})\Phi_i(\boldsymbol{u},\boldsymbol{\rho}).$$

Consequently the space-time components satisfy the Klein-Gordon equation

$$(H^2 - c^2\boldsymbol{P}^2 - m^2c^4)\,\psi_i(t,\boldsymbol{r}) = 0, \tag{1}$$

while the internal structure part obeys

$$S^2\Phi_i(\boldsymbol{u},\boldsymbol{\rho}) = s(s+1)\hbar^2\Phi_i(\boldsymbol{u},\boldsymbol{\rho}), \quad S_3\Phi_i(\boldsymbol{u},\boldsymbol{\rho}) = m_s\hbar\Phi_i(\boldsymbol{u},\boldsymbol{\rho}). \tag{2}$$

Solutions of (2) for $s = 1/2$ imply that Φ_i functions are independent of velocity variables. The functions $\Phi_i^{1/2}(\boldsymbol{\rho})$ are just the four Wigner's functions, so that internal structure Hilbert space is isomorphic to the four-dimensional Hilbert space $\mathbb{C}^4$ [3].

Spin operators and spin projections on the body axis $\boldsymbol{e}_i$, $Z_i = \boldsymbol{e}_i\cdot\boldsymbol{S}$ satisfy the commutation relations

$$[S_i, S_j] = i\hbar\epsilon_{ijk}S_k, \quad [Z_i, Z_j] = -i\hbar\epsilon_{ijk}Z_k, \quad [S_i, Z_j] = 0.$$

Wigner's functions are simultaneous eigenfunctions of the three commuting operators S^2, S_3, and Z_3; in the normal parameterization they are represented by

$$\begin{aligned}
\Phi_1(\alpha,\theta,\phi) &= D^{1/2}_{1/2,1/2}(\alpha,\theta,\phi) = \sqrt{2}(\cos(\alpha/2) + i\cos\theta\sin(\alpha/2)),\\
\Phi_2(\alpha,\theta,\phi) &= D^{1/2}_{-1/2,1/2}(\alpha,\theta,\phi) = i\sqrt{2}\sin(\alpha/2)\,\sin\theta e^{-i\phi},\\
\Phi_3(\alpha,\theta,\phi) &= D^{1/2}_{1/2,-1/2}(\alpha,\theta,\phi) = i\sqrt{2}\sin(\alpha/2)\,\sin\theta e^{i\phi},\\
\Phi_4(\alpha,\theta,\phi) &= D^{1/2}_{-1/2,-1/2}(\alpha,\theta,\phi) = \sqrt{2}(\cos(\alpha/2) - i\cos\theta\sin(\alpha/2)).
\end{aligned}$$

They form an orthonormal set with respect to a normalized invariant measure such that scalar product is defined as

$$<\Phi|\Psi> := \int_0^{2\pi} d\phi \int_0^{\pi} d\theta \int_0^{2\pi} d\alpha\, \Phi^*(\alpha,\theta,\phi)\Psi(\alpha,\theta,\phi)\frac{1}{4\pi^2}\sin\theta\sin^2(\alpha/2).$$

Once this four basis vectors are fixed, when acting on the subspace they span, the differential operators S_i and Z_i have the matrix representation

$$\hat{S} = \frac{\hbar}{2}\begin{pmatrix} \boldsymbol{\sigma} & 0 \\ 0 & \boldsymbol{\sigma} \end{pmatrix}, \tag{3}$$

$$\hat{Z}_1 = \frac{\hbar}{2}\begin{pmatrix} 0 & \mathbb{I} \\ \mathbb{I} & 0 \end{pmatrix}, \quad \hat{Z}_2 = \frac{\hbar}{2}\begin{pmatrix} 0 & i\mathbb{I} \\ -i\mathbb{I} & 0 \end{pmatrix}, \quad \hat{Z}_3 = \frac{\hbar}{2}\begin{pmatrix} \mathbb{I} & 0 \\ 0 & -\mathbb{I} \end{pmatrix}, \tag{4}$$

and the nine components of unit vectors $\boldsymbol{e}_i$, $i = 1,2,3$, are the traceless Hermitian matrices

$$\hat{\boldsymbol{e}}_1 = \frac{1}{3}\begin{pmatrix} 0 & \boldsymbol{\sigma} \\ \boldsymbol{\sigma} & 0 \end{pmatrix}, \quad \hat{\boldsymbol{e}}_2 = \frac{1}{3}\begin{pmatrix} 0 & i\boldsymbol{\sigma} \\ -i\boldsymbol{\sigma} & 0 \end{pmatrix}, \quad \hat{\boldsymbol{e}}_3 = \frac{1}{3}\begin{pmatrix} \boldsymbol{\sigma} & 0 \\ 0 & -\boldsymbol{\sigma} \end{pmatrix}, \tag{5}$$

where $\boldsymbol{\sigma}$ are the three Pauli matrices and $\mathbb{I}$ represents the 2×2 unit matrix.

Since the wave function $\Psi(t, \boldsymbol{r}, \boldsymbol{u}, \boldsymbol{\alpha}) = \sum_i \psi_i(t, \boldsymbol{r})\Phi_i(\alpha, \theta, \phi)$ for spin-1/2 particles, then, once the Φ_i functions that describe the internal structure are identified with the four orthogonal unit vectors of the internal Hilbert space $\mathbb{C}^4$, the wave function becomes a four-component wave function; and the six spin components S_i and Z_j and the nine vector components $(\hat{\boldsymbol{e}}_i)_j$, together the 4×4 unit matrix, completely exhaust the 16 linearly independent 4×4 Hermitian matrices, such that any other internal observable that describes internal structure, for instance internal velocity and acceleration, must necessarily be expressed as a real linear combination of the mentioned sixteen Hermitian matrices.

When Dirac's operator D acts on the wave function, we know that H, $\boldsymbol{P}$, and $\boldsymbol{S}$ are differential operators, but we do not know how to represent the action of velocity $\boldsymbol{u}$ and the $(d\boldsymbol{u}/dt) \times \boldsymbol{u}$ observable. However we know that for the classical particle $\boldsymbol{u}$ and $d\boldsymbol{u}/dt$ are orthogonal vectors and together vector $\boldsymbol{u} \times d\boldsymbol{u}/dt$ they form an orthogonal right handed system, and in the center-of-mass frame the particle describes a circle of radius $R_0 = \hbar/2mc$ for spin-1/2 particles in the plane spanned by $\boldsymbol{u}$ and $d\boldsymbol{u}/dt$. If we relate these vectors with the orthogonal left handed system formed by vectors $\hat{\boldsymbol{e}}_1$, $\hat{\boldsymbol{e}}_2$ and $\hat{\boldsymbol{e}}_3$, as shown in Fig. 1, we have $\boldsymbol{u} = a\hat{\boldsymbol{e}}_1$ and $d\boldsymbol{u}/dt \times \boldsymbol{u} = b\hat{\boldsymbol{e}}_3$, where a and b are positive real numbers. With some minor calculations, we obtain Dirac's operator

$$H - cP \cdot \boldsymbol{\alpha} - \beta mc^2 = 0,$$

where Dirac's matrices are the Pauli-Dirac representation

$$\boldsymbol{\alpha} = \begin{pmatrix} 0 & \boldsymbol{\sigma} \\ \boldsymbol{\sigma} & 0 \end{pmatrix}, \quad \beta = \begin{pmatrix} \mathbb{I} & 0 \\ 0 & -\mathbb{I} \end{pmatrix}.$$

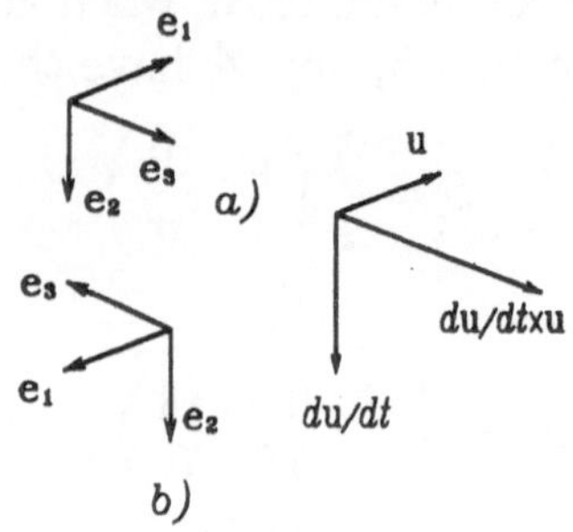

Fig. 1.

The representation suggested by Fig. 2 leads to Weyl's representation:

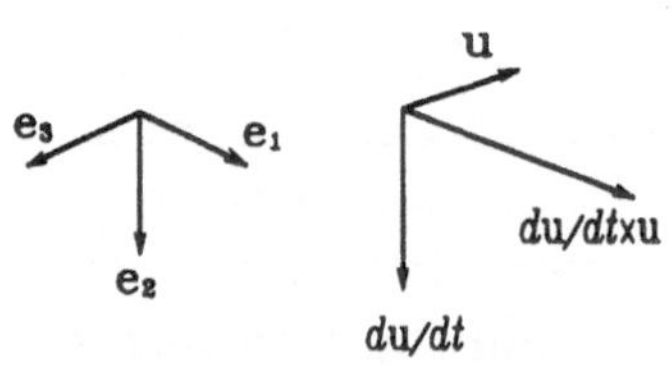

Fig. 2.

where

$$\alpha = \begin{pmatrix} -\sigma & 0 \\ 0 & \sigma \end{pmatrix}, \quad \beta = \begin{pmatrix} 0 & \mathbb{I} \\ \mathbb{I} & 0 \end{pmatrix}$$

Acknowledgment

This work was performed under contract 172.310 E132/91 from Universidad del País Vasco, Euskal Herriko Unibertsitatea.

REFERENCES

[1] M. Rivas, *J. Phys. A* **18**, 1971 (1985); *J. Math. Phys.* **30**, 318 (1989).
[2] M. Rivas, *J. Math. Phys.* **30**, 318 (1989).
[3] A. R. Edmonds, *Angular Momentum in Quantum Mechanics* (Princeton University Press, Princeton, 1957).

QUANTUM TRANSITION AND TEMPORAL DESCRIPTION

Jean Salmon

CNAM, Laboratoire de Physique Générale
292 rue Saint Martin
75141 Paris Cedex 03, France

The author reexamines Crisp's and Jaynes' semi-classical theory and adds a term to the usual Hamiltonian expression revealing irreversibility. The term is derived from the polarization of the sub-quantic medium by the dipolar moment. The result of this is a temporal description of the transition and a satisfactory expression of mean life.

Key words: Hamiltonian, dipolar momentum, quantum transition, sub-quantal medium.

1. INTRODUCTION

The problem of the temporal description of a quantum transition between two levels of an atom to an electron is reexamined. In the Copenhagen school's interpretation such a description is banned. But if one uses Crisp's and Jaynes' semi-classical model and adds a term deriving from the sub-quantic medium, one obtains a Schrödinger modified equation whose solution describes the transition in time.

2. NOTATIONS

A charge Ze can be found at the origin of axes. The electron has a mass m and an electrial charge $-e$; t denotes the time, ε_0 the vacuum premittivity, and $\hbar$ Planck's constant divided by 2π. The position vector is denoted by $\boldsymbol{x}$, the potential energy, V, and usual Hamiltonian by H_0:

$$V = -\frac{Ze^2}{4\pi\varepsilon_0 x}, \quad H_0 = -\frac{\hbar^2}{2m}\Delta + V. \tag{1)(2}$$

The wave function ψ obeys

$$i\hbar\frac{\partial\psi}{\partial t} = H_0\psi. \tag{3}$$

Let us consider a transition between the levels of energy E_1 and E_2 and let us introduce the frequencies ω_1, ω_2, and ω together with

$$E_1 = -h\,\omega_1, \quad E_2 = -h\,\omega_2, \quad \omega = \omega_1 - \omega_2 \qquad (4)(5)(6)$$

The functions ψ_1 and ψ_2 are the Eigenfunctions of H_0 associated wtih E_1 and E_2; thus

$$H_0\,\psi_1 = E_1\,\psi_1, \quad H_0\,\psi_2 - E_2\,\psi_2. \qquad (7)(8)$$

We adopt for ψ the form

$$\psi = c_1(t)\,e^{\imath\omega_1 t}\,\psi_1 + c_2(t)\,e^{\imath\omega_2 t}\psi_2. \tag{9}$$

An adequate temporal description requires c_1 to vary from 0 to 1 and c_2 from 1 to 0. Then, substitution of (9) and (3) shows that either $c_1 = 1, c_2 = 0$ or $c_1 = 0, c_2 = 1$. One cannot describe the transition with H_0.

3. CRISP'S AND JAYNES' FORMULAS [1,2,3,7]

These authors modify the Schrödinger equation by introducing the magnetic potential vector created by the current density $\boldsymbol{j}$ associated with the wave function ψ:

$$\boldsymbol{j} = \frac{ie\hbar}{2m}\left[\psi^+\nabla\,\psi - \psi\nabla\,\psi^+\right]. \tag{10}$$

Schrödinger's equation becomes

$$i\hbar\frac{\partial\psi}{\partial t} = \frac{1}{2m}[-i\hbar\nabla\; + eA]^2\psi + V\psi. \tag{11}$$

Let us define by $\boldsymbol{j}_t$, the transverse current density, and let us adopt the Colomb's gauge. We then have

$$\boldsymbol{A} = \frac{\mu_0}{4\pi}\int |\boldsymbol{x} - \boldsymbol{x}'|^{-1}\,\boldsymbol{j}_t\left(t - \frac{|\boldsymbol{x} - \boldsymbol{x}'|}{c}, \boldsymbol{x}'\right)d^3x', \tag{12}$$

$$\nabla\cdot\boldsymbol{A} = 0. \tag{13}$$

From the system (10)–(13) Crisp and Jaynes obtain the differential equations

$$\frac{d}{dt}[c_1c_1^+] = \frac{\mu_0\omega^3 d^2}{3\pi\hbar c}|c_1c_1^+c_2c_2^+], \tag{14}$$

$$\frac{d}{dt}[c_2c_2^+] = -\frac{\mu_0\omega^3 d^2}{3\pi\hbar c}|c_2c_2^+c_1c_1^+]. \tag{15}$$

in defining the vector $\boldsymbol{d}$ by

$$\boldsymbol{d} = -e\int \psi_1\psi_2 \boldsymbol{x} d^3x, \tag{16}$$

we posit

$$\alpha = \frac{\mu_0\omega^3 d^2}{3\pi\hbar c}, \quad p_1 = c_1c_1^+, \, p_2 = c_2c_2^+. \tag{17)(18)(19}$$

Equations (14) and (15) then become

$$\frac{dp_1}{dt} = \alpha\, p_1p_2, \quad \frac{dp_2}{dt} = -\alpha\, p_1p_2 \tag{20)(21}$$

this gives

$$p_1 = \frac{Ke^{\alpha t}}{1+Ke^{\alpha t}}, \quad p_2 = \frac{1}{1+Ke^{\alpha t}}. \tag{22)(23}$$

Unfortunately, the initial instant corresponding to

$$t = 0, \quad p_1 = 0 \quad p_2 = 1, \tag{24}$$

K is provided the value zero, which is unacceptable. We are therefore going to introduce an external term to the Schrödinger-Maxwell system (4)–(6).

4. ORIGIN AND PART PLAYED BY THE EXTERNAL TERM

The term differs from the usual electromagnetic interaction. In order to justify it, we dare call to mind David Bohm and Louis de Broglie subquantum medium [7].

Let us imagine the dipolar moment of amplitude d and pulsation ω reacts in the sub-quantic medium by creating an electrical field $\boldsymbol{\varepsilon}_s$, not real but purely imaginary, and therefore not measurable. We formulate it by

$$\varepsilon_s = i\int \boldsymbol{\varepsilon}_s(\omega')\sin\omega' t\rho(\omega')d\omega', \tag{25}$$

$\rho(\omega')$ representing a density normalized to unit.

The field $\boldsymbol{\varepsilon}_s$ interacts upon the electron. A purely imaginary potential results from this interaction, formulated as $i\,V_s$. V_s, a real quantity, is given by

$$V_s = -ie\boldsymbol{x} \cdot \boldsymbol{\varepsilon}_s. \tag{26}$$

The term $i\,V_s$ from the optical potential introduces the irreversibility characterizing a radiative transition.

The Schrödinger equation with iV_s becomes

$$i\hbar\frac{\partial\psi}{\partial t} = \frac{1}{2m}[-i\hbar\boldsymbol{\nabla} + e\boldsymbol{A}]^2\psi + iV_s\psi. \tag{27}$$

Let us multiply the two sides by $-\frac{i}{\hbar}c_1\psi_1 e^{-i\omega_1 t}$ and integrate; we get

$$c_1^+\frac{dc_1}{dt} = W_A + W_s \tag{28}$$

W_A is a term already obtained by Crisp and Jaynes. W_s can be defined as follows:

$$\begin{aligned} W_s = \frac{ic_1^*}{2\hbar}\int\int\left[e^{i\omega' t} - e^{-i\omega' t}\right][-e\boldsymbol{x}\cdot\boldsymbol{\varepsilon}_F(\omega')] \\ \cdot\left[c_1\psi_1 + c_2 e^{-i\omega t}\psi_2\right]\psi_1\rho(\omega')d\omega' d^3x \end{aligned} \tag{29}$$

Since

$$\begin{aligned} \int \boldsymbol{x}\psi_1^2 d^3x = \int \boldsymbol{x}\psi_2^2 d^3x = 0, \\ \boldsymbol{d} = -e\int \boldsymbol{x}\psi_1\psi_2 d^3x, \end{aligned} \tag{30}$$

we find

$$\begin{aligned} W_s = \frac{ic_1^+c_2}{2h}\boldsymbol{d}\cdot\int\boldsymbol{\varepsilon}_F(\omega')\left[e^{-i(\omega'-\omega)t} - e^{-i(\omega'+\omega)t}\right]\ldots \\ \ldots\rho(\omega')d\omega'. \end{aligned} \tag{31}$$

Let us assimilate ρ to the Dirac distribution $\delta(\omega' - \omega)$; we obtain

$$W_s = \frac{ic_1^+c_2}{2\hbar}\left[1 - e^{-2i\omega t}\right]\boldsymbol{d}\cdot\boldsymbol{\varepsilon}_F(\omega). \tag{32}$$

Let us take the mean value over the period $\frac{2\pi}{\omega}$ of the factor inside the brackets. The outcome is

$$W_s = \frac{ic_1^+c_2}{2\hbar}\boldsymbol{d}\cdot\boldsymbol{\varepsilon}_F(\omega). \tag{33}$$

We next define $\boldsymbol{\varepsilon}_F$ by supposing that this field is the oscillating dipole field $\boldsymbol{d}e^{i\omega t}$. In the plane of the vectors $\boldsymbol{d}$ and $\boldsymbol{x}$, let us denote by $\boldsymbol{u}$ the unitary vector perpendicular to $\boldsymbol{x}$ and by θ the angle formed by the vectors $\boldsymbol{d}$ and $\boldsymbol{x}$. Let us also introduce a length l_s characterizing the sub-quantum medium.

We then have

$$\boldsymbol{\varepsilon}_F(\omega) = \left[\frac{d\omega^2 \sin\theta}{4\pi\varepsilon_0 c^2 l_s}\right] \boldsymbol{u} \tag{34}$$

Since we have admitted that the sub-quantum medium supports the electromagnetic field, we will assume that l_s is the most natural length that we have at our diposal, $2\pi\frac{c}{\omega}$. So we find

$$\boldsymbol{d} \cdot \boldsymbol{\varepsilon}_F(\omega) = \frac{d^2\omega^2 \sin^2\theta}{8\pi^2 c^3 \varepsilon_0} \tag{35}$$

and thus, on employing the mean value over θ,

$$W_s = (ic_1^+ c_2)\left[\frac{d^2\omega^3}{24\pi^2 c^3 \varepsilon_0 \hbar}\right]. \tag{36}$$

The usual semi-classical theory of radiation uses a mean life θ_0 such that

$$\theta_0 = \frac{12\pi^2 \varepsilon_0 h}{\omega^3 d^2}. \tag{37}$$

Accordingly

$$\alpha = \frac{4\pi}{\theta_0}, \quad W_s = \frac{ic_1^+ c_2}{2\theta_0}, \tag{38)(39}$$

and (28) can be written as

$$c_1^+ \frac{dc_1}{dt} = W_A + i\,\frac{c_1^+ c_2}{2\theta_0}, \tag{40}$$

which has the conjugate form

$$c_1^+ \frac{dc_1^+}{dt} = W_A^+ - i\,\frac{c_1 c_2^+}{2\theta_0}. \tag{41}$$

Crisp and Jaynes theory gives

$$W_A + W_A^+ = \frac{4\pi}{\theta_0} c_1 c_1^+ c_2 c_2^+. \tag{42}$$

We also have

$$p_1 = c_1 c_1^+, \quad p_2 = c_2 c_2^+ = 1 - p_1. \tag{43)(44}$$

One is free to choose c_1 as real and c_2 as purely imaginary; therefore, setting

$$c_2 = ic'_2, \tag{45}$$

where $\dot{c}_2$ is a real quantity, where adding (40) and (41), and taking (42) and (45) into account, one gets

$$\frac{dp_1}{dt} = \frac{1}{\theta_0}\left[4\pi p_1(1-p_1) + p_1^{1/2}(1-p_1)^{1/2}\right]. \tag{46}$$

let us posit

$$p_1 = \sin^2\varphi, \tag{47}$$

so that

$$t = \int \frac{2\theta_0 d\varphi}{4\pi \sin\varphi\cos\varphi + 1}. \tag{48}$$

The evolution of p_1 thus takes place according to the formula

$$t = \frac{\theta_0}{2\pi}\left(1 - \frac{1}{4\pi^2}\right)^{-1/2} \cdot \log \frac{\left[1 + \frac{1}{2\pi}\left(\frac{p_1}{1-p_1}\right)^{1/2} - \left(1 - \frac{1}{4\pi^2}\right)^{1/2}\right]\left[1 + \left(1 - \frac{1}{4\pi^2}\right)^{1/2}\right]}{\left[1 + \frac{1}{2\pi}\left(\frac{p_1}{1-p_1}\right)^{1/2} + \left(1 - \frac{1}{4\pi^2}\right)^{1/2}\right]\left[1 - \left(1 - \frac{1}{4\pi^2}\right)^{1/2}\right]} \tag{49}$$

5. CONCLUSION

The subquantal medium generates a purely imaginary field, whence derives an optical potential which is introduced in the Schrödinger equation. Thus is obtained a description in time of the transition of an electron between two levels.

POSTSCRIPT

If we accept the hypothesis of a pure imaginary field $i\varepsilon_F \sin\omega t$ and suppose that the subquantal medium is made of particles at rest, with mass m_s and electrical charge q_s, sensitive only to a pure imaginary field, then the equation of the motion becomes

$$\frac{d}{dt}\left[m_s\left(1 - \frac{v^2}{c^2}\right)^{-1/2} \boldsymbol{v}\right] = i\, q_s \varepsilon_F \sin\omega t.$$

Since the speed v has to exceed c, we can ask the question: Can the subquantal medium carry only supra-luminal signals?

REFERENCES

1. M. Crisp and E. Jaynes, *Phys. Rev.* **179**, 1253 (1969).
2. A. Barut and J. Kraus, *Found. Phys.* **13**, 189 (1983).
3. A. Barut and J. Van Huele, *Phys. Rev. A* **32**, 3187 (1985).
4. R. Boudet, *Ann. Fond. L. de Broglie* **14(2)**, 119 (1989).
5. R. Blaive and R. Boudet, *Ann. Fond. L. de Broglie* **14(2)**, 147 (1989).
6. R. Boudet, *Ann. Fond. L. de Broglie* **16(2)**, 257 (1991).
7. J. Salmon, *Ann. Fond. L. de Broglie* **15(3)**, 359 (1990).

AUTHOR INDEX

SUBJECT INDEX